Probability and Statistics for Finance

The Frank J. Fabozzi Series

Fixed Income Securities, Second Edition by Frank J. Fabozzi
Focus on Value: A Corporate and Investor Guide to Wealth Creation by James L. Grant and James A. Abate
Handbook of Global Fixed Income Calculations by Dragomir Krgin
Managing a Corporate Bond Portfolio by Leland E. Crabbe and Frank J. Fabozzi
Real Options and Option-Embedded Securities by William T. Moore
Capital Budgeting: Theory and Practice by Pamela P. Peterson and Frank J. Fabozzi
The Exchange-Traded Funds Manual by Gary L. Gastineau
Professional Perspectives on Fixed Income Portfolio Management, Volume 3 edited by Frank J. Fabozzi
Investing in Emerging Fixed Income Markets edited by Frank J. Fabozzi and Efstathia Pilarinu
Handbook of Alternative Assets by Mark J. P. Anson
The Global Money Markets by Frank J. Fabozzi, Steven V. Mann, and Moorad Choudhry
The Handbook of Financial Instruments edited by Frank J. Fabozzi
Collateralized Debt Obligations: Structures and Analysis by Laurie S. Goodman and Frank J. Fabozzi
Interest Rate, Term Structure, and Valuation Modeling edited by Frank J. Fabozzi
Investment Performance Measurement by Bruce J. Feibel
The Handbook of Equity Style Management edited by T. Daniel Coggin and Frank J. Fabozzi
The Theory and Practice of Investment Management edited by Frank J. Fabozzi and Harry M. Markowitz
Foundations of Economic Value Added, Second Edition by James L. Grant
Financial Management and Analysis, Second Edition by Frank J. Fabozzi and Pamela P. Peterson
Measuring and Controlling Interest Rate and Credit Risk, Second Edition by Frank J. Fabozzi, Steven V. Mann, and Moorad Choudhry
Professional Perspectives on Fixed Income Portfolio Management, Volume 4 edited by Frank J. Fabozzi
The Handbook of European Fixed Income Securities edited by Frank J. Fabozzi and Moorad Choudhry
The Handbook of European Structured Financial Products edited by Frank J. Fabozzi and Moorad Choudhry
The Mathematics of Financial Modeling and Investment Management by Sergio M. Focardi and Frank J. Fabozzi
Short Selling: Strategies, Risks, and Rewards edited by Frank J. Fabozzi
The Real Estate Investment Handbook by G. Timothy Haight and Daniel Singer
Market Neutral Strategies edited by Bruce I. Jacobs and Kenneth N. Levy
Securities Finance: Securities Lending and Repurchase Agreements edited by Frank J. Fabozzi and Steven V. Mann
Fat-Tailed and Skewed Asset Return Distributions by Svetlozar T. Rachev, Christian Menn, and Frank J. Fabozzi
Financial Modeling of the Equity Market: From CAPM to Cointegration by Frank J. Fabozzi, Sergio M. Focardi, and Petter N. Kolm
Advanced Bond Portfolio Management: Best Practices in Modeling and Strategies edited by Frank J. Fabozzi, Lionel Martellini, and Philippe Priaulet
Analysis of Financial Statements, Second Edition by Pamela P. Peterson and Frank J. Fabozzi
Collateralized Debt Obligations: Structures and Analysis, Second Edition by Douglas J. Lucas, Laurie S. Goodman, and Frank J. Fabozzi
Handbook of Alternative Assets, Second Edition by Mark J. P. Anson
Introduction to Structured Finance by Frank J. Fabozzi, Henry A. Davis, and Moorad Choudhry
Financial Econometrics by Svetlozar T. Rachev, Stefan Mittnik, Frank J. Fabozzi, Sergio M. Focardi, and Teo Jasic
Developments in Collateralized Debt Obligations: New Products and Insights by Douglas J. Lucas, Laurie S. Goodman, Frank J. Fabozzi, and Rebecca J. Manning
Robust Portfolio Optimization and Management by Frank J. Fabozzi, Peter N. Kolm, Dessislava A. Pachamanova, and Sergio M. Focardi
Advanced Stochastic Models, Risk Assessment, and Portfolio Optimizations by Svetlozar T. Rachev, Stogan V. Stoyanov, and Frank J. Fabozzi
How to Select Investment Managers and Evaluate Performance by G. Timothy Haight, Stephen O. Morrell, and Glenn E. Ross
Bayesian Methods in Finance by Svetlozar T. Rachev, John S. J. Hsu, Biliana S. Bagasheva, and Frank J. Fabozzi
Structured Products and Related Credit Derivatives by Brian P. Lancaster, Glenn M. Schultz, and Frank J. Fabozzi
Quantitative Equity Investing: Techniques and Strategies by Frank J. Fabozzi, CFA, Sergio M. Focardi, Petter N. Kolm

Probability and Statistics for Finance

SVETLOZAR T. RACHEV
MARKUS HÖCHSTÖTTER
FRANK J. FABOZZI
SERGIO M. FOCARDI

John Wiley & Sons, Inc.

Copyright © 2010 by John Wiley & Sons, Inc. All rights reserved.

Published by John Wiley & Sons, Inc., Hoboken, New Jersey.
Published simultaneously in Canada.

No part of this publication may be reproduced, stored in a retrieval system, or transmitted in any form or by any means, electronic, mechanical, photocopying, recording, scanning, or otherwise, except as permitted under Section 107 or 108 of the 1976 United States Copyright Act, without either the prior written permission of the Publisher, or authorization through payment of the appropriate per-copy fee to the Copyright Clearance Center, Inc., 222 Rosewood Drive, Danvers, MA 01923, (978) 750-8400, fax (978) 646-8600, or on the web at www.copyright.com. Requests to the Publisher for permission should be addressed to the Permissions Department, John Wiley & Sons, Inc., 111 River Street, Hoboken, NJ 07030, (201) 748-6011, fax (201) 748-6008, or online at http://www.wiley.com/go/permissions.

Limit of Liability/Disclaimer of Warranty: While the publisher and author have used their best efforts in preparing this book, they make no representations or warranties with respect to the accuracy or completeness of the contents of this book and specifically disclaim any implied warranties of merchantability or fitness for a particular purpose. No warranty may be created or extended by sales representatives or written sales materials. The advice and strategies contained herein may not be suitable for your situation. You should consult with a professional where appropriate. Neither the publisher nor author shall be liable for any loss of profit or any other commercial damages, including but not limited to special, incidental, consequential, or other damages.

For general information on our other products and services or for technical support, please contact our Customer Care Department within the United States at (800) 762-2974, outside the United States at (317) 572-3993, or fax (317) 572-4002.

Wiley also publishes its books in a variety of electronic formats. Some content that appears in print may not be available in electronic books. For more information about Wiley products, visit our web site at www.wiley.com.

Library of Congress Cataloging-in-Publication Data:

Probability and statistics for finance / Svetlozar T. Rachev ... [et al.].
   p. cm. – (Frank J. Fabozzi series ; 176)
   Includes index.
   ISBN 978-0-470-40093-7 (cloth); 978-0-470-90630-9 (ebk); 978-0-470-90631-6 (ebk); 978-0-470-90632-3 (ebk)
   1. Finance–Statistical methods. 2. Statistics. 3. Probability measures. 4. Multivariate analysis. I. Rachev, S. T. (Svetlozar Todorov)
HG176.5.P76 2010
332.01'5195–dc22
                              2010027030

STR
To my grandchildren, Iliana, Zoya, and Svetlozar

MH
To my wife Nadine

FJF
To my sister Lucy

SF
To my mother Teresa
and to the memory of my father Umberto

Contents

Preface — xv
About the Authors — xvii

CHAPTER 1
Introduction — 1
Probability vs. Statistics — 4
Overview of the Book — 5

PART ONE
Descriptive Statistics — 15

CHAPTER 2
Basic Data Analysis — 17
Data Types — 17
Frequency Distributions — 22
Empirical Cumulative Frequency Distribution — 27
Data Classes — 32
Cumulative Frequency Distributions — 41
Concepts Explained in this Chapter — 43

CHAPTER 3
Measures of Location and Spread — 45
Parameters vs. Statistics — 45
Center and Location — 46
Variation — 59
Measures of the Linear Transformation — 69
Summary of Measures — 71
Concepts Explained in this Chapter — 73

CHAPTER 4
Graphical Representation of Data — 75
- Pie Charts — 75
- Bar Chart — 78
- Stem and Leaf Diagram — 81
- Frequency Histogram — 82
- Ogive Diagrams — 89
- Box Plot — 91
- QQ Plot — 96
- Concepts Explained in this Chapter — 99

CHAPTER 5
Multivariate Variables and Distributions — 101
- Data Tables and Frequencies — 101
- Class Data and Histograms — 106
- Marginal Distributions — 107
- Graphical Representation — 110
- Conditional Distribution — 113
- Conditional Parameters and Statistics — 114
- Independence — 117
- Covariance — 120
- Correlation — 123
- Contingency Coefficient — 124
- Concepts Explained in this Chapter — 126

CHAPTER 6
Introduction to Regression Analysis — 129
- The Role of Correlation — 129
- Regression Model: Linear Functional Relationship Between Two Variables — 131
- Distributional Assumptions of the Regression Model — 133
- Estimating the Regression Model — 134
- Goodness of Fit of the Model — 138
- Linear Regression of Some Nonlinear Relationship — 140
- Two Applications in Finance — 142
- Concepts Explained in this Chapter — 149

CHAPTER 7
Introduction to Time Series Analysis — 153
What Is Time Series? 153
Decomposition of Time Series 154
Representation of Time Series with Difference Equations 159
Application: The Price Process 159
Concepts Explained in this Chapter 163

PART TWO
Basic Probability Theory — 165

CHAPTER 8
Concepts of Probability Theory — 167
Historical Development of
    Alternative Approaches to Probability 167
Set Operations and Preliminaries 170
Probability Measure 177
Random Variable 179
Concepts Explained in this Chapter 185

CHAPTER 9
Discrete Probability Distributions — 187
Discrete Law 187
Bernoulli Distribution 192
Binomial Distribution 195
Hypergeometric Distribution 204
Multinomial Distribution 211
Poisson Distribution 216
Discrete Uniform Distribution 219
Concepts Explained in this Chapter 221

CHAPTER 10
Continuous Probability Distributions — 229
Continuous Probability Distribution Described 229
Distribution Function 230
Density Function 232
Continuous Random Variable 237
Computing Probabilities from the Density Function 238
Location Parameters 239
Dispersion Parameters 239
Concepts Explained in this Chapter 245

CHAPTER 11
Continuous Probability Distributions with Appealing Statistical Properties — 247
Normal Distribution — 247
Chi-Square Distribution — 254
Student's t-Distribution — 256
F-Distribution — 260
Exponential Distribution — 262
Rectangular Distribution — 266
Gamma Distribution — 268
Beta Distribution — 269
Log-Normal Distribution — 271
Concepts Explained in this Chapter — 275

CHAPTER 12
Continuous Probability Distributions Dealing with Extreme Events — 277
Generalized Extreme Value Distribution — 277
Generalized Pareto Distribution — 281
Normal Inverse Gaussian Distribution — 283
α-Stable Distribution — 285
Concepts Explained in this Chapter — 292

CHAPTER 13
Parameters of Location and Scale of Random Variables — 295
Parameters of Location — 296
Parameters of Scale — 306
Concepts Explained in this Chapter — 321
Appendix: Parameters for Various Distribution Functions — 322

CHAPTER 14
Joint Probability Distributions — 325
Higher Dimensional Random Variables — 326
Joint Probability Distribution — 328
Marginal Distributions — 333
Dependence — 338
Covariance and Correlation — 341
Selection of Multivariate Distributions — 347
Concepts Explained in this Chapter — 358

Contents   xi

CHAPTER 15
Conditional Probability and Bayes' Rule   **361**
Conditional Probability   362
Independent Events   365
Multiplicative Rule of Probability   367
Bayes' Rule   372
Conditional Parameters   374
Concepts Explained in this Chapter   377

CHAPTER 16
Copula and Dependence Measures   **379**
Copula   380
Alternative Dependence Measures   406
Concepts Explained in this Chapter   412

PART THREE
Inductive Statistics   **413**

CHAPTER 17
Point Estimators   **415**
Sample, Statistic, and Estimator   415
Quality Criteria of Estimators   428
Large Sample Criteria   435
Maximum Likehood Estimator   446
Exponential Family and Sufficiency   457
Concepts Explained in this Chapter   461

CHAPTER 18
Confidence Intervals   **463**
Confidence Level and Confidence Interval   463
Confidence Interval for the Mean of a
    Normal Random Variable   466
Confidence Interval for the Mean of a
    Normal Random Variable with Unknown Variance   469
Confidence Interval for the Variance
    of a Normal Random Variable   471
Confidence Interval for the Variance of a
    Normal Random Variable with Unknown Mean   474
Confidence Interval for the Parameter p
    of a Binomial Distribution   475

    Confidence Interval for the Parameter λ
       of an Exponential Distribution   477
    Concepts Explained in this Chapter   479

CHAPTER 19
Hypothesis Testing   481
    Hypotheses   482
    Error Types   485
    Quality Criteria of a Test   490
    Examples   496
    Concepts Explained in this Chapter   518

PART FOUR
Multivariate Linear Regression Analysis   519

CHAPTER 20
Estimates and Diagnostics for Multivariate Linear Regression Analysis   521
    The Multivariate Linear Regression Model   522
    Assumptions of the Multivariate Linear Regression Model   523
    Estimation of the Model Parameters   523
    Designing the Model   526
    Diagnostic Check and Model Significance   526
    Applications to Finance   531
    Concepts Explained in this Chapter   543

CHAPTER 21
Designing and Building a Multivariate Linear Regression Model   545
    The Problem of Multicollinearity   545
    Incorporating Dummy Variables as Independent Variables   548
    Model Building Techniques   561
    Concepts Explained in this Chapter   565

CHAPTER 22
Testing the Assumptions of the Multivariate Linear Regression Model   567
    Tests for Linearity   568
    Assumed Statistical Properties about the Error Term   570
    Tests for the Residuals Being Normally Distributed   570
    Tests for Constant Variance of the
       Error Term (Homoskedasticity)   573

Absence of Autocorrelation of the Residuals	576
Concepts Explained in this Chapter	581

APPENDIX A
Important Functions and Their Features — 583

Continuous Function	583
Indicator Function	586
Derivatives	587
Monotonic Function	591
Integral	592
Some Functions	596

APPENDIX B
Fundamentals of Matrix Operations and Concepts — 601

The Notion of Vector and Matrix	601
Matrix Multiplication	602
Particular Matrices	603
Positive Semidefinite Matrices	614

APPENDIX C
Binomial and Multinomial Coefficients — 615

Binomial Coefficient	615
Multinomial Coefficient	622

APPENDIX D
Application of the Log-Normal Distribution to the Pricing of Call Options — 625

Call Options	625
Deriving the Price of a European Call Option	626
Illustration	631

REFERENCES   **633**

INDEX   **635**

Preface

In this book, we provide an array of topics in probability and statistics that are applied to problems in finance. For example, there are applications to portfolio management, asset pricing, risk management, and credit risk modeling. Not only do we cover the basics found in a typical introductory book in probability and statistics, but we also provide unique coverage of several topics that are of special interest to finance students and finance professionals. Examples are coverage of probability distributions that deal with extreme events and statistical measures, which are particularly useful for portfolio managers and risk managers concerned with extreme events.

The book is divided into four parts. The six chapters in Part One cover descriptive statistics: the different methods for gathering data and presenting them in a more succinct way while still being as informative as possible. The basics of probability theory are covered in the nine chapters in Part Two. After describing the basic concepts of probability, we explain the different types of probability distributions (discrete and continuous), specific types of probability distributions, parameters of a probability distribution, joint probability distributions, conditional probability distributions, and dependence measures for two random variables. Part Three covers statistical inference: the method of drawing information from sample data about unknown parameters of the population from which the sample was drawn. The three chapters in Part Three deal with point estimates of a parameter, confidence intervals of a parameter, and testing hypotheses about the estimates of a parameter. In the last part of the book, Part Four, we provide coverage of the most widely used statistical tool in finance: multivariate regression analysis. In the first of the three chapters in this part, we begin with the assumptions of the multivariate regression model, how to estimate the parameters of the model, and then explain diagnostic checks to evaluate the quality of the estimates. After these basics are provided, we then focus on the design and the building process of multivariate regression models and finally on how to deal with violations of the assumptions of the model.

There are also four appendixes. Important mathematical functions and their features that are needed primarily in the context of Part Two of this book are covered in Appendix A. In Appendix B we explain the basics of matrix operations and concepts needed to aid in understanding the presentation in Part Four. The construction of the binomial and multinomial coefficients used in some discrete probability distributions and an application of the log-normally distributed stock price to derive the price of a certain type of option are provided in Appendix C and D, respectively.

We would like to thank Biliana Bagasheva for her coauthorship of Chapter 15 (Conditional Probability and Bayes' Rule). Anna Serbinenko provided helpful comments on several chapters of the book.

The following students reviewed various chapters and provided us with helpful comments: Kameliya Minova, Diana Trinh, Lindsay Morriss, Marwan ElChamaa, Jens Bürgin, Paul Jana, and Haike Dogendorf.

We also thank Megan Orem for her patience in typesetting this book and giving us the flexibility to significantly restructure the chapters in this book over the past three years.

Svetlozar T. Rachev
Markus Höchstötter
Frank J. Fabozzi
Sergio M. Focardi
April 2010

About the Authors

Svetlozar (Zari) T. Rachev completed his Ph.D. Degree in 1979 from Moscow State (Lomonosov) University, and his Doctor of Science Degree in 1986 from Steklov Mathematical Institute in Moscow. Currently, he is Chair-Professor in Statistics, Econometrics and Mathematical Finance at the University of Karlsruhe in the School of Economics and Business Engineering, and Professor Emeritus at the University of California, Santa Barbara in the Department of Statistics and Applied Probability. Professor Rachev has published 14 monographs, 10 handbooks, and special-edited volumes, and over 300 research articles. His recently coauthored books published by Wiley in mathematical finance and financial econometrics include *Financial Econometrics: From Basics to Advanced Modeling Techniques* (2007), and *Bayesian Methods in Finance* (2008). He is cofounder of Bravo Risk Management Group specializing in financial risk-management software. Bravo Group was acquired by FinAnalytica for which he currently serves as Chief-Scientist.

Markus Höchstötter is lecturer in statistics and econometrics at the Institute of Statistics, Econometrics and Mathematical Finance at the University of Karlsruhe (KIT). Dr. Höchstötter has authored articles on financial econometrics and credit derivatives. He earned a doctorate (Dr. rer. pol.) in Mathematical Finance and Financial Econometrics from the University of Karlsruhe.

Frank J. Fabozzi is Professor in the Practice of Finance in the School of Management and Becton Fellow at Yale University and an Affiliated Professor at the University of Karlsruhe's Institute of Statistics, Econometrics and Mathematical Finance. Prior to joining the Yale faculty, he was a Visiting Professor of Finance in the Sloan School at MIT. Professor Fabozzi is a Fellow of the International Center for Finance at Yale University and on the Advisory Council for the Department of Operations Research and Financial Engineering at Princeton University. He is the editor of the *Journal of Portfolio Management* and an associate editor of *Quantitative Finance*. He is a trustee for the BlackRock family of closed-end funds. In 2002, he was inducted into the Fixed Income Analysts Society's Hall of Fame and is the 2007 recipient of the C. Stewart Sheppard Award given by the CFA Institute. His recently coauthored books published by Wiley include *Institutional Investment Management* (2009), *Quantitative Equity Investing* (2010), *Bayesian Methods in Finance* (2008), *Advanced Stochastic Models, Risk Assessment, and Portfolio Optimization: The Ideal Risk, Uncertainty, and Performance Measures* (2008), *Financial Modeling of the Equity Market: From CAPM to Cointegration* (2006), *Robust Portfolio Optimization and Management* (2007), and *Financial Econometrics: From Basics to Advanced Modeling Techniques* (2007). Professor Fabozzi earned a doctorate in economics from the City University of New York in 1972. He earned the designation of Chartered Financial Analyst and Certified Public Accountant.

Sergio M. Focardi is Professor of Finance at the EDHEC Business School in Nice and the founding partner of the Paris-based consulting firm The Intertek Group. He is a member of the editorial board of the *Journal of Portfolio Management*. Professor Focardi has authored numerous articles and books on financial modeling and risk management including the following Wiley books: *Quantitative Equity Investing* (2010), *Financial Econometrics* (2007), *Financial Modeling of the Equity Market* (2006), *The Mathematics of Financial Modeling and Investment Management* (2004), *Risk Management: Framework, Methods and Practice* (1998), and *Modeling the Markets: New Theories and Techniques* (1997). He also authored two monographs published by the CFA Institute's monographs: *Challenges in Quantitative Equity Management* (2008) and *Trends in Quantitative Finance* (2006). Professor Focardi has been appointed as a speaker of the CFA Institute Speaker Retainer Program. His research interests include the econometrics of large equity portfolios and the modeling of regime changes. Professor Focardi holds a degree in Electronic Engineering from the University of Genoa and a Ph.D. in Mathematical Finance and Financial Econometrics from the University of Karlsruhe.

CHAPTER 1
Introduction

It is no surprise that the natural sciences (chemistry, physics, life sciences/biology, astronomy, earth science, and environmental science) and engineering are fields that rely on advanced quantitative methods. One of the toolsets used by professionals in these fields is from the branch of mathematics known as probability and statistics. The social sciences, such as psychology, sociology, political science, and economics, use probability and statistics to varying degrees.

There are branches within each field of the natural sciences and social sciences that utilize probability and statistics more than others. Specialists in these areas not only apply the tools of probability and statistics, but they have also contributed to the field of statistics by developing techniques to organize, analyze, and test data. Let's look at examples from physics and engineering (the study of natural phenomena in terms of basic laws and physical quantities and the design of physical artifacts) and biology (the study of living organisms) in the natural sciences, and psychology (the study of the human mind) and economics (the study of production, resource allocation, and consumption of goods and services) in the social sciences.

Statistical physics is the branch of physics that applies probability and statistics for handling problems involving large populations of particles. One of the first areas of application was the explanation of thermodynamics laws in terms of statistical mechanics. It was an extraordinary scientific achievement with far-reaching consequences. In the field of *engineering*, the analysis of risk, be it natural or industrial, is another area that makes use of statistical methods. This discipline has contributed important innovations especially in the study of rare extreme events. The engineering of electronic communications applied statistical methods early, contributing to the development of fields such as queue theory (used in communication switching systems) and introduced the fundamental innovation of measuring information.

Biostatistics and *biomathematics* within the field of biology include many areas of great scientific interest such as public health, epidemiology, demography, and genetics, in addition to designing biological experiments (such as clinical experiments in medicine) and analyzing the results of those experiments. The study of the dynamics of populations and the study of evolutionary phenomena are two important fields in biomathematics. *Biometry* and *biometrics* apply statistical methods to identify quantities that characterize living objects.

Psychometrics, a branch of psychology, is concerned with designing tests and analyzing the results of those tests in an attempt to measure or quantify some human characteristic. Psychometrics has its origins in personality testing, intelligence testing, and vocational testing, but is now applied to measuring attitudes and beliefs and health-related tests.

Econometrics is the branch of economics that draws heavily on statistics for testing and analyzing economic relationships. Within econometrics, there are theoretical econometricians who analyze statistical properties of estimators of models. Several recipients of the Nobel Prize in Economic Sciences received the award as a result of their lifetime contribution to this branch of economics. To appreciate the importance of econometrics to the discipline of economics, when the first Nobel Prize in Economic Sciences was awarded in 1969, the corecipients were two econometricians, Jan Tinbergen and Ragnar Frisch (who is credited for first using the term econometrics in the sense that it is known today). Further specialization within econometrics, and the area that directly relates to this book, is *financial econometrics*. As Jianqing Fan (2004) writes, financial econometrics

> uses statistical techniques and economic theory to address a variety of problems from finance. These include building financial models, estimation and inferences of financial models, volatility estimation, risk management, testing financial economics theory, capital asset pricing, derivative pricing, portfolio allocation, risk-adjusted returns, simulating financial systems, hedging strategies, among others.

Robert Engle and Clive Granger, two econometricians who shared the 2003 Nobel Prize in Economics Sciences, have contributed greatly to the field of financial econometrics.

Historically, the core probability and statistics course offered at the university level to undergraduates has covered the fundamental principles and applied these principles across a wide variety of fields in the natural sciences and social sciences. Universities typically offered specialized courses within these fields to accommodate students who sought more focused applications. The exceptions were the schools of business administration that early on provided a course in probability and statistics with applications to business decision making. The applications cut across finance, marketing, management, and accounting. However, today, each of these areas in business requires specialized tools for dealing with real-world problems in their respective disciplines.

This brings us to the focus of this book. Finance is an area that relies heavily on probability and statistics. The quotation above by Jianqing Fan basically covers the wide range of applications within finance and identifies some of the unique applications. Two examples may help make this clear. First, in standard books on statistics, there is coverage of what one might refer to as "probability distributions with appealing properties." A distribution called the "normal distribution," referred to in the popular press as a "bell-shaped curve," is an example. Considerable space is devoted to this distribution and its application in standard textbooks. Yet, the overwhelming historical evidence suggests that real-world financial data commonly used in financial applications are not normally distributed. Instead, more focus should be on distributions that deal with extreme events, or, in other words, what are known as the "tails" of a distribution. In fact, many market commentators and regulators view the failure of financial institutions and major players in the financial markets to understand non-normal distributions as a major reason for the recent financial debacles throughout the world. This is one of the reasons that, in certain areas in finance, extreme event distributions (which draw from extreme value theory) have supplanted the normal distribution as the focus of attention. The recent financial crisis has clearly demonstrated that because of the highly leveraged position (i.e., large amount of borrowing relative to the value of equity) of financial institutions throughout the world, these entities are very sensitive to extreme events. This means that the management of these financial institutions must be aware of the nature of the tails of distributions, that is, the probability associated with extreme events.

As a second example, the statistical measure of correlation that measures a certain type of association between two random variables may make sense when the two random variables are normally distributed. However, correlation may be inadequate in describing the link between two random variables when a portfolio manager or risk manager is concerned with extreme events that can have disastrous outcomes for a portfolio or a financial institution. Typically models that are correlation based will underestimate the likelihood of extreme events occurring simultaneously. Alternative statistical measures that would be more helpful, the copula measure and the tail dependence, are typically not discussed in probability and statistics books.

It is safe to say that the global financial system has been transformed since the mid-1970s due to the development of models that can be used to value derivative instruments. Complex derivative instruments such as options, caps, floors, and swaptions can only be valued (i.e., priced) using tools from probability and statistical theory. While the model for such pricing was first developed by Black and Scholes (1976) and known as the Black-Scholes option pricing model, it relies on models that can be traced back to the mathematician Louis Bachelier (1900).

In the remainder of this introductory chapter, we do two things. First, we briefly distinguish between the study of probability and the study of statistics. Second, we provide a roadmap for the chapters to follow in this book.

PROBABILITY VS. STATISTICS

Thus far, we have used the terms "probability" and "statistics" collectively as if they were one subject. There is a difference between the two that we distinguish here and which will become clearer in the chapters to follow.

Probability models are theoretical models of the occurrence of uncertain events. At the most basic level, in probability, the properties of certain types of probabilistic models are examined. In doing so, it is assumed that all parameter values that are needed in the probabilistic model are known. Let's contrast this with statistics. Statistics is about empirical data and can be broadly defined as a set of methods used to make inferences from a known sample to a larger population that is in general unknown. In finance and economics, a particular important example is making inferences from the past (the known sample) to the future (the unknown population). In statistics, we apply probabilistic models and we use data and eventually judgment to estimate the parameters of these models. We do not assume that all parameter values in the model are known. Instead, we use the data for the variables in the model to estimate the value of the parameters and then to test hypotheses or make inferences about their estimated values.

Another way of thinking about the study of probability and the study of statistics is as follows. In studying probability, we follow much the same routine as in the study of other fields of mathematics. For example, in a course in calculus, we prove theorems (such as the fundamental theory of calculus that specifies the relationship between differentiation and integration), perform calculations given some function (such as the first derivative of a function), and make conclusions about the characteristics of some mathematical function (such as whether the function may have a minimum or maximum value). In the study of probability, there are also theorems to be proven (although we do not focus on proofs in this book), we perform calculations based on probability models, and we reach conclusions based on some assumed probability distribution. In deriving proofs in calculus or probability theory, deductive reasoning is utilized. For this reason, probability can be considered as a fundamental discipline in the field of mathematics, just as we would view algebra, geometry, and trigonometry. In contrast, statistics is based on inductive reasoning. More specifically, given a sample of data (i.e., observations), we make generalized probabilistic conclusions about the population from which the data are drawn or the process that generated the data.

OVERVIEW OF THE BOOK

The 21 chapters that follow in this book are divided into four parts covering descriptive statistics, probability theory, inductive statistics, and multivariate linear regression.

Part One: Descriptive Statistics

The six chapters in Part One cover descriptive statistics. This topic covers the different tasks of gathering data and presenting them in a more concise yet as informative as possible way. For example, a set of 1,000 observations may contain too much information for decision-making purposes. Hence, we need to reduce this amount in a reasonable and systematic way.

The initial task of any further analysis is to gather the data. This process is explained in Chapter 2. It provides one of the most essential—if not the most essential—assignment. Here, we have to be exactly aware of the intention of our analysis and determine the data type accordingly. For example, if we wish to analyze the contributions of the individual divisions of a company to the overall rate of return earned by the company, we need a completely different sort of data than when we decompose the risk of some investment portfolio into individual risk factors, or when we intend to gain knowledge of unknown quantities in general economic models. As part of the process of retrieving the essential information contained in the data, we describe the methods of presenting the distribution of the data in comprehensive ways. This can be done for the data itself or, in some cases, it will be more effective after the data have been classified.

In Chapter 3, methodologies for reducing the data to a few representative quantities are presented. We refer to these representative quantities as statistics. They will help us in assessing where certain parts of the data are positioned as well as how the data disperse relative to particular positions. Different data sets are commonly compared based on these statistics that, in most cases, proves to be very efficient.

Often, it is very appealing and intuitive to present the features of certain data in charts and figures. In Chapter 4, we explain the particular graphical tools suitable for the different data types discussed in Chapter 2. In general, a graphic uses the distributions introduced in Chapter 2 or the statistics from Chapter 3. By comparing graphics, it is usually a simple task to detect similarities or differences among different data sets.

In Chapters 2, 3, and 4, the analysis focuses only on one quantity of interest and in such cases we say that we are looking at univariate (i.e., one variable) distributions. In Chapter 5, we introduce multivariate distributions; that is, we look at several variables of interest simultaneously. For example, portfolio analysis relies on multivariate analysis. Risk management in general considers the interaction of several variables and the influence that each variable exerts on the others. Most of the aspects from the one-dimensional analysis (i.e., analysis of univariate distributions) can be easily extended to higher dimensions while concepts such as dependence between variables are completely new. In this context, in Chapter 6 we put forward measures to express the degree of dependence between variables such as the covariance and correlation. Moreover, we introduce the conditional distribution, a particular form of distribution of the variables given that some particular variables are fixed. For example, we may look at the average return of a stock portfolio given that the returns of its constituent stocks fall below some threshold over a particular investment horizon.

When we assume that a variable is dependent on some other variable, and the dependence is such that a movement in the one variable causes a known constant shift in the other, we model the set of possible values that they might jointly assume by some straight line. This statistical tool, which is the subject of Chapter 6, is called a *linear regression*. We will present measures of goodness-of-fit to assess the quality of the estimated regression. A popular application is the regression of some stock return on the return of a broad-based market index such as the Standard and Poor's 500. Our focus in Chapter 6 is on the univariate regression, also referred to as a *simple linear regression*. This means that there is one variable (the independent variable) that is assumed to affect some variable (the dependent variable). Part Four of this book is devoted to extending the bivariate regression model to the multivariate case where there is more than one independent variable.

An extension of the regression model to the case where the data set in the analysis is a time series is described in Chapter 7. In time series analysis we observe the value of a particular variable over some period of time. We assume that at each point in time, the value of the variable can be decomposed into several components representing, for example, seasonality and trend. Instead of the variable itself, we can alternatively look at the changes between successive observations to obtain the related difference equations. In time series analysis we encounter the notion of noise in observations. A well-known example is the so-called random walk as a model of a stock price process. In Chapter 7, we will also present the error correction model for stock prices.

Introduction

Part Two: Basic Probability Theory

The basics of probability theory are covered in the nine chapters of Part Two. In Chapter 8, we briefly treat the historical evolution of probability theory and its main concepts. To do so, it is essential that mathematical set operations are introduced. We then describe the notions of outcomes, events, and probability distributions. Moreover, we distinguish between countable and uncountable sets. It is in this chapter, the concept of a random variable is defined. The concept of random variables and their probability distributions are essential in models in finance where, for example, stock returns are modeled as random variables. By giving the associated probability distribution, the random behavior of a stock's return will then be completely specified.

Discrete random variables are introduced in Chapter 9 where some of their parameters such as the mean and variance are defined. Very often we will see that the intuition behind some of the theory is derived from the variables of descriptive statistics. In contrast to descriptive statistics, the parameters of random variables no longer vary from sample to sample but remain constant for all drawings. We conclude Chapter 9 with a discussion of the most commonly used discrete probability distributions: binomial, hypergeometric, multinomial, Poisson, and discrete uniform. Discrete random variables are applied in finance whenever the outcomes to be modeled consist of integer numbers such as the number of bonds or loans in a portfolio that might default within a certain period of time or the number of bankruptcies over some period of time.

In Chapter 10, we introduce the other type of random variables, continuous random variables and their distributions including some location and scale parameters. In contrast to discrete random variables, for continuous random variables any countable set of outcomes has zero probability. Only entire intervals (i.e., uncountable sets) can have positive probability. To construct the probability distribution function, we need the probability density functions (or simply density functions) typical of continuous random variables. For each continuous random variable, the density function is uniquely defined as the marginal rate of probability for any single outcome. While we hardly observe true continuous random variables in finance, they often serve as an approximation to discretely distributed ones. For example, financial derivatives such as call options on stocks depend in a completely known fashion on the prices of some underlying random variable such as the underlying stock price. Even though the underlying prices are discrete, the theoretical derivative pricing models rely on continuous probability distributions as an approximation.

Some of the most well-known continuous probability distributions are presented in Chapter 11. Probably the most popular one of them is the normal distribution. Its popularity is justified for several reasons. First, under certain conditions, it represents the limit distribution of sums of random variables. Second, it has mathematical properties that make its use appealing. So, it should not be a surprise that a vast variety of models in finance are based on the normal distribution. For example, three central theoretical models in finance—the Capital Asset Pricing Model, Markowitz portfolio selection theory, and the Black-Scholes option pricing model—rely upon it. In Chapter 11, we also introduce many other distributions that owe their motivation to the normal distribution. Additionally, other continuous distributions in this chapter (such as the exponential distribution) that are important by themselves without being related to the normal distribution are discussed. In general, the continuous distributions presented in this chapter exhibit pleasant features that act strongly in favor of their use and, hence, explain their popularity with financial model designers even though their use may not always be justified when comparing them to real-world data.

Despite the use of the widespread use of the normal distribution in finance, it has become a widely accepted hypothesis that financial asset returns exhibit features that are not in agreement with the normal distribution. These features include the properties of asymmetry (i.e., skewness), excess kurtosis, and heavy tails. Understanding skewness and heavy tails is important in dealing with risk. The skewness of the distribution of say the profit and loss of a bank's trading desk, for example, may indicate that the downside risk is considerably greater than the upside potential. The tails of a probability distribution indicate the likelihood of extreme events. If adverse extreme events are more likely than what would be predicted by the normal distribution, then a distribution is said to have a heavy (or fat) tail. Relying on the normal distribution to predict such unfavorable outcomes will underestimate the true risk. For this reason, in Chapter 12 we present a collection of continuous distributions capable of modeling asymmetry and heavy tails. Their parameterization is not quite easily accessible to intuition at first. But, in general, each of the parameters of some distribution has a particular meaning with respect to location and overall shape of the distribution. For example, the Pareto distribution that is described in Chapter 12 has a tail parameter governing the rate of decay of the distribution in the extreme parts (i.e., the tails of the distribution).

The distributions we present in Chapter 12 are the generalized extreme value distributions, the log-normal distribution, the generalized Pareto distribution, the normal inverse Gaussian distribution, and the α-stable (or alpha-stable) distribution. All of these distributions are rarely discussed in introductory statistics books nor covered thoroughly in finance books. However, as the overwhelming empirical evidence suggests, especially during volatile periods, the commonly used normal distribution is unsuitable for modeling financial asset returns. The α-stable distributions, a more general class of limiting distributions than the normal distribution, qualifies as a candidate for modeling stock returns in very volatile market environments such as during a financial crisis. As we will explain, some distributions lack analytical closed-form solutions of their density functions, requiring that these distributions have to be approximated using their characteristic functions, which is a function, as will be explained, that is unique to every probability distribution.

In Chapter 13, we introduce parameters of location and spread for both discrete and continuous probability distributions. Whenever necessary, we point out differences between their computation in the discrete and the continuous cases. Although some of the parameters are discussed in earlier chapters, we review them in Chapter 13 in greater detail. The parameters presented in this chapter include quantiles, mean, and variance. Moreover, we explain the moments of a probability distribution that are of higher order (i.e., beyond the mean and variance), which includes skewness and kurtosis. Some distributions, as we will see, may not possess finite values of all of these quantities. As an example, the α-stable distributions only has a finite mean and variance for certain values of their characteristic function parameters. This attribute of the α-stable distribution has prevented it from enjoying more widespread acceptance in the finance world, because many theoretical models in finance rely on the existence of all moments.

The chapters in Part Two thus far have only been dealing with one-dimensional (univariate) probability distributions. However, many fields of finance deal with more than one variable such as a portfolio consisting of many stocks and/or bonds. In Chapter 14, we extend the analysis to joint (or multivariate) probability distributions, the theory of which will be introduced separately for discrete and continuous probability distributions. The notion of random vectors, contour lines, and marginal distributions are introduced. Moreover, independence in the probabilistic sense is defined. As measures of linear dependence, we discuss the covariance and correlation coefficient and emphasize the limitations of their usability. We conclude the chapter with illustrations using some of the most common multivariate distributions in finance.

Chapter 15 introduces the concept of conditional probability. In the context of descriptive statistics, the concept of conditional distributions was explained earlier in the book. In Chapter 15, we give the formal definitions of conditional probability distributions and conditional moments such as the conditional mean. Moreover, we discuss Bayes' formula. Applications in finance include risk measures such as the expected shortfall or conditional value-at-risk, where the expected return of some portfolio or trading position is computed conditional on the fact that the return has already fallen below some threshold.

The last chapter in Part Two, Chapter 16, focuses on the general structure of multivariate distributions. As will be seen, any multivariate distribution can be decomposed into two components. One of these components, the copula, governs the dependence between the individual elements of a random vector and the other component specifies the random behavior of each element individually (i.e., the so-called marginal distributions of the elements). So, whenever the true distribution of a certain random vector representing the constituent assets of some portfolio, for example, is unknown, we can recover it from the copula and the marginal distributions. This is a result frequently used in modeling market, credit, and operational risks. In the illustrations, we demonstrate the different effects various choices of copulae (the plural of copula) have on the multivariate distribution. Moreover, in this chapter, we revisit the notion of probabilistic dependence and introduce an additional dependence measure. In previous chapters, the insufficiency of the correlation measure was pointed out with respect to dependence between asset returns. To overcome this deficiency, we present a measure of tail dependence, which is extremely valuable in assessing the probability for two random variables to jointly assume extremely negative or positive values, something the correlation coefficient might fail to describe.

Part Three: Inductive Statistics

Part Three concentrates on statistical inference as the method of drawing information from sample data about unknown parameters. In the first of the three chapters in Part Three, Chapter 17, the point estimator is presented. We emphasize its random character due to its dependence on the sample data. As one of the easiest point estimators, we begin with the sample mean as an estimator for the population mean. We explain why the sample mean is a particular form of the larger class of linear estimators. The quality of some point estimators as measured by their bias and their mean square error is explained. When samples become very large, estimators may develop certain behavior expressed by their so-called *large sample criteria*. Large sample criteria offer insight into an estimator's behavior as the sample size increases up to infinity. An important large sample criterion is the consistency needed to assure that the estimators will eventually approach the unknown parameter. Efficiency, another large sample criterion, guarantees that this happens faster than for any other unbiased estimator. Also in this chapter, retrieving the best estimator for some unknown parameter, which is usually given by the so-called *sufficient statistic* (if it should exist), is explained. Point estimators are necessary to specify all unknown distributional parameters of models in finance. For example, the return volatility of some portfolio measured by the standard deviation is not automatically known even if we assume that the returns are normally distributed. So, we have to estimate it from a sample of historical data.

In Chapter 18, we introduce the confidence interval. In contrast to the point estimator, a confidence interval provides an entire range of values for the unknown parameter. We will see that the construction of the confidence interval depends on the required confidence level and the sample size. Moreover, the quality criteria of confidence intervals regarding the trade-off between precision and the chance to miss the true parameter are explained. In our analysis, we point out the advantages of symmetric confidence intervals, as well as emphasizing how to properly interpret them. The illustrations demonstrate different confidence intervals for the mean and variance of the normal distribution as well as parameters of some other distributions, such as the exponential distribution, and discrete distributions, such as the binomial distribution.

The final chapter in Part Two, Chapter 19, covers hypothesis testing. In contrast to the previous two chapters, the interest is not in obtaining a single estimate or an entire interval of some unknown parameter but instead in verifying whether a certain assumption concerning this parameter is justified. For this, it is necessary to state the hypotheses with respect to our assumptions. With these hypotheses, one can then proceed to develop a decision rule about the parameter based on the sample. The types of errors made in hypothesis testing—type I and type II errors—are described. Tests are usually designed so as to minimize—or at least bound—the type I error to be controlled by the test size. The often used p-value of some observed sample is introduced in this chapter. As quality criteria, one often focuses on the power of the test seeking to identify the most powerful test for given hypotheses. We explain why it is desirable to have an unbiased and consistent test. Depending on the problem under consideration, a test can be either a one-tailed test or a two-tailed test. To test whether a pair of empirical cumulative relative frequency distributions stem from the same distribution, we can apply the Kolmogorov-Smirnov test. The likelihood-ratio test is presented as the test used when we want to find out whether certain parameters of the distribution are zero or not. We provide illustrations for the most common test situations. In particular, we illustrate the problem of having to find out whether the return volatility of a certain portfolio has increased or not, or whether the inclusion of new stocks into some portfolio increased the overall portfolio return or not.

Part Four: Multivariate Linear Regression

One of the most commonly used statistical tools in finance is regression analysis. In Chapter 6, we introduced the concept of regression for one independent and one dependent variable (i.e., univariate regression or simple linear regression). However, much more must be understand about regression analysis and for this reason in the three chapters in Part Four we extend the coverage to the multivariate linear regression case.

In Chapter 20, we will give the general assumptions of the multivariate linear regression model such as normally and independently distributed errors. Relying on these assumptions, we can lay out the steps of estimating the coefficients of the regression model. Regression theory will rely on some knowledge of linear algebra and, in particular, matrix and vector notation. (This will be provided in Appendix B.) After the model has been estimated, it will be necessary to evaluate its quality through diagnostic checks and the model's statistical significance. The analysis of variance is introduced to assess the overall usefulness of the regression. Additionally, determining the significance of individual independent variables using the appropriate F-statistics is explained. The two illustrations presented include the estimation of the duration of certain sectors of the financial market and the prediction of the 10-year Treasury yield.

In Chapter 21, we focus on the design and the building process of multivariate linear regression models. The three principal topics covered in this chapter are the problem of multicollinearity, incorporating dummy variables into a regression model and model building techniques using stepwise regression analysis. Multicollinearity is the problem that is caused by including in a multivariate linear regression independent variables that themselves may be highly correlated. Dummy variables allow the incorporation of independent variables that represent a characteristic or attribute such as industry sector or a time period within which an observation falls. Because the value of a variable is either one or zero, dummy variables are also referred to as binary variables. A stepwise regression is used for determining the suitable independent variables to be included in the final regression model. The three methods that can be used in a stepwise regression—stepwise inclusion method, stepwise exclusion method, and standard stepwise regression method—are described.

In the introduction to the multivariate linear regression in Chapter 21, we set forth the assumptions about the function form of the model (i.e., that it is linear) and assumptions about the residual or error term in the model (normally distribution, constant variance, and uncorrelated). These assumptions must be investigated. Chapter 22 describes these assumptions in more detail and how to test for any violations. The tools for correcting any violation are briefly described.

Appendixes

Statistics draws on other fields in mathematics. For this reason, we have included two appendices that provide the necessary theoretical background in mathematics to understand the presentations in some of the chapters. In Appendix A, we present important mathematical functions and their features that are needed primarily in the context of Part Two of this book. These functions include the continuous function, indicator function, and monotonic function. Moreover, important concepts from differential and integral calculus are explained. In Appendix B, we cover the fundamentals of matrix operations and concepts needed to understand the presentation in Part Four.

In Appendix C, we explain the construction of the binomial and multinomial coefficients used in some discrete probability distributions covered in Chapter 9. In Appendix D, we present an explicit computation of the price formula for European-style call options when stock prices are assumed to be log-normally distributed.

PART One

Descriptive Statistics

CHAPTER 2
Basic Data Analysis

We are confronted with data every day. Daily newspapers contain information on stock prices, economic figures, quarterly business reports on earnings and revenues, and much more. These data offer observed values of given quantities. In this chapter, we explain the basic data types: qualitative, ordinal, and quantitative. For now, we will restrict ourselves to *univariate data*, that is data of only one dimension. For example, if you follow the daily returns of one particular stock, you obtain a one-dimensional series of observations. If you had observed two stocks, then you would have obtained a two-dimensional series of data, and so on. Moreover, the notions of frequency distributions, empirical frequency distributions, and cumulative frequency distributions are introduced.

The goal of this chapter is to provide the methods necessary to begin data analysis. After reading this chapter, you will learn how to formalize the first impression you obtain from the data in order to retrieve the most basic structure inherent in the data. That is essential for any subsequent tasks you may undertake with the data. Above all, though, you will have to be fully aware of what you want to learn from the data. For example, you may just want to know what the minimum return has been of your favorite stock during the last year before you decide to purchase that stock. Or you may be interested in all returns from last year to learn how this stock typically performs, that is, which returns occur more often than others, and how often. In the latter case, you definitely have to be more involved to obtain the necessary information than just knowing the minimum return. Determining the objective of the analysis is the most important task before getting started in investigating the data.

DATA TYPES
How To Obain Data

Data are gathered by several methods. In the financial industry, we have market data based on regular trades recorded by the exchanges. These data are directly observable. Aside from the regular trading process, there is so-called over-the-counter (OTC) business whose data may be less accessible. Annual reports and quarterly reports are published by companies themselves in print or electronically. These data are available also in the business and finance sections of most major business-oriented print media and the Internet. The fields of marketing and the social sciences employ additional forms of data collection methods such as telephone surveys, mail questionnaires, and even experiments.

If one does research on certain financial quantities of interest, one might find the data available from either free or commercial databases. Hence, one must be concerned with the quality of the data. Unfortunately, very often databases of unrestricted access such as those available on the Internet may be of limited credibility. In contrast, there are many commercial purveyors of financial data who are generally acknowledged as providing accurate data. But, as always, quality has its price.

The Information Contained in the Data

Once the data are gathered, it is the objective of descriptive statistics to visually and computationally convert the information collected into quantities that reveal the essentials in which we are interested. Usually in this context, visual support is added since very often that allows for a much easier grasp of the information.

The field of descriptive statistics discerns different types of data. Very generally, there are two types: nonquantitative (i.e., qualitative and ordinal) and quantitative data.

If certain attributes of an item can only be assigned to categories, these data are referred to as *qualitative data*. For example, stocks listed on the New York Stock Exchange (NYSE) as items can be categorized as belonging to a specific industry sector such as the "banking," "energy," "media and telecommunications," and so on. That way, we assign each item (i.e., stock) as its attribute sector one or possibly more values from the set containing "banking," "energy," "media and telecommunications," and so on.[1] Another example would be the credit ratings assigned to debt obligations by commercial rating companies such as Standard & Poor's, Moody's, and Fitch Ratings. Except for retrieving the value of an attribute, nothing more can be done with qualitative data. One may use a numerical code to indicate the different sectors, e.g, 1 = "banking," 2 = "energy," and so on. However, we cannot perform any computation with these figures since they are simply names of the underlying attribute sector.

---

[1] Instead of attribute, we will most of the time use the term "variable."

On the other hand, if an item is assigned a *quantitative variable*, the value of this variable is numerical. Generally, all real numbers are eligible. Depending on the case, however, one will use discrete values, only, such as integers. Stock prices or dividends, for example, are quantitative data drawing from—up to some digits—positive real numbers. Quantitative data have the feature that one can perform transformations and computations with them. One can easily think of the market capitalization of all companies comprising some index on a certain day while it would make absolutely no sense to do the same with qualitative data.[2]

Data Levels and Scale

In descriptive statistics we group data according to measurement levels. The *measurement level* gives an indication as to the sophistication of the analysis techniques that one can apply to the data collected. Typically, a hierarchy with five levels of measurement is used to group data: nominal, ordinal, interval, ratio, and absolute data. The latter three form the set of quantitative data. If the data are of a certain measurement level, it is said to be scaled accordingly. That is the data are referred to as nominally scaled, and so on.

Nominally scaled data are on the bottom of the hierarchy. Despite the low level of sophistication, this type of data are commonly used. An example of nominally scaled is qualitative data such as the attribute sector of stocks or the credit rating of a debt obligation. We already learned that even though we can assign numbers as proxies to nominal values, these numbers have no numerical meaning whatsoever. We might just as well assign letters to the individual nominal values, for example, "B = banking," "E = energy" and so on.

Ordinally scaled data are one step higher in the hierarchy. One also refers to this type as *rank data* since one can already perform a ranking within the set of values. We can make use of a relationship among the different values by treating them as quality grades. For example, we can divide the stocks comprising a particular stock index according to their market capitalization into five groups of equal size. Let "A" denominate the top 20% of the stocks. Also, let "B" denote the next 20% below, and so on, until we obtain the five groups "A," "B," "C," "D," and "E". After ordinal scaling we can make statements such as group "A" is better than group "C." Hence, we have a natural ranking or order among the values. However, we cannot quantify the difference between them.

---

[2]Market capitalization is the total market value of the common stock of a company. It is obtained by multiplying the number of shares outstanding by the market price per share.

Until now, we can summarize that while we can test the relationship between nominal data for equality only, we can additionally determine a greater or less than relationship between ordinal data.

Data on an *interval scale* are such that they can be reasonably transformed by a linear equation. Suppose we are given values for some variable x. It is now feasible to express a new variable y by the relationship $y = ax + b$ where the x's are our original data. If x has a meaning, then so does y. It is obvious that data have to possess a numerical meaning and therefore be quantitative in order to be measured on an interval scale. For example, consider the temperature F given in degrees Fahrenheit. Then, the corresponding temperature in degrees Celsius C will result from the equation $C = (F - 32)/1.8$. Equivalently, if one is familiar with physics, the same temperature measured in degrees Kelvin, K, will result from $K = C + 273.15$. So, say the temperature on a given day is 55° Fahrenheit for Americans, the same temperature will mean approximately 13° Celsius for Europeans and they will not feel any cooler. Generally, interval data allow for the calculation of differences. For example, (70°–60°) Fahrenheit = 10° Fahrenheit may reasonably express the difference in temperature between Los Angeles and San Francisco. But be careful, the difference in temperature measured in Celsius between the two cities is not the same.

Data measured on a *ratio scale* share all the properties of interval data. In addition, ratio data have a fixed or true zero point. This is not the case with interval data. Their intercept, b, can be arbitrarily changed through transformation. Since the zero point of ratio data is invariable, one can only transform the slope, a. So, for example, $y = ax$ is always a multiple of x. In other words, there is a relationship between y and x given by the ratio a, hence, the name used to describe this type of data. One would not have this feature if one would permit some b different from zero in the transformation. Consider, for example, the stock price, E, of some European stock given in euro units. The same price in U.S. dollars, D, would be D equals E times the exchange rate between euros and U.S. dollars. But if a company's stock price after bankruptcy went to zero, the price in either currency would be zero even at different rates determined by the ratio of U.S. dollar per Euro. This is a result of the invariant zero point.

Absolute data are given by quantitative data measured on a scale even stricter than for ratio data. Here, along with the zero point, the units are invariant as well. Data measured on an absolute scale occurs when transformation would be mathematically feasible but lacked any interpretational implication. A common example is provided by counting numbers. Anybody would agree on the number of stocks comprising a certain stock index. There is no ambiguity as to the zero point and the count increments. If one stock is added to the index, it is immediately clear that the difference to the content of the old index is exactly one unit of stock assuming that no stock is deleted. This absolute scale is the most intuitive and needs no further discussion.

Cross-Sectional Data and Time Series

There is another way of classifying data. Imagine collecting data from one and the same quantity of interest or *variable*. A variable is some quantity that can assume values from a value set. For example, the variable "stock price" can technically assume any nonnegative real number of currency but only one value at a time. Each day, it assumes a certain value that is the day's stock price. As another example, a variable could be the dividend payments from a specific company over some period of time. In the case of dividends, the observations are made each quarter. The set of data then form what is called *time series data*. In contrast, one could pick a particular time period of interest such as the first quarter of the current year and observe the dividend payments of all companies comprising the Standard & Poor's 500 index. By doing so, one would obtain *cross-sectional data* of the universe of stocks in the S&P 500 index at that particular time.

Summarizing, time series data are data related to a variable successively observed at a sequence of points in time. Cross-sectional data are values of a particular variable across some universe of items observed at a unique point in time. This is visualized in Figure 2.1.

FIGURE 2.1 The Relationship between Cross-Sectional and Time Series Data

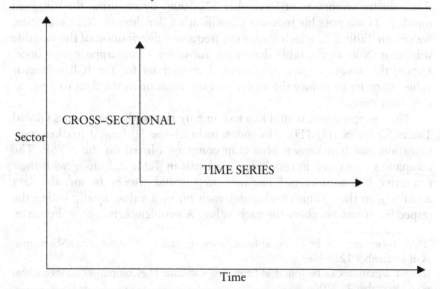

FREQUENCY DISTRIBUTIONS

Sorting and Counting Data

One of the most important aspects when dealing with data is that they are effectively organized and transformed in order to convey the essential information contained in them. This processing of the original data helps to display the inherent meaning in a way that is more accessible for intuition. But before advancing to the graphical presentation of the data, we first describe the methods of structuring data.

Suppose that we are interested in a particular variable that can assume a set of either finite or infinitely many values. These values may be qualitative or quantitative by nature. In either case, the initial step when obtaining a data sample for some variable is to sort the values of each observation and then to determine the frequency distribution of the data set. This is done simply by counting the number of observations for each possible value of the variable. Alternatively, if the variable can assume values on all or part of the real line, the frequency can be determined by counting the number of observations that fall into nonoverlapping intervals partitioning the real line.

In our illustration, we begin with qualitative data first and then move on to the quantitative aspects. For example, suppose we want to analyze the frequency of the industry subsectors of the component stocks in the Dow Jones Industrial Average (DJIA), an index comprised of 30 U.S. stocks.[3] Table 2.1 displays the 30 companies in the index along with their respective industry sectors as of December 12, 2006. By counting the observed number of each possible Industry Classification Benchmark (ICB) subsector, we obtain Table 2.2, which shows the frequency distribution of the variable subsector. Note in the table that many subsector values appear only once. Hence, this might suggest employing a coarser set for the ICB subsector values in order to reduce the amount of information in the data to a necessary minimum.

Now suppose you would like to compare this to the Dow Jones Global Titans 50 Index (DJGTI). Ths index includes the 50 largest market capitalization and best-known blue chip companies listed on the NYSE. The companies contained in this index are listed in Table 2.3 along with their respective ICB subsectors.[4] The next step would also be to sort the data according to their values and count each hit of a value, finally listing the respective count numbers for each value. A problem arises now, however,

---

[3]The information can be found at http://www.dj.com/TheCompany/FactSheets.htm as of December 12, 2006.

[4]The information can be found at http://www.dj.com/TheCompany/FactSheets.htm as of December 12, 2006.

TABLE 2.1 DJIA Components as of December 12, 2006

Company	Industrial Classification Benchmak (ICB) Subsector
3M Co.	Diversified Industrials
Alcoa Inc.	Aluminum
Altria Group Inc.	Tobacco
American Express Co.	Consumer Finance
American International Group Inc.	Full Line Insurance
AT&T Inc.	Fixed Line Telecommunications
Boeing Co.	Aerospace
Caterpillar Inc.	Commercial Vehicles & Trucks
Citigroup Inc.	Banks
Coca-Cola Co.	Soft Drinks
E.I. DuPont de Nemours & Co.	Commodity Chemicals
Exxon Mobil Corp.	Integrated Oil & Gas
General Electric Co.	Diversified Industrials
General Motors Corp.	Automobiles
Hewlett-Packard Co.	Computer Hardware
Home Depot Inc.	Home Improvement Retailers
Honeywell International Inc.	Diversified Industrials
Intel Corp.	Semiconductors
International Business Machines Corp.	Computer Services
Johnson & Johnson	Pharmaceuticals
JPMorgan Chase & Co.	Banks
McDonald's Corp.	Restaurants & Bars
Merck & Co. Inc.	Pharmaceuticals
Microsoft Corp.	Software
Pfizer Inc.	Pharmaceuticals
Procter & Gamble Co.	Nondurable Household Products
United Technologies Corp.	Aerospace
Verizon Communications Inc.	Fixed Line Telecommunications
Wal-Mart Stores Inc.	Broadline Retailers
Walt Disney Co.	Broadcasting & Entertainment

Source: Dow Jones, "The Company, Fact Sheets," http://www.dj.com/.

TABLE 2.2 Frequency Distribution of the Industry Subsectors for the DJIA Components as of December 12, 2006

ICB Subsector	Frequency a_i
Aerospace	2
Aluminum	1
Automobiles	1
Banks	2
Broadcasting & Entertainment	1
Broadline Retailers	1
Commercial Vehicles & Trucks	1
Commodity Chemicals	1
Computer Hardware	1
Computer Services	1
Consumer Finance	1
Diversified Industrials	3
Fixed Line Telecommunications	2
Full Line Insurance	1
Home Improvement Retailers	1
Integrated Oil & Gas	1
Nondurable Household Products	1
Pharmaceuticals	3
Restaurants & Bars	1
Semiconductors	1
Soft Drinks	1
Software	1
Tobacco	1

Source: Dow Jones, "The Company, Fact Sheets," http://www.dj.com/.

when you want to directly compare the numbers with those obtained for the DJIA because the number of stocks contained in each index is not the same. Hence, we cannot compare the respective absolute frequencies. Instead, we have to resort to something that creates comparability of the two data sets. This is done by expressing the number of observations of a particular value as the proportion of the total number of observations in a specific data set. That means we have to compute the ***relative frequency***.

TABLE 2.3 Dow Jones Global Titans 50 Index as of December 12, 2006

Company Name	ICB Subsector
Abbott Laboratories	Pharmaceuticals
Altria Group Inc.	Tobacco
American International Group Inc.	Full Line Insurance
Astrazeneca PLC	Pharmaceuticals
AT&T Inc.	Fixed Line Telecommunications
Bank of America Corp.	Banks
Barclays PLC	Banks
BP PLC	Integrated Oil & Gas
Chevron Corp.	Integrated Oil & Gas
Cisco Systems Inc.	Telecommunications Equipment
Citigroup Inc.	Banks
Coca-Cola Co.	Soft Drinks
ConocoPhillips	Integrated Oil & Gas
Dell Inc.	Computer Hardware
ENI S.p.A.	Integrated Oil & Gas
Exxon Mobil Corp.	Integrated Oil & Gas
General Electric Co.	Diversified Industrials
GlaxoSmithKline PLC	Pharmaceuticals
HBOS PLC	Banks
Hewlett-Packard Co.	Computer Hardware
HSBC Holdings PLC (UK Reg)	Banks
ING Groep N.V.	Life Insurance
Intel Corp.	Semiconductors
International Business Machines Corp.	Computer Services
Johnson & Johnson	Pharmaceuticals
JPMorgan Chase & Co.	Banks
Merck & Co. Inc.	Pharmaceuticals
Microsoft Corp.	Software
Mitsubishi UFJ Financial Group Inc.	Banks
Morgan Stanley	Investment Services
Nestle S.A.	Food Products

TABLE 2.3 (Continued)

Company Name	ICB Subsector
Nokia Corp.	Telecommunications Equipment
Novartis AG	Pharmaceuticals
PepsiCo Inc.	Soft Drinks
Pfizer Inc.	Pharmaceuticals
Procter & Gamble Co.	Nondurable Household Products
Roche Holding AG Part. Cert.	Pharmaceuticals
Royal Bank of Scotland Group PLC	Banks
Royal Dutch Shell PLC A	Integrated Oil & Gas
Samsung Electronics Co. Ltd.	Semiconductors
Siemens AG	Electronic Equipment
Telefonica S.A.	Fixed Line Telecommunications
Time Warner Inc.	Broadcasting & Entertainment
Total S.A.	Integrated Oil & Gas
Toyota Motor Corp.	Automobiles
UBS AG	Banks
Verizon Communications Inc.	Fixed Line Telecommunications
Vodafone Group PLC	Mobile Telecommunications
Wal-Mart Stores Inc.	Broadline Retailers
Wyeth	Pharmaceuticals

Source: Dow Jones, "The Company, Fact Sheets," http://www.dj.com/.

Formal Presentation of Frequency

For a better formal presentation, we denote the (absolute) frequency by a and, in particular, by a_i for the i-th value of the variable. Formally, the relative frequency f_i of the i-th value is then defined by

$$f_i = \frac{a_i}{n}$$

where n is the total number of observations. With k being the number of the different values, the following holds

$$n = \sum_{i=1}^{k} f_i$$

In our illustration, let $n_1 = 30$ be the number of total observations in the DJIA and $n_2 = 50$ the total number of observations in the DJGTI. Table 2.4 shows the relative frequencies for all possible values. Notice that each index has some values that were observed with zero frequency, which still have to be listed for comparison. When we look at the DJIA, we observe that the sectors Diversified Industrials and Pharmaceuticals each account for 10% of all sectors and therefore are the sectors with the highest frequencies. Comparing these two sectors to the DJGTI, we see that Pharmaceuticals play as important a role as a sector with an 18% share, while Diversified Industrials are of minor importance. Instead, Banks are also the largest sector with 18%. A comparison of this sort can now be carried through for all subsectors thanks to the relative frequencies.

Naturally, frequency (absolute and relative) distributions can be computed for all types of data since they do not require that the data have a numerical value.

EMPIRICAL CUMULATIVE FREQUENCY DISTRIBUTION

Accumulating Frequencies

In addition to the frequency distribution, there is another quantity of interest for comparing data that are closely related to the absolute or relative frequency distribution. Suppose that one is interested in the percentage of all large market capitalization stocks in the DJIA with closing prices of less than U.S. $50 on a specific day. One can sort the observed closing prices by their numerical values in ascending order to obtain something like the array shown in Table 2.5 for market prices as of December 15, 2006. Note that since each value occurs only once, we have to assign each value an absolute frequency of 1 or a relative frequency of 1/30, respectively, since there are 30 component stocks in the DJIA. We start with the lowest entry ($20.77) and advance up to the largest value still less than $50, which is $49 (Coca-Cola). Each time we observe less than $50, we added 1/30, accounting for the frequency of each company to obtain an accumulated frequency of 18/30 representing the total share of closing prices below $50. This accumulated frequency is called the *empirical cumulative frequency* at the value $50. If one computes this for all values, one obtains the *empirical cumulative frequency distribution*. The word "empirical" is used because we only consider values that are actually observed. The theoretical equivalent of the cumulative distribution function where all theoretically possible values are considered will be introduced in the context of probability theory in Chapter 8.

TABLE 2.4 Comparison of Relative Frequencies of DJIA and DJGTI.

	Relative Frequencies	
ICB Subsector	DJIA	DJGTI
Aerospace	0.067	0.000
Aluminum	0.033	0.000
Automobiles	0.033	0.020
Banks	0.067	0.180
Broadcasting & Entertainment	0.033	0.020
Broadline Retailers	0.033	0.020
Commercial Vehicles & Trucks	0.033	0.000
Commodity Chemicals	0.033	0.000
Computer Hardware	0.033	0.040
Computer Services	0.033	0.020
Consumer Finance	0.033	0.000
Diversified Industrials	0.100	0.020
Electronic Equipment	0.000	0.020
Fixed Line Telecommunications	0.067	0.060
Food Products	0.000	0.020
Full Line Insurance	0.033	0.020
Home Improvement Retailers	0.033	0.000
Integrated Oil & Gas	0.033	0.140
Investment Services	0.000	0.020
Life Insurance	0.000	0.020
Mobile Telecommunications	0.000	0.020
Nondurable Household Products	0.033	0.020
Pharmaceuticals	0.100	0.180
Restaurants & Bars	0.033	0.000
Semiconductors	0.033	0.040
Soft Drinks	0.033	0.040
Software	0.033	0.020
Telecommunications Equipment	0.000	0.040
Tobacco	0.033	0.020

TABLE 2.5 DJIA Stocks by Share Price in Ascending Order as of December 15, 2006

Company	Share Price
Intel Corp.	$20.77
Pfizer Inc.	25.56
General Motors Corp.	29.77
Microsoft Corp.	30.07
Alcoa Inc.	30.76
Walt Disney Co.	34.72
AT&T Inc.	35.66
Verizon Communications Inc.	36.09
General Electric Co.	36.21
Hewlett-Packard Co.	39.91
Home Depot Inc.	39.97
Honeywell International Inc.	42.69
Merck & Co. Inc.	43.60
McDonald's Corp.	43.69
Wal-Mart Stores Inc.	46.52
JPMorgan Chase & Co.	47.95
E.I. DuPont de Nemours & Co.	48.40
Coca-Cola Co.	49.00
Citigroup Inc.	53.11
American Express Co.	61.90
United Technologies Corp.	62.06
Caterpillar Inc.	62.12
Procter & Gamble Co.	63.35
Johnson & Johnson	66.25
American International Group Inc.	72.03
Exxon Mobil Corp.	78.73
3M Co.	78.77
Altria Group Inc.	84.97
Boeing Co.	89.93
International Business Machines Corp.	95.36

Source: http://www.dj.com/TheCompany/FactSheets.htm, December 15, 2006.

Formal Presentation of Cumulative Frequency Distributions

Formally, the empirical cumulative frequency distribution F_{emp} is defined as

$$F_{emp}(x) = \sum_{i=1}^{k} a_i$$

where k is the index of the largest value observed that is still less than x. In our example, k is 18. When we use relative frequencies, we obtain the empirical relative cumulative frequency distribution defined analogously to the empirical cumulative frequency distribution, this time using relative frequencies. Hence, we have

$$F_{emp}^{f}(x) = \sum_{i=1}^{k} f_i$$

In our example, $F_{emp}^{f}(50) = 18/30 = 0.6 = 60\%$.

Note that the empirical cumulative frequency distribution can be evaluated at any real x even though x need not be an observation. For any value x between two successive observations $x_{(i)}$ and $x_{(i+1)}$, the empirical cumulative frequency distribution as well as the empirical cumulative relative frequency distribution remain at their respective levels at $x_{(i)}$; that is, they are of constant level $F_{emp}(x_{(i)})$ and $F_{emp}^{f}(x_{(i)})$, respectively. For example, consider the empirical relative cumulative frequency distribution for the data shown in Table 2.5. We can extend the distribution to a *function* that determines the value of the distribution at each possible value of the stock price.[5] The function is given in Table 2.6. Notice that if no value is observed more than once, then the empirical relative cumulative frequency distribution jumps by $1/N$ at each observed value. In our illustration, the jump size is 1/30.

In Figure 2.2 the empirical relative cumulative frequency distribution is shown a graph. Note that the values of the function are constant on the extended line between two successive observations, indicated by the solid point to the left of each horizontal line. At each observation, the vertical distance between the horizontal line extending to the right from the preceding observation and the value of the function is exactly the increment, 1/30.

The computation of either form of empirical cumulative distribution function is obviously not intuitive for categorical data unless we assign some meaningless numerical proxy to each value such as "Sector A" = 1, "Sector B" = 2, and so on.

---

[5] A function is the formal way of expressing how some quantity y changes depending on the value of some other quantity x. This leads to the brief functional representation of this relationship such as $y = f(x)$.

Basic Data Analysis

TABLE 2.6 Empirical Relative Cumulative Frequency Distribution of DJIA Stocks from Table 2.5

$$F_{emp}^{f}(x) = \begin{cases} 0.00 & x < 20.77 \\ 0.03 & 20.77 \leq x < 25.56 \\ 0.07 & 25.56 \leq x < 29.77 \\ 0.10 & 29.77 \leq x < 30.07 \\ 0.13 & 30.07 \leq x < 30.76 \\ 0.17 & 30.76 \leq x < 34.72 \\ 0.20 & 34.72 \leq x < 35.66 \\ 0.23 & 35.66 \leq x < 36.09 \\ 0.27 & 36.09 \leq x < 36.21 \\ 0.30 & 36.21 \leq x < 39.91 \\ 0.33 & 39.91 \leq x < 39.97 \\ 0.37 & 39.97 \leq x < 42.69 \\ 0.40 & 42.69 \leq x < 43.60 \\ 0.43 & 43.60 \leq x < 43.69 \\ 0.47 & 43.69 \leq x < 46.52 \\ 0.50 & 46.52 \leq x < 47.95 \\ 0.53 & 47.95 \leq x < 48.40 \\ 0.57 & 48.40 \leq x < 49.00 \\ 0.60 & 49.00 \leq x < 53.11 \\ 0.63 & 53.11 \leq x < 61.90 \\ 0.67 & 61.90 \leq x < 62.06 \\ 0.70 & 62.06 \leq x < 62.12 \\ 0.73 & 62.12 \leq x < 63.35 \\ 0.77 & 63.35 \leq x < 66.25 \\ 0.80 & 66.25 \leq x < 72.03 \\ 0.83 & 72.03 \leq x < 78.73 \\ 0.87 & 78.73 \leq x < 78.77 \\ 0.90 & 78.77 \leq x < 84.97 \\ 0.93 & 84.97 \leq x < 89.93 \\ 0.97 & 89.93 \leq x < 95.36 \\ 1.00 & 95.36 \leq x \end{cases}$$

FIGURE 2.2 Empirical Relative Cumulative Frequency Distribution of DJIA Stocks from Table 2.5

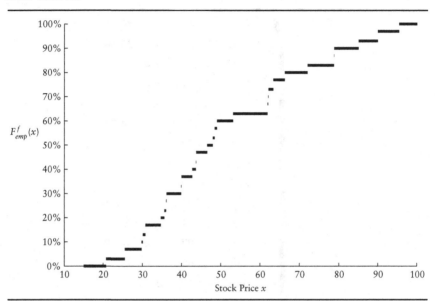

DATA CLASSES

Reasons for Classifying

When quantitative variables are such that the set of values—whether observed or theoretically possible—includes intervals or the entire real numbers, then the variable is said to be a *continuous variable*. This is in contrast to *discrete variables*, which assume values only from a finite or countable set. Variables on a nominal scale cannot be considered in this context. And because of the difficulties with interpreting the results, we will not attempt to explain the issue of classes for rank data either.

When one counts the frequency of observed values of a continuous variable, one notices that hardly any value occurs more than once.[6] Theoretically, with 100% chance, all observations will yield different values. Thus, the method of counting the frequency of each value is not feasible. Instead, the continuous set of values is divided into mutually exclusive intervals. Then for each such interval, the number of values falling within that interval

---

[6]Naturally, the precision given by the number of digits rounded may result in higher occurrences of certain values.

Basic Data Analysis

can be counted again. In other words, one groups the data into classes for which the frequencies can be computed. Classes should be such that their respective lower and upper bounds are real numbers. Moreover, whether the class bounds are elements of the classes or not must be specified. The class bounds of a class have to be bounds of the respective adjacent classes as well, such that the classes seamlessly cover the entire data. The width should be the same for all classes. However, if there are areas where the data are very intensely dense in contrast to areas of lesser density, then the class width can vary according to significant changes in value density. In certain cases, most of the data are relatively evenly scattered within some range while there are extreme values that are located in isolated areas on either end of the data array. Then, it is sometimes advisable to specify no lower bound to the lowest class and no upper bound to the uppermost class. Classes of this sort are called *open classes*. Moreover, one should consider the precision to which the data are given. If values are rounded to the first decimal place but there is the chance that the exact value might vary within half a decimal about the value given, class bounds have to consider this lack of certainty by admitting plus half a decimal on either end of the class.

Formal Procedure for Classifying

Formally, there are four criteria that the classes need to meet.

Criterion #1: Mutual Exclusiveness: Each value can be placed in only one class.
Criterion #2: Completeness: The set of classes needs to cover all values.
Criterion #3: Equidistance: If possible, form classes of equal width.
Criterion #4: Nonemptyness: If possible, avoid forming empty classes.

It is intuitive that the number of classes should increase with an increasing range of values and increasing number of data. Though there are no stringent rules, two rules of thumb are given here with respect to the advised number of classes (first rule) and the best class width (second rule). The first, the so-called *Sturge's rule*, states that for a given set of continuous data of size n, one should use the nearest integer figure to

$$1 + \log_2 n = 1 + 3.222 \log_{10} n$$

Here, $\log_a n$ denotes the logarithm of n to the base a with a being either 2 or 10.

The second guideline is the so called *Freedman-Diaconis rule* for the appropriate class width or bin size. Before turning to the second rule of thumb in more detail, we have to introduce the notion of the *interquartile range* (IQR). This quantity measures the distance between the value where F^f_{emp} is closest to 0.25, that is, the so-called "0.25-quantile," and the value where F^f_{emp} is closest to 0.75, that is, the so-called "0.75-quantile."[7] So, the IQR states how remote the lowest 25% of the observations are from the highest 75%. As a consequence, the *IQR* comprises the central 50% of a data sample. A little more attention will be given to the determination of the above-mentioned quantiles when we discuss sample moments and quantiles since formally there might arise some ambiguity when computing them.[8]

Now we can return to the Freedman-Diaconis rule. It states that a good class width is given by the nearest integer to

$$2 \times IQR \times N^{-1/3}$$

where N is the number of observations in the data set. Note that there is an inverse relationship between the class width and the number of classes for each set of data. That is, given that the partitioning of the values into classes covers all observations, the number of classes n has to be equal to the difference between the largest and smallest value divided by the class width if classes are all of equal size w. Mathematically, that means

$$n = (x_{max} - x_{min}) / w$$

where x_{max} denotes the largest value and x_{min} denotes the smallest value considered.

One should not be intimidated by all these rules. Generally, by mere ordering of the data in an array, intuition produces quite a good feel for what the classes should look like. Some thought can be given to the timing of the formation of the classes. That is, when classes are formed prior to the data-gathering process, one does not have to store the specific values but rather count only the number of hits for each class.

Example of Classifying Procedures

Let's illustrate these rules. Table 2.7 gives the 12-month returns (in percent) of the 235 Franklin Templeton Investments Funds on January 11, 2007. With this many data, it becomes obvious that it cannot be helpful to anyone to know the relative performance for the 235 funds. To obtain an overall impression of the distribution of the data without getting lost in detail, one has to aggregate the information given by classifying the data.

---

[7]The term "percentile" is used interchangeably with quantile.
[8]Note that the IQR cannot be computed for nominal or categorial data in a natural way.

TABLE 2.7 12-Month Returns (in %) for the 235 Franklin Templeton Investment Funds (Luxemburg) on January 11, 2007

Fund	Return	Fund	Return
High Yield A Dis	7.1	US Eqty I Acc USD	8.9
High Yield B Dis	5.6	US Gov A Dis	3.1
High Yield C Acc	6.2	US Gov B Dis	1.8
High Yield I Dis	7.8	US Gov B Acc	1.9
High Yld Eur A Acc	8.3	US Gov C Acc	2.2
High Yld Eur A Dis	8.3	US Gov I Dis	3.8
High Yld Eur I Acc	9.1	US Growth A Acc	3.8
High Yld Eur I Dis	9.1	US Growth B Acc	2.5
Income A Dis	12.8	US Growth C Acc	3.3
Income B Dis	11.4	US Growth I Acc	6.4
Income C Acc	12.1	US Ultra Sh Bd A Dis	3.7
Income C Dis	12.1	US Ultra Sh Bd B Acc	2.5
Income I Acc	13.7	US Ultra Sh Bd B Dis	2.5
India A Acc EUR	29.0	US Ultra Sh Bd C Dis	2.6
India A Acc USD	38.7	US Ultra Sh Bd I Acc	4.2
India A Dis GBP	26.2	US SmMidCapGro A Ac	2.5
India B Acc USD	36.9	US SmMidCapGro B Ac	1.2
India C Acc USD	37.9	US SmMidCapGro C Ac	2.0
India I Acc EUR	30.2	US Tot Rtn A Acc	4.1
India I Acc USD	40.0	US Tot Rtn A Dis	4.2
Mut Beacon AAccEUR	7.4	US Tot Rtn B Acc	2.6
Mut Beacon AAccUSD	15.5	US Tot Rtn B Dis	2.7
Mut Beacon ADisUSD	15.5	US Tot Rtn C Dis	3.1
Mut Beacon Bacc	14.0	US Tot Rtn I Acc	4.8
Mut Beacon Cacc	14.8	Asian Bond A Acc EUR	5.9
Mut Beacon Iacc	16.6	Asian Bond A Acc USD	14.1
Mut Europ AAcc EUR	15.9	Asian Bond A Dis USD	14.0
Mut Europ AAcc USD	24.7	Asian Bond B Dis USD	12.4
Mut Europ Adis EUR	15.9	Asian Bond C Dis USD	13.0
Mut Europ Adis GBP	14.0	Asian Bond I Acc USD	14.6
Mut Europ B Acc	23.1	Asian Grth A Acc EUR	12.7
Mut Europ C Acc USD	23.9	Asian Grth A Acc USD	21.4
Mut Europ C Acc EUR	15.2	Asian Grth A Dis EUR	12.8
Mut Europ I Acc	16.9	Asian Grth A Dis GBP	10.4

TABLE 2.7 (Continued)

Fund	Return	Fund	Return
EmMktBd A Dis EUR	5.2	Glb Inc B Dis	17.2
EmMktBd A Dis USD	13.2	Glb Inc C Dis	17.9
Emg Mkt Bd B Dis	11.9	Glb Inc I Acc	19.4
Emg Mkt Bd C Acc	12.6	Glbl Sm Co A Acc	21.3
Emg Mkt Bd I Acc	14.3	Glbl Sm Co A Dis	21.3
Euro Liq Res A Acc	1.9	Glbl Sm Co C Acc	12.1
Euro Liq Res A Dis	1.9	Glbl Sm Co I Acc	22.4
Euroland Bd A Dis	−1.8	Dlbl Tot Ret A Acc	12.6
Euroland Bd I Acc	−1.2	Dlbl Tot Ret A Dis	12.6
Euroland A Acc	18.5	Dlbl Tot Ret B Acc	10.9
Euroland A Dis	19.8	Dlbl Tot Ret B Dis	10.9
Euroland C Acc	17.8	Dlbl Tot Ret C Dis	11.8
Euroland I Acc	19.6	Dlbl Tot Ret I Acc	13.1
European A Acc USD	24.0	Dlbl Tot Ret I Dis	10.0
European A Acc EUR	15.3	Growth (Euro) A Acc	7.5
European A Dis EUR	15.2	Growth (Euro) A Dis	7.4
European A Dis USD	24.0	Growth (Euro) I Acc	8.4
European C Acc EUR	14.6	Growth (Euro) I Dis	8.4
European I Acc	16.4	Japan A Acc	−8.0
Euro Tot Ret A Acc	−0.4	Korea A Acc	−3.8
Euro Tot Ret A Dis EUR	−0.5	Latin Amer A Acc	35.9
Euro Tot Ret A Dis GBP	−2.6	Latin Amer A Dis GBP	23.6
Euro Tot Ret A Dis USD	7.1	Latin Amer A Dis USD	35.9
Euro Tot Ret C Acc EUR	−1.3	Latin Amer I Acc USD	37.4
Euro Tot Ret C Dis USD	6.2	Thailand A Acc	−11.0
Euro Tot Ret I Acc	−0.3	US$ Liq Res A Acc	4.2
Glbl Bal A Acc EUR	6.5	US$ Liq Res A Dis	4.1
Glbl Bal A Acc USD	14.6	US$ Liq Res B Dis	3.1
Glbl Bal A Dis	14.6	US$ Liq Res C Acc	3.2
Glbl Bal B Acc	13.1	US Value A Acc	14.5
Glbl Bal C Dis	13.9	US Value B Acc	13.0
Glbl Bd A Dis USD	9.2	US Value C Acc	13.8
Glbl Bd A Acc EUR	1.5	US Value I Acc	15.6
Glbl Bd A Dis EUR	1.5		

Basic Data Analysis

For the sake of a better overview, the ordered array is given in Table 2.8. A quick glance at the data sorted in ascending order gives us the lowest (minimum) and largest (maximum) return. Here, we have x_{min} = −18.3% and x_{max} = 41.3%, yielding a range of 59.6% to cover.

We first classify the data according to Sturge's rule. For the number of classes, n, we obtain the nearest integer to $1 + \log_2 235 = 8.877$, which is 9. The class width is then determined by the range divided by the number of classes, 59.6%/9, yielding a width of roughly 6.62%. This is not a nice number to deal with, so we may choose 7% instead without deviating noticeably from the exact numbers given by Sturge's rule. We now cover a range of 9 × 7% = 63%, which is slightly larger than the original range of the data.

TABLE 2.8 Ordered Array of the 235 12-Month Returns for the Franklin Templeton Investment Funds (Luxemburg)

Obs. (i)	Value	Obs. (i)	Value	Obs. (i)	Value	Obs. (i)	Value	Obs. (i)	Value
$x(1)$	−18.3	$x(24)$	0.2	$x(47)$	2.7	$x(70)$	6.4	$x(93)$	9.2
$x(2)$	−17.6	$x(25)$	0.6	$x(48)$	2.9	$x(71)$	6.4	$x(94)$	9.3
$x(3)$	−12.6	$x(26)$	0.9	$x(49)$	3.1	$x(72)$	6.5	$x(95)$	9.9
$x(4)$	−12.2	$x(27)$	1.2	$x(50)$	3.1	$x(73)$	6.9	$x(96)$	10.0
$x(5)$	−11.4	$x(28)$	1.4	$x(51)$	3.1	$x(74)$	7.1	$x(97)$	10.4
$x(6)$	−11.0	$x(29)$	1.4	$x(52)$	3.2	$x(75)$	7.1	$x(98)$	10.4
$x(7)$	−8.3	$x(30)$	1.5	$x(53)$	3.3	$x(76)$	7.1	$x(99)$	10.7
$x(8)$	−8.0	$x(31)$	1.5	$x(54)$	3.7	$x(77)$	7.4	$x(100)$	10.9
$x(9)$	−6.9	$x(32)$	1.5	$x(55)$	3.8	$x(78)$	7.4	$x(101)$	10.9
$x(10)$	−5.9	$x(33)$	1.8	$x(56)$	3.8	$x(79)$	7.5	$x(102)$	11.0
$x(11)$	−3.8	$x(34)$	1.9	$x(57)$	4.1	$x(80)$	7.5	$x(103)$	11.1
$x(12)$	−2.6	$x(35)$	1.9	$x(58)$	4.1	$x(81)$	7.7	$x(104)$	11.1
$x(13)$	−1.8	$x(36)$	1.9	$x(59)$	4.2	$x(82)$	7.8	$x(105)$	11.4
$x(14)$	−1.8	$x(37)$	1.9	$x(60)$	4.2	$x(83)$	8.1	$x(106)$	11.4
$x(15)$	−1.4	$x(38)$	2.0	$x(61)$	4.2	$x(84)$	8.2	$x(107)$	11.8
$x(16)$	−1.3	$x(39)$	2.0	$x(62)$	4.8	$x(85)$	8.3	$x(108)$	11.9
$x(17)$	−1.2	$x(40)$	2.2	$x(63)$	4.9	$x(86)$	8.3	$x(109)$	12.1
$x(18)$	−0.7	$x(41)$	2.5	$x(64)$	5.2	$x(87)$	8.4	$x(110)$	12.1
$x(19)$	−0.5	$x(42)$	2.5	$x(65)$	5.6	$x(88)$	8.4	$x(111)$	12.1
$x(20)$	−0.4	$x(43)$	2.5	$x(66)$	5.9	$x(89)$	8.9	$x(112)$	12.1
$x(21)$	−0.4	$x(44)$	2.5	$x(67)$	5.9	$x(90)$	9.1	$x(113)$	12.4
$x(22)$	−0.4	$x(45)$	2.6	$x(68)$	6.2	$x(91)$	9.1	$x(114)$	12.6
$x(23)$	−0.3	$x(46)$	2.6	$x(69)$	6.2	$x(92)$	9.1	$x(115)$	12.6

TABLE 2.8 (Continued)

Obs. (i)	Value	Obs. (i)	Value	Obs. (i)	Value	Obs. (i)	Value	Obs. (i)	Value
x(116)	12.6	x(140)	14.4	x(164)	16.6	x(188)	20.5	x(212)	23.9
x(117)	12.6	x(141)	14.5	x(165)	16.7	x(189)	20.5	x(213)	24.0
x(118)	12.7	x(142)	14.6	x(166)	16.9	x(190)	20.5	x(214)	24.0
x(119)	12.8	x(143)	14.6	x(167)	17.1	x(191)	20.6	x(215)	24.7
x(120)	12.8	x(144)	14.6	x(168)	17.2	x(192)	20.7	x(216)	25.3
x(121)	13.0	x(145)	14.6	x(169)	17.4	x(193)	21.2	x(217)	26.2
x(122)	13.0	x(146)	14.7	x(170)	17.8	x(194)	21.3	x(218)	29.0
x(123)	13.0	x(147)	14.8	x(171)	17.8	x(195)	21.3	x(219)	30.2
x(124)	13.1	x(148)	14.9	x(172)	17.9	x(196)	21.3	x(220)	33.1
x(125)	13.1	x(149)	15.2	x(173)	18.4	x(197)	21.3	x(221)	33.2
x(126)	13.2	x(150)	15.2	x(174)	18.5	x(198)	21.4	x(222)	33.3
x(127)	13.3	x(151)	15.3	x(175)	18.7	x(199)	21.4	x(223)	34.1
x(128)	13.3	x(152)	15.3	x(176)	18.7	x(200)	21.5	x(224)	34.8
x(129)	13.7	x(153)	15.5	x(177)	18.9	x(201)	21.9	x(225)	35.9
x(130)	13.7	x(154)	15.5	x(178)	19.3	x(202)	22.1	x(226)	35.9
x(131)	13.8	x(155)	15.5	x(179)	19.4	x(203)	22.1	x(227)	36.1
x(132)	13.9	x(156)	15.6	x(180)	19.6	x(204)	22.4	x(228)	36.4
x(133)	14.0	x(157)	15.7	x(181)	19.7	x(205)	22.6	x(229)	36.9
x(134)	14.0	x(158)	15.8	x(182)	19.7	x(206)	22.7	x(230)	37.4
x(135)	14.0	x(159)	15.9	x(183)	19.8	x(207)	23.1	x(231)	37.6
x(136)	14.0	x(160)	15.9	x(184)	19.9	x(208)	23.1	x(232)	37.9
x(137)	14.1	x(161)	16.0	x(185)	19.9	x(209)	23.1	x(233)	38.7
x(138)	14.3	x(162)	16.4	x(186)	20.0	x(210)	23.6	x(234)	40.0
x(139)	14.4	x(163)	16.4	x(187)	20.4	x(211)	23.7	x(235)	41.3

Selecting a value for the lower-class bound of the lowest class slightly below our minimum, say −20.0%, and an upper-class bound of the highest class, say 43.0%, we spread the surplus of the range (3.4%) evenly. The resulting classes can be viewed in Table 2.9, where in the first row, the indexes of the class are given. The second row contains the class bounds. Brackets indicate that the value belongs to the class whereas parentheses exclude given values. So, we obtain a half-open interval for each class containing all real numbers between the lower bound and just below the upper bound, thus excluding that value. In the third row, we have the number of observations that fall into the respective classes.

We can check for compliance with the four criteria given earlier. Because we use half-open intervals, we guarantee that Criterion #1 is fulfilled. Since

TABLE 2.9 Classes for the 235 Fund Returns According to Sturge's Rule

Class Index I	$[a_i; b_i)$	a_I
1	[−20,−13)	2
2	[−13,−6)	7
3	[−6,1)	17
4	[1,8)	56
5	[8,15)	66
6	[15,22)	53
7	[22,29)	16
8	[29,36)	9
9	[36,43)	9

the lowest class starts at −20%, and the highest class ends at 43%, Criterion #2 is satisfied. All nine classes are of width 7%, which complies with Criterion #3. Finally, compliance with Criterion #4 can be checked easily.

Next, we apply the Freedman-Diaconis rule. With our ordered array of data, we can determine the 0.25-quartile by selecting the observation whose index is the first to exceed $0.25 \times N = 0.25 \times 235 = 58.75$. This yields the value of observation 59, which is 4.2%. Accordingly, the 0.75-quartile is given by the value whose index is the first to exceed $0.75 \times 235 = 176.25$. For our return data, it is x_{177}, which is 18.9%. The IQR is computed as 18.9% − 4.2% = 14.7% such that the class width (or bin size) of the classes is now determined according to

$$w = 2 \times IQR \times 1/\sqrt[3]{235} = 4.764\%$$

Taking the data range of 59.6% from the previous calculation, we obtain as the suggested number of classes 59.6%/4.764 = 12.511. Once again, this is not a neat looking figure. We stick with the initial class width of $w = 4.764\%$ as closely as possible by selecting the next integer, say 5%. And, without any loss of information, we extend the range artificially to 60%. So, we obtain for the number of classes 60%/5 = 12, which is close to our original real number 12.511 computed according to the Freedman-Diaconis rule but much nicer to handle. We again spread the range surplus of 0.4% (60% − 59.6%) evenly across either end of the range such that we begin our lowest class at −18.5% and end our highest class at 41.5%. The classes are given in Table 2.10. The first row of the table indicates the indexes of the class while the

TABLE 2.10 Classes for the 235 Fund Returns According to the Freedman-Diaconis Rule

I	$[a_i; b_i)$	a_I
1	[−18.5;−13.5)	2
2	[−13.5;−8.5)	4
3	[−8.5;−3.5)	5
4	[−3.5;1.5)	18
5	[1.5;6.5)	42
6	[6.5;11.5)	35
7	[11.5;16.5)	57
8	[16.5;21.5)	36
9	[21.5;26.5)	18
10	[26.5;31.5)	2
11	[31.5;36.5)	9
12	[36.5;41.5)	7

second row gives the class bounds. The number of observations that fall into each class is shown in the last row.[9]

Let us next compare Tables 2.9 and 2.10. We observe a finer distribution when the Freedman-Diaconis rule is employed because this rule generates more classes for the same data. However, it is generally difficult to judge, which rule provides us with the better information because, as can be seen in our illustration, the two rules set up completely different classes. But the choice of class bounds is essential. By just slightly shifting the bounds between two adjacent classes, many observations may fall from one class into the other due to this alteration. As a result, this might produce a totally different picture about the data distribution. So, we have to be very careful when we interpret the two different results.

For example, class 7, that is, [22;29) in Table 2.9, contains 16 observations.[10] Classes 9 and 10 of Table 2.10 cover approximately the same range, [21.5;31.5). Together they account for 20 observations. We could now easily present two scenarios that would provide rather different conceptions

---
[9]One can easily check that the four requirements for the classes are again met.
[10]The notation $[x_1,x_2)$ is the short way of stating that we have selected all real numbers between x_1 and x_2 but not including the latter one as indicated by the parenthesis (")") rather than the bracket ("]"). This is a so-called *half-open interval* in contrast to an open interval indicated by all parentheses or a closed interval as indicated by all brackets. Parentheses indicate exclusion while brackets indicate inclusion of the respective bounds.

about the frequency. In scenario one, suppose, one assumes that two observations are between 21.5 and 22.0. Then there would have to be 16 observations between 22.0 and 26.5 to add up to 18 observations in class 9 of Table 2.10. This, in return, would mean that the 16 observations of class 7 from Table 2.9 would all have to lie between 22.0 and 26.5 as well. Then the two observations from class 10 of Table 2.10 must lie beyond 29.0. The other scenario could assume that we have four observations between 21.5 and 22.0. Then, for similar reasons as before, we would have 14 observations between 22.0 and 26.5. The two observations from class 10 of Table 2.10 would now have to be between 26.5 and 29.0 so that the total of 16 observations in class 7 of Table 2.9 is met. See how easily slightly different classes can lead to an ambiguous interpretation? Looking at all classes at once, many of these puzzles can be solved. However, some uncertainty remains. As can be seen, the choice of the number of classes, and thus the class, bounds can have a significant impact on the information that the data conveys when condensed into classes.

CUMULATIVE FREQUENCY DISTRIBUTIONS

In contrast to the empirical cumulative frequency distributions, in this section we will introduce functions that convey basically the same information, that is, the frquency distribution, but rely on a few more assumptions. These cumulative frequency distributions introduced here, however, should not be confused with the theoretical definitions given in the chapters on probability theory to follow, even though one will clearly notice that the notion is akin to both.

The *absolute cumulative frequency* at each class bound states how many observations have been counted up to this particular class bound. However, we do not exactly know how the data are distributed within the classes. On the other hand, when relative frequencies are used, the cumulative relative frequency distribution states the overall proportion of all values up to a certain lower or upper bound of some class.

So far, things are not much different from the definition of the empirical cumulative frequency distribution and empirical cumulative relative frequency distribution. At each bound, the empirical cumulative frequency distribution and cumulative frequency coincide. However, an additional assumption is made regarding the distribution of the values between bounds of each class when computing the cumulative frequency distribution. The data are thought of as being continuously distributed and equally spread between the particular bounds.[11] Hence, both forms of the cumulative

---

[11]This type of assumed behavior will be defined in Chapter 11 when we cover continuous rectangular probability distributions as a uniform distribution of data.

frequency distributions increase in a linear fashion between the two class bounds. So for both forms of cumulative distribution functions, one can compute the accumulated frequencies at values inside of classes.

For a more thorough analysis of this, let's use a more formal presentation. Let I denote the set of all class indexes i with i being some integer value between 1 and $n_I = |I|$ (i.e., the number of classes). Moreover, let a_j and f_j denote the (absolute) frequency and relative frequency of some class j, respectively. The cumulative frequency distribution at some upper bound, x_u^i, of a given class i is computed as

$$F(x_u^i) = \sum_{j: x_u^j \leq x_u^i} a_j = \sum_{j: x_u^j \leq x_l^i} a_j + a_i \qquad (2.1)$$

In words, this means that we sum up the frequencies of all classes whose upper bound is less than x_u^i plus the frequency of class i itself. The corresponding cumulative relative frequency distribution at the same value is then,

$$F^f(x_u^i) = \sum_{j: x_u^j \leq x_u^i} f_j = \sum_{j: x_u^j \leq x_l^i} f_j + f_i \qquad (2.2)$$

This describes the same procedure as in equation (2.1) using relative frequencies instead of frequencies. For any value x in between the boundaries of, say, class i, x_l^i and x_u^i, the cumulative relative frequency distribution is defined by

$$F^f(x) = F^f(x_l^i) + \frac{x - x_l^i}{x_u^i - x_l^i} f_i \qquad (2.3)$$

In words, this means that we compute the cumulative relative frequency distribution at value x as the sum of two things. First, we take the cumulative relative frequency distribution at the lower bound of class i. Second, we add that share of the relative frequency of class i that is determined by the part of the whole interval of class i that is covered by x.

Figure 2.3 might appeal more to intuition. At the bounds of class i, we have values of the cumulative relative frequency given by $F^f(x_l^i)$ and $F^f(x_u^i)$. We assume that the cumulative relative frequency increases linearly along the line connecting $F^f(x_l^i)$ and $F^f(x_u^i)$. Then at any value x^* inside of class i, we find the corresponding value $F^f(x^*)$ by the intersection of the dash-dotted line and the vertical axis as shown in the figure. The dash-dotted line is obtained by extending a horizontal line through the intersection of the vertical line through x^* and the line connecting $F^f(x_l^i)$ and $F^f(x_u^i)$ with slope $[F^f(x^*) - F^f(x_l^i)]/(x^* - x_l^i)$.

FIGURE 2.3 Determination of Frequency Distribution within Class Bounds

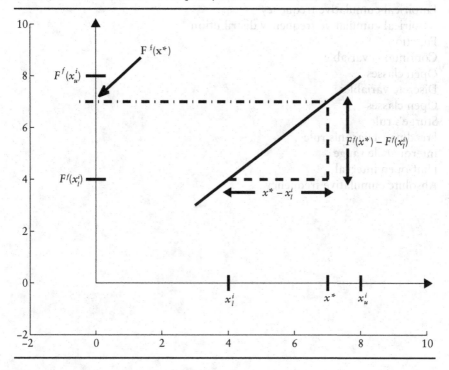

CONCEPTS EXPLAINED IN THIS CHAPTER (IN ORDER OF PRESENTATION)

Univariate data
Qualitative data
Quantitative data
Measurement level
Nominally scaled data
Ordinally scaled data
Rank data
Interval scale
Ratio scale
Absolute data
Variable
Time series data
Cross-sectional data

Relative frequency
Empirical cumulative frequency
Empirical cumulative frequency distribution
Function
Continuous variable
Open classes
Discrete variables
Open classes
Sturge's rule
Freedman-Diaconis rule
Interquartile range
Half-open interval
Absolute cumulative frequency

CHAPTER 3

Measures of Location and Spread

Now that we have the data at our disposal, maybe classified, and possibly the corresponding frequency distributions computed as well, it is time to retrieve the information using some concise measures. Generally, there are two possible ways to do this. In this chapter, one of the two ways is presented: the computation of key numbers conveying specific information about the data. The alternative, the graphical representation, will be presented in the next chapter. As key numbers we will introduce measures for the center and location of the data as well as measures for the spread of the data.

PARAMETERS VS. STATISTICS

Before we go further, however, we have to introduce a distinction that is valid for any type of data. We have to be aware of whether we are analyzing the entire population or just a sample from that population. The key numbers when dealing with populations are called *parameters* while we refer to *statistics* when we observe only a sample. Parameters are commonly denoted by Greek letters while statistics are usually assigned Roman letters.

The difference between these two measures is that parameters are values valid for the entire population or universe of data and, hence, remain constant throughout whereas statistics may vary with every different sample even though they each are selected from the very same population. This is easily understood using the following example. Consider the average return of all stocks listed in the S&P 500 index during a particular year. This quantity is a parameter μ, for example, since it represents all these stocks. If one randomly selects 10 stocks included in the S&P 500 stocks, however, one may end up with an average return for this sample that deviates from the population average, μ. The reason would be that by chance one has picked stocks that do not represent the population very well. For example, one might by chance select the top 10 performing stocks included in the S&P

500. Their returns will yield an average (statistic) that is above the average of all 500 stocks (parameter). The opposite analog arises if one had picked the 10 worst performers. In general, deviations of the statistics from the parameters are the result of one selecting the sample.

CENTER AND LOCATION

The measures we present first are those revealing the center and the location of the data. The center and location are expressed by three different measures: mean, median, and mode.

Mean

The *mean* is the quantity given by the sum of all values divided by the *size* of the data set. The size is the number of possible values or observations.[1] Note that the size of a population is usually indicated by N. The size of a sample is given by the number of data contained in the sample, commonly symbolized by n. Hence, the population mean is defined by

$$\mu = \frac{\sum_{i=1}^{N} x_i}{N} \qquad (3.1)$$

In contrast, the sample mean is defined by

$$\bar{x} = \frac{\sum_{i=1}^{n} x_{s,i}}{n} \qquad (3.2)$$

Note that the summation in equation (3.2) is only over the elements in the sample set, $x_{s,i}$. In the following, we will not use the indication s (as done here for $x_{s,i}$) when we consider samples since it will be obvious from the context.

The interpretation of the mean is as follows: The mean gives an indication as to which value the data are scattered about. Moreover, on average, one has to expect a data value equal to the mean when selecting an observation at random. However, one incurs some loss of information that is not insignificant. Given a certain data size, a particular mean can be obtained from different values. One extreme would be that all values are equal to the mean. The other extreme could be that half of the observations are

---

[1]Commonly, the size or *magnitude* of a set S is denoted by $|S|$. So, if $S = \{x_1, \ldots, x_{100}\}$, $|S| = 100$.

extremely to the left and half of the observations are extremely to the right of the mean, thus, leveling out, on average.

For example, consider a set S with only two possible values (i.e., x_1 and x_2) such that $|S| = 2$. Let the first scenario be $x_1 = x_2 = 0$. The resulting mean is $\mu = 0.5 \times (x_1 + x_2) = 0$. A second scenario could be $x_1 = -1,000$ and $x_2 = 1,000$, again resulting in a mean $\mu = 0.5 \times (x_1 + x_2) = 0$. So, both scenarios produce equivalent means even though they result from completely different data. This loss of information is the cost of consolidating data. It is inherent in all sorts of reduction of data to key figures.

The requirements on the data such that the mean can be computed should be noted. The data have to be quantitative; otherwise, computation of the mean is infeasible. As a consequence, both nominal and rank scale data have no means. So, one has to be careful when attempting to calculate means.

If we are dealing with classified data, the mean is computed in a slightly different manner. Since we have no information as to the true values, we choose the centers of the classes to represent the values within the classes. The centers of the classes are each weighted with the classes' respective relative frequencies and summed to compute the mean. Formally, we will use the following notation:

n = the number of observations
$I \in \{1, \ldots, n_c\}$ = the class index
$c_I = b_I - a_I$ = the center of class I
p_I = the relative frequency of class I
h_I = the absolute frequency of class I

Then, the sample mean of classified data is defined by

$$\bar{x}^c = \frac{1}{n}\sum_{I=1}^{n_c} c_I \times h_I = \sum_{I=1}^{n_c} c_I \times p_I \qquad (3.3)$$

To illustrate, recall the classes as given in Table 2.9 of Chapter 2, which lists the classes according to Sturge's rule. We obtain as class centers the values in Table 3.1 The mean is

$$\bar{x}_C = \frac{1}{235}\left(-16.5 \times 2 - 9.5 \times 7 - \ldots + 39.5 \times 9\right) = 12.364$$

TABLE 3.1 Class Centers of Fund Returns

I	1	2	3	4	5	6	7	8	9
c_I	−16.5	−9.5	−2.5	4.5	11.5	18.5	25.5	32.5	39.5

Note that if we had computed the mean from the original data according to the definition given by equation (3.2), we would have obtained $\bar{x} = 12.323$, which is almost the same number. We can see that in this example, classification results in little to no loss as far as the mean is concerned. This might be an indication of an equal distribution of the data inside of the classes or, alternatively, of lopsided data in some classes being canceled out by data tilted in the opposite direction in other classes.

Median

A second measure of the center of a distribution is the *median*. In symbols, we denote the population median by $\hat{\mu}$ and the sample median by m_d, respectively. Roughly speaking, the median divides data by value into a lower half and an upper half. A more rigorous definition for the median is that we require that at least half of the data are no greater and at least half of the data are no smaller than the median itself.[2]

This can be stated in a formal way as follows. For the median, it has to be true that

$$\frac{1}{n}\left|\{i \mid x_{(i)} \leq m_d\}\right| \geq 0.5$$

and

$$\frac{1}{n}\left|\{i \mid x_{(i)} \geq m_d\}\right| \geq 0.5$$

In words, this means that both the number of indexes whose corresponding values are no greater than m_d and the number of indexes whose corresponding values are no less than m_d have to account for, at least, 50% of the data.

To calculate either one, it is imperative that we have the data presented in an array of ascending values. Now, for a set containing an uneven number n of values or observations, we obtain the median by the data point with index $(n-1)/2 + 1$. This value can be expressed by the so-called *ceiling function* $\lceil n/2 \rceil$, which is the smallest integer number greater than or equal to $n/2$. Our median thus obtained is then

$$x_{(\lceil n/2 \rceil)} \qquad (3.4)$$

---

[2] As we will see in subsequent chapters, the median of a distribution can be one single value as well as a set of values.

The feature of this median, now, is that the number of values that is not greater than it is exactly equal to the number of values that is not smaller than it, namely $(n-1)/2$. When we look at the corresponding empirical relative cumulative distribution function, F_{emp}^f, we will notice that the function first crosses the level of 0.5 at the median; that is, $F_{emp}^f(m_d) > 0.5$.

We will try to make this clear with a simple example. Suppose that we are analyzing seven companies (C_1 through C_7) with respect to their percentage stock price gains and their 2006 credit rating as assigned by Standard and Poor's (S&P). The data are shown in Table 3.2. According to (3.4), the median of the percentage stock price gains is the value $x_{(\lceil 7/2 \rceil)} = x_{(4)} = 0.063$, which is the return of company C_2. By the same definition, the median of the S&P ratings is $x_{(\lceil 7/2 \rceil)}^{S\&P} = x_{(4)}^{S\&P} = BB$. Note that because companies C_2 and C_4 have a BB rating, the third and fourth positions are shared by them. So the ordered array of S&P ratings may place either company C_2 or company C_4 in position 4 and, hence, make its value (BB) the median. Additionally, the implication of the empirical cumulative relative frequency distribution on the selection of the median is demonstrated. In the fourth column, the empirical cumulative relative frequency distribution is given for both the returns and the S&P ratings of the seven companies. The graphs of the empirical cumulative relative frequency distributions are shown in Figures 3.1 and 3.2.

In Figure 3.1, since each value occurs once only, we have seven different entries of which one, namely 0.063, is exactly in the middle having three values smaller and larger than it. This is indicated in the figure by the cumulative frequency function first crossing the value of 0.5 at the return of 0.063. In Figure 3.2, there are six frequency increments only, despite the fact that there are seven companies. This is due to companies C_2 and C_4 having the same rating, BB. So, for the rating of BB, the cumulative frequency of 0.5 is crossed for the first time. There is no ambiguity with respect to the median.

TABLE 3.2 Empirical Distribution of Percentage Returns and S&P Rating

Company	Return	S&P Rating	F_{emp}^f of Return/Rating
C_1	0.071	CCC	0.714 / 0.286
C_2	0.063	BB	0.571 / 0.429
C_3	0.051	AA	0.429 / 0.857
C_4	0.047	BB	0.286 / 0.571
C_5	0.027	A	0.143 / 0.714
C_6	0.073	AAA	0.857 / 1.000
C_7	0.092	R[a]	1.000 / 0.143

[a]R means "under regulatory supervision due to its financial condition."

FIGURE 3.1 Empirical Distribution of Percentage Stock Price Returns

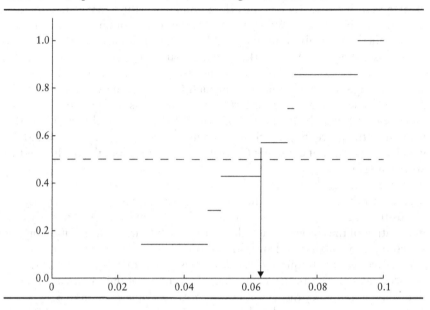

FIGURE 3.2 Empirical Distribution of S&P Credit Ratings

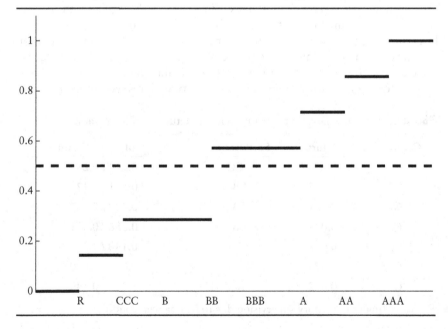

TABLE 3.3 Empirical Distribution of Percentage Returns and S&P Rating with Additional Company

Company	Return	S&P Rating	F_{emp}^f of Return/Rating
C_1	0.071	A	0.625 / 0.750
C_2	0.063	B	0.500 / 0.375
C_3	0.051	AA	0.375 / 0.875
C_4	0.047	BB	0.250 / 0.625
C_5	0.027	CCC	0.125 / 0.250
C_6	0.073	AAA	0.750 / 1.000
C_7	0.092	R[a]	0.875 / 0.125
C_8	0.096	B	1.000 / 0.500

[a] R means "under regulatory supervision due to its financial condition."

If, in contrast, we have a set containing an even number of observations n^e, the median is calculated as the average of the value with index $n^e/2$ and the value with index $n^e/2 + 1$.[3] Thus, we have

$$m_d = \frac{1}{2}\left(x_{(n/2)} + x_{(n/2+1)}\right) \qquad (3.5)$$

In this case where there is an even n^e, the empirical relative cumulative distribution function has a value of at least 0.5 at the median; that is, $F_{emp}^f(m_d) \geq 0.5$. Why not take $x_{(n/2)}$? The reason is that at least half of the values are at least as large and at least half of the values (including the value itself) are not greater than $x_{(n/2)}$. But, the same is true for $x_{(n/2+1)}$. So, there is some ambiguity in this even case. Hence, the definition given by equation (3.5).

To also convey the intuition behind the definition in the case where there is an even number of observations, let's extend the previous example by using eight rather than seven companies. We have added company C_8. Also, company C_2 has received a credit downgrade to B. The new data set is displayed in Table 3.3.

We now use the definition given by equation (3.5) to compute the median of the returns. Thus, we obtain the following median

$$m_d = \frac{1}{2}\left(x_{(8/2)} + x_{(8/2+1)}\right) = 0.5 \times (0.063 + 0.071) = 0.067$$

We display this in Figure 3.3 where we depict the corresponding cumulative relative empirical frequency distribution.

[3] In a similar context in probability theory, we will define the median as the set of values between $x_{(n/2)}$ and $x_{(n/2+1)}$.

FIGURE 3.3 Empirical Distribution of Percentage Stock Price Returns; Even Number of Companies

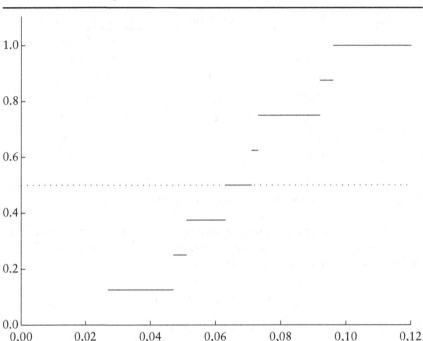

For the ratings, we naturally obtain no number for the median since the ratings are given as rank scaled values. The definition for the median given by equation (3.5) therefore does not work in this case. As a solution, we define the median to be the value between $x_{(8/2)}^{S\&P}$ and $x_{(8/2+1)}^{S\&P}$; that is, it is between B and BB, which is a theoretical value since it is not an observable value. However, if $x_5^{S\&P}$ had been B, for example, the median would be B. The cumulative relative empirical frequency distribution is shown in Figure 3.4.

Note that one can determine the median in exactly the same way by starting with the highest values and working through to the lowest (i.e., in the opposite direction of how it is done here). The resulting order would be descending but produce the same medians.

One common mistake is to confuse the object with its value of a certain attribute when stating the median. In the example above, the median is a certain return or rating (i.e., a certain value of some attribute) but not the company itself.

For classified data, we have to determine the median by some different method. If the cumulative relative frequency distribution is 0.5 at some class

FIGURE 3.4 Empirical Distribution of S&P Ratings; Even Number of Companies

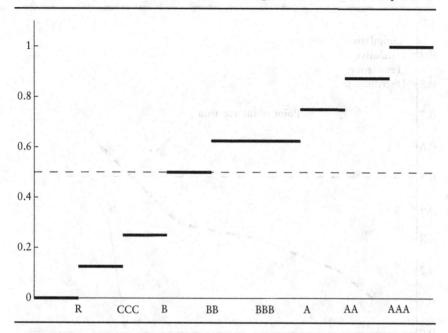

bound, then this class bound is defined to be the median. If, however, such a class bound does not exist, we have to determine the class whose lower bound is less than 0.5 and whose upper class bound is greater than 0.5; that is, we have to find what is referred to as the *class of incident*. The median is determined by linear interpolation. For a visual presentation of this procedure, see Figure 3.5 where the class of incidence is bounded by a and b. Formally, if it is not a class bound, the population median of classified data is

$$\hat{\mu}^C = a_1 + \frac{0.5 - F^f(a_1)}{F^f(b_1) - F^f(a_1)} \times (b_1 - a_1) \qquad (3.6)$$

where $a_1 < 0.5$ and $b_1 > 0.5$. The corresponding sample median is computed using the empirical relative cumulative distribution function, instead, and denoted by m_d^C. For our data from Table 2.9 in Chapter 2, this would lead to a median equal to

$$m_d^C = 8 + \frac{0.5 - 0.349}{0.630 - 0.349}(15 - 8) = 11.765$$

By nature of the median, it should be intuitive that the data have to be at least of rank scale. Thus, we cannot compute this measure for nominal

FIGURE 3.5 Interpolation Method to Retrieve Median of Classified Data

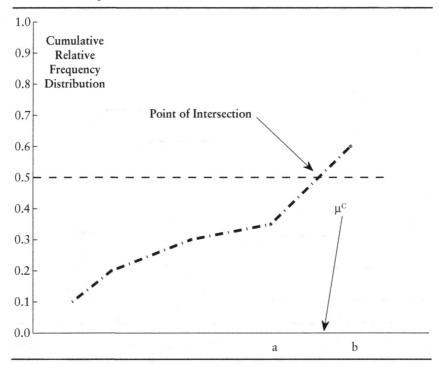

data, even though it is occasionally done, producing meaningless results and resulting in drawing wrong conclusions.

Mode

A third measure of location presented here is the *mode*. Its definition is simple. It is the value that occurs most often in a data set. If the distribution of some population or the empirical distribution of some sample are known, the mode can be determined to be the value corresponding to the highest frequency. For populations, the mode is denoted by M and for the sample the mode is denoted by m. Formally, the mode is defined by

$$M = \max_i f(x_i) \text{ or } m = \max_i f^{emp}(x_i) \quad (3.7)$$

In our earlier example using the S&P ratings of the eight companies, the mode is the value B since $f^{emp}(B) = 0.25 = \max$.

If it should be the case that the maximum frequency is obtained by more than one value, one speaks of a *multimodal* data set. In the case of only one mode, the data set is referred to as *unimodal*.

When we have class data, we cannot determine a mode. Instead, we determine a *mode class*. It is the class with the greatest absolute or relative frequency. In other words, it is the class with the most observations of some sample in it. Formally, it is

$$I_M = \left\{ I \middle| h_I = \max_J h_J \right\} = \left\{ I \middle| p_I = \max_J p_J \right\} \qquad (3.8)$$

For the class data in Table 2.9 in Chapter 2, the mode class is class $I = 5$.

Of the three measures of central tendency, the mode is the measure with the greatest loss of information. It simply states which value occurs most often and reveals no further insight. This is the reason why the mean and median enjoy greater use in descriptive statistics. While the mean is sensitive to changes in the data set, the mode is absolutely invariant as long as the maximum frequency is obtained by the same value. The mode, however, is of importance, as will be seen, in the context of the shape of the distribution of data. A positive feature of the mode is that it is applicable to all data levels.

Weighted Mean

Both the mean and median are special forms of some more general location parameters and statistics.

The mean is a particular form of the more general *weighted mean* in that it assigns equal weight of to all values, that is, $1/N$ for the parameter and $1/n$ for the statistic. The weighted mean, on the other hand, provides the option to weight each value individually. Let w_i denote the weight assigned to value x_i, we then define the population weighted mean by

$$\mu_w = \sum_{i=1}^{N} w_i x_i \Big/ \sum_{i=1}^{N} w_i \qquad (3.9)$$

and the sample weighted mean by

$$\bar{x}_w = \sum_{i=1}^{n} w_i x_i \Big/ \sum_{i=1}^{n} w_i \qquad (3.10)$$

respectively. One reason for using a weighted mean rather than an *arithmetic mean* might be that extreme values might be of greater importance than other values. Or, if we analyze observations of some phenomenon over some period of time (i.e., we obtain a time series), we might be interested in the more recent values compared to others stemming from the past.[4]

As an example, consider once again the eight companies from the previous example with their attributes "return" and "S&P rating." To compute the weighted average of all eight returns, we might weight the returns according to the respective company's S&P credit rating in the following manner. The best performer's return is weighted by 8, the second best performer's returns by 7, and so on. If two or more companies have the same rating, they are assigned the weights equal to the average of the positions they occupy. Hence, the weighted mean is computed as

$$\bar{x}_m = \frac{8 \times 7.3 + 7 \times 5.1 + 6 \times 7.1 + 5 \times 4.7 + 3.5 \times 9.6 + 3.5 \times 6.3 + 2 \times 2.7 + 1 \times 9.2}{8 + 7 + 6 + 5 + 3.5 \times 2 + 2 + 1} \times 0.01$$
$$= 0.064$$

Quantiles

Now let us turn to the generalization of the median. The median is a particular case of a *quantile* or *percentile*. Analogously to the median, a percentile divides the ordered data array, whether population or sample, into two parts. The lower part represents α percent of the data while the upper part accounts for the remaining $(1 - \alpha)$ percent of the data for some given share $\alpha \in (0, 1)$.[5]

Formally, we define the population α-percentile μ_α for countable data by

$$\frac{1}{N}\left|\left\{i \mid x_{(i)} \leq \mu_\alpha\right\}\right| \geq \alpha \quad \text{and} \quad \frac{1}{N}\left|\left\{i \mid x_{(i)} \geq \mu_\alpha\right\}\right| \geq 1 - \alpha \qquad (3.11)$$

and the sample percentile q_α by

$$\frac{1}{N}\left|\left\{i \mid x_{(i)} \leq q_\alpha\right\}\right| \geq \alpha \quad \text{and} \quad \frac{1}{N}\left|\left\{i \mid x_{(i)} \geq q_\alpha\right\}\right| \geq 1 - \alpha \qquad (3.12)$$

That is, in equations (3.11) and (3.12) the portion of values no greater than either μ_α or q_α is at least α, while the share of values no less than either μ_α or q_α is at least $(1 - \alpha)$. At the α-percentiles, the cumulative distribution functions assume values of at least α for the first time.

---

[4]The term arithmetic mean is the formal expression for equally weighted average.
[5]Hence, the percentiles corresponding to some share α are more specifically called α-*percentiles*, expressing the partitioning with respect to this particular share.

FIGURE 3.6 0.25- and 0.30-Percentiles of Returns

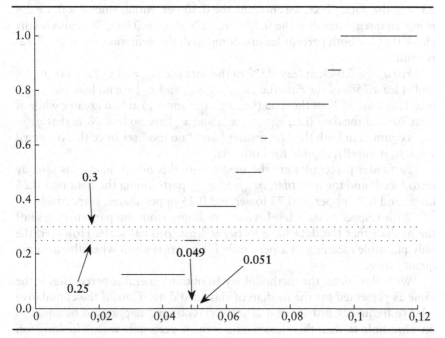

The following convention is used for the same reason as with the median: to avoid ambiguity.[6] That is, if $n \times \alpha$ is an integer, that is, there exists some index (i) with exactly the value of $n \times \alpha$, then the α-percentiles is defined to be $0.5 \times [x_{(i)} + x_{(i+1)}]$. Again, the percentile defined as the arithmetic mean of $x_{(i)}$ and $x_{(i+1)}$ is just by convention. The α-percentile could just as well be defined to be any value between these two numbers. In this case, the corresponding cumulative distribution function assumes the value α at $x_{(i)}$. If, however, $n \times \alpha$ is not an integer, then the requirements of equations (3.11) and (3,12) are met by the value in the array with the smallest index greater than $n \times \alpha$ (i.e., $x_{([n \times \alpha])}$).

We illustrate the idea behind α-percentiles with an example given in Figure 3.6. Consider again the eight companies given in Table 3.3 and their two attributes "return" and "S&P rating." Assume that we are interested in the worst 25% and 30% returns. Thus, we have to compute the 0.25- and 0.30-percentiles. First, we turn our attention to the 0.25-percentile. With n equal to 8, then $8 \times 0.25 = 2$. Hence, we compute $q_{0.25} = 0.5(x_{(2)} + x_{(3)}) = 0.5(0.047 + 0.051) = 0.049$. So, the 0.25-percentile is not an observed value.

---

[6]In the following, we will present the theory in the sample case only. The ideas are analogous in the population setting.

The meaning of this is that any return no greater than 0.049 is in the lower 25% of the data. Next, we compute the 0.30-percentile. Since $8 \times 0.3 = 2.4$ is not an integer number, the 0.30-percentile is $x_{(3)} = 0.051$. We can visually check that for both percentiles just computed, the definition given by (3.12) is valid.

For $q_{0.25} = 0.049$, at least 25% of the data (i.e, $x_{(1)}$ and $x_{(2)}$) are no greater and at least 75% of the data (i.e., $x_{(3)}, \ldots, x_{(7)}$, and $x_{(8)}$) are no less. For $q_{0.30} = 0.051$, at least 30% of the data (i.e., $x_{(1)}, x_{(2)}$, and $x_{(3)}$) are no greater while at least 70% of the data (i.e., $x_{(3)}, \ldots, x_{(7)}$, and $x_{(8)}$) are no less. Note that $q_{0.3} = x_{(3)}$ is counted in both the "no greater" and "no less" set since this observed value is, naturally, eligible for both sets.

Particular percentiles are the median—in this notation, $q_{0.5}$ (as already introduced) and the *quartiles*, $q_{0.25}$ and $q_{0.75}$, partitioning the data into 0.25 lower and 0.75 upper or 0.75 lower and 0.25 upper shares, respectively.

With respect to data level issues, we know from the particular case of the median that the data have to be, at least, ordinal scale. However, the only plausible meaning of a percentile for any α is given when the data are quantitative.

With class data, the methodology to obtain general α-percentiles is the same as presented for the median of classified data. Thus, if the cumulative relative frequency distribution at some class bound happens to be equal to our threshold α, then this class bound is the α-percentile we are looking for. If, however, this is not the case, we have to determine the class of incidence. We can restate the definition of the class of incidence as follows: It is the class whose lower bound is less than and whose upper bound is greater than α. Again, through linear interpolation, we obtain the particular percentile.

Formally then, we obtain the population α-percentile by

$$\mu_\alpha^c = a_I + \frac{\alpha - F^f(a_I)}{F^f(b_I) - F^f(a_I)} \tag{3.13}$$

with $F(a_I) < \alpha$ and $F(b_I) > \alpha$. The sample percentile is obtained by

$$q_\alpha^c = a_I + \frac{\alpha - F_{emp}^f(a_I)}{F_{emp}^f(b_I) - F_{emp}^f(a_I)} \tag{3.14}$$

in that we use the empirical cumulative relative frequency distribution of the sample rather than the population cumulative relative frequency distribution from equation (3.13). As an example, consider the data from Table 2.9 in Chapter 2. Suppose that we are interested in the 0.3-percentiles. The class of incidence is class 4 since $F(a_4) = 0.111 < 0.3$ and $F(b_4) = 0.349 > 0.3$. Hence, according to equation (3.14), the 0.3-quantile is given by

$$q_{0.3} = 1 + \frac{0.300 - 0.111}{0.349 - 0.111} \times (8-1) = 6.563$$

VARIATION

Rather than measures of the center or one single location, we now discuss measures that capture the way the data are spread either in absolute terms or relative terms to some reference value such as, for example, a measure of location. Hence, the measures introduced here are *measures of variation*. We may be given the average return, for example, of a selection of stocks during some period. However, the average value alone is incapable of providing us with information about the variation in returns. Hence, it is insufficient for a more profound insight in the data. Like almost everything in real life, the individual returns will most likely deviate from this reference value, at least to some extent. This is due to the fact that the driving force behind each individual object will cause it to assume a value for some respective attribute that is inclined more or less in some direction away from the standard.

While there are a great number of measures of variation that have been proposed in the finance literature, we limit our coverage to those that are more commonly used in finance.

Range

Our first measure of variation is the *range*. It is the simplest measure since it merely computes the difference between the maximum and the minimum of the data set. Formally, let x_{min} and x_{max} denote the minimum and maximum values of a data set, respectively. Then, the range is defined by

$$R = x_{max} - x_{min} \tag{3.15}$$

As an example, let's use the return data of the eight companies from Table 3.3. The maximum return is $x_{max} = x_{(8)} = 0.096$ while the minimum return is $x_{min} = x_{(1)} = 0.027$. Thus, the range is $r = 0.096 - 0.027 = 0.069$.

What does this value tell us? While the mean is known to be 0.065, the values extend over some interval that is wider than the value of the average (i.e., wider than 0.065). It seems like the data might be very scattered. But can we really assume that? Not really, since we are only taking into consideration the extreme values of either end of the data. Hence, this measure is pretty sensitive to shifts in these two values while the rest of the data in between may remain unchanged. The same range is obtained from two data

TABLE 3.4 Alternative Set of Returns of Eight Companies

Company	Return
A*	0.027
D*	0.060
B*	0.064
C*	0.065
H*	0.067
G*	0.068
F*	0.070
E*	0.096

sets that have the same extremes but very different structures within the data. For example, suppose that we have a second set of returns for the eight companies as shown in Table 3.4 with the asterisks indicating the names of the companies in our new data set. As we can easily see for this new data set, the range is the same. However, the entire structure within the data set is completely different. The data in this case are much less scattered between the extremes. This is not indicated by the range measure, however. Hence, the range is the measure of variation with the most limited usefulness, in this context, due to its very limited insight into the data structure.

For data classes, we obtain the range by the span between the upper bound of the uppermost class and the lower bound of the lowest class. Formally, the range of class data is given by

$$R^C = b_{n^C} - a_1 = \max_I b_I - \min_I a_I \qquad (3.16)$$

where n^C is the number of classes and the class indexes $I \in \{1, \ldots, n^C\}$. For our data from Table 2.9 in Chapter 2, we obtain as the range

$$R^C = b_9 - a_1 = 43 - (-20) = 63$$

This, of course, is larger than the range for the underlying data, $R = 59.6$, since the classes are required to cover the entire data set.

Interquartile Range

A solution to the range's sensitivity to extremes is provided by the *interquartile range* (IQR) in that the most extreme 25% of the data at both ends are discarded. Hence, the IQR is given by the difference between the upper (i.e., 25%) and lower (i.e., 75%) quartiles, respectively. Formally, the population IQR is defined by

$$IQR = \mu_{0.75} - \mu_{0.25} \qquad (3.17)$$

The sample IQR is defined analogously with the quartiles replaced by the corresponding sample quartiles.

As an example, consider the return data given in Table 3.3. With $q_{0.25} = 0.049$ and $q_{0.75} = 0.083$ (rounded from 0.0825), the IQR is 0.034. Note that only the values $x_{(2)}$, $x_{(3)}$, $x_{(6)}$, and $x_{(7)}$ enter the computation. For the remaining data, it only has to be maintained that the numerical *order* of their values is kept to obtain the same value for the IQR.

When the data are classified, the sample IQR is analogously defined by

$$IQR^C = q^C_{0.75} - q^C_{0.25} \qquad (3.18)$$

For the class data from Table 2.9 of Chapter 2, we obtain as quartiles $q^C_{0.25} = 5.094$ and $q^C_{0.75} = 18.731$.[7] Hence, the IQR is given by $IQR^C = q^C_{0.75} - q^C_{0.25} = 13.637$. The actual IQR of the original data is 14.7.

The IQR represents the body of the distribution. The influence of rare extremes is deleted. But still, the IQR uses only a fraction of the information contained in the data. It conveys little about the entire variation. Naturally, if the IQR is large, the outer segments of the data are bound to be further away from some center than would be feasible if the IQR was narrow. But as with the range, the same value for the IQR can be easily obtained analytically by many different data sets.

Absolute Deviation

To overcome the shortcomings of the range and IQR as measures of variation, we introduce a third measure that accounts for all values in the data set. It is the so-called **mean absolute deviation** (MAD). The MAD is the average deviation of all data from some reference value.[8] The deviation is usually measured from the median. So, for a population, the MAD is defined to be

$$\partial_{MAD} = \frac{1}{N} \sum_{i=1}^{N} |x_i - \hat{\mu}| \qquad (3.19)$$

whereas for a sample, it is

---

[7] The computation of the quartiles is left as an exercise for the reader.
[8] The reference value is usually a measure of the center.

$$d_{MAD} = \frac{1}{n}\sum_{i=1}^{n}|x_i - m_d| \tag{3.20}$$

The MAD measure takes into consideration every data value. Due to the absolute value brackets, only the length of the deviation enters the calculation since the direction, at least here, is not of interest.[9]

For the data in Table 3.3, the MAD is computed to be

$$d_{MAD} = \frac{1}{8}\big[|0.071 - 0.067| + |0.063 - 0.067| + \ldots + |0.096 - 0.067|\big] = 0.018$$

So, on average, each return deviates from the median by 1.8% per year.

For class data, the same problem arises as with the mean; that is, we do not have knowledge of the true underlying data. So we cannot compute the distance between individual data from the median. Therefore, we seek an alternative in that we use the class centers representing the data inside of the classes instead. Then, the mean average deviation is the weighted sum of the deviations of each class's central value from the classified data median where the weights represent the relative class weights. So, if the central value of class I is denoted by c_I, then the formal definition of the class data MAD of a population of size N is given by

$$MAD^C = \frac{1}{N}\sum_{I=1}^{n_C}|c_I - \hat{\mu}^C| \times h_I = \sum_{I=1}^{n_C}|c_I - \hat{\mu}^C| \times f_I \tag{3.20}$$

where the number of classes is given by n_C. The MAD of sample classified data is given by

$$MAD^C = \frac{1}{n}\sum_{I=1}^{n_C}|c_I - m_d^C| \times h_I = \sum_{I=1}^{n_C}|c_I - m_d^C| \times f_I \tag{3.21}$$

for a sample of size n with, again, n_C classes.

For example, using the sample in Table 2.9 in Chapter 2, the MAD is computed as[10]

$$\begin{aligned}MAD^C &= \frac{1}{235}\big(|-16.5 - 11.765| \times 2 + |-9.5 - 11.765| \times 7 + \ldots + |39.5 - 11.765| \times 9\big) \\ &= |-16.5 - 11.765| \times 0.009 + |-9.5 - 11.765| \times 0.029 + \ldots + |39.5 - 11.765| \times 0.038 \\ &= 8.002\end{aligned}$$

---

[9]The absolute value of x, denoted $|x|$, is the positive value of x neglecting its sign. For example, if $x = -5$, then $|x| = |-5| = 5 = |5|$.
[10]Precision is three decimals. However, we will not write down unnecessary zeros.

If we compute the MAD for the original data, we observe a value of 8.305, which is just slightly higher then MAD^C. Thus, the data are relatively well represented by the class centers with respect to the MAD. This is a result for this particular data set and, of course, does not have to be like this in general.

Variance and Standard Deviation

The next measure we introduce, the *variance*, is the measure of variation used most often. It is an extension of the MAD in that it averages not only the absolute but the squared deviations. The deviations are measured from the mean. The square has the effect that larger deviations contribute even more to the measure than smaller deviations as would be the case with the MAD. This is of particular interest if deviations from the mean are more harmful the larger they are. In the conext of the variance, one often speaks of the averaged squared deviations as risk.

Formally, the population variance is defined by

$$\sigma^2 = \frac{1}{N}\sum_{i=1}^{N}(x_i - \mu)^2 \qquad (3.22)$$

One can show that equation (3.22) can be alternatively written as

$$\sigma^2 = \frac{1}{N}\left(\sum_{i=1}^{N} x_i^2\right) - \mu^2 \qquad (3.23)$$

which is sometimes preferable to the form in equation (3.22).

The sample variance is defined by

$$s^2 = \frac{1}{n}\sum_{i=1}^{n}(x_i - \bar{x})^2 \qquad (3.24)$$

using the sample mean instead. If, in equation (3.24), we use the divisor $n-1$ rather than just n, we obtain the *corrected sample variance*, which we denote s^{*2}. This is due to some issue to be introduced in Chapter 17.

As an example to illustrate the calculation of equation (3.24), we use as our sample the revenue for nine European banks based on investment banking fees generated from initial public offerings, bond underwriting, merger deals, and syndicated loan underwriting from January through March 2007. The data are shown in Table 3.5. With a sample mean of $\bar{x}$ = $178 million, the computation of the sample variance, then, yields

TABLE 3.5 Ranking of European Banks by Investment Banking Revenue from January through March 2007

Bank	Revenue (in million $)
Deutsche Bank	284
JP Morgan	188
Royal Bank of Scotland	173
Citigroup	169
BNP Paribas	169
Merrill Lynch	157
Credit Suisse	157
Morgan Stanley	155
Lalyon	153

Source: Dealogic published in the European edition of the *Wall Street Journal*, March 20, 2007.

$$s^2 = \frac{1}{8}\left[(284-178)^2 + (188-178)^2 + \ldots + (153-178)^2\right] \times 1,000,000^2$$
$$= 1,694 \times 1,000,000^2$$

or, in words, roughly $1.5 billion. This immense figure represents the average that a bank's revenue deviates from the mean squared. Definitely, the greatest chunk is contributed by Deutsche Bank's absolute deviation amounting to roughly $100 million. By squaring this amount, the effect on the variance becomes even more pronounced The variance would reduce significantly if, say, Deutsche Bank's deviation were of the size of the remaining bank's average deviation.

Related to the variance is the even more commonly stated measure of variation, the ***standard deviation.*** The reason is that the units of the standard deviation correspond to the original units of the data whereas the units are squared in the case of the variance. The standard deviation is defined to be the positive square root of the variance. Formally, the population standard deviation is

$$\sigma = \sqrt{\frac{1}{N}\sum_{i=1}^{N}(x_i - \mu)^2} \tag{3.25}$$

with the corresponding definition for the sample standard deviation being

$$s = \sqrt{\frac{1}{n-1}\sum_{i=1}^{n}(x_i - \bar{x})^2} \tag{3.26}$$

Hence, from the European bank revenue example, we obtain for the sample standard deviation

$$s = \sqrt{1694 \times 1,000,000^2} = \$42 \text{ million}$$

This number can serve as an approximate average deviation of each bank's revenue from the nine bank's mean revenue.

As discussed several times so far, when we are dealing with class data, we do not have access to the individual data values. So, the intuitive alternative is, once again, to use the class centers as representatives of the values within the classes. The corresponding sample variance of the classified data of size n is defined by

$$s_C^2 = \frac{1}{n}\sum_{I=1}^{n^C}(c_I - \bar{x}^C)^2 \times h_I = \sum_{I=1}^{n^C}(c_I - \bar{x}^C)^2 \times f_I$$

for n^C classes.[11] Alternatively, we can give the class variance by

$$s_C^2 = \frac{1}{n}\left(\sum_{I=1}^{n^C} c_I^2 \times h_I\right) - \bar{x}^2 = \sum_{I=1}^{n^C} c_I^2 \times f_I - \bar{x}^2 \qquad (3.27)$$

As an example, we use the data in Table 2.9 in Chapter 2. The variance is computed using equation (3.27).[12] So, with the mean roughly equal to $\bar{x}^C = 12.4$, we obtain

$$s_C^2 = (-16.5)^2 \times \frac{2}{235} + (-9.5)^2 \times \frac{7}{235} + \ldots + (39.5)^2 \times \frac{9}{235} - (12.4)^2 = 116.2$$

The corresponding standard deviation is $s_C = \sqrt{116.2} = 10.8$, which is higher than the MAD. This is due to large deviations being magnified by squaring them compared to the absolute deviations used for the computation of the MAD.

Skewness

The last measure of variation we describe in this chapter is the *skewness*. There exist several definitions for this measure. The **Pearson skewness** is defined as three times the difference between the median and the mean divided by the standard deviation.[13] Formally, the population Pearson skewness is

---

[11]The corresponding population variance is defined analogously with the sample size n replaced by the population size N.
[12]In this example, we round to one decimal place.
[13]To be more precise, this is only one of Pearson's skewness coefficients. Another one not represented here employs the mode instead of the mean.

$$\sigma_p = \frac{(\hat{\mu} - \mu)}{\sigma} \qquad (3.28)$$

and the sample skewness is

$$s_p = \frac{(m_d - \bar{x})}{s} \qquad (3.29)$$

As can be easily seen, for symmetrically distributed data skeweness is zero. For data with the mean being different from the median and, hence, located in either the left or the right half of the data, the data are skewed. If the mean is in the left half, the data are skewed to the left (or left skewed) since there are more extreme values on the left side compared to the right side. The opposite (i.e., skewed to the right (or right skewed)), is true for data whose mean is further to the right than the median. In contrast to the MAD and variance, the skewness can obtain positive as well as negative values. This is because not only is some absolute deviation of interest but the direction, as well.

Consider the data in Table 3.5 and let's compute equation (3.29). The median is given by $m_d = x_{(5)} = x_{(6)} = \$169$ million and the sample mean is $\bar{x} = \$178$ million. Hence, the Pearson skewness turns out to be $s_p = (\$169 - \$178)/\$41 = -0.227$, indicating left-skewed data. Note how the units (millions of dollars) vanish in this fraction since the standard deviation has the same units as the original data.

A different definition of skewness is presented by the following

$$\sigma_3 = \frac{1}{N} \sum_{i=1}^{N} (x_i - \mu)^3 / \sigma^3 \qquad (3.30)$$

Here, large deviations are additionally magnified by the power three and, akin to the previous definition of Pearson's skewness given by equation (3.29), the direction is expressed. In contrast to equation (3.29) where just two measures of center enter the formula, each data value is considered here.

As an example, assume that the nine banks in Table 3.5 represents the entire population. Thus, we have the population standard deviation given as $\sigma = \$39$ million. The skewness as in equation (3.30) is now

$$\frac{1}{9 \times 39^3} ((284-178)^3 + (188-178)^3 + \ldots + (153-178)^3) = 2.148$$

Interestingly, the skewness is now positive. This is in contrast to the Pearson skewness for the same data set. The reason is that equation (3.30) takes into consideration the deviation to the third power of every data value, whereas in equation (3.29) the entire data set enters the calculation only through the mean. Hence, large deviations have a much bigger impact in equation (3.30) than in equation (3.29). Once again, the big gap between Deutsche Bank's revenue and the average revenue accounts for the strong warp of the distribution to the right that is not given enough credit in equation (3.30).

Data Levels and Measures of Variation

A final note on data level issues with respect to the measures of variation is in order. Is is intuitive that nominal data are unsuitable for the computation of any of the measures of variation just introduced. Again, the answer is not so clear for rank data. The range might give some reasonable result in this case if the distance between data points remains constant so that two different data sets can be compared. The more sophisticated a measure is, however, the less meaningful the results are. Hence, the only scale that all these measures can be reasonably applied to are quantitative data.

Empirical Rule

Before continuing, we mention one more issue relevant to all types of distributions of quantitative data. If we know the mean and standard deviation of some distribution, by the so called *empirical rule*, we can assess that at least 75% of the data are within two standard deviations about the mean and at least 89% of the data are within three standard deviations about the mean.[14]

Coefficient of Variation and Standardization

As mentioned previously, the standard deviation is the most popular measure of variation. However, some difficulty might arise when one wants to compare the variation of different data sets. To overcome this, a relative measure is introduced using standard deviation relative to the mean. This measure is called the *coefficient of variation* and defined for a population by

$$v = \frac{\sigma}{\mu} \qquad (3.31)$$

and for samples, by

$$v = \frac{s}{\bar{x}} \qquad (3.32)$$

---

[14]This is the result of Chebychev's Theorem that is introduced in the context of probability theory in Chapter 11.

TABLE 3.6 Investment Banking Revenue Ranking of a Hypothetical Sample

Bank	Revenue (in million $)
A	59
B	55
C	51
D	50
E	48
F	42
G	37
H	35
I	20

The advantage of this measure over the mere use of the standard deviation will be apparent from the following example. Suppose we have a sample of banks with revenues shown in Table 3.6. Computing the mean and standard deviation for this new group, we obtain $\bar{x}$ = $44 million and s = $12 million, respectively. At first glance, this second group of banks appears to vary much less than the European sample of banks given in Table 3.5 when comparing the standard deviations. However, the coefficient of variation offers a different picture. For the first group, we have v_1 = 0.231 and for the second group, we have v_2 = 0.273. That is, relative to its smaller mean, the data in the second sample has greater variation.

Generally, there is a problem of comparison of data. Often, the data are transformed to overcome this. The data are said to be ***standardized*** if the transformation is such that the data values are reduced by the mean and divided by the standard deviation. Formally, the population and sample data are standardized by

$$z = \frac{x - \mu}{\sigma} \tag{3.33}$$

and

$$\tilde{z} = \frac{x - \bar{x}}{s} \tag{3.34}$$

respectively. So, for the two groups from Tables 3.5 and 3.6, we obtain Table 3.7 of standardized values. Note that the units ($ million) have vanished as a result of the standardization process since both the numerator and denominator are of the same units, hence, canceling out each other. Thus, we can create comparability by simply standardizing data.

TABLE 3.7 Standardized Values of Investment Banking Revenue Data for Nine European Banks

European Bank	Standardize Value	Hypothetical Bank	Standardized Value
Deutsche Bank	2.567	A	1.238
JP Morgan	0.235	B	0.905
Royal Bank of Scotland	−0.130	C	0.573
Citigroup	−0.227	D	0.490
BNP Paribas	−0.227	E	0.323
Merrill Lynch	−0.518	F	−0.176
Credit Suisse	−0.518	G	−0.591
Morgan Stanley	−0.567	H	−0.758
Lalyon	−0.615	I	−2.005

TABLE 3.8 Sample Return Data

x_i	0.027	0.035	0.039	0.046	0.058	0.062	0.080	0.096
$y_i = a \times x_i + b$	−1.199	−0.867	−0.697	−0.400	0.111	0.281	1.046	1.726

$a = 18.051, b = -2.355$

MEASURES OF THE LINEAR TRANSFORMATION

The linear transformation of some form $y = a \times x + b$ of quantitative data x as used, for example, in the standardization process is a common procedure in data analysis. For example, through standardization it is the goal to obtain data with zero mean and unit variance.[15] So, one might wonder what the impact of this is on the measures presented in this chapter.

First, we will concentrate on the measures of center and location. The transformation has no effect other than a mere shift and a rescaling as given by the equation. Hence, for example, the new population mean is $\mu^t = a \times \mu + b$, and the new mode is $m^t = a \times m + b$. We demonstrate this for the median of the sample annual return data in Table 3.8.

The median of the original data $(x_{(i)}, i = 1, \ldots, 8)$ is $m_d = 0.5 \times (x_{(4)} + x_{(5)})$ = 0.052. In the second row of the table, we have the transformed data $(y_{(i)}, i = 1, \ldots, 8)$. If we compute the median of the transformed data the way we are accustomed to we obtain

---

[15]This sort of transformation is monotone. If a is positive, then it is even direction preserving.

$$m_d^t = 0.5 \times (y_{(4)} + y_{(5)}) = -0.144$$

Instead, we could have obtained this value by just transforming the x-data median, which would have yielded

$$m_d^t = 18.051 \times m_d - 2.355 = -0.144$$

as well.

For the measures of variation, things are different. We will present here the effect of the linear transformation on the range, the MAD, and the variance. Let us once again denote the transformed data by $y_i = a \times x_i + b$ where the original data are the x_i.

We begin with the range of y_i, R_y, which is given by

$$\begin{aligned} R_y &= \max(y_i) - \min(y_i) \\ &= \max(a \times x_i + b) - \min(a \times x_i + b) \\ &= a \times \max(x_i) + b - a \times \min(x_i) - b \\ &= a \times R_x \end{aligned}$$

where R_x denotes the range of the original data. Hence, the transformation has a mere rescaling effect of the size a on the original range. Since the b cancel out each other, the transformation's change of location has no effect on the range. In equation (3.35), we used the fact that for a linear transformation with positive a, the maximum of the original x produces the maximum of the transformed data and the minimum of the original x produces the minimum of the transformed data. In the case of a negative a, the range would become negative. As a consequence, by convention we only consider the absolute value of a when we compute the range of transformed data (i.e., $R_y = |a| \times R_x$).

Next, we analyze the MAD. Let MAD_x and MAD_y denote the original and the transformed data's MAD, respectively. Furthermore, let m_d and m_d^t denote the original and the transformed data's median, respectively. The transformed MAD is then

$$\begin{aligned} MAD_y &= \frac{1}{n}\sum_{i=1}^{n}|y_i - m_d^t| = \frac{1}{n}\sum_{i=1}^{n}|a \times x_i + b - a \times m_d - b| \\ &= \frac{1}{n}\sum_{i=1}^{n}|a \times x_i - a \times m_d| = |a| \times \frac{1}{n}\sum_{i=1}^{n}|x_i - m_d| = |a| \times MAD_x \quad (3.36) \end{aligned}$$

where we have used the fact already known to us that the original median is translated into the median of the transformed data. Hence, the original

MAD is merely rescaled by the absolute value of a. Again, the shift in location by b has no effect on the MAD.

Finally, we examine the population variance.[16] Let σ_x^2 and σ_y^2 denote the variance of the original and the transformed data, respectively. Moreover, the population means of the x_i and the y_i are denoted by μ_x and μ_y, respectively. Then the variance of the y_i is

$$\begin{aligned}
\sigma_y^2 &= \frac{1}{N}\sum_{i=1}^{N}(y_i - \mu_y)^2 \\
&= \frac{1}{N}\sum_{i=1}^{N}(a \times x_i + b - a \times \mu_x - b)^2 \\
&= \frac{1}{N}\sum_{i=1}^{N}(a \times x_i - a \times \mu_x)^2 \\
&= a^2 \times \frac{1}{N} \times \sum_{i=1}^{N}(x_i - \mu_x)^2 \\
&= a^2 \times \sigma_x^2
\end{aligned} \qquad (3.37)$$

In equation (3.37), we used the fact that the mean as a measure of center is transformed through multiplication by a and a shift by b as described before. Thus, the variance of the transformed data is obtained by multiplying the original variance by the square of a. That is, once again, only the scale factor a has an effect on the variance while the shift can be neglected.

In general, we have seen that the linear transformation affects the measures of variation through the rescaling by a whereas any shift b has no consequence.

SUMMARY OF MEASURES

We introduced a good number of concepts in this chapter. For this reason, in Table 3.9 the measures of location/center and variation are listed with their data level qualifications and transformation features.

---

[16]The transformation of the sample variance is analogous with the population replaced with the sample mean.

TABLE 3.9 Measures of Center/Location and Variation: Data Level Qualification and Sensitivity to Linear Transformation

Measure of Location[a]	Data Level[b]	Sensitivity to Transformation $y = a \times x + b$
Mean	MQ	$\mu_y = a \cdot \mu_x + b \qquad \bar{y} = a \cdot \bar{x} + b$
Median	(R), MQ	$\hat{\mu}_y = a \cdot \hat{\mu} + b \qquad m_{d,y} = a \cdot m_{d,x} + b$
Mode	N, R, MQ	$M_y = a \cdot M_x + b \qquad m_y = a \cdot m_x + b$
α-percentile	(R), MQ	$\mu_{\alpha,y} = a \cdot \mu_{\alpha,x} + b \qquad q_{\alpha,y} = a \cdot q_{\alpha,x} + b$

Measure of Variation	Data Level	Sensitivity to Transformation $y = a \times x + b$
Range	MQ	$R_y = |a| \cdot R_x$
MAD	MQ	$MAD_y = |a| \cdot MAD_x$
Variance	MQ	$\sigma_y^2 = a^2 \cdot \sigma_x^2 \qquad s_y^2 = a^2 \cdot s_x^2$
Pearson – Skewness[c]	MQ	$\sigma_{P,y} = sign(a) \cdot \sigma_{P,x} \qquad s_{P,y} = sign(a) \cdot s_{P,x}$
Generalized Skewness	MQ	$\sigma_{3,y} = a^{\frac{3}{2}} \cdot \sigma_{3,y} \qquad s_{3,y} = a^{\frac{3}{2}} \cdot s_{3,x}$
Coefficient of Variation	MQ	$\upsilon_y = \dfrac{|a| \cdot \sigma_x}{a \cdot \mu_x + b} \qquad v_y = \dfrac{|a| \cdot s_x}{a \cdot \bar{x} + b}$

[a] If two formulae are given for a measure, the left formula is for the population while the right one is for the sample.
[b] N = nominal scale (qualitative); R = rank scale (ordinal); MQ = metric or cardinal scale (quantitative, i.e., interval, ratio, and absolute scale).
[c] The sign of a number x, denoted by $sign(x)$, is the indicator whether a number is negative, positive, or zero. For negative numbers, it is equal to -1, for positive numbers, it is equal to 1, and for zero, it is equal to 0. For example, $sign(-5) = -1$.

CONCEPTS EXPLAINED IN THIS CHAPTER
(IN ORDER OF PRESENTATION)

Parameters
Statistics
Mean
Median
Ceiling function
Class of incident
Mode
Multimodal
Unimodal
Mode class
Weighted mean
Arithmetic mean
Quantile
Percentile
Quartiles
Measures of variation
Range
Interquartile range
Mean absolute deviation
Variance
Corrected sample variance
Standard deviation
Skewness
Pearson skewness
Empirical rule
Coefficient of variation
Standardized

CHAPTER 4

Graphical Representation of Data

In this chapter, we describe various ways of representing data graphically. Typically, graphs are more intuitively appealing than a table of numbers. As a result, they are more likely to leave an impression on the user. In general, the objective when using graphs and diagrams is to provide a high degree of information efficiency and with greater clarity. The intention is to present the information contained in the data as attractively as possible. Though there are numerous graphical tools available, the coverage in this chapter is limited to the presentation of the most commonly used types of diagrams. Some diagrams are suited for relative as well as absolute frequencies. Wherever possible, this is taken into consideration by thinking about the purposes for using one or the other. In general, diagrams cannot be used for all data levels. So with the introduction of each diagram, the data level issue is clarified.

We begin with the graphs suitable and most commonly used with data of categorical or rank scale. The use of diagrams is then extended to quantitative data with a countable value set so that individual values are clearly disjoint from each other by some given step or unit size, and to class data.

PIE CHARTS

The first graphical tool to be introduced is the *pie chart*, so-named because of the circular shape with slices representing categories or values. The size of the slices is related to the frequencies of the values represented by them. One speaks of this relationship as proportionality between the size and the frequency.[1]

One further component of the pie chart is also attributed meaning, the radius. That is, if several data sets of possibly different size are compared with each other, the radius of each pie chart is proportional to the magnitude of the data set it represents. More precisely, if one pie represents a set A of size $S_A = |A|$ and a second pie represents a set B of size $S_B = |B|$, which is

---
[1]Proportionality of the quantities X and Y indicates that X is some real multiple of Y.

TABLE 4.1 Berkshire Hathaway Third Quarter Reports: Revenues of 1996 and 2006

Revenues (in million $)	1996	2006
Insurance and Other		
Insurance premiums earned	$971	$6,359
Sales and service revenues	722	13,514
Interest, dividend and other investment income	220	1,117
Investment gains/losses	96	278
Utilities and Energy		
Operating revenues		2,780
Other revenues		69
Finance and Financial Products		
Income from finance businesses	6	
Interest income		400
Investment gains/losses		
Derivative gains/losses		−11
Other		854
Total	$2,015	$25,360

Source: Berkshire Hathaway, Inc., http://www.berkshirehathaway.com/ (April 4, 2007).

$k = S_A/S_B$ times that of A, then the pie of B has to have a radius r_B, which is $\sqrt{k}$ times the length of radius r_A of pie A, tht is, $r_B = r_A \cdot \sqrt{k}$.[2]

As an example, we consider the (unaudited) third quarter reports of Berkshire Hathaway, Inc. for the years 1996 and 2006 as shown in Table 4.1. In particular, we analyze the revenues of both quarters. Note that some positions appear in the report for one year while they do not in the other year. However, we are more interested in positions that appear in both. In the table, we have the corresponding revenue positions listed as given by the individual reports. From the table, we can see that in 2006, the position "derivative gains and losses" has a negative value that creates a problem since we can only represent positive—or, at least, with the same sign—val-

---

[2] The size or area of a circle is given by $r^2 \times \pi$ where π is a constant roughly equal to 3.14. This is the reason why the radius enters into the formula as a squared term. Consequently, the ratio of two areas is equal to the ratio of the two individual radii squared. The square root of the ratio expresses the proportionality between the two radii.

ues. So, we just charge it, here, against the position "Other" to obtain a value of this position equal to $843 million. This procedure, however, is most likely not appreciated if one is interested in each individual position.

We can now construct two separate pie charts representing 1996 and 2006. We depict them in Figure 4.1 and Figure 4.2, respectively.

Note that the two largest positions in 1996 remain the two largest positions in 2006. However, while "insurance premiums earned" was the largest position in 1996, accounting for 48% of revenues, it ranked second in 2006, accounting for 25% of revenue. In absolute figures, though, the value is about seven times as large in 2006 as in 1996. Since the total of the revenues for 2006 exceeds 10 times that of 1996, this has to be reflected in the radius. As mentioned, the radius of 2006 has to be

$$r_{2006} = r_{1996} \times \sqrt{\$25,360 / \$2,015} = r_{1996} \times 3.546$$

For a size comparison, we display both pie charts jointly in Figure 4.3.

FIGURE 4.1 Pie Chart of Third Quarter Berskhire Revenues, 1996

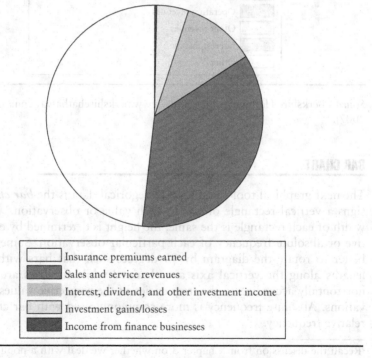

Insurance premiums earned
Sales and service revenues
Interest, dividend, and other investment income
Investment gains/losses
Income from finance businesses

Source: Berkshire Hathaway, Inc., http://www.berkshirehathaway.com/ (April 4, 2007).

FIGURE 4.2 Pie Chart of Third Quarter Berkshire Revenues, 2006.

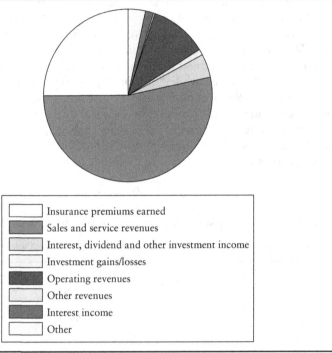

Source: Berkshire Hathaway, Inc., http://www.berkshirehathaway.com/ (April 4, 2007).

BAR CHART

The next graphical tool suitable for categorical data is the ***bar chart***. It assigns a vertical rectangle or bar to each value or observation.[3] While the width of each rectangle is the same, the height is determined by either relative or absolute frequency of each particular observation. Sometimes it is better to rotate the diagram by 90 degrees such that bars with identical lengths along the vertical axis are obtained while the bars are extended horizontally according to the frequencies of the respective values or observations. Absolute frequency is more commonly used with bar charts than relative frequency.

---

[3]Recall the discussion from Chapter 3 on whether we deal with a population with all possible values from the value set or just a sample with a certain selection from the value set.

Graphical Representation of Data

FIGURE 4.3 Comparison of 1996 and 2006 Berkshire Pie Charts

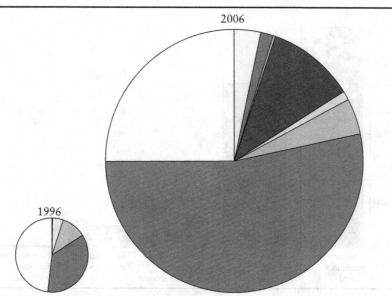

Source: Berkshire Hathaway, Inc., http://www.berkshirehathaway.com/ (April 4, 2007).

The use of bar charts is demonstrated using the Berkshire third quarter data from the previous example. We perform the task for both years, 1996 and 2006, again correcting for the negative values in position "Derivative gains/losses" in 2006 in the same fashion. The bar charts are shown in Figures 4.4 and 4.5. The two largest positions of both reports become immediately apparent again. In contrast to the pie chart, their respective performance relative to the other revenue positions of each report may become more evident using bar charts. While the bar of "insurance premiums earned" towers over the bar of "sales and service revenues" in 1996, this is exactly the opposite in the bar chart for 2006. One must be aware of the absolute frequency units of both diagrams. Even though the bars seem to be about the same height, the vertical axis is different by a factor of 10.

In Figure 4.6, we list each observation without netting. This means we do not net the negative "Derivative gains/losses" against "Other". Even though the first position of the two is negative (i.e., –$11,000,000), we still prefer to assign it a bar extending upward with height equal to the absolute value of this position (i.e., +$11,000,000).[4] In some way, we have to indicate that the value is negative. Here, it is done by the text arrow.

---

[4]Alternatively, we might have extended the bar below the horizontal axis.

FIGURE 4.4 Bar Chart of Berkshire Third Quarter Revenue ($ million), 1996

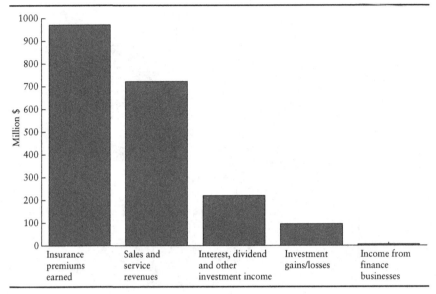

Source: Berkshire Hathaway, Inc., http://www.berkshirehathaway.com/ (April 4, 2007).

FIGURE 4.5 Bar Chart of Berkshire Third Quarter Revenue ($ million), 2006

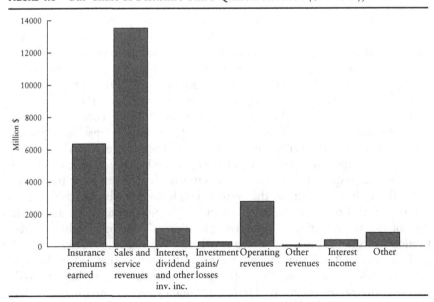

Source: Berkshire Hathaway, Inc., http://www.berkshirehathaway.com/ (April 4, 2007).

FIGURE 4.6 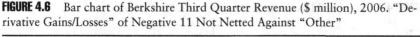Bar chart of Berkshire Third Quarter Revenue ($ million), 2006. "Derivative Gains/Losses" of Negative 11 Not Netted Against "Other"

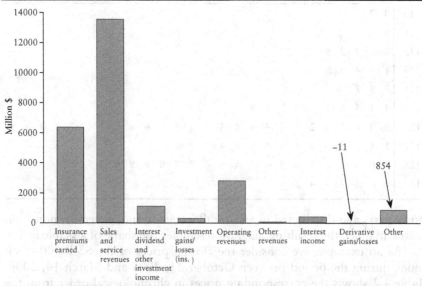

Source: Berkshire Hathaway, Inc., http://www.berkshirehathaway.com/ (April 4, 2007).

The width of the bars has been arbitrary selected in both charts. Usually, they are chosen such that the diagram fits optically well with respect to the height of the bars and such that data values can be easily identified. But beyond this, the width has no meaning.

STEM AND LEAF DIAGRAM

While the the pie chart and bar chart are graphical tools intended for use with categorical data, the next four tools are intended for use with quantitative data. The first such graphical tool we will explain is the so-called *stem and leaf diagram*. The data are usually integer-valued and positive. The diagram is constructed from a numerically ordered array of data with leading one or more digits of the observed values listed along a vertical axis in ascending order. For each observed value, the row with the corresponding leading digit(s) is selected. The remaining specifying digit of each observation is noted to the right of the vertical axis along the row, again, in numerically ascending order. How this works exactly will be shown in the next example. Hence, with the data more frequent around the median than in the

FIGURE 4.7 Stem and Leaf Diagram of S&P 500 Index Prices

Stem	Leaves
120	3 7
121	5 9
122	0 0 1 3 9
123	1 1 4 5
124	3 8 8 9
125	4 5 5 6 7 7 7 7 8 9
126	0 0 0 1 1 2 3 3 4 4 4 4 5 5 5 5 6 6 7 7 7 7 8 8 9 9
127	1 1 2 3 3 3 4 6 6 8 8 8
128	0 0 1 2 2 3 3 4 4 5 5 5 6 7 7 8 8 9 9 9
129	0 0 1 3 4 4 7

outer parts and distributed in an overall irregular and, generally, non-symmetric manner, the resulting diagram resembles the shape of a maple leaf.

As an example, we consider the closing prices of the S&P 500 stock index during the period between October 31, 2005 and March 14, 2006. Table 4.2 shows the corresponding prices in chronological order from the top left to the bottom right. We next have to order the individual prices according to size. This is easy and, hence, not shown here. The resulting stem and leaf diagram is shown in Figure 4.7. As we can see, we have selected a period where the S&P 500 stock index was between 1200 and 1299. So, all prices have two leading digits in common, that is, 12, while the third digit indicates the row in the diagram. The last (i.e., fourth) digit is the number that is put down to the right along the row in numerical order. If you look at the diagram, you will notice that most prices in this sample are between 1260 and 1299. In general, the data are skewed to the right.

FREQUENCY HISTOGRAM

The *frequency histogram* (or simply *histogram*) is a graphical tool used for quantitative data with class data. On the horizontal axis are the class bounds while the vertical axis represents the class frequencies divided by their respective class widths. We call this quantity the *frequency density* since it is proportional to the class frequency. The frequency can be either relative or absolute. The concept of density is derived from the following. Suppose, one has two data sets and one uses the same classes for both data. Without loss of generality, we have a look at the first class. Suppose for the first data set, we have twice as many observations falling into the first class

TABLE 4.2 S&P 500 Index Prices (U.S. dollars, rounded) for the Period October 31, 2005 through March 14, 2006

Price	Date	Price	Date	Price	Date	Price	Date
1207	20051031	1265	20051202	1285	20060106	1264	20060209
1203	20051101	1262	20051205	1290	20060109	1267	20060210
1215	20051102	1264	20051206	1290	20060110	1263	20060213
1220	20051103	1257	20051207	1294	20060111	1276	20060214
1220	20051104	1256	20051208	1286	20060112	1280	20060215
1223	20051107	1259	20051209	1288	20060113	1289	20060216
1219	20051108	1260	20051212	1283	20060117	1287	20060217
1221	20051109	1267	20051213	1278	20060118	1283	20060221
1231	20051110	1273	20051214	1285	20060119	1293	20060222
1235	20051111	1271	20051215	1261	20060120	1288	20060223
1234	20051114	1267	20051216	1264	20060123	1289	20060224
1229	20051115	1260	20051219	1267	20060124	1294	20060227
1231	20051116	1260	20051220	1265	20060125	1281	20060228
1243	20051117	1263	20051221	1274	20060126	1291	20060301
1248	20051118	1268	20051222	1284	20060127	1289	20060302
1255	20051121	1269	20051223	1285	20060130	1287	20060303
1261	20051122	1257	20051227	1280	20060131	1278	20060306
1266	20051123	1258	20051228	1282	20060201	1276	20060307
1268	20051125	1254	20051229	1271	20060202	1278	20060308
1257	20051128	1248	20051230	1264	20060203	1272	20060309
1257	20051129	1269	20060103	1265	20060206	1282	20060310
1249	20051130	1273	20060104	1255	20060207	1284	20060313
1265	20051201	1273	20060105	1266	20060208	1297	20060314

Note: Date column is of format yyyymmdd where y = year, m = month, and d = day. S&P 500 index price levels.

compared to the second data set. This then leads to a higher data concentration in class one in case of the first data set; in other words, the data are more densely located in class one for data set one compared to data set two.

The diagram is made up of rectangles above each class, horizontally confined by the class bounds and with a height determined by the class's frequency density. Thus, the histogram's appearance is similar to the bar chart. But it should not be confused with a bar chart because this can lead to misinterpretation.[5] If relative frequencies are used, then the area confined by the rectangles adds up to one. On the other hand, if absolute frequencies are used, then the area is equal to the number of data values, n. That follows because if we add up all the individual rectangles above each class, we obtain for

$$\text{Relative frequencies: } A = \sum_{I=1}^{n_C} \frac{f_I}{\Delta I} \times \Delta I = \sum_{I=1}^{n_C} f_I = 1$$

$$\text{Absolute frequencies: } A = \sum_{I=1}^{n_C} \frac{a_I}{\Delta I} \times \Delta I = \sum_{I=1}^{n_C} a_I = n$$

for a sample of size n where we have n_C classes, and where ΔI is the width of class I.

The histogram helps in determining the approximate center, spread, and overall shape of the data. We demonstrate the features of the histogram using six exemplary data sets. In Figure 4.8 we have three identical data sets except that data set 2 is shifted with respect to data set 1 and data set 3 is shifted by some additional amount. That is, all three data sets have measures of center and location that are different from each other by some shift. Apart from this, the spread and overall shape of the individual histograms are equivalent.

In Figure 4.9, we show three additional data sets. Data set 4 is still symmetric but, in contrast to the three previous data sets, it has varying class widths. Besides being shifted to the right, data set 5 has an overall shape that is different from data set 4 in that it appears more compact. It is still symmetric, but the spread is smaller. And finally, data set 6 is shifted even further to the right compared to data sets 4 and 5. Unlike the other five histograms, this one is not symmetric but skewed to the right. Furthermore, the spread seems to exceed the histogram of all the others data sets.

---

[5]Recall that the bar chart is intended for use with qualitative data while the histogram is for use with class data. Thus, they are used for different purposes. With quantitative data and constant class width, the histogram can be used as a bar chart where, in that case, the height represents plain frequencies rather than frequency densities.

Graphical Representation of Data

FIGURE 4.8 Exemplary Histograms: Same Shape, Same Spread, Different Medians

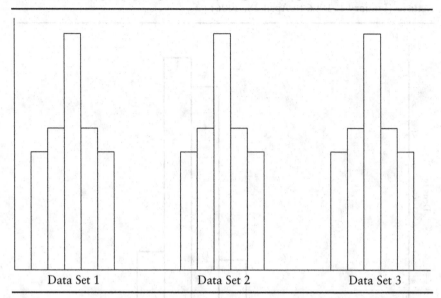

FIGURE 4.9 Exemplary Histograms: Different Median, Spread, and Shape

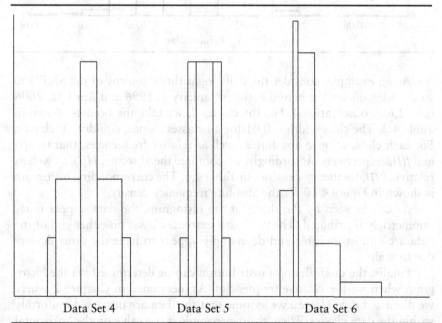

FIGURE 4.10 Histogram with Absolute Frequency Density *H(I)* of Daily S&P 500 Returns (January 1, 1996 to April 28, 2006)

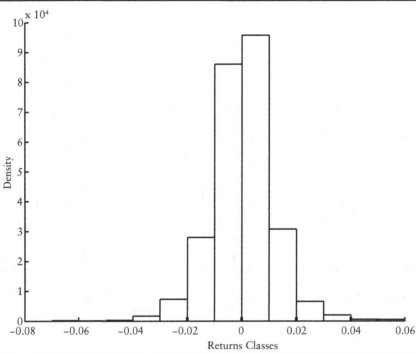

As an example, consider the daily logarithmic returns of the S&P 500 stock index during the period between January 2, 1996 and April 28, 2006 (i.e., 2,600 observations). For the classes *I*, we take the bounds shown in Table 4.3. The class width is 0.01 for all classes, hence, equidistant classes. For each class, we give absolute as well as relative frequencies, that is, *a(I)* and *f(I)*, respectively. Accordingly, we compute the absolute, *H(I)*, as well as relative, *h(I)*, frequency densities in Table 4.4. The corresponding histogram is shown in Figure 4.10 for the absolute frequency density.

As can be seen by the shape of the histogram, the data appear fairly symmetricly distributed. The two most extreme classes on either end of the data are almost invisible and deceivingly appear to have the same density due to scale.

Finally, the quantiles of a distribution can be determined via the histogram when we use relative frequencies. As mentioned in Chapter 2 where we discussed data classes, we assume that the data are dispersed uniformly within the data classes. Then, the α-percentile is the value on the horizontal

TABLE 4.3 Class Data of Daily S&P 500 Logarithmic Returns from January 2, 1996 through April 28, 2006 with $a(I)$ Absolute and $f(I)$ Relative Frequency of Class I

I	[−0.07,−0.06)	[−0.06,−0.05)	[−0.05,−0.04)	[−0.04,−0.03)	[−0.03,−0.02)	[−0.02,−0.01)	[−0.01,0.00)
$a(I)$	2	1	3	17	73	280	861
$f(I)$	0.00	0.00	0.00	0.01	0.03	0.11	0.33
I	[0,0.01)	[0.01,0.02)	[0.02,0.03)	[0.03,0.04)	[0.04,0.05)	[0.05,0.06)	
$a(I)$	958	308	66	20	6	5	
$f(I)$	0.37	0.12	0.03	0.01	0.00	0.00	

TABLE 4.4 Frequency Densities of S&P 500 Logarithmic Returns from January 2, 1996 through April 28, 2006 with $H(I)$ Absolute and $h(I)$ Relative Frequency Density

I	[−0.07,−0.06)	[−0.06,−0.05)	[−0.05,−0.04)	[−0.04,−0.03)	[−0.03,−0.02)	[−0.02,−0.01)	[−0.01,0.00)
$H(I)$	200	100	300	1700	7300	28000	86100
$h(I)$	0.08	0.04	0.12	0.65	2.81	10.77	33.12
I	[0,0.01)	[0.01,0.02)	[0.02,0.03)	[0.03,0.04)	[0.04,0.05)	[0.05,0.06)	
$H(I)$	95800	30800	6600	2000	600	500	
$h(I)$	36.85	11.85	2.54	0.77	0.23	0.19	

axis to the left of which the area covered by the histogram accounts for α% of the data. In our example, we may be interested in the 30% lower share of the data, that is, we are looking for the α 0.3-quantile.

Since class 1 represents 60% of the data already, the 0.3-quantile is to be within this class. Because of the uniform distribution, the median of class 1 is equal to this percentile, that is, $q_{0.3} = 4.5$.

Using the methods described for computation of quantiles for class data, we determine class seven to be the incidence class since the empirical cumulative relative frequency distribution is equal to 0.14 at α_7 and 0.48 at β_7, respectively. The 0.3-quantile is then computed to be $q_{0.3} = -0.005$.

We double check this by computing the area covered by the histogram for values less than or equal to -0.005. This area is

$$A = 0.01 \times (h_1 + h_2 + h_3 + h_4 + h_5 + h_6) + (-0.005 - (-0.01))/(0.01) \times h_7 \times 0.01$$
$$= 0.3 = 30\%$$

The procedure is visualized in Figure 4.11.

FIGURE 4.11 Determination of 30% Quantile Using the Histogram for Relative Frequency Densities, $f(I)$

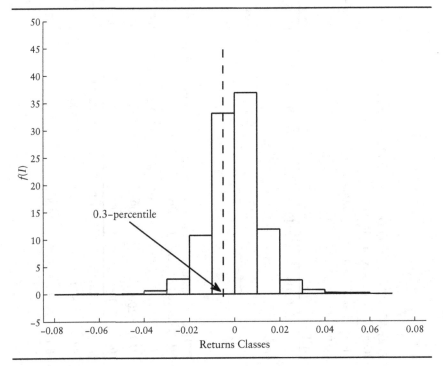

Graphical Representation of Data

OGIVE DIAGRAMS

From our coverage of cumulative frequency distributions in Chapter 2, we know that at some class bound b the cumulative frequency distribution function is equal to the sum of the absolute frequencies of all classes to the left of that particular bound while the cumulative relative frequency distribution function equals the sum of the relative frequencies of all classes to the left of the same bound. With histograms, this equals the area covered by all rectangles up to bound b. For continuous data, we assume that inside the classes, the cumulative frequency increases linearly. If we plot the cumulative relative frequency distribution at each class bound and interpolate linearly, this diagram is called an *ogive*.[6]

As an example, we use the returns from Table 4.5 along with the classes in Table 4.6. In Figure 4.12, we display the empirical cumulative relative frequency distribution of the returns.[7] In Figure 4.13, we match the empirical cumulative relative frequency distribution with the ogive diagram obtained from the data classes.

TABLE 4.5 Daily S&P 500 Logarithmic Returns of Period March 17, 2006 through April 28, 2006

Price	Date	Price	Date	Price	Date
0.0015	20060317	−0.0042	20060331	−0.0029	20060417
−0.0017	20060320	0.0023	20060403	0.0174	20060418
−0.0060	20060321	0.0063	20060404	0.0017	20060419
0.0060	20060322	0.0043	20060405	0.0012	20060420
−0.0026	20060323	−0.0019	20060406	−0.0001	20060421
0.0010	20060324	−0.0103	20060407	−0.0024	20060424
−0.0010	20060327	0.0008	20060410	−0.0049	20060425
−0.0064	20060328	−0.0077	20060411	0.0028	20060426
0.0075	20060329	0.0012	20060412	0.0033	20060427
−0.0020	20060330	0.0008	20060413	0.0007	20060428

Note: Date column is of format *yyyymmdd* where y = year, m = month, and d = day. S&P 500 returns.

---

[6]Note that for intermediate classes, the upper and lower bounds of adjacent classes coincide.

[7]The dashed line extending to the right at the return value 0.0174 indicates that the empirical cumulative relative frequency distribution remains constant at one since all observations have been accounted for.

TABLE 4.6 Classes of Daily S&P 500 Stock Index Returns for Period March 17, 2006 through April 28, 2006

Class	Bounds	Ogive
I	[−0.015, −0.008)	0.033
II	[−0.008, −0.001)	0.400
III	[−0.001, 0.006)	0.867
IV	[0.006, 0.013)	0.967
V	[0.013, 0.020)	1.000

Note: Right column contains values of ogive at upper-class bounds.

FIGURE 4.12 Empirical Cumulative Relative Frequency Distribution of Daily S&P 500 Returns for the Period March 17, 2006 through April 28, 2006

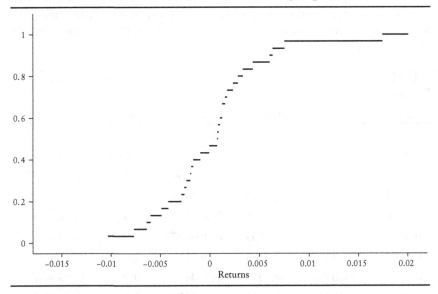

Notice that at each upper-class bound, ogive and F^f_{emp} intersect. This is because at each upper-class bound, all values less than or equal to the respective bounds have been already considered by the empirical cumulative relative frequency distribution. However, the ogive keeps increasing in a linear manner until it reaches the cumulative relative frequency evaluated at the respective upper-class bounds.[8] Hence, the ogive assumes the value of F^f_{emp} at each class bound.

---

[8]Remember that, in contrast, the empirical cumulative distribution functions increase by jumps only.

FIGURE 4.13 Empirical Cumulative Relative Frequency Distribution of the Daily S&P 500 Stock Index Returns versus Ogive of Classes of Same Data for the Period March 17, 2006 through April 28, 2006

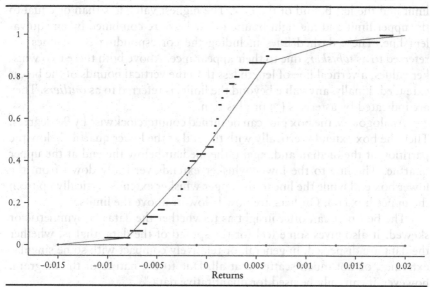

Looking at classes I and V in Figure 4.13, one can see that the ogive attributes frequency to areas outside the data range. In the case of class I starting at −0.015, the ogive ascends even though the first observation does not occur until −0.0103. Analogously, the ogive keeps ascending between the values 0.0174 and 0.020 until it assumes the value one. This is despite the fact that no more values could be observed beyond 0.0174. It is due to the assumption of a continuous uniform distribution of the data within classes already mentioned.

BOX PLOT

The *box plot* or *box and whisker plot* manages in a simple and efficient way to present both measures of location and variation. It is commonly used in the context of testing the quality of estimators of certain parameters.

To construct a box plot, the median and the lower and upper quartiles are needed. The interquartile range (IQR) representing the middle 50% of the data is indicated by a horizontal box with its left and right bounds given by lines extending vertically above the lower and upper quartile, respectively. Another vertical line of equivalent length extends above the median.

The values at 1.5 times the IQR to either the left and right of the lower and upper quartiles, define the *lower* and *upper limit*, respectively. Dashed horizontal lines combine the lowest value greater than or equal to the lower limit and the left bound of the box. The highest value less than or equal to the upper limit and the right bound of the box are combined by an equivalent line. These dashed lines including the corresponding two values are referred to as **whiskers** due to their appearance. Above both these two whisker values, a vertical line of length less than the vertical bounds of the box is extended. Finally, any value beyond the limits is referred to as **outliers**. They are indicated by asterisks (*) or plus sign.

Analogously, the box plot can be turned counterclockwise by 90 degrees. Then the box extends vertically with the end at the lower quartile below the partition at the median and, again, the median below the end at the upper quartile. The line to the lower whisker extends vertically down from the lower box end while the line to the upper whisker extends vertically up from the upper-box end. Outliers are now below or above the limits.

The box plot can offer insight as to whether the data are symmetric or skewed. It also gives some feel for the spread of the data, that is, whether the data are dispersed, in general, or relatively compact with some singular extremes, or not much scattered at all. Due to the nature of the diagram, however, it can only be used for quantitative data.[9]

To illustrate the procedure for generating the box plot, we first analyze the daily logarithmic returns of the Euro-U.S. dollar exchange rate. The period of analysis is January 1, 1999 through June 29, 2007 ($n = 2{,}216$ observations). The next step is to determine the quartiles and the median. They are given to be

$$q_{0.25} = -0.0030$$

$$m_d = 0.0001$$

$$q_{0.75} = 0.0030$$

From the quartiles, we can compute the IQR. This enables us to then determine the lower and upper limit. For our data, we obtain

$$LL = q_{0.25} - 1.5 \cdot IQR = -0.0120$$

$$UL = q_{0.75} + 1.5 \cdot IQR = 0.0120$$

---

[9]Even though quantiles were shown to exist for ordinal-scale data, the computation of the IQR is infeasible.

where LL denotes lower limit and UL denotes upper limit. With our data, we find that the values of the limits are actually observed values. Thus, we obtain as whisker ends

$$LW = LL = -0.0120$$

$$UW = UL = 0.0120$$

with LW and UW denoting the lower- and upper-whisker ends, respectively.

By construction, any value beyond these values is an outlier denoted by an asterisk. The resulting box plot is depicted in Figure 4.14.

As a second illustration, we use the daily S&P 500 stock index returns between January 4, 2007 and July, 20, 2007 ($n = 137$ logarithmic returns). The quantities of interest in this example are

$$q_{0.25} = -0.0015$$

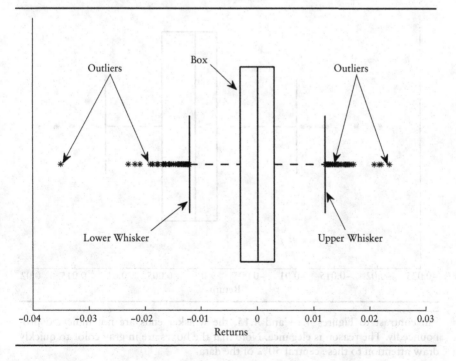

FIGURE 4.14 Box Plot of the Daily Logarithmic Returns of the EUR-USD Exchange Rate from January 1, 1999 through June 29, 2007

$$q_{0.75} = 0.0049$$
$$m_d = 0.0010$$
$$IQR = 0.0063$$
$$LL = -0.0011$$
$$UL = 0.0144$$
$$LW = -0.0107$$
$$UW = 0.0114$$

The resulting box plot is displayed in Figure 4.15. It can be seen in the figure that neither the lower nor upper limits are assumed by observations. Consequently, it follows that the whiskers do not extend across the full length between quartiles and limits. In general, the second plot reveals some skewness in the data that can be confirmed by checking the corresponding statistics.

For illustrative purposes only, the plots are also displayed turned counterclockwise by 90 degrees. This can be observed in Figures 4.16 and 4.17.[10]

FIGURE 4.15 Box Plot of the Daily S&P 500 Stock Index Logarithmic Returns Over the Period January 4, 2007 through July 20, 2007

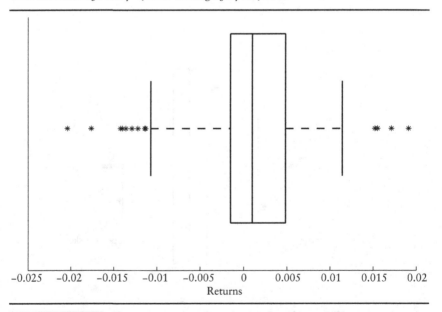

[10]In contrast to Figures 4.14 and 4.15, the whisker ends are not indicated pronouncedly. The reason is elegance. Note that the boxes are in gray color to quickly draw attention to these central 50% of the data.

Graphical Representation of Data

FIGURE 4.16 Box Plot of Daily EUR-USD Exchange Rate Returns (counterclockwise by 90 degrees)

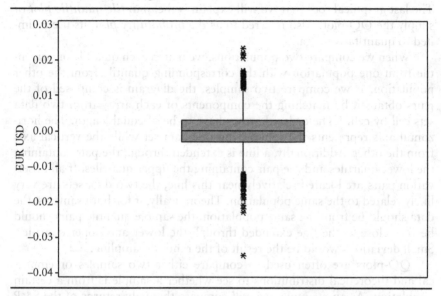

FIGURE 4.17 Box Plot of Daily S&P 500 Stock Index Returns (counterclockwise by 90 degrees)

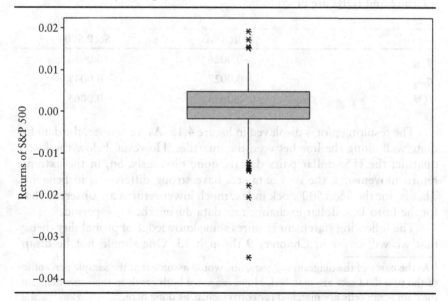

QQ PLOT

The last graphical tool that we will explain is the *quantile-quantile plot* or simply the QQ-plot. Also referred to as the *probability plot*, its use is limited to quantitative data.

When we compare two populations, we match each quantile or percentile from one population with the corresponding quantile from the other population. If we compare two samples, the diagram is composed of the pairs obtained by matching the components of each array from two data sets cell by cell.[11] The ordered arrays have to be of equal length. The horizontal axis represents the values from one data set while the vertical axis from the other. Additionally, a line is extended through the pair containing the lower quartiles and the pair containing the upper quartiles. If all observation pairs are located relatively near this line, the two data sets are very likely related to the same population. Theoretically, if for both samples, the data should be from the same population, the sample quantile pairs should be very close to the line extended through the lower and upper quartiles. Small deviations would be the result of the random sampling.

QQ-plots are often used to compare either two samples or empirical and theoretical distributions to see whether a sample is from a certain population. As an example, we will compare the daily returns of the S&P 500 stock index and the Euro-U.S. dollar exchange rate during the period between January 4, 2007 and June 26, 2007. The corresponding sample quartiles and IQRs are given to be

	EUR-USD	S&P 500
$q_{0.25}$	−0.0026	−0.0016
$q_{0.75}$	0.0027	0.0049
IQR	0.0053	0.0065

The resulting plot is displayed in Figure 4.18. As we can see, the data fits quite well along the line between the quartiles. However, below the lower quartile, the U.S. dollar pairs deviate quite obviously. So, in the extreme return movements, the two data sets have strong differences in behavior. That is, for the S&P 500 stock index, much lower returns are observed than for the Euro-U.S. dollar exchange rate data during the same period.

The following statement requires some knowledge of probability theory that we will cover in Chapters 9 through 13. One should not be disap-

---

[11] As the name of the diagram suggests, one would assume that the sample percentiles of the two data sets are matched. However, since both result in the same diagram, the data array cells are matched for convenience, as done here.

FIGURE 4.18 QQ Plot of Daily S&P 500 Returns versus EUR-USD Exchange Rate Returns for the Period January 4, 2007 through June 26, 2007

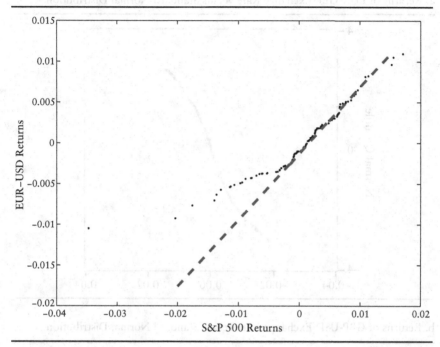

pointed if the following remark is not quite intuitive, at this stage. Typically, an empirical distribution obtained from a sample is compared to a theoretical distribution from a population from which this sample might be drawn. Very often in financial applications, the continuous normal (or Gaussian distribution) is the initial choice distribution.[12] Then, if the quantile pairs should deviate to some degree from the imaginary line through the quartile, the hypothesis of a normal distribution for the analyzed data sample is rejected and one will have to look for some other distribution. We will not go into detail at this point, however. In Figure 4.19, we have two QQ plots. The left one displays the empirical quantiles of the daily Euro-British pound (GBP) exchange rate returns (horizontal axis) matched with the theoretical quantiles from the standard normal distribution (vertical axis). The relationship looks curved rather than linear. Hence, the normal distribution ought to be rejected in favor of some alternative distribution. If we look at the right plot displaying the sample quantiles of the GBP-USD exchange rate returns matched with the standard normal quantiles, we might notice that

---

[12]The normal distribution will be discussed in Chapter 11.

FIGURE 4.19 Comparison of Empirical and Theoretical Distributions Using Box Plots

a. Returns of EUR-GBP Exchange Rate versus Standard Normal Distribution

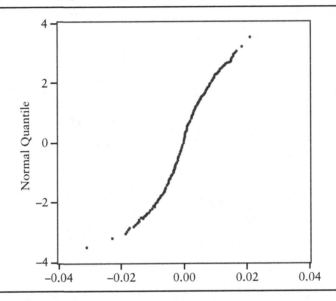

b. Returns of GBP-USD Exchange Rate versus Standard Normal Distribution

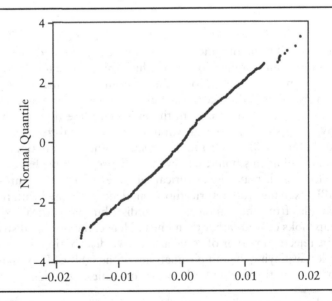

in this case the relationship looks fairly linear. So, at a first glance, the GBP-USD exchange rate returns might be modeled by the normal distribution.

CONCEPTS EXPLAINED IN THIS CHAPTER
(IN ORDER OF PRESENTATION)

Pie chart
Bar chart
Stem and leaf diagram
Frequency histogram
Frequency density
Ogive
Box plot (box and whisker plot)
Whiskers outliers
Quantile-quantile plot
Probability plot

CHAPTER 5
Multivariate Variables and Distributions

In previous chapters, we examined samples and populations with respect to one variable or attribute only. That is, we restricted ourselves to one-dimensional analysis.[1] However, for many applications of statistics to problems in finance, there is typically less of a need to analyze one variable in isolation. Instead, a typical problem faced by practitioners is to investigate the common behavior of several variables and joint occurrences of events. In other words, there is the need to establish joint frequency distributions. Along with measures determining the extent of dependence between variables, we will also introduce graphical tools for higher dimensional data to obtain a visual conception of the underlying data structure.

DATA TABLES AND FREQUENCIES

As in the one-dimensional case, we first gather all joint observations of our variables of interest. For a better overview of occurrences of the variables, it might be helpful to set up a table with rows indicating observations and columns representing the different variables. This table is called the *table of observations*. Thus, the cell of, say, row i and column j contains the value that observation i has with respect to variable j. Let us express this relationship between observations and variables a little more formally by some functional representation.

In the following, we will restrict ourselves to observations of pairs, that is, $k = 2$. In this case, the observations are *bivariate variables* of the form $x = (x_1, x_2)$.[2] The first component x_1 assumes values in the set V of possible

---
[1] The word *variable* will be used often instead of feature or characteristic. It has the same meaning, however, indicating the variable nature of the observed values.
[2] In this chapter's context, a variable consists of two components that, in turn, are one-dimensional variables.

values while the second component x_2 takes values in W, that is, the set of possible values for the second component.

Consider the Dow Jones Industrial Average over some period, say one month (roughly 22 trading days). The index includes the stock of 30 companies. The corresponding table of observations could then, for example, list the roughly 22 observation dates in the columns and the individual company names row-wise. So, in each column, we have the stock prices of all constituent stocks at a specific date. If we single out a particular row, we have narrowed the observation down to one component of the joint observation at that specific day.

Since we are not so much interested in each particular observation's value with respect to the different variables, we condense the information to the degree where we can just tell how often certain variables have occurred.[3] In other words, we are interested in the frequencies of all possible pairs with all possible combinations of first and second components. The task is to set up the so-called *joint frequency distribution*. The *absolute joint frequency* of the components x and y is denoted by

$$a_{x,y}(v,w) \qquad (5.1)$$

which is the number of occurrences counted of the pair (v,w). The *relative joint frequency distribution* is denoted by[4]

$$f_{x,y}(v,w) \qquad (5.2)$$

The relative frequency is obtained by dividing the absolute frequency by the number of observations.

While joint frequency distributions exist for all data levels, one distinguishes between qualitative data, on the one hand, and rank and quantitative data, on the other hand, when referring to the table displaying the joint frequency distribution. For qualitative (nominal scale) data, the corresponding table is called a *contingency table* whereas the table for rank (ordinal) scale and quantitative data is called a *correlation table*.

As an example, consider the daily returns of the S&P 500 stock index between January 2 and December 31, 2003. There are 252 observations (i.e., daily returns); that is, $n = 252$. To see whether the day of the week influences the sign of the stock returns (i.e., positive or negative), we sort the returns according to the day of the week as done in Table 5.1. Accumulating the 252 returns categorized by sign, for each weekday, we obtain

---

[3]This is reasonable whenever the components assume certain values more than once.
[4]Note the index refers to both components.

TABLE 5.1 Contingency Table: Absolute Frequencies of Sign (v) of Returns Sorted by Weekday (w)

		\multicolumn{5}{c}{w}					$a(w_j)$
		Mon	Tue	Wed	Thu	Fri	$a(w_j)$
	Positive	31	26	27	30	23	137
v	Negative	17	25	25	21	27	115
	$a(v_j)$	48	51	52	51	50	252

Note: Period of observation is January 2 through December 31, 2003.

the absolute frequencies as given in Table 5.1. We see that while there have been more positive returns than negative returns (i.e., 137 versus 115), the difference between positive and negative returns is greatest on Mondays. On Fridays, as an exception, there have been more negative returns than positive ones.

As another example, and one that we will used throughout this chapter, consider the bivariate monthly logarithmic return data of the S&P 500 stock index and the General Electric (GE) stock for the period between January 1996 and December 2003, 96 observation pairs. The original data are given in Table 5.2. We slightly transform the GE returns x by rounding them to two digits. Furthermore, we separate them into two sets where one set of returns coincides with negative S&P 500 stock index returns and the other set coincides with non-negative S&P 500 stock index returns. Thus, we obtain a new bivariate variable of which the first component, x, is given by the GE returns with values v and the second component, y, is the sign of the S&P 500 stock index returns with values $w = -$ and $w = +$.[5] The resulting contingency table of the absolute frequencies according to equation (5.1) is given in Table 5.2. The relative frequencies according to equation (5.2) are given in Table 5.3.

If we look at the extreme values of the GE returns, we notice that the minimum return of $v = -0.38$ occurred simultaneously with a negative S&P 500 return. On the other hand, the maximum GE return of $v = 0.23$ occurred on a day when the index return was positive.

In general, it should be intuitively obvious from the GE returns under the two different regimes (i.e., positive or negative index returns) that the stock returns behave quite differently depending on the sign of the index return.

---

[5] The (−) sign indicates negative returns and the (+) sign indicates non-negative returns. Note that, in contrast to the original bivariate returns, the new variable is not quantitative since its second component is merely a coded nominal or, at best, rank scale variable.

TABLE 5.2 Absolute Frequencies $a_{x,y}(v,w)$ of Rounded Monthly GE Stock Returns x versus Sign of Monthly S&P 500 Stock Index Returns y

GE Return (x)	S&P 500 (y) −	S&P 500 (y) +
−0.38	1	
−0.30	1	
−0.26	1	
−0.22		1
−0.16	1	
−0.13	2	1
−0.12	1	
−0.11	1	1
−0.10		1
−0.09	1	2
−0.08	2	
−0.07	3	1
−0.06	1	2
−0.05	2	1
−0.03	2	1
−0.02	1	4
−0.01	3	4
0.00	1	8
0.01		4
0.02	2	5
0.03	1	1
0.04	4	1
0.05		2
0.06	1	1
0.07	3	1
0.08		4
0.09		3
0.10	1	1
0.11	1	2
0.12		1
0.14		1
0.15	1	
0.17		2
0.19		1
0.23		1

Note: Column 2: w = negative sign, Column 3: w = positive sign. Zero frequency is denoted by a blank.

TABLE 5.3 Relative Frequencies $f_{x,y}(v,w)$ of Rounded Monthly GE Stock Returns (x) versus Sign of Monthly S&P 500 Stock Index Returns (y)

GE Return (x)	S&P 500 (y) −	S&P 500 (y) +
−0.38	0.0104	
−0.30	0.0104	
−0.26	0.0104	
−0.22		0.0104
−0.16	0.0104	
−0.13	0.0208	0.0104
−0.12	0.0104	
−0.11	0.0104	0.0104
−0.10		0.0104
−0.09	0.0104	0.0208
−0.08	0.0208	
−0.07	0.0313	0.0104
−0.06	0.0104	0.0208
−0.05	0.0208	0.0104
−0.03	0.0208	0.0104
−0.02	0.0104	0.0417
−0.01	0.0313	0.0417
0.00	0.0104	0.0833
0.01		0.0417
0.02	0.0208	0.0521
0.03	0.0104	0.0104
0.04	0.0417	0.0104
0.05		0.0208
0.06	0.0104	0.0104
0.07	0.0313	0.0104
0.08		0.0417
0.09		0.0313
0.10	0.0104	0.0104
0.11	0.0104	0.0208
0.12		0.0104
0.14		0.0104
0.15	0.0104	
0.17		0.0208
0.19		0.0104
0.23		0.0104

Note: Column 2: w = negative sign, Column 3: w = positive sign. Zero frequency is denoted by a blank.

CLASS DATA AND HISTOGRAMS

As in the univariate (i.e., one-dimensional) case, it is sometimes useful to transform the original data into class data. The requirement is that the data are quantitative. The reasons for classing are the same as before. Instead of single values, rows and columns may now contain intervals representing the class bounds. Note that either both variables can be class data or just one. The joint frequency of classed data is, again, best displayed using histograms of the corresponding density, a concept that will be defined now. In the bivariate (i.e., two-dimensional) case, the histogram is a three-dimensional object where the two-dimensional base plane is formed by the two axes representing the values of the two variables. The axis in the third dimension represents the *frequency density* of each combination class I and class J defined by

$$H_{I,J} = \frac{a(I,J)}{\Delta I \times \Delta J} \qquad (5.3)$$

In words, the absolute frequency of pairs denoted by $a(I,J)$ with the first component in class I and the second component in class J is divided by the area with length ΔI and width ΔJ. Using relative frequencies, we obtain the equivalent definition of density

$$h_{I,J} = \frac{f(I,J)}{\Delta I \times \Delta J} \qquad (5.4)$$

Note that definitions (5.3) and (5.4) do not yield the same values, so one has to clarify which form of frequency is applied, relative or absolute.

Rather than an area, the bivariate histogram now represents a volume. When using absolute joint frequencies, as done in definition (5.3), the entire volume covered by the histogram is equal to the total number of observations (n). When using relative frequencies, as done in definition (5.4), the entire volume under the histogram is equal to one or 100%.

In our GE-S&P 500 return example,[6] we have 96 pairs of joint observations where component 1 is the return of the index in a certain month and component 2 is the stock's return in the same month. With the ranges determined by the respective minima (i.e., $x_{\min}^{S\&P500} = -0.1576$ and $x_{\min}^{GE} = -0.3803$) and the respective maxima (i.e., $x_{\max}^{S\&P500} = 0.0923$ and $x_{\max}^{GE} = 0.2250$), reasonable lowest classes are given with lower bounds of –0.40 for both return samples. We can thus observe that the range of the index is less than half the

---

[6] The returns are listed in Table 5.2.

Multivariate Variables and Distributions

FIGURE 5.1 Histogram of Relative Joint Frequency of S&P 500 Index and GE Stock Returns

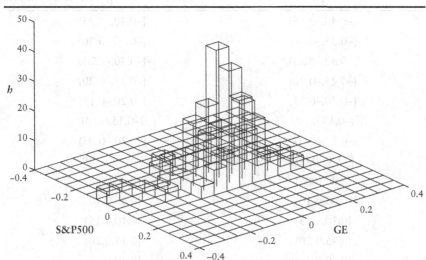

range of the stock, which is in line with the hypothesis that an index is less likely to suffer from extreme movements than individual stocks.

We choose two-dimensional classes of constant width of 0.05 in each dimension.[7] Thus, we obtain the class bounds as given in Table 5.4. First, we determine the absolute bivariate frequencies $a(I,J)$ of each class (I,J) by counting the respective return pairs that fall into the class with index I in the first and J in the second component. Then, we divide the frequencies by the total number of observations (i.e., 96) to obtain the relative frequencies $f(I,J)$. To compute the relative densities according to equation (5.2), we divide the relative frequencies by the area $0.05 \times 0.05 = 0.0025$, which is the product of the respective constant class widths. The resulting values for formula (5.3) are given in Table 5.5. Plotting the histogram, we obtain the graphic in Figure 5.1.

MARGINAL DISTRIBUTIONS

Observing bivariate data, one might be interested in only one particular component. In this case, the joint frequency in the contingency or correla-

---

[7]Note that, for simplicity, we formed the same classes for both variables despite the fact that the individual ranges of the two components are different. Moreover, the rules for univariate classes (as discussed in Chapter 2) would have suggested no more than 10 classes in each dimension. However, we prefer to distinguish more thoroughly to reveal more of the bivariate return structure inherent in the data.

TABLE 5.4 Class Bounds of Classes I (S&P 500) and J (GE)

I	J
[−0.40,−0.35)	[−0.40,−0.35)
[−0.35,−0.30)	[−0.35,−0.30)
[−0.30,−0.25)	[−0.30,−0.25)
[−0.25,−0.20)	[−0.25,−0.20)
[−0.20,−0.15)	[−0.20,−0.15)
[−0.15,0.10)	[−0.15,0.10)
[−0.10,−0.05)	[−0.10,−0.05)
[−0.05,0.00)	[−0.05,0.00)
[0.00,0.05)	[0.00,0.05)
[0.05,0.10)	[0.05,0.10)
[0.10,0.15)	[0.10,0.15)
[0.15,0.20)	[0.15,0.20)
[0.20,0.25)	[0.20,0.25)
[0.25,0.30)	[0.25,0.30)
[0.30,0.35)	[0.30,0.35)
[0.35,0.40)	[0.35,0.40)

tion table can be aggregated to produce the univariate distribution of the one variable of interest. In other words, the joint frequencies are projected into the frequency dimension of that particular component. This distribution so obtained is called the *marginal distribution*. The marginal distribution treats the data as if only the one component was observed while a detailed joint distribution in connection with the other component is of no interest.

The frequency of certain values of the component of interest is measured by the *marginal frequency*. For example, to obtain the marginal frequency of the first component whose values v are represented by the rows of the contingency or correlation table, we add up all joint frequencies in that particular row, say i. Thus, we obtain the row sum as the marginal frequency of this component v_i. That is, for each value v_i, we sum the joint frequencies over all pairs (v_i, w_j) where v_i is held fix.

To obtain the marginal frequency of the second component whose values w are represented by the columns, for each value w_j, we add up the joint frequencies of that particular column j to obtain the column sum. This time

TABLE 5.5 Histogram Values of S&P 500 Stock Index and GE Stock Returns

	[-0.40, -0.35]	[-0.35, -0.30]	[-0.30, -0.25]	[-0.25, -0.20]	[-0.20, -0.15]	[-0.15, -0.10]	[-0.10, -0.05]	[-0.05, 0.00]	[0.00, 0.05]	[0.05, 0.10]	[0.10, 0.15]	[0.15, 0.20]	[0.20, 0.25]	[0.25, 0.30]	[0.30, 0.35]	[0.35, 0.40]
[-0.40, -0.35]	0	0	0	0	0	0	0	0	0	0	0	0	0	0	0	0
[-0.35, -0.30]	0	0	0	0	0	0	0	0	0	0	0	0	0	0	0	0
[-0.30, -0.25]	0	0	0	0	0	0	0	0	0	0	0	0	0	0	0	0
[-0.25, -0.20]	0	0	0	0	0	0	0	0	0	0	0	0	0	0	0	0
[-0.20, -0.15]	0	0	0	0	0	0	4.1667	0	0	0	0	0	0	0	0	0
[-0.15, 0.10]	0	0	0	0	0	4.1667	0	0	0	0	0	0	0	0	0	0
[-0.10, -0.05]	4.1667	4.1667	0	0	0	8.3333	4.1667	8.3333	0	0	0	0	0	0	0	0
[-0.05, 0.00]	0	0	0	4.1667	4.1667	4.1667	20.8333	25.0000	16.6667	8.3333	0	0	0	0	0	0
[0.00, 0.05]	0	0	0	0	0	4.1667	16.6667	45.8333	37.5000	8.3333	4.1667	0	0	0	0	0
[0.05, 0.10]	0	0	4.1667	4.1667	4.1667	4.1667	8.3333	12.5000	29.1667	20.8333	8.3333	4.1667	0	0	0	0
[0.10, 0.15]	0	0	0	0	0	0	0	0	0	4.1667	8.3333	0	0	0	0	0
[0.15, 0.20]	0	0	0	0	0	0	0	0	0	0	0	0	0	0	0	0
[0.20, 0.25]	0	0	0	0	0	0	0	0	0	0	0	0	0	0	0	0
[0.25, 0.30]	0	0	0	0	0	0	0	0	0	0	0	0	0	0	0	0
[0.30, 0.35]	0	0	0	0	0	0	0	0	0	0	0	0	0	0	0	0
[0.35, 0.40]	0	0	0	0	0	0	0	0	0	0	0	0	0	0	0	0

Note: Rows are classes of the index and columns are classes of the stock.

we sum over all pairs (v_i, w_j) keeping w_j fix. Formally, the relative marginal frequency at value v_i of component variable x is defined by

$$f_x(v_i) = \sum_j f(v_i, w_j) \qquad (5.5)$$

where the sum is over all values w_j of the component y. The converse case, that is, the relative marginal frequency at value w_j of the component variable y, is given by the following definition

$$f_y(w_j) = \sum_i f(v_i, w_j) \qquad (5.6)$$

where summation is over all values v_i of component variable x.

For example, consider the bivariate variable where the sign of the S&P 500 stock index returns is one component and the other component is the GE stock returns. From the relative frequencies in Table 5.3, we compute the marginal frequencies defined by equations (5.5) and (5.6). The results are given in Table 5.4. From the table, we see that in $f_y(w = +) = 60\%$ of the cases, the returns of the S&P 500 were positive. We also learn that the most common GE return value is zero, occurring with a frequency of $f_x(v = 0) = 0.0937$; that is, in $0.0937 \times 96 = 9$ months of that period, the S&P 500 stock index remained unchanged.

GRAPHICAL REPRESENTATION

A common graphical tool used with bivariate data arrays is given by the so-called *scatter diagram* or *scatter plot*. In this diagram, the values of each pair are displayed. Along the horizontal axis, usually the values of the first component are displayed while along the vertical axis, the values of the second component are displayed. The scatter plot is helpful in visualizing whether the variation of one component variable somehow affects the variation of the other. If, for example, the points in the scatter plot are dispersed all over in no discernible pattern, the variability of each component may be unaffected by the other. This is visualized in Figure 5.2. The other extreme is given if there is a functional relationship between the two variables. Here, two cases are depicted. In Figure 5.3, the relationship is linear whereas in Figure 5.4, the relationship is of some higher order.[8] When two (or more) variables are observed at a certain point in time, one speaks of *cross-sectional analysis*. In contrast, analyzing one and the same variable at different

---

[8] As a matter of fact, in Figure 5.3, we have $y = 0.3 + 1.2x$. In Figure 5.4, we have $y = 0,2 + x^3$.

Multivariate Variables and Distributions 111

FIGURE 5.2 Scatter Plot: Extreme 1—No Relationship of Component Variables x and y

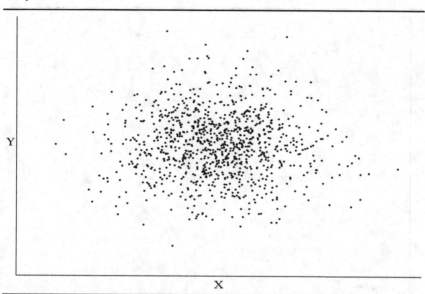

FIGURE 5.3 Scatter Plot: Extreme 2—Perfect Linear Relationship between Component Variables x and y

FIGURE 5.4 Scatter Plot: Extreme 3—Perfect Cubic Functional Relationship between Component Variables *x* and *y*

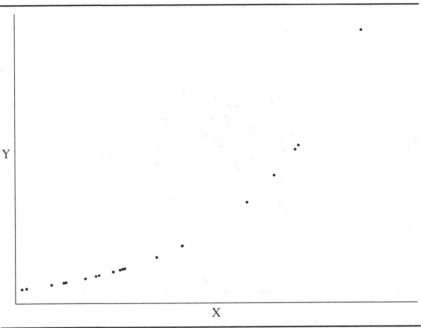

points in time, one refers to as *time series analysis*. We will come back to the analysis of various aspects of joint behavior in more detail in the subsections that follow this discussion.

Once again, consider the bivariate monthly return data of the S&P 500 stock index and the GE stock from the histogram example. We plot the pairs of returns such that the GE returns are the horizontal components while the index returns are the vertical components. The resulting plot is displayed in Figure 5.5. By observing the plot, we can roughly assess, at first, that there appears to be no distinct structure in the joint behavior of the data. However, by looking a little bit more thoroughly, one might detect a slight linear relationship underlying the two returns series. That is, the observations appear to move around some invisible line starting from the bottom left corner and advancing to the top right corner. This would appear quite reasonable since one might expect some link between the GE stock and the overall index.

FIGURE 5.5 Scatter Plot of Monthly S&P 500 Stock Index Returns versus Monthly GE Stock Returns

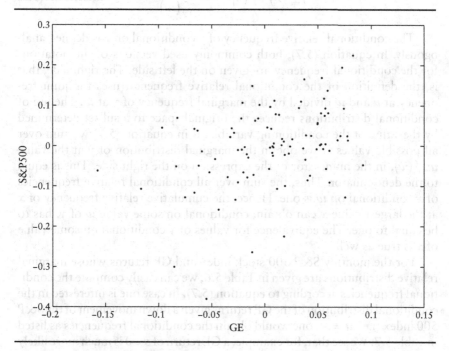

CONDITIONAL DISTRIBUTION

With the marginal distribution as previously defined, we obtain the frequency of component x at a certain value v, for example. We treat variable x as if variable y did not exist and we only observed x. Hence, the sum of the marginal frequencies of x has to be equal to one. The same is true in the converse case for variable y. Looking at the contingency or correlation table, the joint frequency at the fixed value v of the component x may vary in the values w of component y. Then, there appears to be some kind of influence of component y on the occurrence of value v of component x. The influence, as will be shown later in equations (5.14) and (5.15), is mutual. Hence, one is interested in the distribution of one component given a certain value for the other component. This distribution is called the *conditional frequency distribution*.[9] The conditional relative frequency *of* x conditional on w is defined by

---

[9]We will give definitions and demonstrate the issue of conditional frequencies using relative frequencies only. The definitions and intuition can easily be extended to the use of absolute frequencies by replacing respective quantities where necessary.

$$f_{x|w}(v) = f(x|w) = \frac{f_{x,y}(v,w)}{f_y(w)} \tag{5.7}$$

The conditional relative frequency of y conditional on v is defined analogously. In equation (5.7), both commonly used versions of the notations for the conditional frequency are given on the left side. The right side, that is, the definition of the conditional relative frequency, uses the joint frequency at v and w divided by the marginal frequency of y at w. The use of conditional distributions reduces the original space to a subset determined by the value of the conditioning variable. If in equation (5.7) we sum over all possible values v, we obtain the marginal distribution of y at the value w, $f_y(w)$, in the numerator of the expression on the right side. This is equal to the denominator. Thus, the sum over all conditional relative frequencies of x conditional on w is one. Hence, the cumulative relative frequency of x at the largest value x can obtain, conditional on some value w of y, has to be equal to one. The equivalence for values of y conditional on some value of x is true as well.

For the monthly S&P 500 stock index and GE returns whose marginal relative distributions are given in Table 5.6, we can easily compute the conditional frequencies according to equation (5.7). In case one is interested in the conditional distribution of the GE returns given a down movement of the S&P 500 index (i.e., $w = -$), one would obtain the conditional frequencies as listed in Table 5.7. We see that, for example, a GE return of $v = 0$ is much more likely under an up-scenario of the index (i.e, $f(0|+) = 0.1379$) than under a down-scenario (i.e., $f(0|-) = 0.0263$). Under a down-scenario, the most frequent GE return value is $v = 0.04$ with conditional frequency $f(0.04|-) = 0.1054$. However, in an unconditional setting of Table 5.6, the joint occurrence of $v = 0.04$ and $w = -$ happens only with $f(0.04,-) = 0.0417$.

CONDITIONAL PARAMETERS AND STATISTICS

Analogous to univariate distributions, it is possible to compute measures of center and location for conditional distributions. For example, the sample mean of x conditional on some value w of y is given by

$$\bar{x}_{|y} = \sum_i x_i f^{emp}(x_i|w) \tag{5.8}$$

The corresponding population mean is given by

$$\mu_{x|w} = \sum_i v_i f(v_i|w) \tag{5.9}$$

Multivariate Variables and Distributions

TABLE 5.6 Marginal Relative Frequencies of Rounded Monthly GE Stock Returns x, $f(v)$, and of Sign of Monthly S&P 500 Stock Index Returns, $f(w)$

GE Return x	S&P 500 y −	S&P 500 y +	$f(v)$
−0.38	0.0104		0.0104
−0.30	0.0104		0.0104
−0.26	0.0104		0.0104
−0.22		0.0104	0.0104
−0.16	0.0104		0.0104
−0.13	0.0208	0.0104	0.0312
−0.12	0.0104		0.0104
−0.11	0.0104	0.0104	0.0208
−0.10		0.0104	0.0104
−0.09	0.0104	0.0208	0.0312
−0.08	0.0208		0.0208
−0.07	0.0313	0.0104	0.0417
−0.06	0.0104	0.0208	0.0312
−0.05	0.0208	0.0104	0.0312
−0.03	0.0208	0.0104	0.0312
−0.02	0.0104	0.0417	0.0521
−0.01	0.0313	0.0417	0.0730
0.00	0.0104	0.0833	0.0937
0.01		0.0417	0.0417
0.02	0.0208	0.0521	0.0729
0.03	0.0104	0.0104	0.0208
0.04	0.0417	0.0104	0.0521
0.05		0.0208	0.0208
0.06	0.0104	0.0104	0.0208
0.07	0.0313	0.0104	0.0417
0.08		0.0417	0.0417
0.09		0.0313	0.0313
0.10	0.0104	0.0104	0.0208
0.11	0.0104	0.0208	0.0312
0.12		0.0104	0.0104
0.14		0.0104	0.0104
0.15	0.0104		0.0104
0.17		0.0208	0.0208
0.19		0.0104	0.0104
0.23		0.0104	0.0104
$f(w)$	0.40	0.60	1.00

Note: A zero joint frequency is denoted by a blank.

TABLE 5.7 Conditional Frequencies—$f_{x|w}(v) = f(v|w)$ of Rounded Monthly GE Stock Returns (x) Conditional on Sign of Monthly S&P 500 Stock Index Returns (y)

GE Return (x)	S&P 500 (y)	
	−	+
−0.38	0.0263	
−0.30	0.0263	
−0.26	0.0263	
−0.22		0.0172
−0.16	0.0263	
−0.13	0.0526	0.0172
−0.12	0.0263	
−0.11	0.0263	0.0172
−0.10		0.0172
−0.09	0.0263	0.0344
−0.08	0.0526	
−0.07	0.0791	0.0172
−0.06	0.0263	0.0344
−0.05	0.0526	0.0172
−0.03	0.0526	0.0172
−0.02	0.0263	0.0691
−0.01	0.0791	0.0691
0.00	0.0263	0.1379
0.01		0.0691
0.02	0.0526	0.0863
0.03	0.0263	0.0172
0.04	0.1054	0.0172
0.05		0.0344
0.06	0.0263	0.0172
0.07	0.0791	0.0172
0.08		0.0691
0.09		0.0518
0.10	0.0263	0.0172
0.11	0.0263	0.0344
0.12		0.0172
0.14		0.0172
0.15	0.0263	
0.17		0.0344
0.19		0.0172
0.23		0.0172
$\sum_{v} f(v|w)$	1.0000	1.0000

Note: A zero frequency is denoted by a blank.

In equation (5.8), we sum over all empirical relative frequencies of x conditional on w whereas in equation (5.9), we sum over the relative frequencies for all possible population values of x given w. Also, the conditional variance of x given w can be computed in an analogous fashion. The conditional population variance is given by

$$\sigma^2_{x|w} = f(v_i \mid w) \sum_i (v_i - \mu_{x|w})^2 \qquad (5.10)$$

The conditional sample variance is given by

$$s^2_{x|w} = f^{emp}(v_i \mid w) \sum_i (v_i - \bar{x}_{|w})^2 \qquad (5.11)$$

In contrast to definition (5.11) where we just sum over observed values, in definition (5.10), we sum over all possible values that x can assume (i.e., the entire set of feasible value).

For the computation of the conditional sample means of the GE returns, we use equation (5.8). A few intermediate results are listed in Table 5.8. The statistics are, then, $\bar{x}_{|w=-} = -0.0384$ and $\bar{x}_{|w=+} = 0.0214$. Comparison of the two shows that when we have negative index returns, the conditional sample average of the GE returns reflects this, in that we have a negative value. The opposite holds with positive index returns. A comparison of the two statistics in absolute values indicates that, on average, the negative returns are larger given a negative index scenario than the positive returns given a positive index scenario.

For the conditional sample variances of the GE returns we use definition (5.11) and obtain

$$s^2_{x|w=-} = 0.0118 \text{ and } s^2_{x|w=-} = 0.0154$$

These statistics reveal quite similar spread behavior within the two sets.[10]

INDEPENDENCE

The previous discussion raised the issue that a component may have influence on the occurrence of values of the other component. This can be analyzed by comparison of the joint frequencies of x and y with the value in one component fixed, say $x = v$. If these frequencies vary for different values of y, then the occurrence of values x is not independent of the value of y. It

---

[10]Note that in our examples we will consistently use the incorrect notation f despite the fact that we are dealing with empirical frequencies throughout.

TABLE 5.8 Conditional Sample Means of Rounded Monthly GE Stock Returns Conditional on Sign v of Monthly S&P 500 Stock Index Returns

GE Return x	S&P 500 y					
	$w=-$		$w=+$		$f(v\|w=-) \times (v-\bar{x}_{w=-})^2$	$f(v\|w=+) \times (v-\bar{x}_{w=+})^2$
	$f(v\|w=-)$	$v \times f(v\|w=-)$	$f(v\|w=+)$	$v \times f(v\|w=+)$		
−0.38	0.0263	−0.0100			0.0031	0.0042
−0.30	0.0263	−0.0079			0.0018	0.0027
−0.26	0.0263	−0.0068			0.0013	0.0021
−0.22			0.0172	−0.0038	0.0000	0.0000
−0.16	0.0263	−0.0042			0.0004	0.0009
−0.13	0.0526	−0.0068	0.0172	−0.0022	0.0004	0.0012
−0.12	0.0263	−0.0032			0.0002	0.0005
−0.11	0.0263	−0.0029	0.0172	−0.0019	0.0001	0.0005
−0.10			0.0172	−0.0017	0.0000	0.0000
−0.09	0.0263	−0.0024	0.0344	−0.0031	0.0001	0.0003
−0.08	0.0526	−0.0042			0.0001	0.0005
−0.07	0.0791	−0.0055	0.0172	−0.0012	0.0001	0.0007
−0.06	0.0263	−0.0016	0.0344	−0.0021	0.0000	0.0002
−0.05	0.0526	−0.0026	0.0172	−0.0009	0.0000	0.0003
−0.03	0.0526	−0.0016	0.0172	−0.0005	0.0000	0.0001
−0.02	0.0263	−0.0005	0.0691	−0.0014	0.0000	0.0000
−0.01	0.0791	−0.0008	0.0691	−0.0007	0.0001	0.0001
0.00	0.0263		0.1379		0.0000	0.0000
0.01			0.0691	0.0007	0.0000	0.0000
0.02	0.0526	0.0011	0.0863	0.0017	0.0002	0.0000
0.03	0.0263	0.0008	0.0172	0.0005	0.0001	0.0000
0.04	0.1054	0.0042	0.0172	0.0007	0.0006	0.0000
0.05			0.0344	0.0017	0.0000	0.0000
0.06	0.0263	0.0016	0.0172	0.0010	0.0003	0.0000
0.07	0.0791	0.0055	0.0172	0.0012	0.0009	0.0002
0.08			0.0691	0.0055	0.0000	0.0000
0.09			0.0518	0.0047	0.0000	0.0000
0.10	0.0263	0.0026	0.0172	0.0017	0.0005	0.0002
0.11	0.0263	0.0029	0.0344	0.0038	0.0006	0.0002
0.12			0.0172	0.0021	0.0000	0.0000
0.14			0.0172	0.0024	0.0000	0.0000
0.15	0.0263	0.0039			0.0009	0.0004
0.17			0.0344	0.0059	0.0000	0.0000
0.19			0.0172	0.0033	0.0000	0.0000
0.23			0.0172	0.0040	0.0000	0.0000
	$\bar{x}_{\|w=-} = -0.0384$		$\bar{x}_{\|w=+} = 0.0214$			

Multivariate Variables and Distributions

is equivalent to check whether a certain value of x occurs more frequently given a certain value of y, that is, check the conditional frequency of x conditional on y, and compare this conditional frequency with the marginal frequency at this particular value of x. The formal definition of *independence* is if for all v,w

$$f_{x,y}(v,w) = f_x(v) \cdot f_y(w) \qquad (5.12)$$

that is, for any pair (v,w), the joint frequency is the mathematical product of their respective marginals. By the definition of the conditional frequencies, we can state an equivalent definition as in the following

$$f_x(v) = f(v \mid w) = \frac{f_{x,y}(v,w)}{f_y(w)} \qquad (5.13)$$

which, in the case of independence of x and y, has to hold for all values v and w. Conversely, the analogous of equation (5.13) has to be true for the marginal frequency of y, $f_y(w)$, at any value w. In general, if one can find one pair (v,w) where either equations (5.12) or (5.13) and, hence, both do not hold, then x and y are *dependent*. So, it is fairly easy to show that x and y are dependent by simply finding a pair violating equations (5.12) and (5.13).

Now we show that the concept of influence of x on values of y is analogous. Thus, the feature of statistical dependence of two variables is mutual. This will be shown in a brief formal way by the following. Suppose that the frequency of the values of x depends on the values of y, in particular,[11]

$$f_x(v) \neq \frac{f_{x,y}(v,w)}{f_y(w)} = f(v \mid w) \qquad (5.14)$$

Multiplying each side of equation (5.14) by $f_y(w)$ yields

$$f_{x,y}(v,w) \neq f_x(v) \cdot f_y(w) \qquad (5.15)$$

which is just the definition of dependence. Dividing each side of equation (5.15) by $f_x(v) > 0$ gives

$$\frac{f_{x,y}(v,w)}{f_x(v)} = f(w \mid v) \neq f_y(w)$$

---

[11] This holds provided that $f_y(w) > 0$.

showing that the values of y depend on x. Conversely, one can demonstrate the mutuality of the dependence of the components.

Let's use the conditional frequency data from the GE returns. An easy counterexample to show that the data are not independent is, for example,

$$f_x(-0.38)f_y(-) = 0.0104 \times 0.3956 = 0.0042 \neq 0.0104 = f_{X,Y}(-0.38,-)$$

Thus, since the joint frequency of a GE return of –0.38 and a negative index return does not equal the product of the marginal frequencies, we can conclude that the component variables are not independent.

COVARIANCE

In this bivariate context, we introduce a measure of joint variation for quantitative data. It is the (sample) *covariance* defined by

$$s_{x,y} = \text{cov}(x,y) = \frac{1}{n}\sum_{i=1}^{n}(x_i - \bar{x})(y_i - \bar{y}) \qquad (5.16)$$

In definition (5.16), for each observation, the deviation of the first component from its mean is multiplied by the deviation of the second component from its mean. The sample covariance is then the average of all joint deviations. Some tedious calculations lead to an equivalent representation of definition (5.16)

$$s_{x,y} = \text{cov}(x,y) = \frac{1}{n}\sum_{i=1}^{n}v_i w_i - \overline{xy}$$

which is a transformation analogous to the one already presented for variances.

Using relative frequency distributions, definition (5.16) is equivalent to

$$\text{cov}(x,y) = \sum_{i=1}^{r}\sum_{j=1}^{s}f_{x,y}(v_i,w_j)(v_i - \bar{x})(w_j - \bar{y}) \qquad (5.17)$$

In equation (5.17), the value set of component variable x has r values while that of y has s values. For each pair (v,w), the product of the joint deviations from the respective means is weighted by the joint relative frequency.[12] From equation (5.17) we can see that, in case of independence of x

---

[12] The definition for absolute frequencies is analogous with the relative frequencies replaced by the absolute frequencies and the entire expression divided by the number of observations n.

and y, the covariance can be split up into the product of two terms. One term is variable only in x values while the other term is variable only in y values.

$$\begin{aligned}
\operatorname{cov}(x,y) &= \sum_{i=1}^{r}\sum_{j=1}^{s} f_x(v_i)f_y(w_j)(v_i - \bar{x})(w_j - \bar{y}) \\
&= \sum_{i=1}^{s} f_x(v_i)(v_i - \bar{x})\sum_{j=1}^{r} f_y(w_j)(w_j - \bar{y}) \\
&= \left[\underbrace{\sum_{i=1}^{r} f_x(v_i)v_i}_{\bar{x}} - \bar{x}\underbrace{\sum_{i=1}^{r} f_x(v_i)}_{1}\right]\left[\underbrace{\sum_{j=1}^{s} f_y(w_j)w_j}_{\bar{y}} - \bar{y}\underbrace{\sum_{j=1}^{s} f_y(w_j)}_{1}\right] \\
&= 0
\end{aligned} \qquad (5.18)$$

The important result of equation (5.18) is that the covariance of independent variables is equal to zero. The converse, however, is not generally true; that is, one cannot automatically conclude independence from zero covariance. This statement is one of the most important results in statistics and probability theory. Technically, if the covariance of x and y is zero, the two variables are said to be uncorrelated. For any value of $\operatorname{cov}(x,y)$ different from zero, the variables are correlated. Since two variables with zero covariance are uncorrelated but not automatically independent, it is obvious that independence is a stricter criterion than no correlation.[13]

This concept is exhibited in Figure 5.6. In the plot, the two sets representing correlated and uncorrelated variables are separated by the dashed line. Inside of the dashed circle, we have uncorrelated variables while the correlated variables are outside. Now, as we can see by the dotted line, the set of independent variables is completely contained within the dashed circle of uncorrelated variables. The complementary set outside the dotted circle (i.e., the dependent variables) contains all of the correlated as well as part of the uncorrelated variables. Since the dotted circle is completely inside of the dashed circle, we see that independence is a stricter requirement than uncorrelatedness.

The concept behind Figure 5.6 of zero covariance with dependence can be demonstrated by a simple example. Consider two hypothetical securities, x and y, with the payoff pattern given in Table 5.9. In the left column below y, we have the payoff values of security y while in the top row we have the payoff values of security x. Inside of the table are the joint frequencies of the pairs (x,y). As we can see, each particular value of x occurs in combination with only one particular value of y. Thus, the two variables (i.e., the payoffs

---
[13] The reason is founded in the fact that the terms in the sum of the covariance can cancel out each other even though the variables are not independent.

FIGURE 5.6 Relationship between Correlation and Dependence of Bivariate Variables

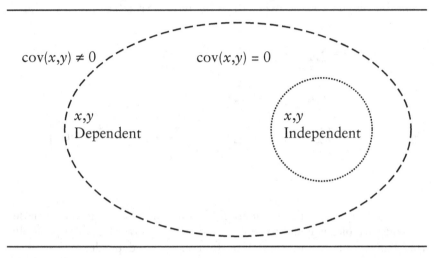

TABLE 5.9 Payoff Table of the Hypothetical Variables x and y with Joint Frequencies

y \ x	7/6	13/6	−5/6	−11/6
1	1/3			
−2		1/6		
2			1/6	
−1				1/3

of x and y) are dependent. We compute the means of the two variables to be $\bar{x} = 0$ and $\bar{y} = 0$, respectively. The resulting sample covariance according to equation (5.17) is then

$$s_{x,y} = \frac{1}{3}\cdot\left(\frac{7}{6}-0\right)\cdot\left(1 \ -0\right)+\ldots+\frac{1}{3}\left(-\frac{11}{6}-0\right)\left(1 \ -0\right) = 0$$

which indicates zero correlation. Note that despite the fact that the two variables are obviously dependent, the joint occurrence of the individual values is such that, according to the covariance, there is no relationship apparent.

The previous example is a very simple and artificial example to demonstrate this effect. As another example, consider the monthly returns of the S&P 500 stock index and GE stock from Table 5.2. With the respective

Multivariate Variables and Distributions

means given as $\bar{x}_{S\&P500} = 0.0062$ and $\bar{x}_{GE} = -0.0022$, according to equation (5.16), we obtain $s_{S\&P500,GE} = cov(r_{S\&P500}, r_{GE}) = 0.0018$.

CORRELATION

If the covariance of two variables is non-zero we know that, formally, the variables are dependent. However, the degree of correlation is not uniquely determined.

This problem is apparent from the following illustration. Suppose we have two variables, x and y, with cov(x,y) of certain value. A linear transformation of, at least, one variable, say $ax + b$, will generally lead to a change in value of the covariance due to the following property of the covariance

$$cov(ax + b, y) = a\,cov(x, y)$$

This does not mean, however, that the transformed variable is more or less correlated with y than x was. Since the covariance is obviously sensitive to transformation, it is not a reasonable measure to express the degree of correlation.

This shortcoming of the covariance can be circumvented by dividing the joint variation as defined by equation (5.16) by the product of the respective variations of the component variables. The resulting measure is the *Pearson correlation coefficient* or simply the *correlation coefficient* defined by

$$r_{x,y} = \frac{cov(x,y)}{s_x \cdot s_y} \quad (5.19)$$

where the covariance is divided by the product of the standard deviations of x and y. By definition, $r_{x,y} \in [-1,1]$ for any bivariate quantitative data. Hence, we can compare different data with respect to the correlation coefficient equation (5.19). Generally, we make the following distinction

$r_{x,y} < 0$ Negative correlation

$r_{x,y} = 0$ No correlation

$r_{x,y} > 0$ Positive correlation

to indicate the possible direction of joint behavior.

In contrast to the covariance, the correlation coefficient is invariant with respect to linear transformation. That is, it is said to be *scaling invariant*. For example, if we translate x to $ax + b$, we still have

$$r_{ax+b,y} = \text{cov}(ax+b,y)/(s_{ax+b} \cdot s_y) = a\,\text{cov}(x,y)/as_x \cdot s_y = r_{x,y}$$

For example, using the monthly bivariate return data from the S&P 500 and GE, we compute $s_{S\&P500} = Var(r_{S\&P500}) = 0.0025$ and $s_{GE} = Var(r_{GE}) = 0.0096$ such that, according to (5.19), we obtain as the correlation coefficient the value $r_{S\&P500,GE} = 0.0018/(0.0497 \cdot 0.0978) = 0.3657$. This indicates a noticeable correlation despite a covariance close to zero. The reason is that, while the covariance is influenced by the small size of the returns, the correlation coefficient is invariant to the scale and, hence, detects the true linear dependence.

In standard statistical analysis of financial and economic data, one often resorts to functions of the original variables such as squares or higher powers to detect dependence structures even though the correlations are zero. In other words, if the data should yield a correlation of zero (i.e., $r_{x,y} = 0$), we could, for example, look at x^2 and y^2 instead. Very often, correlation is then detected between x^2 and y^2, which is in favor of dependence of the variables x and y.

We issue the warning that the correlation statistic measured is a result of each individual sample and, hence, influenced by the data even though the data of different samples stems from the same population. This sensitivity needs to be kept in mind as with all statistics.[14] We repeat the warning in Chapter 6 regarding the insufficiency of the correlation coefficient as a measure of general dependence.

CONTINGENCY COEFFICIENT

So far, we could only determine the correlation of quantitative data. To extend this analysis to any type of data, we introduce another measure. The so-called *chi-square test statistic* denoted by χ^2 using relative frequencies is defined by

$$\chi^2 = n\sum_{i=1}^{r}\sum_{j=1}^{s}\frac{\left(f_{x,y}(v_i,w_j) - f_x(v_i)f_y(w_j)\right)^2}{f_x(v_i)f_y(w_j)} \quad (5.20)$$

and using absolute frequencies by

---

[14]Some statistics are robust against certain untypical behavior in samples with respect to the population. That is, they capture the "true" relationships quite well regardless of single outliers or extreme values of any type. However, this general sensitivity cannot be neglected *a priori* (i.e., in advance).

$$\chi^2 = \frac{1}{n}\sum_{i=1}^{r}\sum_{j=1}^{s}\frac{\left(n \cdot a_{x,y}(v_i,w_j) - a_x(v_i)a_y(w_j)\right)^2}{a_x(v_i)a_y(w_j)} \qquad (5.21)$$

The intuition behind equations (5.20) and (5.21) is to measure the average squared deviations of the joint frequencies from what they would be in case of independence. When the components are, in fact, independent, then the chi-square test statistic is zero. However, in any other case, we have the problem that, again, we cannot make an unambiguous statement to compare different data sets. The values of the chi-square test statistic depend on the data size n. For increasing n, the statistic can grow beyond any bound such that there is no theoretical maximum. The solution to this problem is given by the *Pearson contingency coefficient* or simply *contingency coefficient* defined by

$$C = \sqrt{\frac{\chi^2}{n+\chi^2}} \qquad (5.22)$$

The contingency coefficient by definition (5.22) is such that $0 \leq C < 1$. Consequently, it assumes values that are strictly less than one but may become arbitrarily close to one. This is still not satisfactory enough for our purpose to design a measure that can uniquely determine the respective degrees of dependence of different data sets.

There is another coefficient that can be used based on the following. Suppose we have bivariate data where the value set of the first component variable contains r different values and the value set of the second component variable contains s different values. In the extreme case of total dependence of x and y, each variable will assume a certain value if and only if the other variable assumes a particular corresponding value. Hence, we have $k = \min\{r,s\}$ unique pairs that occur with positive frequency whereas any other combination does not occur at all (i.e., has zero frequency). Then one can show that

$$C = \sqrt{\frac{k-1}{k}}$$

such that, generally, $0 \leq C \leq \sqrt{(k-1)/k} < 1$. Now, the standardized coefficient can be given by

$$C_{corr} = \sqrt{\frac{k}{k-1}}C \qquad (5.23)$$

which is called the *corrected contingency coefficient* with $0 \leq C \leq 1$. With the measures (5.20) through (5.23) and the corrected contingency coefficient, we can determine the degree of dependence for any type of data.

The products of the marginal relative or absolute frequencies, that is, $f_x(v_i)f_y(w_j)$ and $a_x(v_i)a_y(w_j)$, used in equations (5.20) and (5.21), respectively, form the so-called *indifference table* consisting of the frequencies as if the variables were independent.

Using GE returns, from the joint as well as the marginal frequencies of Table 5.6, we can compute the chi-square test statistic according to equation (5.20) to obtain $\chi^2 = 9.83$ with $n = 96$. The intermediate results are listed in Table 5.10. According to equation (5.22), the contingency coefficient is given by $C = 0.30$. With $k = \min\{2,35\} = 2$, we obtain as the corrected contingency coefficient $C_{corr} = 0.43$. Though not perfect, there is clearly dependence between the monthly S&P 500 stock index and GE stock returns.

CONCEPTS EXPLAINED IN THIS CHAPTER (IN ORDER OF PRESENTATION)

Table of observations
Bivariate variables
Joint frequency distribution
Absolute joint frequency
Relative joint frequency distribution
Contingency table
Correlation table
Frequency density
Marginal distribution
Marginal frequency
Scatter diagram
Scatter plot
Cross-sectional analysis
Time series analysis
Conditional frequency distribution
Independence
Dependent
Covariance
Pearson correlation coefficient (correlation coefficient)
Scaling invariant
Chi-square test statistic
Pearson contingency coefficient (contingency coefficient)
Corrected contingency coefficient
Indifference table

TABLE 5.10 Indifference Table of Rounded Monthly GE Stock Returns x versus Sign y of Monthly S&P 500 Stock Index Returns

GE Return x	S&P 500 y −	S&P 500 y +	$f_x(v)$	$\left(f_{x,y}(v,w) - f_x(v) \cdot f_y(w)\right)^2$	$\dfrac{\left(f_{x,y}(v,w) - f_x(v) \cdot f_y(w)\right)^2}{f_x(v) \cdot f_y(w)}$
−0.38	0.0104		0.0104	0.0015	0.0022
−0.30	0.0104		0.0104	0.0015	0.0022
−0.26	0.0104		0.0104	0.0015	0.0022
−0.22		0.0104	0.0104	0.0007	0.0010
−0.16	0.0104		0.0104	0.0015	0.0022
−0.13	0.0208	0.0104	0.0312	0.0009	0.0013
−0.12	0.0104		0.0104	0.0015	0.0022
−0.11	0.0104	0.0104	0.0208	0.0001	0.0001
−0.10		0.0104	0.0104	0.0007	0.0010
−0.09	0.0104	0.0208	0.0312	0.0001	0.0001
−0.08	0.0208		0.0208	0.0030	0.0045
−0.07	0.0313	0.0104	0.0417	0.0021	0.0031
−0.06	0.0104	0.0208	0.0312	0.0001	0.0001
−0.05	0.0208	0.0104	0.0312	0.0009	0.0013
−0.03	0.0208	0.0104	0.0312	0.0009	0.0013
−0.02	0.0104	0.0417	0.0521	0.0008	0.0013
−0.01	0.0313	0.0417	0.0730	0.0000	0.0000
0.00	0.0104	0.0833	0.0937	0.0031	0.0047
0.01		0.0417	0.0417	0.0027	0.0040
0.02	0.0208	0.0521	0.0729	0.0004	0.0006
0.03	0.0104	0.0104	0.0208	0.0001	0.0001
0.04	0.0417	0.0104	0.0521	0.0033	0.0050
0.05		0.0208	0.0208	0.0013	0.0020
0.06	0.0104	0.0104	0.0208	0.0001	0.0001
0.07	0.0313	0.0104	0.0417	0.0021	0.0031
0.08		0.0417	0.0417	0.0027	0.0040
0.09		0.0313	0.0313	0.0020	0.0030
0.10	0.0104	0.0104	0.0208	0.0001	0.0001
0.11	0.0104	0.0208	0.0312	0.0001	0.0001
0.12		0.0104	0.0104	0.0007	0.0010
0.14		0.0104	0.0104	0.0007	0.0010
0.15	0.0104		0.0104	0.0015	0.0022
0.17		0.0208	0.0208	0.0013	0.0020
0.19		0.0104	0.0104	0.0007	0.0010
0.23		0.0104	0.0104	0.0007	0.0010
$f_y(w)$	0.40	0.60	1.00		

Note: The second and third columns contain the indifference values $f_x(v) \cdot f_y(w)$.

CHAPTER 6

Introduction to Regression Analysis

In this chapter and the one to follow, we introduce methods to express joint behavior of bivariate data. It is assumed that, at least to some extent, the behavior of one variable is the result of a functional relationship between the two variables. In this chapter, we introduce the linear regression model including its ordinary least squares estimation, and the goodness-of-fit measure for a regression.

Regression analysis is important in order to understand the extent to which, for example, a security price is driven by some more global factor. Throughout this chapter, we will only consider quantitative data since most of the theory presented does not apply to any other data level.

Before advancing into the theory of regression, we note the basic idea behind a regression. The essential relationship between the variables is expressed by the measure of scaled linear dependence, that is, correlation.

THE ROLE OF CORRELATION

In many applications, how two entities behave together is of interest. Hence, we need to analyze their joint distribution. In particular, we are interested in the joint behavior of them, say x and y, linearly. The appropriate tool is given by the covariance of x and y. More exactly, we are interested in their correlation expressed by the correlation coefficient explained in Chapter 5. Generally, we know that correlation assumes values between −1 and 1 where the sign indicates the direction of the linear dependence. So, for example, a correlation coefficient of −1 implies that all pairs (x,y) are located perfectly on a line with negative slope. This is important for modeling the regression of one variable on the other. The strength of the intensity of dependence, however, is unaffected by the sign. For a general consideration, only the absolute value of the correlation is of importance. This is essential in assessing the extent of usefulness of assuming a linear relationship between the two variables.

When dealing with regression analysis, a problem may arise from seemingly correlated data even though they are not. This is expressed by accidental comovements of components of the observations. This effect is referred to as a *spurious regression*.

Stock Return Example

As an example, we consider monthly returns of the S&P 500 stock index for the period January 31, 1996 and December 31, 2003. The data are provided in Table 6.1. This time span includes 96 observations. To illustrate the linear dependence between the index and individual stocks, we take the monthly stock returns of an individual stock, General Electric (GE), covering the same period. The data are also given in Table 6.1. The correlation coefficient of the two series is $r_{S\&P500,GE}^{monthly} = 0.7125$ using the formula shown in Chapter 5. This indicates a fairly strong correlation in the same direction between the stock index and GE. So, we can expect with some certainty that GE's stock moves in the same direction as the index. Typically, there is a positive correlation between stock price movement and a stock index.

For comparison, we also compute the correlation between these two series using weekly as well as daily returns from the same period. (The data are not shown here.) In the first case, we have $r_{S\&P500,GE}^{weekly} = 0.7616$ while in the latter, we have $r_{S\&P500,GE}^{daily} = 0.7660$. This difference in value is due to the fact that while the true correlation is some value unknown to us, the correlation coefficient as a statistic depends on the sample data.

Correlation in Finance

Let's focus on the inadequacy of using the correlation as an expression of the dependence of variables. This issue is of extreme relevance in finance. Historically, returns have been modeled with distributions where it was sufficient to know the correlations in order to fully describe the dependence structure between security returns. However, empirical findings have revealed that, in reality, the correlation as a measure of linear dependence is insufficient to meaningfully express the true dependence between security returns. For this reason, the statistical concept of *copula* has been introduced in finance to help overcome this deficiency. Often, extreme joint movements of returns occur that cannot be explained by the correlation. In addition, a measure of *tail dependence* is used to express the degree to which one has to expect a security to falter once another security is already on its way to hit bottom. We will cover this topic in Chapter 16.

Introduction to Regression Analysis

TABLE 6.1 Monthly Returns of the S&P 500 Stock Index and General Electric During the Period January 31, 1996 and December 31, 2003

	S&P500	GE		S&P500	GE		S&P500	GE
Jan 31, '96	0.0321	0.0656	Sep 30, '98	0.0605	0.0077	May 31, '01	0.0050	0.0124
Feb 29, '96	0.0069	−0.015	Oct 30, '98	0.0772	0.1019	Jun 29, '01	−0.025	−0.002
Mar 29, '96	0.0078	0.0391	Nov 30, '98	0.0574	0.0343	Jul 31, '01	−0.010	−0.105
Apr 30, '96	0.0133	−0.006	Dec 31, '98	0.0548	0.1296	Aug 31, '01	−0.066	−0.056
May 31, '96	0.0225	0.0709	Jan 29, '99	0.0401	0.0307	Sep 28, '01	−0.085	−0.073
Jun 28, '96	0.0022	0.0480	Feb 26, '99	−0.032	−0.041	Oct 31, '01	0.0179	−0.017
Jul 31, '96	−0.046	−0.045	Mar 31, '99	0.0380	0.1050	Nov 30, '01	0.0724	0.0603
Aug 30, '96	0.0186	0.0121	Apr 30, '99	0.0372	−0.042	Dec 31, '01	0.0075	0.0474
Sep 30, '96	0.0527	0.0968	May 28, '99	−0.025	−0.032	Jan 31, '02	−0.015	−0.072
Oct 31, '96	0.0257	0.0622	Jun 30, '99	0.0530	0.1084	Feb 28, '02	−0.020	0.0443
Nov 29, '96	0.0708	0.0738	Jul 30, '99	−0.032	−0.030	Mar 28, '02	0.0360	−0.024
Dec 31, '96	−0.021	−0.042	Aug 31, '99	−0.006	0.0330	Apr 30, '02	−0.063	−0.163
Jan 31, '97	0.0595	0.0489	Sep 30, '99	−0.028	0.0597	May 31, '02	−0.009	−0.005
Feb 28, '97	0.0059	−0.004	Oct 29, '99	0.0606	0.1373	Jun 28, '02	−0.075	−0.058
Mar 31, '97	−0.043	−0.028	Nov 30, '99	0.0188	−0.037	Jul 31, '02	−0.082	0.1194
Apr 30, '97	0.0567	0.1154	Dec 31, '99	0.0562	0.1786	Aug 30, '02	0.0048	−0.056
May 30, '97	0.0569	0.0867	Jan 31, '00	−0.052	−0.141	Sep 30, '02	−0.116	−0.185
Jun 30, '97	0.0425	0.0757	Feb 29, '00	−0.020	−0.003	Oct 31, '02	0.0829	0.0390
Jul 31, '97	0.0752	0.0822	Mar 31, '00	0.0923	0.1754	Nov 29, '02	0.0555	0.0791
Aug 29, '97	−0.059	−0.109	Apr 28, '00	−0.031	0.0160	Dec 31, '02	−0.062	−0.097
Sep 30, '97	0.0517	0.0918	May 31, '00	−0.022	0.0099	Jan 31, '03	−0.027	−0.046
Oct 31, '97	−0.035	−0.044	Jun 30, '00	0.0236	0.0108	Feb 28, '03	−0.017	0.0488
Nov 28, '97	0.0436	0.1373	Jul 31, '00	−0.016	−0.023	Mar 31, '03	0.0083	0.0645
Dec 31, '97	0.0156	0.0002	Aug 31, '00	0.0589	0.1354	Apr 30, '03	0.0779	0.1469
Jan 30, '98	0.0100	0.0569	Sep 29, '00	−0.054	−0.012	May 30, '03	0.0496	−0.024
Feb 27, '98	0.0680	0.0038	Oct 31, '00	−0.004	−0.041	Jun 30, '03	0.0112	0.0077
Mar 31, '98	0.0487	0.1087	Nov 30, '00	−0.083	−0.096	Jul 31, '03	0.0160	−0.006
Apr 30, '98	0.0090	−0.010	Dec 29, '00	0.0040	−0.023	Aug 29, '03	0.0177	0.0407
May 29, '98	−0.019	−0.019	Jan 31, '01	0.0340	−0.028	Sep 30, '03	−0.012	0.0164
Jun 30, '98	0.0386	0.0881	Feb 28, '01	−0.096	0.0159	Oct 31, '03	0.0535	−0.025
Jul 31, '98	−0.011	−0.010	Mar 30, '01	−0.066	−0.087	Nov 28, '03	0.0071	−0.010
Aug 31, '98	−0.157	−0.105	Apr 30, '01	0.0740	0.1569	Dec 31, '03	0.0495	0.0848

REGRESSION MODEL: LINEAR FUNCTIONAL RELATIONSHIP BETWEEN TWO VARIABLES

So far, we have dealt with cross-sectional bivariate data understood as being coequal variables, x and y. Now we will present the idea of treating one variable as a reaction to the other where the other variable is considered to be exogenously given. That is, y as the *dependent variable* depends on the realization of the *regressor* or *independent variable* x. In this context, the joint behavior described in the previous section is now thought of as y being some function of x and possibly some additional quantity. In other words, we assume a functional relationship between the two variables given by the equation

$$y = f(x) \tag{6.1}$$

which is an exact deterministic relationship. However we admit that the variation of y will be influenced by other quantities. Thus, we allow for some additional quantity representing a *residual term* that is uncorrelated with x which is assumed to account for any movement of y unexplained by (6.1). Since these residuals are commonly assumed to be normally distributed—a concept to be introduced in Chapter 11—assuming that residuals are uncorrelated is equivalent to assuming that residuals are independent of x. Hence, we obtain a relationship as modeled by the following equation

$$y = f(x) + \varepsilon \tag{6.2}$$

where the residual or error is given by ε.

In addition to being independent of anything else, the residual is modeled as having zero mean and some constant variance, σ_e^2. A disturbance of this sort is considered to be some unforeseen information or shock. Assume a linear functional relationship,

$$f(x) = \alpha + \beta x \tag{6.3}$$

where the population parameters α and β are the vertical axis intercept and slope, respectively. With this assumption, equation (6.2) is called a *simple linear regression* or a *univariate regression*.[1] The parameter β determines how much y changes with each unit change in x. It is the average change in y dependent on the average change in x one can expect. This is not the case when the relationship between x and y is not linear.

---

[1] We refer to the simple linear regression as a univariate regression because there is only one independent variable whereas a multivariate regression (the subject of Part Four of this book) as a regression with more than one independent variable. In the regression literature, however, a simple linear regression is sometimes referred to as a "bivariate regression" because there are two variables, one dependent and one independent. In this book we will use the term *univariate regression*.

DISTRIBUTIONAL ASSUMPTIONS OF THE REGRESSION MODEL

The independent variable can be a deterministic quantity or a random variable. The first case is typical of an experimental setting where variables are controlled. The second case is typical in finance where we regress quantities over which we do not have any direct control, for example the returns of a stock and of an index

The error terms (or residuals) in (6.2) are assumed to be independently and identically distributed (denoted by i.i.d.). The concept of independence and identical distribution means the following. First, independence guarantees that each error assumes a value that is unaffected by any of the other errors. So, each error is absolutely unpredictable from knowledge of the other errors. Second, the distributions of all errors are the same. Consequently, for each pair (x,y), an error or residual term assumes some value independently of the other residuals in a fashion common to all the other errors, under equivalent circumstances. The i.i.d. assumption is important if we want to claim that all information is contained in the function (6.1) and deviations from (6.1) are purely random. In other words, the residuals are *statistical noise* such that they cannot be predicted from other quantities.[2]

The distribution identical to all residuals is assumed to have zero mean and constant variance such that the mean and variance of y conditional on x are

$$\mu_{y|x} = f(x) = \alpha + \beta x$$
$$\sigma^2_{y|x} = \sigma^2_e \qquad (6.4)$$

In words, once a value of x is given, we assume that, on average, y will be exactly equal to the functional relationship. The only variation in (6.4) stems from the residual term. This is demonstrated in Figure 6.1. We can see the ideal line given by the linear function. Additionally, the disturbance terms are shown taking on values along the dash-dotted lines for each pair x and y. For each value of x, ε has the mean of its distribution located on the line $\alpha + \beta \cdot x$ above x. This means that, on average, the error term will have no influence on the value of y, $\bar{y} = \bar{f}(x)$ where the bar above a term denotes the average. The x is either exogenous and, hence, known such that $\bar{f}(x) = f(x)$ or x is some endogenous variables and, thus, $\bar{f}(x)$ is the expected value of $f(x)$.

---

[2] If the errors do not seem to comply with the i.i.d. requirement, then something would appear to be wrong with the model. Moreover, in that case a lot of the estimation results would be faulty.

FIGURE 6.1 Linear Functional Relationship between *x* and *y* with Distribution of Disturbance Term

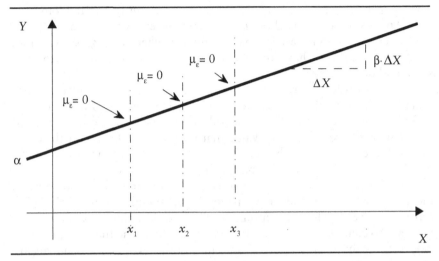

The distributions of all ε are identical. Typically, these distributions are assumed to follow *normal distributions*, a distribution that will be discussed in Chapter 11.[3] Consequently, the error terms are continuous variables that are normally distributed with zero mean and constant variance. Formally, this is indicated by

$$\varepsilon \stackrel{iid}{\sim} N(0, \sigma^2)$$

We will not, however, discuss this any further here but instead postpone our discussion to Chapter 8 where we cover random variables and their respective probability distributions.

ESTIMATING THE REGRESSION MODEL

Even if we assume that the linear assumption (6.2) is plausible, in most cases we will not know the population parameters. We have to estimate the population parameters to obtain the sample regression parameters. An initial approach might be to look at the scatter plot of *x* and *y* and iteratively draw a line through the points until one believes the best line has been found. This

---

[3]The normal distribution is a symmetric distribution of continuous variables with values on the entire real line.

Introduction to Regression Analysis

FIGURE 6.2   Scatter Plot of Data with Two Different Lines as Linear Fits

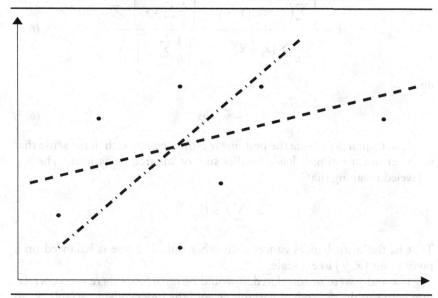

approach is demonstrated in Figure 6.2. We have five pairs of bivariate data. While at first glance both lines appear reasonable, we do not know which one is optimal. There might very well exist many additional lines that will look equally suited if not better. The intuition behind retrieving the best line is to balance it such that the sum of the vertical distances of the y-values from the line is minimized.

What we need is a formal criterion that determines optimality of some linear fit. Formally, we have to solve

$$\min_{a,b} \sum_{i=1}^{n}(y_i - a - bx_i)^2 \qquad (6.5)$$

That is, we need to find the estimates a and b of the parameters α and β, respectively, that minimize the total of the squared errors. Here, the error is given by the disturbance between the line and the true observation y. By taking the square, we penalize larger disturbances more strongly than smaller ones. This approach given by (6.5) is called the ***ordinary least squares*** (OLS) ***regression*** or methodology.[4] Here, the minimum is obtained analytically through first derivatives with respect to α and β, respectively. The resulting estimates are, then, given by

---

[4]Although several authors are credited with the concept of least squares regression, it was originally conceived by C.F. Gauss.

$$b = \frac{\frac{1}{n}\sum_{i=1}^{n}(x_i - \bar{x})(y_i - \bar{y})}{\frac{1}{n}\sum_{i=1}^{n}(x_i - \bar{x})^2} = \frac{\left(\frac{1}{n}\sum_{i=1}^{n}x_i y_i\right) - \overline{xy}}{\left(\frac{1}{n}\sum_{i=1}^{n}x_i^2\right) - \bar{x}^2} \qquad (6.6)$$

and

$$a = \bar{y} - b\bar{x} \qquad (6.7)$$

Least squares provide the best linear estimate approach in the sense that no other linear estimate has a smaller sum of squared deviations.[5] The line is leveled meaning that

$$\sum_{i=1}^{n} e_i = 0$$

That is, the disturbances cancel each other out. The line is balanced on a pivot point $(\bar{x}, \bar{y})$ like a scale.

If x and y were uncorrelated, b would be zero. Since there is no correlation between the dependent variable, y, and the independent variable, x, all variation in y would be purely random, that is, driven by the residuals, ε. The corresponding scatter plot would then look something like Figure 6.3 with the regression line extending horizontally. This is in agreement with a regression coefficient $\beta = 0$.

Application to Stock Returns

As an example, consider again the monthly returns from the S&P 500 index (indicated by X) and the GE stock (indicated by Y) from the period between January 31, 1996 and December 31, 2003. Below we list the intermediate results of regressing the index returns on the stock returns.

$\bar{x}$ = 0.0062

$\bar{y}$ = 0.0159

$\frac{1}{96}\sum_{i=1}^{96} x_i y_i$ = 0.0027

---

[5]For functional relationships higher than of linear order, there is often no analytical solution. The optima have to be determined numerically or by some trial and error algorithms.

Introduction to Regression Analysis

FIGURE 6.3 Regression of Uncorrelated Variables x and y

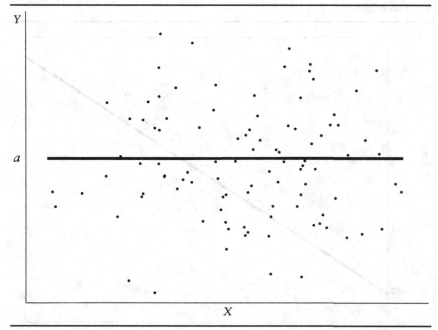

$$\frac{1}{96}\sum_{i=1}^{96} x_i^2 = 0.0025$$

$$\bar{x}^2 = 0.00004$$

(Here, we chose to present $\bar{x}^2$ with the more precise five digits since the rounded number of 0.0000 would lead to quite different results in the subsequent calculations.) Putting this into (6.6) and (6.7), we obtain

$$b = \frac{0.0027 - 0.0062 \cdot 0.0159}{0.0025 - 0.00004} = 1.0575$$
$$a = 0.0159 - 1.0575 \cdot 0.0062 = 0.0093$$

The estimated regression equation is then

$$\hat{y} = 0.0093 + 1.0575x$$

The scatter plot of the observation pairs and the resulting least squares regression line are shown in Figure 6.4.

FIGURE 6.4 Scatter Plot of Observations and Resulting Least Squares Regression Line

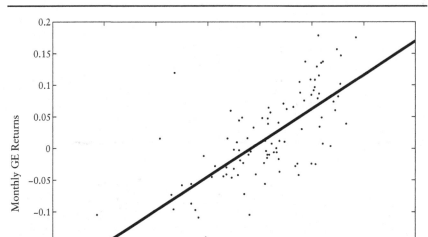

From both the regression parameter b as well as the graphic, we see that the two variables tend to move in the same direction. This supports the previous finding of a positive correlation coefficient. This can be interpreted as follows. For each unit return in the S&P 500 index value, one can expect to encounter about 1.06 times a unit return in the GE stock return. The equivalent values for the parameters using weekly and daily returns are $b = 1.2421$ and $a = 0.0003$ and $b = 1.2482$ and $a = 0.0004$, respectively.

GOODNESS OF FIT OF THE MODEL

As explained in the previous chapter, the correlation coefficient $r_{x,y}$ is a measure of the relative amount of the linear relationship between x and y. We need to find a related measure to evaluate how suitable the line is that has been derived from least squares estimation. For this task, the *coefficient of determination*, or R^2, is introduced. This *goodness-of-fit* measure calculates how much of the variation in y is caused or explained by the variation in x.

Introduction to Regression Analysis

If the percentage explained by the coefficient of determination is small, the fit might not be a too overwhelming one. Before introducing this measure formally, we present some initial considerations.

Consider the variance of the observations y by analyzing the *total sum of squares* of y around its means as given by

$$SST = \sum_{i=1}^{n}(y_i - \bar{y})^2$$

The total sum of squares (denoted by SST) can be decomposed into the *sum of squares explained by the regression* (denoted by SSR) and the *sum of squared errors* (denoted by SSE). That is,[6]

$$SST = SSR + SSE$$

with

$$SSR = \sum_{i=1}^{n}(\hat{y}_i - \bar{y})^2$$

and

$$SSE = \sum_{i}^{n}(y_i - \hat{y}_i)^2 = \sum_{i=1}^{n}e_i^2 = \sum_{i=1}^{n}y_i^2 - a\sum_{i=1}^{n}y_i - b\sum_{i=1}^{n}x_i y_i$$

where $\hat{y}$ is the estimated value for y from the regression. The SSR is that part of the total sum of squares that is explained by the regression term $f(x)$. The SSE is the part of the total sum of squares that is unexplained or equivalently the sum of squares of the errors. Now, the coefficient of determination is defined by[7]

$$R^2 = \frac{Var(f(x))}{s_y^2} = \frac{\frac{1}{n}\sum_{i=1}^{n}(a+bx_i - \bar{y})^2}{s_y^2} = \frac{\frac{1}{n}\sum_{i=1}^{n}(\hat{y}_i - \bar{y})^2}{\frac{1}{n}\sum_{i=1}^{n}(y_i - \bar{y})^2}$$

$$= \frac{SSR}{SST} = \frac{SST - SSE}{SST} = 1 - \frac{SSE}{SST}$$

---

[6] The notation in books explaining the R^2 differs. In some books, SSR denotes sum of squares of the residuals (where R represents residuals, i.e., the errors) and SSE denotes sum of sum of squares explained by the regression (where E stands for explained). Notice that the notation is just the opposite of what we used above.

[7] Note that the average means of y and $\hat{y}$ are the same (i.e., they are both equal to $\bar{y}$).

R^2 takes on values in the interval [0,1]. The meaning of $R^2 = 0$ is that there is no discernable linear relationship between x and y. No variation in y is explained by the variation in x. Thus, the linear regression makes little sense. If $R^2 = 1$, the fit of the line is perfect. All of the variation in y is explained by the variation in x. In this case, the line can have either a positive or negative slope and, in either instance, expresses the linear relationship between x and y equally well.[8] Then, all points (x_i, y_i) are located exactly on the line.

As an example, we use the monthly return data from the previous example. Employing the parameters $b = 1.0575$ and $a = 0.0093$ for the regression $\hat{y}_t$ estimates, we obtain $SST = 0.5259$, $SSR = 0.2670$, and $SSE = 0.2590$. The $R^2 = 0.5076$ (0.2670/0.5259). For the weekly fit, we obtain, $SST = 0.7620$, $SSR = 0.4420$, and $SSE = 0.3200$ while got daily fit we have $SST = 0.8305$, $SSR = 0.4873$, and $SSE = 0.3432$. The coefficient of determination is $R^2 = 0.5800$ for weekly and $R^2 = 0.5867$ for daily.

Relationship between Coefficient of Determination and Correlation Coefficient

Further analysis of the R^2 reveals that the coefficient of determination is just the squared correlation coefficient, $r_{x,y}$, of x and y. The consequence of this equality is that the correlation between x and y is reflected by the goodness-of-fit of the linear regression. Since any positive real number has a positive and a negative root with the same absolute value, so does R^2. Hence, the extreme case of $R^2 = 1$ is the result of either $r_{x,y} = -1$ or $r_{x,y} = 1$. This is repeating the fact mentioned earlier that the linear model can be increasing or decreasing in x. The extent of the dependence of y on x is not influenced by the sign. As stated earlier, the examination of the absolute value of $r_{x,y}$ is important to assess the usefulness of a linear model.

With our previous example, we would have a perfect linear relationship between the monthly S&P 500 (i.e., x) and the monthly GE stock returns (i.e., y), if say, the GE returns were $y = 0.0085 + 1.1567x$. Then $R^2 = 1$ since all residuals would be zero and, hence, the variation in them (i.e., SSE would be zero, as well).

LINEAR REGRESSION OF SOME NONLINEAR RELATIONSHIP

Sometimes, the original variables do not allow for the concept of a linear relationship. However, the assumed functional relationship is such that a

---

[8] The slope has to different from zero, however, since in that case, there would be no variation in the y-values. As a consequence, any change in value in x would have no implication on y.

FIGURE 6.5 Least Squares Regression Fit for Exponential Functional Relationship

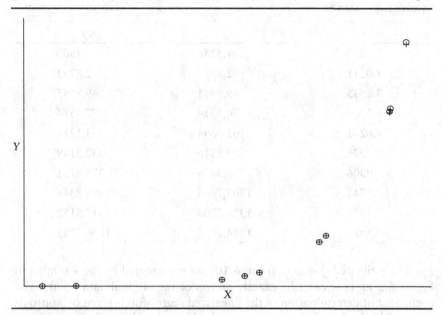

transformation $h(y)$ of the dependent variable y might lead to a linear functionality between x and the transform, h. This is demonstrated by some hypothetical data in Figure 6.5 where the y-values appear to be the result of some exponential function of the x-values. The original data pairs in Table 6.2 are indicated by the ○ symbols in Figure 6.5.

We assume that the functional relationship is of the form

$$y = \alpha e^{\beta \cdot x} \qquad (6.8)$$

To linearize (6.8), we have the following natural logarithm transformation of the y-values to perform

$$\ln y = \ln \alpha + \beta x \qquad (6.9)$$

Linear Regression of Exponential Data

We estimate using OLS the $\ln y$ on the x-values to obtain $\ln a = 0.044$ and $b = 0.993$. Retransformation yields the following functional equation

$$\hat{y} = a \cdot e^{b \cdot x} = 1.045 \cdot e^{0.993 \cdot x}$$

TABLE 6.2 Values of Exponential Relationship Between x and y Including Least Squares Regression Fit, $\hat{y}$

x	y	$\hat{y}$
0.3577	1,5256	1.4900
1.0211	2,8585	2.8792
3.8935	49,1511	49.8755
4.3369	76,5314	77.4574
4.6251	102,0694	103.1211
5.7976	329,5516	330.3149
5.9306	376,3908	376.9731
7.1745	1305,7005	1296.2346
7.1917	1328,3200	1318.5152
7.5089	1824,2675	1806.7285

The estimated $\hat{y}$-values from (6.10) are represented by the + symbol in Figure 6.5 in most cases lie exactly on top of the original data points. The coefficient of determination of the linearized regression is given by approximately $R^2 = 1$ which indicates a perfect fit. Note that this is the least squares solution to the linearized problem (6.9) and not the originally assumed functional relationship. The regression parameters for the original problem obtained in some other fashion than via linearization may provide an even tighter fit with an R^2 even closer to one.[9]

TWO APPLICATIONS IN FINANCE

In this section, we provide two applications of regression analysis to finance.

Characteristic Line

We discuss now a model for security returns. This model suggests that security returns are decomposable into three parts. The first part is the return of a risk-free asset. The second is a security specific component. And finally, the third is the return of the market in excess of the risk-free asset (i.e., *excess return*) which is then weighted by the individual security's covariance with the market relative to the market's variance Formally, this is

---
[9]As noted earlier, for functional relationships higher than of linear order, there is often no analytical solution, the optima having to be determined numerically or by some trial-and-error algorithms.

Introduction to Regression Analysis

$$R_S = R_f + \alpha_S + \beta_{S,M}(R_M - R_f) \qquad (6.11)$$

where

R_S = the individual security's return
R_f = the risk-free return
α_S = the security specific term
$\beta_{S,M} = Cov(R_S, R_M) / Var(R_M)$ = the so-called beta factor

The beta factor measures the sensitivity of the security's return to the market. Subtracting the risk-free interest rate R_f from both sides of equation (6.11) we obtain the expression for excess returns

$$R_S - R_f = \alpha_S + \beta_{S,M}(R_M - R_f)$$

or equivalently

$$r_S = \alpha_S + \beta_{S,M} r_M \qquad (6.12)$$

which is called the *characteristic line* where $r_S = R_S - R_f$ and $r_M = R_M - R_f$ denote the respective excess returns of the security and the market.

This form provides for a version similar to (6.3). The model given by (6.12) implies that at each time t, the observed excess return of some security $r_{S,t}$ is the result of the functional relationship

$$r_{S,t} = \alpha_S + \beta_{S,M} r_{M,t} + \varepsilon_{S,t} \qquad (6.13)$$

So, equation (6.13) states that the actual excess return of some security S is composed of its specific return and the relationship with the market excess return, that is, $\alpha_S + \beta_{S,M} r_{M,t}$, and some error $\varepsilon_{S,t}$ from the exact model at time t. The term α_S is commonly interpreted as a measure of performance of the security above or below its performance that is attributed to the market performance. It is often referred to as the average abnormal performance of the stock.

While we have described the characteristic line for a stock, it also applies to any portfolio or funds. To illustrate, we use the monthly returns between January 1995 and December 2004 shown in Table 6.3 for two actual mutual funds which we refer to as fund A and fund B. Both are large capitalization stock funds. As a proxy for the market, we use the S&P 500 stock index.[10] For the estimation of the characteristic line in excess return form given by equation (6.12), we use the excess return data in Table 6.3. We employ the

---

[10]The data were provided by Raman Vardharaj. The true funds names cannot be revealed.

TABLE 6.3 Data to Estimate the Characteristic Line of Two Large-Cap Mutual Funds

Month	Market Excess Return	Excess Return for Fund A	Excess Return for Fund B
01/31/1995	2.18	0.23	0.86
02/28/1995	3.48	3.04	2.76
03/31/1995	2.50	2.43	2.12
04/30/1995	2.47	1.21	1.37
05/31/1995	3.41	2.12	2.42
06/30/1995	1.88	1.65	1.71
07/31/1995	2.88	3.19	2.83
08/31/1995	−0.20	−0.87	0.51
09/30/1995	3.76	2.63	3.04
10/31/1995	−0.82	−2.24	−1.10
11/30/1995	3.98	3.59	3.50
12/31/1995	1.36	0.80	1.24
01/31/1996	3.01	2.93	1.71
02/29/1996	0.57	1.14	1.49
03/31/1996	0.57	0.20	1.26
04/30/1996	1.01	1.00	1.37
05/31/1996	2.16	1.75	1.78
06/30/1996	0.01	−1.03	−0.40
07/31/1996	−4.90	−4.75	−4.18
08/31/1996	1.71	2.32	1.83
09/30/1996	5.18	4.87	4.05
10/31/1996	2.32	1.00	0.92
11/30/1996	7.18	5.68	4.89
12/31/1996	−2.42	−1.84	−1.36
01/31/1997	5.76	3.70	5.28
02/28/1997	0.42	1.26	−1.75
03/31/1997	−4.59	−4.99	−4.18
04/30/1997	5.54	4.20	2.95
05/31/1997	5.65	4.76	5.56
06/30/1997	4.09	2.61	2.53
07/31/1997	7.51	5.57	7.49
08/31/1997	−5.97	−4.81	−3.70
09/30/1997	5.04	5.26	4.53
10/31/1997	−3.76	−3.18	−3.00
11/30/1997	4.24	2.81	2.52
12/31/1997	1.24	1.23	1.93
01/31/1998	0.68	−0.44	−0.70
02/28/1998	6.82	5.11	6.45
03/31/1998	4.73	5.06	3.45
04/30/1998	0.58	−0.95	0.64

TABLE 6.3 (Continued)

Month	Market Excess Return	Excess Return for Fund A	Excess Return for Fund B
05/31/1998	−2.12	−1.65	−1.70
06/30/1998	3.65	2.96	3.65
07/31/1998	−1.46	−0.30	−2.15
08/31/1998	−14.89	−16.22	−13.87
09/30/1998	5.95	4.54	4.40
10/31/1998	7.81	5.09	4.24
11/30/1998	5.75	4.88	5.25
12/31/1998	5.38	7.21	6.80
01/31/1999	3.83	2.25	2.76
02/28/1999	−3.46	−4.48	−3.36
03/31/1999	3.57	2.66	2.84
04/30/1999	3.50	1.89	1.85
05/31/1999	−2.70	−2.46	−1.66
06/30/1999	5.15	4.03	4.96
07/31/1999	−3.50	−3.53	−2.10
08/31/1999	−0.89	−1.44	−2.45
09/30/1999	−3.13	−3.25	−1.72
10/31/1999	5.94	5.16	1.90
11/30/1999	1.67	2.87	3.27
12/31/1999	5.45	8.04	6.65
01/31/2000	−5.43	−4.50	−1.24
02/29/2000	−2.32	1.00	2.54
03/31/2000	9.31	6.37	5.39
04/30/2000	−3.47	−4.50	−5.01
05/31/2000	−2.55	−3.37	−4.97
06/30/2000	2.06	0.14	5.66
07/31/2000	−2.04	−1.41	1.41
08/31/2000	5.71	6.80	5.51
09/30/2000	−5.79	−5.24	−5.32
10/31/2000	−0.98	−2.48	−5.40
11/30/2000	−8.39	−7.24	−11.51
12/31/2000	−0.01	2.11	3.19
01/31/2001	3.01	−0.18	4.47
02/28/2001	−9.50	−5.79	−8.54
03/31/2001	−6.75	−5.56	−6.23
04/30/2001	7.38	4.86	4.28
05/31/2001	0.35	0.15	0.13
06/30/2001	−2.71	−3.76	−1.61
07/31/2001	−1.28	−2.54	−2.10
08/31/2001	−6.57	−5.09	−5.72

TABLE 6.3 (Continued)

Month	Market Excess Return	Excess Return for Fund A	Excess Return for Fund B
09/30/2001	−8.36	−6.74	−7.55
10/31/2001	1.69	0.79	2.08
11/30/2001	7.50	4.32	5.45
12/31/2001	0.73	1.78	1.99
01/31/2002	−1.60	−1.13	−3.41
02/28/2002	−2.06	−0.97	−2.81
03/31/2002	3.63	3.25	4.57
04/30/2002	−6.21	−4.53	−3.47
05/31/2002	−0.88	−1.92	−0.95
06/30/2002	−7.25	−6.05	−5.42
07/31/2002	−7.95	−6.52	−7.67
08/31/2002	0.52	−0.20	1.72
09/30/2002	−11.01	−9.52	−6.18
10/31/2002	8.66	3.32	4.96
11/30/2002	5.77	3.69	1.61
12/31/2002	−5.99	−4.88	−3.07
01/31/2003	−2.72	−1.73	−2.44
02/28/2003	−1.59	−0.57	−2.37
03/31/2003	0.87	1.01	1.50
04/30/2003	8.14	6.57	5.34
05/31/2003	5.18	4.87	6.56
06/30/2003	1.18	0.59	1.08
07/31/2003	1.69	1.64	3.54
08/31/2003	1.88	1.25	1.06
09/30/2003	−1.14	−1.42	−1.20
10/31/2003	5.59	5.23	4.14
11/30/2003	0.81	0.67	1.11
12/31/2003	5.16	4.79	4.69
01/31/2004	1.77	0.80	2.44
02/29/2004	1.33	0.91	1.12
03/31/2004	−1.60	−0.98	−1.88
04/30/2004	−1.65	−2.67	−1.81
05/31/2004	1.31	0.60	0.77
06/30/2004	1.86	1.58	1.48
07/31/2004	−3.41	−2.92	−4.36
08/31/2004	0.29	−0.44	−0.11
09/30/2004	0.97	1.09	1.88
10/31/2004	1.42	0.22	1.10
11/30/2004	3.90	4.72	5.53
12/31/2004	3.24	2.46	3.27

estimators (6.6) and (6.7). For fund A, the estimated regression coefficients are $a_A = -0.21$ and $b_{A,S\&P500} = 0.84$, and therefore $r_A = -0.21 + 0.84 \cdot r_{S\&P500}$. For fund B we have $a_B = 0.01$ and $b_{B,S\&P500} = 0.82$, and therefore $r_B = 0.01 + 0.82 \cdot r_{S\&P500}$.

Interpreting the results of the performance measure estimates a, we see that for fund A there is a negative performance relative to the market while for it appears that fund B outperformed the market. For the estimated betas (i.e., b) we obtain for fund A that with each expected unit return of the S&P 500 index, fund A yields, on average, a return of 84% of that unit. This is roughly equal for fund B where for each unit return to be expected for the index, fund B earns a return of 82% that of the index. So, both funds are, as expected, positively related to the performance of the market.

The goodness-of-fit measure (R^2) is 0.92 for the characteristic line for fund A and 0.86 for fund B. So, we see that the characteristic lines for both mutual funds provide good fits.

Application to Hedging[11]

As another application of regression, let's see how it is used in *hedging*. Portfolio managers and risk managers use hedging to lock in some price of an asset that is expected to be sold at a future date. The concern is obviously that between the time a decision is made to sell an asset and the asset is sold there will be an adverse movement in the price of the asset. When hedging is used to protect against a decline in an asset's price, the particular hedge used is called a *short hedge*. A short hedge involves selling the hedging instrument. When hedging, the manager must address the following questions:

1. What hedging instrument should be used?
2. How much of the hedging instrument should be shorted?

The hedging instrument can be a cash market instrument or a derivative instrument such as a futures contract or a swap. Typically a derivative instrument is used. A primary factor in determining which derivative instrument will provide the best hedge is the degree of correlation between the price movement of the asset to be hedged and the price movement of the derivative instrument that is a candidate for hedging. (Note here we see an application of correlation to hedging.)

For example, consider a risk manager seeking to hedge a position in a long-term corporate bond. The price risk is that interest rates will rise in the future when it is anticipated that the corporate bond will be sold and as a result, the price of the corporate bond will decline. There are no corporate

---
[11] This example is taken from Fabozzi (2008).

bond derivative contracts that can be used to hedge against this interest rate risk. Let's suppose that the manager decides that a futures contract, a type of derivative instrument, should be used. There are different futures contracts available: Treasury bonds futures, Treasury bill futures, municipal bond futures, and stock index futures. Obviously, stock index futures would not be a good candidate given that the correlation of stock price movements and interest rates that affect corporate bond prices may not be that strong. Municipal bond futures involve tax-exempt interest rates and would not be a good candidate for a hedging instrument. Treasury bills involve short-term interest rate and hence the correlation between short-term and long-term interest rates would not be that strong. The most suitable futures contract would be Treasury bond futures contract. When using a hedging instrument that is not identical to the instrument to be hedged, the hedge is referred to as a *cross hedge*.

Given the hedging instrument, the amount of that instrument to be shorted (sold) must be determined. This amount is determined by the hedge ratio. For example, let's suppose that we are going to hedge a long-term corporate bond using a Treasury bond futures. Suppose that the hedge ratio is 1.157. This means that for every $1 million par value of the instrument to be hedged, $1.157 million par value of the hedging instrument (i.e., Treasury bond futures contract) should be sold (i.e., shorted). In the case of a Treasury bond futures contract, there is really not one Treasury bond that can be delivered to satisfy the future contract. Instead, there are many eligible Treasury bonds that can be delivered. The one that is assumed to be delivered and that is what is known as the "cheapest to deliver" (CTD) Treasury bond. A discussion of how the CTD is determined is beyond the scope of this text. Now here is where regression comes into play.

In cross hedging, the hedge ratio must be refined to take into account the relationship between the yield levels and yield spreads. More specifically, because of cross hedging, the hedge ratio is adjusted by multiplying it by what is referred to as the *yield beta* which is found by estimating the following regression:

$$\text{Yield change of corporate bond to be hedged} = \alpha + \beta \text{ yield change in CTD issue} + \varepsilon \quad (6.14)$$

where the estimated β is the yield beta.

To illustration, suppose that on December 24, 2007, a portfolio manager owned $10 million par value of the Procter & Gamble (P&G) 5.55% issue maturing March 5, 2037 and trading to yield 5.754%. The manager plans to sell the issue in March 2008. To hedge this position, suppose that the portfolio manager used U.S. Treasury bond futures maturing on the last business day of March 2008. The CTD issue for this futures contract is the

Treasury bond 6.25% issue maturing on 8/15/2003. The hedge ratio for this hedge without the adjustment for yield beta is 1.157. Table 6.4 shows the yield and yield change for the P&G bond and the CTD Treasury issue. The yield beta using the data in Table 6.4 to estimate the regression given by (6.8) and the adjusted hedge ratio are given below:

Number of Trading Days	Yield Beta	Adjusted Hedge Ratio
Prior 30 trading days ending 12/21/2007	0.906	1.048
Prior 90 trading days ending 12/21/2007	0.894	1.034

As can be seen, the adjusted hedge ratio is considerably different from the hedge ratio without adjusting for the cross hedge using the regression to compute the yield beta.

CONCEPTS EXPLAINED IN THIS CHAPTER (IN ORDER OF PRESENTATION)

Spurious regression
Copula
Tail dependence
Dependent variable
Regressor
Independent variable
Residual term
Simple linear regression
Univariate regression
Statistical noise
Normal distribution
Ordinary least squares regression
Coefficient of determination
Goodness of fit
Total sum of squares
Sum of squares regression
Sum of squares
Excess return
Characteristic line
Hedging
Short hedge
Cross hedge
Yield beta

TABLE 6.4 Yield and Yield Change for each Trading Day for the P&G 5.55% 3/5/2037 and Treasury 6.25% 8/15/2023 (CTD issue): 2/28/2007–12/21/2007

Trading Day	P&G 5.55% 3/5/2037			Treasury 6.25% 8/15/2023 (CTD issue)		
	Price	Yield	Yield Change	Price	Yield	Yield Change
8/15/2007	93.822	5.999	0.013	112.821	5.071	0.020
8/16/2007	94.982	5.911	−0.088	113.916	4.978	−0.093
8/17/2007	93.970	5.987	0.076	113.100	5.047	0.069
8/20/2007	94.396	5.955	−0.032	113.542	5.009	−0.038
8/21/2007	94.770	5.927	−0.028	114.011	4.969	−0.040
8/22/2007	93.650	6.012	0.085	113.777	4.989	0.020
8/23/2007	94.070	5.980	−0.032	114.068	4.964	−0.025
8/24/2007	94.095	5.978	−0.002	114.220	4.951	−0.013
8/27/2007	94.526	5.945	−0.033	114.715	4.910	−0.041
8/28/2007	94.951	5.914	−0.031	114.989	4.887	−0.023
8/29/2007	95.492	5.873	−0.041	114.814	4.901	0.014
8/30/2007	95.492	5.873	0.000	114.814	4.901	0.000
8/31/2007	96.118	5.827	−0.046	115.195	4.869	−0.032
9/4/2007	96.078	5.830	0.003	115.050	4.881	0.012
9/5/2007	95.406	5.880	0.050	115.915	4.809	−0.072
9/6/2007	95.198	5.895	0.015	115.722	4.825	0.016
9/7/2007	96.560	5.795	−0.100	117.060	4.715	−0.110
9/10/2007	97.272	5.743	−0.052	117.754	4.658	−0.057
9/11/2007	97.156	5.751	0.008	117.548	4.675	0.017
9/12/2007	95.307	5.887	0.136	117.098	4.711	0.036
9/13/2007	94.495	5.948	0.061	116.326	4.774	0.063
9/14/2007	94.775	5.927	−0.021	116.500	4.760	−0.014
9/17/2007	94.902	5.917	−0.010	116.530	4.757	−0.003
9/18/2007	94.306	5.962	0.045	116.174	4.786	0.029
9/19/2007	93.665	6.011	0.049	115.308	4.857	0.071
9/20/2007	92.145	6.129	0.118	113.770	4.985	0.128
9/21/2007	93.149	6.051	−0.078	114.270	4.943	−0.042
9/24/2007	93.026	6.060	0.009	114.396	4.932	−0.011
9/25/2007	94.718	5.931	−0.129	114.301	4.940	0.008
9/26/2007	94.613	5.939	0.008	114.199	4.948	0.008

TABLE 6.4 (Continued)

	P&G 5.55% 3/5/2037			Treasury 6.25% 8/15/2023 (CTD issue)		
Trading Day	Price	Yield	Yield Change	Price	Yield	Yield Change
9/27/2007	95.402	5.880	−0.059	114.877	4.891	−0.057
9/28/2007	95.281	5.889	0.009	114.877	4.891	0.000
10/1/2007	95.806	5.850	−0.039	115.284	4.857	−0.034
10/2/2007	96.055	5.832	−0.018	115.572	4.833	−0.024
10/3/2007	96.059	5.831	−0.001	115.381	4.849	0.016
10/4/2007	96.335	5.811	−0.020	115.632	4.828	−0.021
10/5/2007	94.915	5.916	0.105	114.224	4.945	0.117
10/8/2007	94.915	5.916	0.000	114.224	4.945	0.000
10/9/2007	94.915	5.916	0.000	114.224	4.945	0.000
10/10/2007	96.247	5.818	−0.098	114.224	4.944	−0.001
10/11/2007	95.952	5.839	0.021	114.039	4.960	0.016
10/12/2007	95.630	5.863	0.024	113.761	4.983	0.023
10/15/2007	95.643	5.862	−0.001	113.815	4.978	−0.005
10/16/2007	95.523	5.871	0.009	113.769	4.982	0.004
10/17/2007	97.233	5.746	−0.125	115.112	4.869	−0.113
10/18/2007	97.677	5.714	−0.032	115.608	4.827	−0.042
10/19/2007	98.915	5.625	−0.089	116.738	4.734	−0.093
10/22/2007	99.170	5.607	−0.018	116.916	4.719	−0.015
10/23/2007	98.903	5.626	0.019	116.698	4.737	0.018
10/24/2007	99.369	5.593	−0.033	117.488	4.672	−0.065
10/25/2007	99.155	5.608	0.015	117.329	4.685	0.013
10/26/2007	98.780	5.635	0.027	117.026	4.710	0.025
10/29/2007	99.034	5.617	−0.018	117.180	4.697	−0.013
10/30/2007	99.185	5.606	−0.011	117.105	4.703	0.006
10/31/2007	98.068	5.686	0.080	116.050	4.789	0.086
11/1/2007	99.589	5.578	−0.108	117.353	4.682	−0.107
11/2/2007	100.343	5.526	−0.052	118.040	4.626	−0.056
11/5/2007	99.978	5.551	0.025	117.731	4.651	0.025
11/6/2007	99.454	5.587	0.036	117.265	4.688	0.037
11/7/2007	98.263	5.672	0.085	117.235	4.691	0.003
11/8/2007	98.356	5.665	−0.007	117.585	4.662	−0.029

TABLE 6.4 (Continued)

	P&G 5.55% 3/5/2037			Treasury 6.25% 8/15/2023 (CTD issue)		
Trading Day	Price	Yield	Yield Change	Price	Yield	Yield Change
11/9/2007	98.264	5.672	0.007	118.216	4.611	−0.051
11/12/2007	98.264	5.672	0.000	118.216	4.611	0.000
11/13/2007	99.111	5.612	−0.060	117.976	4.629	0.018
11/14/2007	98.893	5.627	0.015	117.977	4.629	0.000
11/15/2007	99.922	5.555	−0.072	119.051	4.543	−0.086
11/16/2007	100.081	5.544	−0.011	119.201	4.530	−0.013
11/19/2007	100.754	5.497	−0.047	119.890	4.475	−0.055
11/20/2007	100.504	5.514	0.017	119.916	4.473	−0.002
11/21/2007	100.727	5.499	−0.015	120.180	4.452	−0.021
11/23/2007	101.189	5.467	−0.032	120.565	4.421	−0.031
11/26/2007	103.537	5.310	−0.157	122.554	4.265	−0.156
11/27/2007	102.331	5.390	0.080	121.322	4.361	0.096
11/28/2007	101.554	5.443	0.053	120.608	4.417	0.056
11/29/2007	101.863	5.422	−0.021	121.492	4.347	−0.070
11/30/2007	101.140	5.471	0.049	120.804	4.401	0.054
12/3/2007	101.802	5.426	−0.045	121.638	4.335	−0.066
12/4/2007	101.881	5.420	−0.006	121.670	4.332	−0.003
12/5/2007	99.350	5.595	0.175	121.262	4.364	0.032
12/6/2007	98.063	5.686	0.091	120.151	4.451	0.087
12/7/2007	96.671	5.787	0.101	118.645	4.571	0.120
12/10/2007	96.317	5.813	0.026	118.244	4.603	0.032
12/11/2007	98.093	5.684	−0.129	120.015	4.461	−0.142
12/12/2007	98.142	5.680	−0.004	119.156	4.529	0.068
12/13/2007	96.989	5.764	0.084	117.981	4.624	0.095
12/14/2007	96.284	5.815	0.051	117.245	4.684	0.060
12/17/2007	96.824	5.776	−0.039	117.764	4.641	−0.043
12/18/2007	98.203	5.676	−0.100	118.801	4.557	−0.084
12/19/2007	98.796	5.634	−0.042	119.325	4.515	−0.042
12/20/2007	99.466	5.587	−0.047	119.987	4.462	−0.053
12/21/2007	97.757	5.708	0.121	118.364	4.592	0.130

CHAPTER 7

Introduction to Time Series Analysis

In this chapter, we introduce the element of time as an index of a series of univariate observations. Thus, we treat observations as being obtained successively rather than simultaneously. We present a simple time series model and its components. In particular, we focus on the trend, the cyclical, and seasonal terms as well as the error or disturbance of the model. Furthermore, we introduce the random walk and error correction models as candidates for modeling security price movements. Here the notion of innovation appears. Time series are significant in modeling price processes as well as the dynamics of economic quantities.

WHAT IS TIME SERIES?

So far, we have either considered two-component variables cross-sectionally coequal, which was the case in correlation analysis, or we have considered one variable to be, at least partially, the functional result of some other quantity. The intent of this section is to analyze variables that change in time, in other words, the objects of the analysis are *time series*. The observations are conceived as compositions of functions of time and other exogenous and endogenous variables as well as lagged values of the series itself or other quantities. These latter quantities may be given exogenously or also depend on time.

To visualize this, we plot the graph of 20 daily closing values of the German stock market index, the DAX, in Figure 7.1. The values are listed in Table 7.1. The time points of observation t with equidistant increments are represented by the horizontal axis while the DAX index values are represented by the vertical axis.

TABLE 7.1 DAX Values of the Period May 3 to May 31, 2007

Date	t	Level
5/3/2007	1	7883.04
5/4/2007	2	7764.97
5/7/2007	3	7781.04
5/8/2007	4	7739.20
5/9/2007	5	7697.38
5/10/2007	6	7735.88
5/11/2007	7	7659.39
5/14/2007	8	7619.31
5/15/2007	9	7607.54
5/16/2007	10	7499.50
5/17/2007	11	7481.25
5/18/2007	12	7505.35
5/21/2007	13	7459.61
5/22/2007	14	7479.34
5/23/2007	15	7415.33
5/24/2007	16	7475.99
5/25/2007	17	7442.20
5/29/2007	18	7525.69
5/30/2007	19	7516.76
5/31/2007	20	7476.69

Source: Deutsche Börse, http://deutsche-boerse.com/.

DECOMPOSITION OF TIME SERIES

Each point in Figure 7.1 is a pair of the components, time and value. In this section, the focus is on the dynamics of the observations; that is, one wants to know what the values are decomposable into at each point in time. A time series with observations x_t, $t = 1, 2, \ldots, n$ is usually denoted by $\{x\}_t$.[1] In the context of time series analysis, for any value x_t, the series is thought of as a composition of several quantities. The most traditional decomposition is of the form

$$x_t = T_t + Z_t + S_t + U_t \qquad (7.1)$$

---

[1]The number of dates n may theoretically be infinite. We will restrict ourselves to finite lengths.

FIGURE 7.1 DAX Index Values: May 3 to May 31, 2007

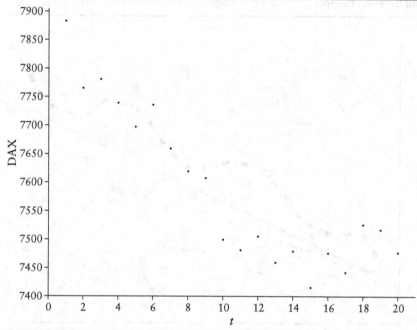

where

T_t = trend
Z_t = cyclical term
S_t = seasonal term
U_t = disturbance (or error)

While the trend and seasonal terms are assumed to be deterministic functions of time (i.e., their respective values at some future time t are known at any lagged time $t - d$, which is d units of time prior to t), the cyclical and disturbance terms are random. One also says that the last two terms are **stochastic**.[2] Instead of the cyclical term Z_t and the disturbance U_t, one sometimes incorporates the so-called *irregular term* of the form $I_t = \phi \cdot I_{t-1} + U_t$ with $0 < \phi \leq 1$. That is, instead of equation (7.1), we have now

$$x_t = T_t + S_t + I_t \tag{7.2}$$

---

[2]The case where all four components of the time series are modeled as stochastic quantities is not considered here.

FIGURE 7.2 Decomposition of Time Series into Trend T, Seasonal Component S, and Irregular Component I

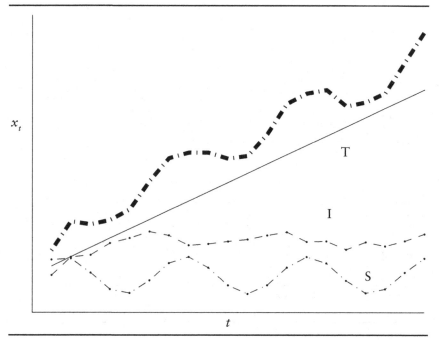

With the coefficient ϕ, we control how much of the previous time's irregular value is lingering in the present. If ϕ is close to zero, the prior value is less significant than if ϕ were close to one or even equal to one.

Note that U_t and I_{t-1} are independent. Since I_t depends on the prior value I_{t-1} scaled by ϕ and disturbed only by U_t, this evolution of I_t is referred to as *autoregressive of order one*.[3] As a consequence, there is some relation between the present and the previous level of I. Thus, these two are correlated to an extent depending on ϕ. This type of correlation between levels at time t and different times from the same variable is referred to as *autocorrelation*.[4]

In Figure 7.2, we present the decomposition of some hypothetical time series. The straight solid line T is the linear trend. The irregular component I is represented by the dashed line, and the seasonal component S is the dash-dotted line at the bottom of the figure. The resulting thick dash-dotted line is the time series $\{x\}_t$ obtained by adding all components.

---

[3]Order 1 indicates that the value of the immediately prior period is incorporated into the present period's value.
[4]The concept of correlation was introduced in Chapter 5.

Introduction to Time Series Analysis

Application to S&P 500 Index Returns

As an example, we use the daily S&P 500 stock index returns from the same period January 2, 1996 to December 31, 2003. To obtain an initial impression of the data, we plot them in the scatter plot in Figure 7.3. At first glance, it is kind of difficult to detect any structure within the data. However, we will decompose the returns according to equation (7.2). A possible question might be, is there a difference in the price changes depending on the day of the week? For the seasonality, we consider a period of length five since there are five trading days within a week. The seasonal components, $S_t(weekday)$, for each weekday (i.e., Monday through Friday) are given below:

Monday	−0.4555
Tuesday	0.3814
Wednesday	0.3356
Thursday	−0.4723
Friday	0.1759

FIGURE 7.3 Daily Returns of S&P 500 Stock Index Between January 2, 1996 and December 31, 2003

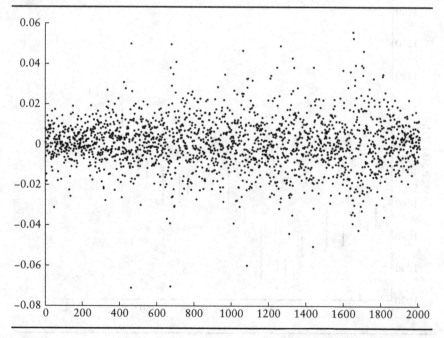

The coefficient of the irregular term is ϕ = 0.2850 indicating that the previous period's value is weighted by about one third in the computation of this period's value. The overall model of the returns, then, looks like

$$y_t = T_t + S_t + I_t = T_t + S_t(weekday) - 0.2850 I_{t-1} + U_t$$

The technique used to estimate the times series model is the *moving average* method. Since it is beyond the scope of this chapter, we will not discuss it here.

As can be seen by Figure 7.4, it might appear difficult to detect a linear trend, at least, when one does not exclude the first 15 observations. If there really is no trend, most of the price is contained in the other components rather than any deterministic term. Efficient market theory that is central in financial theory does not permit any price trend since this would indicate that today's price does not contain all information available. By knowing that the price grows deterministically, this would have to be already embodied into today's price.

FIGURE 7.4 Daily S&P 500 Stock Index Prices with Daily Changes Extending Vertically

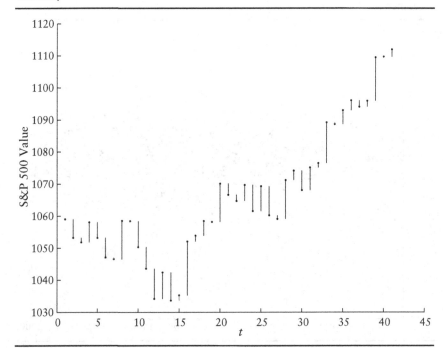

REPRESENTATION OF TIME SERIES WITH DIFFERENCE EQUATIONS

Rather than representing $\{x\}_t$ by (7.1) or (7.2), often the dynamics of the components of the series are given. So far, the components are considered as quantities at certain points in time. However, it may sometimes be more convenient to represent the evolution of $\{x\}_t$ by *difference equations* of its components. The four components in difference equation form could be thought of as

$$\Delta x_t = x_t - x_{t-1} = \Delta T_t + \Delta I_t + \Delta S_t \qquad (7.3)$$

with the change in the linear trend $\Delta T_t = c$ where c is a constant, and

$$\Delta I_t = \phi\left(I_{t-1} - I_{t-2}\right) + \xi_t$$

where ξ are disturbances themselves, and

$$\Delta T_t + \Delta S_t = h(t)$$

where $h(t)$ is some deterministic function of time. The symbol Δ indicates change in value from one period to the next.

The concept that the disturbance terms are *i.i.d.* means that the ξ behave in a manner common to all ξ (i.e., *identically distributed*) though *independently* of each other. The concept of statistical independence was introduced in Chapter 5 while for random variables, this will be done in Chapter 14.

In general, difference equations are some functions of lagged values, time, and other stochastic variables. In time series analysis, one most often encounters the task of estimating difference equations such as the type above, for example. The original intent of time series analysis was to provide some reliable tools for forecasting.[5]

By forecasting, we assume that the change in value of some quantity, say x, from time t to time $t + 1$ occurs according to the difference equation (7.3). However, since we do not know the value of the disturbance in $t + 1$, ξ_{t+1}, at time t, we incorporate its expected value, that is, zero. All other quantities in equation (7.3) are deterministic and, thus, known in t. Hence, the forecast really is the expected value in $t + 1$ given the information in t.

APPLICATION: THE PRICE PROCESS

Time series analysis has grown more and more important for verifying financial models. Price processes assume a significant role among these models.

---
[5]See, for example, Enders (1995).

Below we discuss two commonly encountered models for price processes given in a general setting: random walk and error correction.[6] The theory behind them is not trivial. In particular, the error correction model applies expected values computed conditional on events (or information), which is a concept to be introduced in Chapter 15. One should not be discouraged if these models appear somewhat complicated at this early stage of one's understanding of statistics.

Random Walk

Let us consider some price process given by the series $\{S\}_t$.[7] The dynamics of the process are given by

$$S_t = S_{t-1} + \varepsilon_t \qquad (7.4)$$

or, equivalently, $\Delta S_t = \varepsilon_t$.

In words, tomorrow's price, S_{t+1}, is thought of as today's price plus some *random shock* that is independent of the price. As a consequence, in this model, known as the *random walk*, the increments $S_t - S_{t-1}$ from $t-1$ to t are thought of as completely undeterministic. Since the ε_t have a mean of zero, the increments are considered fair.[8] An increase in price is as likely as a downside movement. At time t, the price is considered to contain all information available. So at any point in time, next period's price is exposed to a random shock.

Consequently, the best estimate for the following period's price is this period's price. Such price processes are called *efficient* due to their immediate information processing.

A more general model, for example, AR(p), of the form

$$S_t = \alpha_0 + \alpha_1 S_{t-1} + \ldots + \alpha_p S_{t-p} + \varepsilon_t$$

with several lagged prices could be considered as well. This price process would permit some slower incorporation of lagged prices into current prices. Now for the price to be a random walk process, the estimation would have to produce $a_0 = 0$, $a_1 = 1$, $a_2 = \ldots = a_p = 0$.

---

[6]In Parts Two and Three of this book we will introduce an additional price process using logarithmic returns.

[7]Here the price of some security at time t, S_t, ought not be confused with the seasonal component in equation (7.1).

[8]Note that since the ε assumes values on the entire real number line, the stock price could potentially become negative. To avoid this problem, logarithmic returns are modeled according to equation (7.4) rather than stock prices.

Application to S&P 500 Index Returns

As an example to illustrate equation (7.4), consider the daily S&P 500 stock index prices between November 3, 2003 and December 31, 2003. The values are given in Table 7.2 along with the daily price changes. The resulting plot is given in Figure 7.4. The intuition given by the plot is roughly that, on each day, the information influencing the following day's price is unpredictable and, hence, the price change seems completely arbitrary. Hence, at first glance much in this figure seems to support the concept of a random walk. Concerning the evolution of the underling price process, it looks reasonable to assume that the next day's price is determined by the previous day's price plus some random change. From Figure 7.4, it looks as if the changes occur independently of each other and in a manner common to all changes (i.e., with identical distribution).

Error Correction

We next present a price model that builds on the relationship between spot and forward markets. Suppose we extend the random walk model slightly by introducing some forward price for the same underlying stock S. That is, at time t, we agree by contract to purchase the stock at $t + 1$ for some price determined at t. We denote this price by $F(t)$. At time $t + 1$, we purchase the stock for $F(t)$. The stock, however, is worth S_{t+1} at that time and need not—and most likely will not—be equal to $F(t)$. It is different from the agreed forward price by some random quantity ε_{t+1}. If this disturbance has zero mean, as defined in the random walk model, then the price is fair. Based on this assumption, the reasonable forward price would equal[9]

$$F(t) = E[S_{t+1} | t] = E[S_t + \varepsilon_t | t] = S_t$$

So, on average, the difference between S and F should fulfill the following condition:

$$\Delta \equiv S_{t+1} - F(t) \approx 0$$

If, however, the price process permits some constant terms such as some upward trend, for example, the following period's price will no longer be equal to this period's price plus some random shock. The trend will spoil

---

[9] Note that we employ expected values conditional on time t to express that we base our forecast on all information available at time t. The expected value—or mean—as a parameter of a random variable will be introduced in Chapter 13. The conditional expected value will be introduced in Chapter 15.

TABLE 7.2 Daily S&P 500 Stock Index Values and Daily Changes between 11/3/2003 and 12/31/2003

Date	P_t	Δ_t
12/31/2003	1111.92	2.28
12/30/2003	1109.64	0.16
12/29/2003	1109.48	13.59
12/26/2003	1095.89	1.85
12/24/2003	1094.04	−1.98
12/23/2003	1096.02	3.08
12/22/2003	1092.94	4.28
12/19/2003	1088.66	−0.52
12/18/2003	1089.18	12.70
12/17/2003	1076.48	1.35
12/16/2003	1075.13	7.09
12/15/2003	1068.04	−6.10
12/12/2003	1074.14	2.93
12/11/2003	1071.21	12.16
12/10/2003	1059.05	−1.13
12/09/2003	1060.18	−9.12
12/08/2003	1069.30	7.80
12/05/2003	1061.50	−8.22
12/04/2003	1069.72	4.99
12/03/2003	1064.73	−1.89
12/02/2003	1066.62	−3.50
12/01/2003	1070.12	11.92
11/28/2003	1058.20	−0.25
11/26/2003	1058.45	4.56
11/25/2003	1053.89	1.81
11/24/2003	1052.08	16.80
11/21/2003	1035.28	1.63
11/20/2003	1033.65	−8.79
11/19/2003	1042.44	8.29
11/18/2003	1034.15	−9.48
11/17/2003	1043.63	−6.72
11/14/2003	1050.35	−8.06
11/13/2003	1058.41	−0.12
11/12/2003	1058.53	11.96
11/11/2003	1046.57	−0.54
11/10/2003	1047.11	−6.10
11/07/2003	1053.21	−4.84
11/06/2003	1058.05	6.24
11/05/2003	1051.81	−1.44
11/04/2003	1053.25	−5.77
11/03/2003	1059.02	

Introduction to Time Series Analysis

the fair price, and the forward price designed as the expected value of the following period's stock price conditional on this period's information will contain a systematic error. The model to be tested is, then,

$$S_{t+1} = \alpha_0 + \alpha_1 F(t) + \varepsilon_t$$

with a potential nonzero linear trend captured by α_0. A fair price would be if the estimates are $a_0 = 0$ and $a_1 = 1$. Then, the markets would be in approximate equilibrium. If not, the forward prices have to be adjusted accordingly to prohibit predictable gains from the differences in prices.

The methodology to do so is the so-called *error correction model* in the sense that today's (i.e., this period's) deviations from the equilibrium price have to be incorporated into tomorrow's (i.e., the following period's) price to return to some long-term equilibrium. The model is given by the equations system

$$S_{t+2} = S_{t+1} - \alpha\left(S_{t+1} - F(t)\right) + \varepsilon_{t+2}, \quad \alpha > 0$$
$$F(t+1) = F(t) + \beta\left(S_{t+1} - F(t)\right) + \xi_{t+1}, \quad \beta > 0$$

with

$$E[\varepsilon_{t+2} \mid t+1] = 0$$
$$E[\xi_{t+1} \mid t] = 0$$

At time $t + 2$, the term $\alpha\left(S_{t+1} - F(t)\right)$ in the price of S_{t+2} corrects for deviations from the equilibrium $\left(S_{t+1} - F(t)\right)$ stemming from time $t + 1$. Also, we adjust our forward price $F(t + 1)$ by the same deviation scaled by β. Note that, now, the forward price, too, is affected by some *innovation*, ξ_{t+1}, unknown at time t. In contrast to some disturbance or error term ε, which simply represents some deviation from an exact functional relationship, the concept of innovation such as in connection with the ξ_{t+1} is that of an independent quantity with a meaning such as, for example, new information or shock.

In general, the random walk and error correction models can be estimated using least squares regression introduced in Chapter 6. However, this is only legitimate if the regressors (i.e., independent variables) and disturbances are uncorrelated.

CONCEPTS EXPLAINED IN THIS CHAPTER (IN ORDER OF PRESENTATION)

Time series
Stochastic

Irregular term
Autoregressive of order one
Autocorrelation
Difference equations
Random shock
Random walk
Efficient
Error correction model
Innovation

PART TWO

Basic Probability Theory

CHAPTER 8
Concepts of Probability Theory

In this chapter, we introduce the general concepts of probability theory. Probability theory serves as the quantification of risk in finance. To estimate probabilistic models, we have to gather and process empirical data. In this context, we need the tools provided by statistics. We will see that many concepts from Part One can be extended to the realm of probability theory.

We begin by introducing a few preliminaries such as formal set operations, right-continuity, and nondecreasing functions. We then explain probability, randomness, and random variables, providing both their formal definitions and the notation used in this field.

HISTORICAL DEVELOPMENT OF ALTERNATIVE APPROACHES TO PROBABILITY

Before we introduce the formal definitions, we provide a brief outline of the historical development of probability theory and the alternative approaches since probability is, by no means, unique in its interpretation. We will describe the two most common approaches: relative frequencies and axiomatic system.

Probability as Relative Frequencies

The relative frequencies approach to probability was conceived by Richard von Mises in 1928 and as the name suggests formulates probability as the relative frequencies $f(x_i)$ introduced in Chapter 2. This initial idea was extended by Hans Reichenbach. Given large samples, it was understood that $f(x_i)$ was equal to the true probability of value x_i. For example, if $f(x_i)$ is small, then the true probability of value x_i occurring should be small, in general. However, $f(x_i)$ itself is subject to uncertainty. Thus, the relative frequencies might deviate from the corresponding probabilities. For example, if the sample is not large enough, whatever large may be, then, it is likely that we obtain a rare set of observations and draw the wrong conclusion with respect to the underlying probabilities.

This point can be illustrated with a simple example. Consider throwing a six-sided dice 12 times.[1] Intuitively, one would expect the numbers 1 through 6 to occur twice, each since this would correspond to the theoretical probabilities of 1/6 for each number. But since so many different outcomes of this experiment are very likely possible, one might observe relative frequencies of these numbers different from 1/6. So, based on the relative frequencies, one might draw the wrong conclusion with respect to the true underlying probabilities of the according values. However, if we increase the repetitions from 12 to 1,000, for example, with a high degree of certainty, the relative frequency of each number will be pretty close to 1/6.

The reasoning of von Mises and Reichenbach was that since extreme observations are unlikely given a reasonable sample size, the relative frequencies will portray the true probabilities with a high degree of accuracy. In other words, probability statements based on relative frequencies were justifiable since, in practice, highly unlikely events could be ruled out.

In the context of our dice example, they would consider as unlikely that certain numbers appeared significantly more often than others if the series of repetitions is, say, 1,000. But still, who could guarantee that we do not accidentally end up throwing 300 1s, 300 2s, 400 3s, and nothing else?

We see that von Mises' approach becomes relevant, again, in the context of estimating and hypothesis testing. For now, however, we will not pay any further attention to it but turn to the alternative approach to probability theory.

Axiomatic System

Introduced by Andrei N. Kolmogorov in 1933, the axiomatic system abstracted probability from relative frequencies as obtained from observations and instead treated probability as purely mathematical. The variables were no longer understood as the quantities that could be observed but rather as some theoretical entities "behind the scenes." Strict rules were set up that controlled the behavior of the variables with respect to their likelihood of assuming values from a predetermined set. So, for example, consider the price of a stock, say General Electric (GE). GE's stock price as a variable is not what you can observe but a theoretical quantity obeying a particular system of probabilities. What you observe is merely realizations of the stock price with no implication on the true probability

---

[1]There is always the confusion about whether a single dice should be referred to as a "die." In the United Kingdom, die is used; in the United States, dice is the accepted term.

Concepts of Probability Theory

of the values since the latter is given and does not change from sample to sample. The relative frequencies, however, are subject to change depending on the sample.

We illustrate the need for an axiomatic system due to the dependence of relative frequencies on samples using our dice example. Consider the chance of occurrence of the number 1. Based on intuition, since there are six different "numbers of dots" on a dice, the number 1 should have a chance of 1/6, right? Suppose we obtain the information based on two samples of 12 repetitions each, that is, $n_1 = n_2 = 12$. In the following table, we report the absolute frequencies, a_i, representing how many times the individual numbers of dots 1 through 6 were observed.

Number of Dots	Absolute Frequencies a_i	
	Sample 1	Sample 2
1	4	1
2	1	1
3	3	1
4	0	1
5	1	1
6	3	7
Total	12	12

That is, in sample 1, 1 dot was observed 4 times while, in sample 2, 1 dot was observed only once, and so on.

From the above observations, we obtain the following relative frequencies

Number of Dots	Relative Frequencies $f(x_i)$	
	Sample 1	Sample 2
1	0.3333	0.0833
2	0.0833	0.0833
3	0.2500	0.0833
4	0.0000	0.0833
5	0.0833	0.0833
6	0.2500	0.5833
Total	1.0000	1.0000

That is, in sample 1, 1 dot was observed 33.33% of the time while in sample 2, 1 dot was observed 8.33% of the time, and so on. We see that both samples lead to completely different results about the relative frequencies for the number of dots. But, as we will see, the theoretical probability is 1/6 = 0.1667, for each value 1 through 6. So, returning to our original question of the chance of occurrence of 1 dot, the answer is still 1/6 = 0.1667.

In finance, the problem arising with this concept of probability is that, despite the knowledge of the axiomatic system, we do not know for sure what the theoretical probability is for each value. We can only obtain a certain degree of certainty as to what it approximately might be. This insight must be gained from estimation based on samples and, thus, from the related relative frequencies. So, it might appear reasonable to use as many observations as possible. However, even if we try to counteract the sample-dependence of relative frequencies by using a large number of observations, there might be a change in the underlying probabilities exerting additional influence on the sample outcome. For example, during the period of a bull market, the probabilities associated with an upward movement of some stock price might be higher than under a bear market scenario.

Despite this shortcoming, the concept of probability as an abstract quantity as formulated by Kolmogorov has become the standard in probability theory and, hence, we will resort to it in Part Two of the book.

SET OPERATIONS AND PRELIMINARIES

Before proceeding to the formal definition of probability, randomness, and random variables we need to introduce some terminology.

Set Operations

A set is a combination of **elements**. Usually, we denote a set by some capital (upper-case) letter, e.g. S, while the elements are denoted by lower-case letters such as a, b, c, ... or a_1, a_2, ... To indicate that a set S consists of exactly the elements a, b, c, we write $S = \{a,b,c\}$. If we want to say that element a belongs to S, the notation used is that $a \in S$ where $\in$ means "belongs to." If, instead, a does not belong to S, then the notation used is $a \notin S$ where $\notin$ means does not belong to.

A type of set such as $S = \{a,b,c\}$ is said to be **countable** since we can actually count the individual elements a, b, and c. A set might also consist of all real numbers inside of and including some bounds, say a and b. Then, the set is equal to the interval from a to b, which would be expressed in mathematical notation as $S = [a,b]$. If either one bound or both do not belong to the set, then this would be written as either $S = (a,b]$, $S = [a,b)$, or $S = (a,b)$, respectively, where the parentheses denote that the value is excluded. An interval is an ***uncountable*** set since, in contrast to a countable set $S = \{a,b,c\}$, we cannot count the elements of an interval.[2]

We now present the operators used in the context of sets. The first is *equality* denoted by $=$ and intuitively stating that two sets are equal, that is, $S_1 = S_2$, if they consist of the same elements. If a set S consists of no elements, it is referred to as an *empty set* and is denoted by $S = \emptyset$. If the elements of S_1 are all contained in S_2, the notation used is $S_1 \subset S_2$ or $S_1 \subseteq S_2$. In the first case, S_2 also contains additional elements not in S_1, while, in the second case, the sets might also be equal. For example, let $S_1 = \{a,b\}$ and $S_2 = \{a,b,c\}$, then $S_1 \subset S_2$. The operator $\subseteq$ would indicate that S_2 consists of, at least, a and b. Or, let $M_1 = [0,1]$ and $M_2 = [0.5,1]$, then $M_2 \subset M_1$.

If we want to join a couple of sets, we use the *union operator* denoted by $\cup$. For example, let $S_1 = \{a,b\}$ and $S_2 = \{b,c,d\}$, then the union would be $S_1 \cup S_2 = \{a,b,c,d\}$. Or, let $M_1 = [0,1]$ and $M_2 = [0.5,1]$, then $M_2 \cup M_1 = [0,1] = M_1$.[3] If we join n sets $S_1, S_2, ..., S_n$ with $n \geq 2$, we denote the union by $\cup_{i=1}^{n} S_i$.

The opposite operator to the union is the difference denoted by the "\" symbol. If we take the difference between set S_1 and set S_2, that is, $S_1 \setminus S_2$, we discard from S_1 all the elements that are common to both, S_1 and set S_2. For example, let $S_1 = \{a,b\}$ and $S_2 = \{b,c,d\}$, then $S_1 \setminus S_2 = \{a\}$.

To indicate that we want to single out elements that are contained in several sets simultaneously, then we use the *intersection operator* $\cap$. For example, with the previous sets, the intersection would be $S_1 \cap S_2 = \{b\}$. Or, let $M_1 = [0,1]$ and $M_2 = [0.5,1]$, then $M_1 \cap M_2 = [0.5,1] = M_2$.[4] Instead of the $\cap$ symbol, one sometimes writes $S_1 S_2$ to indicate intersection.

If two sets contain no common elements (i.e., the intersection is the empty set), then the sets are said to be *pairwise disjoint*. For example, the sets $S_1 = \{a,b\}$ and $S_2 = \{c,d\}$ are pairwise disjoint since $S_1 \cap S_2 = \emptyset$. Or, let $M_1 = [0,0.5)$ and $M_2 = [0.5, 1]$, then $M_1 \cap M_2 = \emptyset$. If we intersect n sets $S_1, S_2, ..., S_n$ with $n \geq 2$, we denote the intersection by $\cap_{i=1}^{n} S_i$.

The *complement* to some set S is denoted by $\overline{S}$. It is defined as $S \cap \overline{S} = \emptyset$ and $S \cup \overline{S} = \Omega$. That is, the complement $\overline{S}$ is the remainder of Ω that is not contained in S.

---

[2]Suppose we have the interval $[1,2]$, that is all real numbers between 1 and 2. We cannot count all numbers inside of this interval since, for any two numbers such as, for example, 1 and 1.001, 1.0001, or even 1.000001, there is always infinitely many more numbers that lie between them.

[3]Note that in a set we do not consider an element more than once.

[4]Note that we can even join an infinite number of sets since the union is countable, that is, $\cup_{i=1}^{\infty} S_i$. The same is true for intersections with $\cap_{i=1}^{\infty} S_i$.

Right-Continuous and Nondecreasing Functions

Next we introduce two concepts of functions that should be understood in order to appreciate probability theory: right-continuous function and nondecreasing function.

A function f is right-continuous at $\tilde{x}$ if the limit from the right of the function values coincides with the actual value of f at $\tilde{x}$. Formally, that is $\lim_{x > \tilde{x}} f(x) = f(\tilde{x})$. We illustrate this in Figure 8.1. At the abscissae $x1$ and $x2$, the function f jumps to $f(x1)$ and $f(x2)$ respectively.[5] After each jump, the function remains at the new level, for some time. Hence, approaching $x1$ from the right, that is, for higher x-values, the function f approaches $f(x1)$ smoothly. This is not the case when approaching $x1$ from the left since f jumps at $x1$ and, hence, deviates from the left-hand limit. The same reasoning applies to f at abscissa $x2$. A function is said to be a *right-continuous function* if it is right-continuous at every value on the x-axis.

FIGURE 8.1 Demonstration of Right-Continuity of Some Hypothetical Function f at Values $x1$ and $x2$

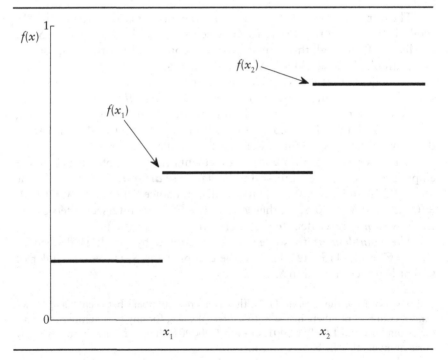

---

[5]By *abscissa* we mean a value on the horizontal x-axis.

FIGURE 8.2 Hypothetical Nondecreasing Function f

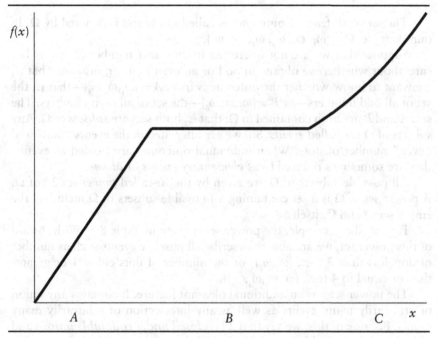

A function f is said to be a ***nondecreasing function*** if it never assumes a value smaller than any value to the left. We demonstrate this using Figure 8.2. We see that while, in the different sections A, B, and C, f might grow at different rates, it never decreases. Even for x-values in section B, f has zero and thus a nonnegative slope.

An example of a right-continuous and nondecreasing function is the empirical relative cumulative distribution function, $F_{emp}^f(x)$, introduced in Chapter 2.

Outcome, Space, and Events

Before we dive into the theory, we will use examples that help illustrate the concept behind the definitions that follow later in this section.

Let us first consider, again the number of dots of a dice. If we throw it once, we observe a certain value, that is, a realization of the abstract number of dots, say 4. This is a one particular outcome of the random experiment. We will denote the outcomes by ω and a particular outcome i will be denoted by ω_i. We might just as well have realized 2, for example, which would represent another outcome. All feasible outcomes, in this experiment, are given by

$$\omega_1 = 1 \quad \omega_2 = 2 \quad \omega_3 = 3 \quad \omega_4 = 4 \quad \omega_5 = 5 \quad \omega_6 = 6$$

The set of all feasible outcomes is called *space* and is denoted by Ω. In our example, $\Omega = \{\omega_1, \omega_2, \omega_3, \omega_4, \omega_5, \omega_6\}$.

Suppose that we are not interested in the exact number of points but care about whether we obtain an odd or an even number, instead. That is, we want to know whether the outcome is from $A = \{\omega_1, \omega_3, \omega_5\}$—that is, the set of all odd numbers—or $B = \{\omega_2, \omega_4, \omega_6\}$—the set of all even numbers. The sets A and B are both contained in Ω; that is, both sets are *subsets* of Ω. Any subsets of Ω are called *events*. So, we are interested in the events "odd" and "even" number of dots. When individual outcomes are treated as events, they are sometimes referred to as *elementary events* or *atoms*.

All possible subsets of Ω are given by the so-called *power set* 2^Ω of Ω. A power set of Ω is a set containing all possible subsets of Ω including the empty set $\emptyset$ and Ω, itself.[6]

For our dice example, the power set is given in Table 8.1. With the aid of this power set, we are able to describe all possible events such as number of dots less than 3 (i.e., $\{\omega_1, \omega_2\}$) or the number of dots either 1 or greater than or equal to 4 (i.e., $\{\omega_1, \omega_4, \omega_5, \omega_6\}$).

The power set has an additional pleasant feature. It contains any union of arbitrarily many events as well as any intersection of arbitrarily many events. Because of this, we say that 2^Ω is *closed under countable unions* and *closed under countable intersections*. Unions are employed to express that

TABLE 8.1 The Power Set of the Example Number of Dots of a Dice

$$
\begin{aligned}
2^\Omega = \{&\emptyset, \{\omega_1\}, \{\omega_2\}, \{\omega_3\}, \{\omega_4\}, \{\omega_5\}, \{\omega_6\}, \{\omega_2, \omega_3\}, \{\omega_2, \omega_4\}, \{\omega_2, \omega_5\}, \{\omega_2, \omega_6\}, \ldots \\
&\{\omega_3, \omega_4\}, \{\omega_3, \omega_5\}, \{\omega_3, \omega_6\}, \{\omega_4, \omega_5\}, \{\omega_4, \omega_6\}, \{\omega_5, \omega_6\}, \{\omega_1, \omega_2, \omega_3\}, \\
&\{\omega_1, \omega_2, \omega_4\}, \{\omega_1, \omega_2, \omega_5\}, \{\omega_1, \omega_2, \omega_6\}, \{\omega_1, \omega_3, \omega_4\}, \{\omega_1, \omega_3, \omega_5\}, \{\omega_1, \omega_3, \omega_6\}, \\
&\{\omega_1, \omega_4, \omega_5\}, \{\omega_1, \omega_4, \omega_6\}, \{\omega_1, \omega_5, \omega_6\}, \{\omega_2, \omega_3, \omega_4\}, \{\omega_2, \omega_3, \omega_5\}, \{\omega_2, \omega_3, \omega_6\}, \ldots \\
&\{\omega_2, \omega_4, \omega_5\}, \{\omega_2, \omega_4, \omega_6\}, \{\omega_2, \omega_5, \omega_6\}, \{\omega_3, \omega_4, \omega_5\}, \{\omega_3, \omega_4, \omega_6\}, \{\omega_3, \omega_5, \omega_6\}, \\
&\{\omega_4, \omega_5, \omega_6\}, \{\omega_1, \omega_2, \omega_3, \omega_4\}, \{\omega_1, \omega_2, \omega_3, \omega_5\}, \{\omega_1, \omega_2, \omega_3, \omega_6\}, \{\omega_1, \omega_2, \omega_4, \omega_5\}, \\
&\{\omega_1, \omega_2, \omega_4, \omega_6\}, \{\omega_1, \omega_2, \omega_5, \omega_6\}, \{\omega_1, \omega_3, \omega_4, \omega_5\}, \{\omega_1, \omega_3, \omega_4, \omega_6\}, \\
&\{\omega_1, \omega_3, \omega_5, \omega_6\}, \{\omega_1, \omega_4, \omega_5, \omega_6\}, \{\omega_2, \omega_3, \omega_4, \omega_5\}, \{\omega_2, \omega_3, \omega_4, \omega_6\}, \\
&\{\omega_2, \omega_3, \omega_5, \omega_6\}, \{\omega_2, \omega_4, \omega_5, \omega_6\}, \{\omega_3, \omega_4, \omega_5, \omega_6\}, \{\omega_1, \omega_2, \omega_3, \omega_4, \omega_5\}, \\
&\{\omega_1, \omega_2, \omega_3, \omega_4, \omega_6\}, \{\omega_1, \omega_2, \omega_3, \omega_5, \omega_6\}, \{\omega_1, \omega_2, \omega_4, \omega_5, \omega_6\}, \\
&\{\omega_1, \omega_3, \omega_4, \omega_5, \omega_6\}, \{\omega_1, \omega_3, \omega_4, \omega_5, \omega_6\}\} \\
= &\Omega
\end{aligned}
$$

Note: The notation $\{\omega_i\}$ for $i = 1, 2, \ldots, 6$ indicates that the outcomes are treated as events.

---

[6] For example, let $\Omega = \{1, 2, 3\}$, then the power set $2^\Omega = \{\emptyset, \{1\}, \{2\}, \{3\}, \{1, 2\}, \{1, 3\}, \{2, 3\}, \Omega\}$. That is, we have included all possible combinations of the original elements of Ω.

at least one of the events has to occur. We use intersections when we want to express that the events have to occur simultaneously. The power set also contains the complements to all events.

As we will later see, all these properties of the power set are features of a σ-*algebra*, often denoted by $\mathbb{A}$.

Now consider an example were the space Ω is no longer countable. Suppose that we are analyzing the daily logarithmic returns for a common stock or common stock index. Theoretically, any real number is a feasible outcome for a particular day's return.[7] So, events are characterized by singular values as well as closed or open intervals on the real line.[8] For example, we might be interested in the event E that the S&P 500 stock index return is "at least 1%." Using the notation introduced earlier, this would be expressed as the half-open interval $E = [0.01, \infty)$.[9] This event consists of the uncountable union of all outcomes between 0.01 and ∞. Now, as the sets containing all feasible events, we might take, again, the power set of the real numbers, that is, 2^Ω with $\Omega = (-\infty, \infty) = \mathbb{R}$.[10] But, for theoretical reasons beyond the scope of this text, that might cause trouble.

Instead, we take a different approach. To design our set of events of the uncountable space Ω, we begin with the inclusion of the events "any real number," which is the space Ω, itself, and "no number at all," which is the empty set $\emptyset$. Next, we include all events of the form "less than or equal to a", for any real number a, that is, we consider all half-open intervals $(-\infty, a]$, for any $a \in \mathbb{R}$. Now, for each of these $(-\infty, a]$, we add its complement $(-\infty, a]$ = $\Omega \setminus (-\infty, a] = (a, \infty)$, which expresses the event "greater than a." So far, our set of events contains $\emptyset$, Ω, all sets $(-\infty, a]$, and all the sets (a, ∞). Furthermore, we include all possible unions and intersections of everything already in the set of events as well as of the resulting unions and intersections themselves.[11] By doing this, we guarantee that any event of practical relevance of an uncountable space is considered by our set of events.

With this procedure, we construct the Borel σ-algebra, $\mathbb{B}$. This is the collection of events we will use any time we deal with real numbers.

---

[7]Let us assume, for now, that we are not restricted to a few digits due to measurement constraints or quotes conventions in the stock market. Instead, we consider being able to measure the returns to any degree of precision.

[8]By real line, we refer to all real numbers between minus infinity ($-\infty$) and plus infinity (∞). We think of them as being arrayed on a line such as the horizontal axis of a graph.

[9]By convention, we never include ∞ since it is not a real number.

[10]The symbol R is just a mathematical abbreviation for the real numbers.

[11]For example, through the intersection of $(-\infty, 2]$ and $(1, \infty)$, we obtain $(1, 2]$ while the intersection of $(-\infty, 4]$ and $(3, \infty)$ yields $(3, 4]$. These resulting intervals, in turn, yield the union $(1,2] \cup (3,4]$, which we wish to have in our set of events also.

The events from the respective σ-*algebra* of the two examples can be assigned probabilities in a unique way as we will see.

The Measurable Space

Let us now express the ideas from the previous examples in a formal way. To describe a random experiment, we need to formulate

1. Outcomes ω
2. Space Ω
3. σ-algebra 𝔸

Definition 1—Space: The space Ω contains all outcomes. Depending on the outcomes ω, the space Ω is either countable or uncountable.

Definition 2—σ-algebra: The σ-algebra 𝔸 is the collection of events (subsets of Ω) with the following properties:

a. $\Omega \in \mathbb{A}$ and $\emptyset \in \mathbb{A}$.
b. If event $E \in \mathbb{A}$ then $\bar{E} \in \mathbb{A}$.
c. If the countable sequence of events $E_1, E_2, E_3, \ldots \in \mathbb{A}$ then $\bigcup_{i=1}^{\infty} E_i \in \mathbb{A}$ and $\bigcap_{i=1}^{\infty} E_i \in \mathbb{A}$.

Definition 3—Borel σ-algebra: The σ-algebra formed by $\emptyset$, $\Omega = \mathbb{R}$, intervals $(\infty, a]$ for some real a, and countable unions and intersections of these intervals is called a ***Borel σ-algebra*** and denoted by $\mathbb{B}$.

Note that we can have several σ-algebrae for some space Ω. Depending on the events we are interested in, we can think of a σ-algebra 𝔸 that contains fewer elements than 2^{Ω} (i.e., countable Ω), or the Borel σ-algebra (i.e., uncountable Ω). For example, we might think of $\mathbb{A} = \{\emptyset, \Omega\}$, that is, we only want to know whether any outcome occurs or nothing at all.[12] It is easy to verify that this simple 𝔸 fulfills all requirements a, b, and c of Definition 2.

Definition 4—Measurable space: The tuple (Ω,𝔸) with 𝔸 being a σ-algebra of Ω is a ***measurable space***.[13]

Given a measureable space, we have enough to describe a random experiment. All that is left is to assign probabilities to the individual events. We will do so next.

---

[12] The empty set is interpreted as the *improbable event*.
[13] A *tuple* is the combination of several components. For example, when we combine two values a and b, the resulting tuple is (a,b), which we know to be a pair. If we combine three values a, b, and c, the resulting tuple (a,b,c) is known as a triplet.

PROBABILITY MEASURE

We start with a brief discussion of what we expect of a probability or *probability measure*, that is, the following properties:

Property 1: A probability measure should assign each event E from our σ-algebra a nonnegative value corresponding to the chance of this event occurring.

Property 2: The chance that the empty set occurs should be zero since, by definition, it is the improbable event of "no value."

Property 3: The event that "any value" might occur (i.e., Ω) should be 1 or, equivalently, 100% since some outcome has to be observable.

Property 4: If we have two or more events that have nothing to do with one another that are pairwise disjoint or ***mutually exclusive***, and create a new event by uniting them, the probability of the resulting union should equal the sum of the probabilities of the individual events.

To illustrate, let:

- The first event state that the S&P 500 log return is "maximally 5%," that is, $E_1 = (-\infty, 0.05]$.
- The second event state that the S&P 500 log return is "at least 10%," that is, $E_2 = [0.10, \infty)$.

Then, the probability of the S&P log return *either* being no greater than 5% *or* no less than 10% should be equal to the probability of E_1 plus the probability of E_2.

Let's proceed a little more formally. Let $(\Omega, \mathbb{A})$ be a measurable space. Moreover, consider the following definition.

Definition 5—Probability measure: A function P on the σ-algebra $\mathbb{A}$ of Ω is called a *probability measure* if it satisfies:

a. $P(\emptyset) = 0$ and $P(\Omega) = 1$.
b. For a countable sequence of events $E_1, E_2, \ldots$ in $\mathbb{A}$ that are pairwise disjoint (i.e., $E_i \cap E_j = \emptyset \ldots, i \neq j$), we have[14]

$$P\left(\bigcup_{i=1}^{\infty} E_i\right) = \sum_{i=1}^{\infty} P(E_i)$$

---

[14]This property is referred to as *countable additivity*.

Then we have everything we need to model randomness and chance, that is, we have the space Ω, the σ-algebra $\mathbb{A}$ of Ω, and the probability measure P. This triplet $(\Omega,\mathbb{A},P)$ forms the so called *probability space*.

At this point, we introduce the notion of **P-almost surely (P-a.s.)** occurring events. It is imaginable that even though $P(\Omega) = 1$, not all of the outcomes in Ω contribute positive probability. The entire positive probability may be contained in a subset of Ω while the remaining outcomes form the *unlikely* event with respect to the probability measure P. The event accounting for the entire positive probability with respect to P is called the *certain event with respect to P*.[15] If we denote this event by E_{as}, then we have $P(E_{as}) = 1$ yielding $P(\Omega \backslash E_{as}) = 0$.

There are certain peculiarities of P depending on whether Ω is countable or not. It is essential to analyze these two alternatives since this distinction has important implications for the determination of the probability of certain events. Here is why.

Suppose, first, that Ω is countable. Then, we are able to assign the event $\{\omega_i\}$ associated with an individual outcome, ω_i, a nonnegative probability $p_i = P(\{\omega_i\})$, for all $\omega_i \in \Omega$. Moreover, the probability of any event E in the σ-algebra $\mathbb{A}$ can be computed by adding the probabilities of all outcomes associated with E. That is,

$$P(E) = \sum_{\omega_i \in E} p_i$$

In particular, we have

$$P(\Omega) = \sum_{\omega_i \in \Omega} p_i = 1$$

Let us resume the six-sided dice tossing experiment. The probability of each number of dots 1 through 6 is 1/6 or formally,

$$P(\{\omega_1\}) = P(\{\omega_2\}) = ... = P(\{\omega_6\}) = 1/6$$

or equivalently,

$$p_1 = p_2 = ... = p_6 = 1/6$$

Suppose, instead, we have $\Omega = \mathbb{R}$. That is, Ω is uncountable and our σ-algebra is given by the Borel σ-algebra, $\mathbb{B}$. To give the probability of the events E in $\mathbb{B}$, we need an additional device, given in the next definition.

---

[15] We have to specify with respect to which probability measure this event is almost surely. If it is obvious, however, as is the case in this book, we omit this specification.

Definition 6—Distribution function: A function F is a ***distribution function of the probability measure*** P if it satisfies the following properties:

a. F is right-continuous.
b. F is nondecreasing.
c. $\lim_{x \to -\infty} F(x) = 0$ and $\lim_{x \to \infty} F(x) = 1$.
d. For any $x \in \mathbb{R}$, we have $F(x) = P((-\infty,x])$.

It follows that, for any interval $(x,y]$, we compute the associated probability according to

$$F(y) - F(x) = P((x,y]) \qquad (8.1)$$

So, in this case we have a function F uniquely related to P from which we derive the probability of any event in $\mathbb{B}$. Note that in general F is only right-continuous, that is the limit of $F(y)$, when $y > x$ and $y > x$, is exactly $F(x)$. At point x we might have a jump of the distribution $F(x)$. The size of this jump equals $P(\{x\})$. This distribution function can be interpreted in a similar way to the relative empirical cumulative distribution function $F_{emp}^{f}(x)$ in Chapter 2. That is, we state the probability of our quantity of interest being less than or equal to x.

To illustrate, the probability of the S&P 500 log return being, at most 1%, $E = (-\infty, 0.01]$, is given by $F^{S\&P\ 500}(0.01) = P((-\infty, 0.01])$,[16] while the probability of it being between –1 and 1% is

$$F^{S\&P\ 500}(0.01) - F^{S\&P\ 500}(-0.01) = P((-0.01, 0.01])$$

RANDOM VARIABLE

Now time has come to introduce the concept of a ***random variable***. When we refer to some quantity as being a random variable, we want to express that its value is subject to uncertainty, or randomness. Technically, the variable of interest is said to be ***stochastic***. In contrast to a deterministic quantity whose value can be determined with certainty, the value of a random variable is not known until we can observe a realized outcome of the random experiment. However, since we know the probability space $(\Omega, \mathbb{A}, P)$, we are aware of the possible values it can assume.

One way we can think of a random variable denoted by X is as follows. Suppose we have a random experiment where some outcome ω from the

---

[16] We use the index in $F^{S\&P\ 500}$ to emphasize that this distribution function is unique to the probability of events related to the S&P 500 log returns.

space Ω occurs. Then, depending on this ω, the random variable X assumes some value $X(\omega) = x$, where ω can be understood as input to X. What we observe, finally, is the value x, which is only a consequence of the outcome ω of the underlying random experiment.

For example, we can think of the price of a 30-year Treasury bond as a random variable assuming values at random. However, expressed in a somewhat simple fashion, the 30-year Treasury bond depends completely on the prevailing market interest rate (or yield) and, hence, is a function of it. So, the underlying random experiment concerns the prevailing market interest rate with some outcome ω while the price of the Treasury bond, in turn, is merely a function of ω.

Consequently, a random variable is a function that is completely deterministic and depends on the outcome ω of some random experiment. In most applications, random variables have values that are real numbers.

So, we understand random variables as functions from some space into an image or *state* space. We need to become a little more formal at this point. To proceed, we will introduce a certain type of function, the ***measurable function***, in the following

Definition 7—Measureable function: Let $(\Omega, \mathbb{A})$ and $(\Omega', \mathbb{A}')$ be two measurable spaces. That is Ω, Ω' are spaces and $\mathbb{A}$, $\mathbb{A}'$ their σ-algebrae, respectively. A function $X: \Omega \to \Omega'$ is $\mathbb{A}$-$\mathbb{A}'$-*measurable* if, for any set $E' \in \mathbb{A}'$, we have[17]

$$X^{-1}(E') \in \mathbb{A}$$

In words, this means that a function from one space to another is measurable if the origin with respect to this function of each image in the σ-algebra of the state space can be traced in the σ-algebra of the domain space.

We illustrate this in Figure 7.3. Function X creates images in Ω' by mapping outcomes ω from Ω with values $X(\omega) = x$ in Ω'. In reverse fashion, for each event E' in the state space with σ-algebra $\mathbb{A}'$, X^{-1} finds the corresponding origin of E' in σ-algebra $\mathbb{A}$ of the probability space.

Now, we define a random variable X as a measurable function. That means for each event in the state space σ-algebra, $\mathbb{A}'$, we have a corresponding event in the σ-algebra of the domain space, $\mathbb{A}$.

To illustrate this, let us consider the example with the dice. Now we will treat the "number of points" as a random variable X. The possible outcome

---

[17]Instead of *A-A'-measurable,* we will, henceforth, use simply *measureable* since, in our statements, it is clear which σ-algebrae are being referred to.

FIGURE 7.3 Relationship between Image E' and $X^{-1}(E')$ through the Measurable Function X

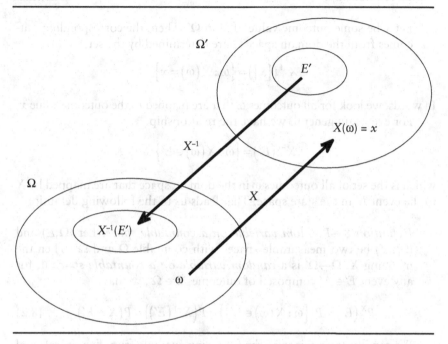

values of X are given by the state space Ω', namely, $\Omega' = \{1,2,3,4,5,6\}$.[18] The origin or domain space is given by the set of outcomes $\Omega = \{\omega_1, \omega_2, \omega_3, \omega_4, \omega_5, \omega_6\}$. Now, we can think of our random variable X as the function $X: \Omega \to \Omega'$ with the particular map $X(\omega_i) = i$ with $i = 1, 2, \ldots, 6$.

Random Variables on a Countable Space

We will distinguish between random variables on a countable space and on an uncountable space. We begin with the countable case.

The random variable X is a function mapping the countable space Ω into the state space Ω'. The state space Ω' contains all outcomes or values that X can obtain.[19] Thus, all outcomes in Ω' are countable images of the

---

[18] Note that we do not define the outcomes of number of dots as nominal or even rank data anymore, but as numbers. That is 1 is 1, 2 is 2, and so on.

[19] Theoretically, Ω' does not have to be countable, that is it could contain more elements that X can assume values. But we restrict ourselves to countable state spaces Ω' consisting of exactly all the values of X.

outcomes ω in Ω. Between the elements of the two spaces, we have the following relationship.

Let x be some outcome value of X in Ω'. Then, the corresponding outcomes from the domain space Ω are determined by the set

$$X^{-1}(\{x\}) = \{\omega : X(\omega) = x\}$$

In words, we look for all outcomes ω that are mapped to the outcome value x. For events, in general, we have the relationship

$$X^{-1}(E') = \{\omega : X(\omega) \in E'\}$$

which is the set of all outcomes ω in the domain space that are mapped by X to the event E' in the state space. That leads us to the following definition:

Definition 8—Random variable on a countable space: Let $(\Omega, \mathbb{A})$ and $(\Omega', \mathbb{A}')$ be two measurable spaces with countable Ω and Ω'. Then the mapping $X: \Omega \to \Omega'$ is a *random variable on a countable space* if, for any event $E' \in \mathbb{A}'$ composed of outcomes $x \in \Omega'$, we have

$$P^X(E') = P(\{\omega : X(\omega) \in E'\}) = P(X^{-1}(E')) = P(X \in E') \quad (8.2)$$

We can illustrate this with the following example from finance referred to as the "binomial stock price model." The random variable of interest will be the price of some stock. We will denote the price of the stock by S. Suppose at the beginning of period t, the price of the stock is \$20 (i.e., $S_t = \$20$). At the beginning of the following period, $t + 1$, the stock price is either $S_{t+1} = \$18$ or $S_{t+1} = \$22$. We model this in the following way.
Let:

- $(\Omega, \mathbb{A})$ and $(\Omega', \mathbb{A}')$ be two measurable spaces with $\Omega' = \{\$18, \$22\}$, (i.e., the state space of the period $t + 1$ stocks price) and $\mathbb{A}$ (i.e., the corresponding σ-algebra of all events with respect to the stock price in $t + 1$).
- Ω be the space consisting of the outcomes of some random experiment completely influencing the $t + 1$ stock price.
- $\mathbb{A}$ be the corresponding σ-algebra of Ω with all events in the origin space.

Now, we can determine the origin of the event that

$$S_{t+1} = \$18 \text{ by } E_{\text{down}} = \{\omega : S(\omega) = \$18\}$$

and

$$S_{t+1} = \$22 \text{ by } E_{up} = \{\omega : S(\omega) = \$22\}$$

Thus, we have partitioned Ω into the two events, E_{down} and E_{up}, related to the two period $t + 1$ stock prices. With the probability measure P on Ω, we have the probability space $(\Omega, \mathbb{A}, P)$. Consequently, due to equation (8.2), we are able to compute the probability $P^S(\$18) = P(E_{down})$ and $P^S(\$22) = P(E_{up})$, respectively.

We will delve into this, including several examples in Chapter 9, where we cover discrete random variables.

Random Variables on an Uncountable Space

Now let's look at the case when the probability space $(\Omega, \mathbb{A}, P)$ is no longer countable. Recall the particular way in which events are assigned probabilities in this case.

While for a countable space any outcome ω can have positive probability, that is, $p_\omega > 0$, this is not the case for individual outcomes of an uncountable space. On an uncountable space, we can have the case that only events associated with intervals have positive probability. These probabilities are determined by distribution function $F(x) = P(X \leq x) = P(X < x)$ according to equation (8.1).

This brings us to the following definition:

Definition 9—Random variable on a general possibly uncountable space: Let $(\Omega, \mathbb{A})$ and $(\Omega', \mathbb{A}')$ be two measurable spaces with, at least, Ω uncountable. The map $X: \Omega \to \Omega'$ is a *random variable on the uncountable space* $(\Omega, \mathbb{A}, P)$ if it is measurable. That is, if, for any $E' \in \mathbb{A}'$, we have

$$X^{-1}(E') \in \mathbb{A}$$

probability from $(\Omega, \mathbb{A}, P)$ on $(\Omega', \mathbb{A}')$ by

$$P^X(E') = P(\{\omega : X(\omega) \in E'\}) = P(X^{-1}(E')) = P(X \in E')$$

We call this the *probability law* or *distribution* of X. Typically, the probability of $X \in E'$ is written using the following notation:

$$P^X(E') = P(X \in E')$$

Very often, we have the random variable X assume values that are real numbers (i.e., $\Omega' = \mathbb{R}$ and $\mathbb{B}' = \mathbb{B}$). Then, the events in the state space are characterized by countable unions and intersections of the intervals $(-\infty, a]$ corresponding to the events $\{X \leq a\}$, for real numbers a. In this case, we require that to be a random variable, X satisfies

$$\{\omega : X(\omega) \leq a\} = X^{-1}((\infty, a]) \in \mathbb{B}$$

for any real a.

To illustrate, let's use a call option on a stock. Suppose in period t we purchase a call option on a certain stock expiring in the next period $T = t + 1$. The strike price, denoted by K, is \$50. Then as the buyer of the call option, in $t + 1$ we are entitled to purchase the stock for \$50 no matter what the market price of the stock (S_{t+1}) might be. The value of the call option at time $t + 1$, which we denote by C_{t+1}, depends on the market price of the stock at $t + 1$ relative to the strike price (K). Specifically,

- If S_{t+1} is less than K, then the value of the option is zero, that is, $C_{t+1} = 0$
- If S_{t+1} is greater than K, then the value of the option is equal to $S_{t+1} - K$

Let $(\Omega, \mathbb{A}, P)$ be the probability space with the stock price in $t + 1$; that is, $S_{t+1} = s$ representing the uncountable real-valued outcomes. So, we have the uncountable probability space $(\Omega, \mathbb{A}, P) = (\mathbb{R}, \mathbb{B}, P)$. Assume that the price at $t + 1$ can take any nonnegative value. Assume further that the probability of exactly s is zero (i.e., $P(S_{t+1} = s) = 0$), that is, the distribution function of the price at $T = 1$ is continuous. Let the value of the call option in $T = t + 1$, C_{t+1}, be our random variable mapping from Ω to Ω'. Since the possible values of the call option at $t + 1$ are real numbers, the state space is uncountable as well. Hence, we have $(\Omega', \mathbb{A}') = (\mathbb{R}, \mathbb{B})$. C_{t+1}, to be a random variable, is a $\mathbb{B}$-$\mathbb{B}'$-measurable function.

Now, the probability of the call becoming worthless is determined by the event in the origin space that the stock price falls below K. Formally, that equals

$$P^{C_{t+1}}(0) = P(C_{t+1} \leq 0) = P(S_{t+1} \leq K) = P((-\infty, K])$$

since the corresponding event in $\mathbb{A}$ to a 0 value for the call option is $(-\infty, K]$. Equivalently, $C_{t+1}^{-1}(\{0\}) = (-\infty, K]$. Any positive value c of C_{t+1} is associated with zero probability since we have

$$P^{C_{t+1}}(c) = P(C_{t+1} = c) = P(S_{t+1} = c + K) = 0$$

due to the relationship $C_{t+1} = S_{t+1} - K$ for $S_{t+1} > K$.

CONCEPTS EXPLAINED IN THIS CHAPTER (IN ORDER OF PRESENTATION)

Elements
Countable
Uncountable
Empty set
Union operator
Intersection operator
Pairwise disjoint
Complement
Right continuous function
Nondecreasing function
Space
Subsets
Events
Elementary events
Atoms
Power set
Closed under countable unions
Closed under countable intersections
σ-algebra
Borel σ-algebra
Measurable space
Probability measure
Mutually exclusive
Probability space
P-almost surely (P-a.s.)
Unlikely
Certain event with respect to P
Distribution function
Random variable
Stochastic
Measurable function
Random variable on a countable space
Random variable on an uncountable space
Probability law

CHAPTER 9
Discrete Probability Distributions

In this chapter, we learn about random variables on a countable space and their distributions. As measures of location and spread, we introduce their mean and variance. The random variables on the countable space will be referred to as *discrete random variables*. We present the most important discrete random variables used in finance and their *probability distribution* (also called *probability law*):

- Bernoulli
- Binomial
- Hypergeometric
- Multinomial
- Poisson
- Discrete uniform

The operators used for these distributions will be derived and explained in Appendix C of this book. Operators are concise expressions that represent particular, sometimes lengthy, mathematical operations. The appendix to this chapter provides a summary of the discrete distributions covered.

DISCRETE LAW

In order to understand the distributions discussed in this chapter, we will explain the general concept of a *discrete law*. Based on the knowledge of countable probability spaces, we introduce the random variable on the countable space as the discrete random variable. To fully comprehend the discrete random variable, it is necessary to become familiar with the process of assigning probabilities to events in the countable case. Furthermore, the cumulative distribution function will be presented as an important representative of probability. It is essential to understand the mean and variance parameters. Wherever appropriate, we draw analogies to descriptive statistics for a facilitation of the learning process.

Random Variable on the Countable Space

Recall that the probability space $(\Omega, \mathbb{A}, P)$ where Ω is a countable space. The probability of any event E is given by

$$P(E) = \sum_{\omega_i \in E} p_i$$

with the p_i being the probabilities of the individual outcomes ω_i in the event E. Remember that the random variable X is the mapping from Ω into Ω' such that the state space Ω' is countable.[1] Thus, we found out that the probability of any event E' in the state space has probability

$$P(X \in E') = P^X(E') = \sum_{\omega_i : X(\omega_i) \in E'} p_i$$

since E' is associated with the set

$$\{\omega_i : X(\omega_i) \in E'\}$$

through X. The probability of each individual outcome of X yields the discrete probability law of X. It is given by $P(X = x_i) = p_i^X$, for all $x_i \in \Omega'$.

Only for individual discrete values x is the probability p^X positive. This is similar to the empirical frequency distribution with positive relative frequency f_i at certain observed values. If we sort the $x_i \in \Omega$ in ascending order, analogous to the empirical relative cumulative frequency distribution

$$F_{emp}^f(x) = \sum_{x_i \leq x} f_i$$

we obtain the *discrete cumulative distribution* (*cdf*) of X,

$$F^X(x) = P(X \leq x) = \sum_{x_i \leq x} p_i^X$$

That is, we express the probability that X assumes a value no greater than x.

Suppose we want to know the probability of obtaining at most 3 dots when throwing a dice. That is, we are interested in the *cdf* of the random variable number of dots, at the value $x = 3$. We obtain it by

$$F^X(3) = p_1 + p_2 + p_3 = 1/3 + 1/3 + 1/3 = 0.5$$

---

[1] We denote random variables by capital letters, such as X, whereas the outcomes are denoted by small letters, such as x_i.

FIGURE 9.1 Cumulative Distribution Function of Number of Dots Appearing from Tossing a Dice

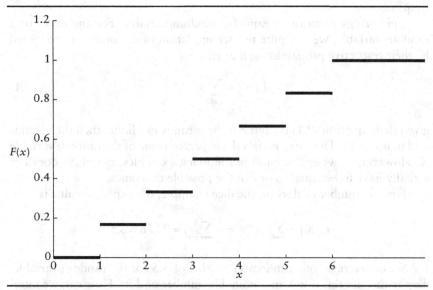

where the p_i denote the respective probabilities of the number of dots less than or equal to 3. A graph of the *cdf* is shown in Figure 9.1.

Mean and Variance

In Chapter 3, we introduced the sample mean and variance. While these are sample dependent statistics, the notion of a parameter for the entire population had been conveyed as well. Here we present the mean and variance of the distribution as parameters where the probability space can be understood as the analog to the population.

To illustrate, we use the random variable number of dots obtained by tossing a dice. Since we treat the numbers as numeric values, we are able to perform transformations and computations with them. By throwing a dice several times, we would be able to compute a sample average based on the respective outcome. So, a question could be: What number is theoretically expected? In our discussion below, we see how to answer that question.

Mean

The mean is the population equivalent to the sample average of a quantitative variable. In order to compute the sample average, we sum up all observations and divide the resulting value by the number of observations, which we will denote by n. Alternatively, we sum over all values weighted by their relative frequencies.

This brings us to the mean of a random variable. For the mean of a random variable, we compute the accumulation of the outcomes weighted by their respective probabilities; that is,

$$E(X) = \sum_{x_i \in \Omega} x_i \cdot p_i^X \tag{9.1}$$

given that equation (9.1) is finite.[2] If the mean is not finite, then the mean is said to not exist. The mean equals the *expected value* of the random variable X. However, as we will see in the following examples, the mean does not actually have to be equal to one of the possible outcomes.

For the number of dots on the dice example, the expected value is

$$E(X) = \sum_{i=1}^{6} i \cdot p_i = \frac{1}{6} \sum_{i=1}^{6} i = 21/6 = 3.5$$

So, on average, one can expect a value of 3.5 for the random variable, despite the fact this is not an obtainable number of dots. How can we interpret this? If we were to repeat the dice tossing many times, record for each toss the number of dots observed, then, if we averaged over all numbers obtained, we would end up with an average very close if not identical to 3.5.

Let's move from the dice tossing example to look at the binomial stock price model that we introduced in the previous chapter. With the stock price S at the end of period 1 being either $S_1 = \$18$ or $S_1 = \$22$, we have only these two outcomes with positive probability each. We denote the probability measure of the stock price at the end of period 1 by $P^S(\cdot)$. At the beginning of the period, we assume the stock price to be $S_0 = \$20$. Furthermore, suppose that up- and down-movements are equally likely; that is, $P^S(18) = \frac{1}{2}$ and $P^S(22) = \frac{1}{2}$. So we obtain

$$E(S) = \frac{1}{2} \cdot \$18 + \frac{1}{2} \cdot \$22 = \$20$$

This means on average, the stock price will remain unchanged even though $20 is itself not an obtainable outcome.

We can think of it this way. Suppose, we observed some stock over a very long period of time and the probabilities for up- and down-movements did not change. Furthermore suppose that each time the stock price was $20 at the beginning of some period, we recorded the respective end-of-period

---

[2]Often, the mean is denoted as the parameter μ.

Discrete Probability Distributions

price. Then, we would finally end up with an average of these end-of-period stock prices very close to if not equal to $20.

Variance

Just like in the realm of descriptive statistics, we are interested in the dispersion or spread of the data. For this, we introduce the *variance* as a measure. While in Part One we analyzed the sample variance, in this chapter our interest is focused on the variance as a parameter of the random variable's distribution.

Recall, that a sample measure of spread gives us information on the average deviation of observations from their sample mean. With the help of the variance, we intend to determine the magnitude we have to theoretically expect of the squared deviation of the outcome from the mean. Again, we use squares to eliminate the effect from the signs of the deviations as well as to emphasize larger deviations compared to smaller ones, just as we have done with the sample variance.

For the computation of the expected value of the squared deviations, we weight the individual squared differences of the outcomes from the mean with the probability of the respective outcome. So, formally, we define the variance of some random variable X to be

$$\sigma_X^2 = Var(X) = \sum_{x_i \in \Omega} (x_i - E(X))^2 p_i^X \qquad (9.2)$$

For example, for the number of dots obtained from tossing a dice, we obtain the variance

$$\begin{aligned}\sigma_X^2 = Var(X) &= \sum_{i=1}^{6} (i - E(X))^2 p_i^X \\ &= \frac{1}{6}\left[(1-3.5)^2 + (2-3.5)^2 + \ldots + (6-3.5)^2\right] \\ &= 2.9167\end{aligned}$$

Thus, on average, we have to expect a squared deviation from the mean by roughly 2.9.

Standard Deviation

We know from Part One there is a problem in interpreting the variance. A squared quantity is difficult to relate to the original random variable. For this reason, just as we have done with descriptive statistics, we use the *standard deviation*, which is simply the square root of the variance. Formally, the standard deviation is given by

$$\sigma_X = \sqrt{Var(X)}$$

The standard deviation appeals to intuition because it is a quantity that is of the same scale as the random variable X. In addition, it helps in assessing where the probability law assigns its probability mass. A rule of thumb is that at least 75% of the probability mass is assigned to a vicinity of the mean that extends two standard deviations in each direction. Furthermore, this rule states that in at least 89% of the times a value will occur that lies in a vicinity of the mean of three standard deviations in each direction.

For the number of dots obtained from tossing a dice, since the variance is 2.9167, the standard deviation is

$$\sigma_X = \sqrt{2.9167} = 1.7078$$

In Figure 9.2, we display all possible outcomes 1 through 6 indicated by the ○ symbol, including the mean of $E(X) = 3.5$. We extend a vicinity about the mean of length $\sigma_X = 1.7078$, indicated by the "+" symbol, to graphically relate the magnitude of the standard deviation to the possible values of X.

BERNOULLI DISTRIBUTION

In the remainder of this chapter, we introduce the most common discrete distributions used in finance. We begin with the simplest one, the ***Bernoulli distribution***.

Suppose, we have a random variable X with two possible outcomes. That is, we have the state space $\Omega' = \{x_1, x_2\}$. The distribution of X is given by the probability for the two outcomes, that is,

$$p_1^X = p \text{ and } p_2^X = 1 - p$$

Now, to express the random experiment of drawing a value for X, all we need to know is the two possible values in the state space and parameter p representing the probability of x_1. This situation is represented concisely by the Bernoulli distribution. This distribution is denoted $B(p)$ where p is the probability parameter.

Formally, the Bernoulli distribution is associated with random variables that assume the values $x_1 = 1$ and $x_2 = 0$, or $\Omega' = \{0,1\}$. That is why this

FIGURE 9.2 Relation Between Standard Deviation (σ = 1.7078) and Scale of Possible Outcomes 1, 2, ..., 6 Indicated by the ○ Symbol

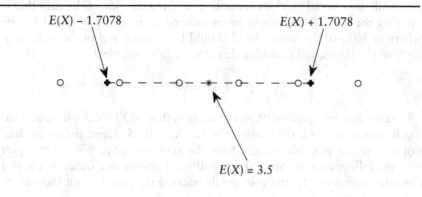

distribution is sometimes referred to as the "zero-one distribution." One usually sets the parameter p equal to the probability of x_1 such that

$$p = P(X = x_1) = P(X = 1)$$

The mean of a Bernoulli distributed random variable is

$$E(X) = 0 \cdot (1-p) + 1 \cdot p = p \qquad (9.3)$$

and the variance is

$$\begin{aligned} Var(X) &= (0-p)^2 \cdot (1-p) + (1-p)^2 \cdot p \\ &= p \cdot (1-p) \end{aligned} \qquad (9.4)$$

The Bernoulli random variable is commonly used when one models the random experiment where some quantity either satisfies a certain criterion or not. For example, it is employed when it is of interest whether an item is intact or broke. In such applications, we assign the outcome "success" the numerical value 1 and the outcome "failure" the numerical value 0, for example. Then, we model the random variable X describing the state of the item as Bernoulli distributed.

Consider the outcomes when flipping a coin: head or tail. Now we set head equal to the numerical value 0 and tail equal to 1. We take X as the Bernoulli distributed random variable describing the side of the coin that is up after the toss. What should be considered a *fair* coin? It would be one where in 50% of the tosses, head should be realized and in the remaining 50% of the tosses, tail should realized. So, a fair coin yields

$$p = 1 - p = 0.5$$

According to equation (9.3), the mean is then $E(X) = 0.5$ while, according to equation (9.4), the variance is $Var(X) = 0.25$. Here, the mean does not represent a possible value x from the state space Ω'. We can interpret it in the following way: Since 0.5 is halfway between one outcome (0) and the other outcome (1), the coin is fair because the mean is not inclined to either outcome.

As another example, we will take a look at credit risk modeling by considering the risk of default of a corporation. Default occurs when the corporation is no longer able to meet its debt obligations, A priori, default occurring during some period is uncertain and, hence, is treated as random. Here we view the corporation's failure within the next year as a Bernoulli random variable X. When the corporation defaults, $X = 0$ and in the case of survival $X = 1$. For example, a corporation may default within the next year with probability

$$P(X = 0) = 1 - p = 1 - e^{-0.04} = 0.0392$$

and survive with probability

$$P(X = 1) = p = e^{-0.04} = 0.9608$$

The reason we used the exponential function in this situation for the probability distribution will be understood once we introduce the exponential distribution in Chapter 11.

We can, of course, extend the prerequisites of the Bernoulli distribution to a more general case, That is, we may choose values for the two outcomes, x_1 and x_2 of the random variable X different from 0 and 1. Then, we set the parameter p equal to either one of the probabilities $P(X = x_1)$ or $P(X = x_2)$. The distribution yields mean

$$E(X) = x_1 \cdot p + x_2 \cdot (1 - p)$$

and variance

Discrete Probability Distributions

$$Var(X) = (x_1 - E(X))^2 \cdot p + (x_2 - E(X))^2 \cdot (1-p)$$

where we set $p = P(X = x_1)$.

We illustrate this generalization of the Bernoulli distribution in the case of the binomial stock price model. Again, we denote the random stock price at time period 1 by S_1. Recall that the state space $\Omega' = \{\$18, \$22\}$ containing the two possible values for S_1. The probability of S_1 assuming value $\$18$ can be set to

$$P(S_1 = \$18) = p$$

so that

$$P(S_1 = \$22) = 1 - p$$

Hence, we have an analogous situation to a Bernoulli random experiment, however, with $\Omega' = \{\$18, \$22\}$ instead of $\Omega' = \{0,1\}$.

Suppose, that

$$P(S_1 = \$18) = p = 0.4 \text{ and } P(S_1 = \$22) = 1 - p = 0.6$$

Then the mean is

$$E(S_1) = 0.4 \cdot \$18 + 0.6 \cdot \$22 = \$20.4$$

and the variance

$$Var(S_1) = (\$18 - \$20.4)^2 \cdot 0.4 + (\$22 - \$20.4)^2 \cdot 0.6 = (\$3.84)^2$$

BINOMIAL DISTRIBUTION

Suppose that we are no longer interested in whether merely one single item satisfies a particular requirement such as success or failure. Instead, we want to know the number of items satisfying this requirement in a sample of n items. That is, we form the sum over all items in the sample by adding 1 for each item that is success and 0 otherwise. For example, it could be the number of corporations that satisfy their debt obligation in the current year from a sample of 30 bond issues held in a portfolio. In this case, a corporation would be assigned 1 if it satisfied its debt obligation and 0 if it did not. We would then sum up over all 30 bond issues in the portfolio.

Now, one might realize that this is the linking of n single Bernoulli trials. In other words, we perform a random experiment with n "independent" and identically distributed Bernoulli random variables, which we denote by $B(p)$. Note that we introduced two important assumptions: independent random variables and identically distributed random variables. Independent random variables or independence is an important statistical concept that requires a formal definition. We will provide one but not until Chapter 14. However, for now we will simply relate independence to an intuitive interpretation such as uninfluenced by another factor or factors. So in the Bernoulli trials, we assume independence, which means that the outcome of a certain item does not influence the outcome of any others. By identical distribution we mean that the two random variables' distributions are the same. In our context, it implies that for each item, we have the same $B(p)$ distribution.

This experiment is as if one draws an item from a bin and replaces it into the bin before drawing the next item. Thus, this experiment is sometimes referred to as *drawing with replacement*. All we need to know is the number of trials, n, and the parameter p related to each single drawing. The resulting sum of the Bernoulli random variables is distributed as a *binomial distribution* with parameters n and p and denoted by $B(n,p)$.

Let X be distributed $B(n,p)$. Then, the random variable X assumes values in the state space $\Omega' = \{0,1,2,\ldots,n\}$. In words, the total X is equal to the number of items satisfying the particular requirement (i.e., having a value of 1). X has some integer value i of at least 0 and at most n.

To determine the probability of X being equal to i, we first need to answer the following question: How many different samples of size n are there to yield a total of i hits (i.e., realizations of the outcome i)? The notation to represent realizing i hits out of a sample of size n is

$$\binom{n}{i} \tag{9.5}$$

The expression in equation (9.5) is called the *binomial coefficient* and is explained in Appendix C of this book.

Since in each sample the n individual $B(p)$ distributed items are drawn independently, the probability of the sum over these n items is the product of the probabilities of the outcomes of the individual items.[3] We illustrate this in the next example.

---

[3] The definition of independence in probability theory will not be given until Chapter 14. However, the concept is similar to that introduced in Chapter 5 for relative frequencies.

Discrete Probability Distributions

Suppose, we flip a fair coin 10 times (i.e., $n = 10$) and denote by Y_i the result of the i-th trial. We denote by $Y_i = 1$ that the i-th trial produced head and by $Y_i = 0$ that it produced tail. Assume, we obtain the following result

Y_1	Y_2	Y_3	Y_4	Y_5	Y_6	Y_7	Y_8	Y_9	Y_{10}
1	1	0	0	0	1	0	1	1	0

So, we observe $X = 5$ times head. For this particular result that yields $X = 5$, the probability is

$$P(Y_1 = 1, Y_2 = 1, \ldots, Y_{10} = 0) = P(Y_1 = 1) \cdot P(Y_2 = 1) \cdot \ldots \cdot P(Y_{10} = 0)$$
$$= p \cdot p \cdot \ldots \cdot (1-p)$$
$$= p^5 \cdot (1-p)^5$$

Since we are dealing with a fair coin (i.e., $p = 0.5$), the above probability is

$$P(Y_1 = 1, Y_2 = 1, \ldots, Y_{10} = 0) = 0.5^5 \cdot 0.5^5 = 0.5^{10} \approx 0.0010$$

With

$$\binom{10}{5} = 252$$

different samples leading to $X = 5$, we compute the probability for this value of the total as

$$P(X = 5) = \binom{10}{5} p^5 \cdot (1-p)^5 = 252 \cdot 0.5^{10} = 0.2461$$

So, in roughly one fourth of all samples of $n = 10$ independent coin tosses, we obtain a total of $X = 5$ 1s or heads.

From the example, we see that the exponent for p is equal to the value of the total X (i.e., $i = 5$), and the exponent for $1 - p$ is equal to $n - i = 5$.

Let p be the parameter from the related Bernoulli distribution (i.e., $P(X = 1) = p$). The probability of the $B(n,p)$ random variable X being equal to some $i \in \Omega'$ is given by

$$P(X = i) = \binom{n}{i} \cdot p^i \cdot (1-p)^{n-i}, \quad i = 1, 2, \ldots, n \qquad (9.6)$$

For a particular selection of parameters, the probability distribution at certain values can be found in the four tables in the appendix to this chapter. The mean of a $B(n,p)$ random variable is

$$E(X) = n \cdot p \qquad (9.7)$$

and its variance is

$$Var(X) = n \cdot p \cdot (1-p) \qquad (9.8)$$

Below we will apply what we have just learned to be the binomial stock price model and two other applications.

Application to the Binomial Stock Price Model

Let's extend the binomial stock price model in the sense that we link T successive periods during which the stock price evolves.[4] In each period $(t, t+1]$, the price either increases or decreases by 10%. For now, it is not important how we obtained 10%. However, intituively the 10% can be thought of as the volatility of the stock price S. Thus, the corresponding factor by which the price will change from the previous period is 0.9 (down movement) and 1.1 (up movement). Based on this assumption about the price movement for the stock each period, at the end of the period $(t, t+1]$, the stock price is

$$S_{t+1} = S_t \cdot Y_{t+1}$$

where the random variable Y_{t+1} assumes a value from $\{0.9, 1.1\}$, with 0.9 representing a price decrease of 10% and 1.1 a price increase of 10%. Consequently, in the case of $Y_{t+1} = 1.1$, we have

$$S_{t+1} = S_t \cdot 1.1$$

while, in case of $Y_{t+1} = 0.9$, we have

$$S_{t+1} = S_t \cdot 0.9$$

For purposes of this illustration, let's assume the following probabilities for the down movement and up movement, respectively,

---

[4]The entire time span of length T is subdivided into the adjacent period segments $(0,1], (1,2], \ldots, (T-1, T]$.

$$P(Y_{t+1} = 1.1) = p = 0.6$$

and

$$P(Y_{t+1} = 0.9) = 1 - p = 0.4$$

After T periods, we have a random total of X up movements; that is, for all periods $(0,1], (1,2], \ldots,$ and $(T-1,T]$, we increment X by 1 if the period related factor $Y_{t+1} = 1.1$, $t = 0, 1, \ldots, T-1$. So, the result is some $x \in \{1,2,\ldots,T\}$. The total number of up movements, X, is a binomial distributed $B(T,p)$ random variable on the probability space $(\Omega',\mathbb{A}',P^X)$ where

1. The state space is $\Omega' = \{1,2,\ldots,T\}$.
2. σ-algebra $\mathbb{A}'$ is given by the power set $2^{\Omega'}$ of Ω'.
3. P^X is denoted by the binomial probability distribution given by

$$P(X = k) = \binom{T}{k} p^k (1-p)^{T-k}, \ k = 1,2,\ldots,T$$

with $p = 0.6$.

Consequently, according to equations (9.7) and (9.8), we have

$$E(X) = 2 \cdot 0.6 = 1.2$$

and

$$Var(X) = 2 \cdot 0.6 \cdot 0.4 = 0.48$$

By definition of S_T and X, we know that the evolution of the stock price is such that

$$S_T = S_0 \cdot 1.1^X \cdot 0.9^{T-X}$$

Let us next consider a random variable that is not binomial itself, but related to a binomial random variable. Now, instead of considering the $B(T,p)$ distributed total X, we could introduce as a random variable, the stock price at T (i.e., S_T). Using an illustration, we will derive the stock price independently of X and, then, emphasize the relationship between S_T and X. Note that S_T is not a binomial random variable.

Let us set $T = 2$. We may start with an initial stock price of $S_0 = \$20$. At the end of the first period, that is, $(0,1]$, we have

$$S_1 = S_0 \cdot Y_1$$

either equal to

$$S_1 = \$20 \cdot 1.1 = \$22$$

or

$$S_1 = \$20 \cdot 0.9 = \$18$$

At the end of the second period, that is, $(1,2]$, we have

$$S_2 = S_1 \cdot Y_2 = \$22 \cdot 1.1 = \$24.20$$

or

$$S_2 = S_1 \cdot Y_2 = \$22 \cdot 0.9 = \$19.80$$

In the case where $S_1 = \$22$, and

$$S_2 = S_1 \cdot Y_2 = \$18 \cdot 1.1 = \$19.80$$

or

$$S_2 = S_1 \cdot Y_2 = \$18 \cdot 0.9 = \$16.20$$

in the case where $S_1 = \$18$.

That is, at time $t + 1 = T = 2$, we have three possible values for S_2, namely, \$24.20, \$19.80, and \$16.20. Hence, we have a new state space that we will denote by $\Omega_S' = \{\$16.2, \$19.8, \$24.2\}$. Note that $S_2 = \$19.80$ can be achieved in two different ways: (1) $S_1 = S_0 \cdot 1.1 \cdot 0.9$ and (2) $S_1 = S_0 \cdot 0.9 \cdot 1.1$. The evolution of this pricing process, between time 0 and $T = 2$, can be demonstrated using the ***binomial tree*** given in Figure 9.3.

As σ-algebra, we use $\mathbb{A} = 2^{\Omega_S'}$, which is the power set of the state space Ω_S'. It includes events such as, for example, "stock price in $T = 2$ no greater than \$19.80," defined as $E' = \{S_2 \leq \$19.80\}$.

The probability distribution of S_2 is given by the following

$$P(S_2 = \$24.20) = P(Y_1 = 1.1) \cdot P(Y_2 = 1.1)$$
$$= \binom{2}{2} p^2 = 0.6^2 = 0.36$$

FIGURE 9.3 Binomial Stock Price Model with Two Periods

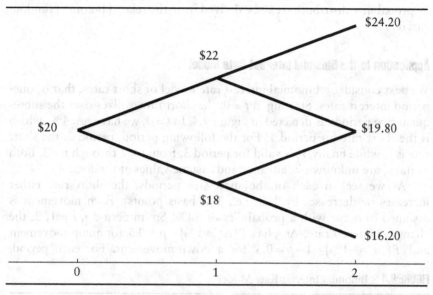

Note: Starting price $S_0 = \$20$. Upward factor $u = 1.1$, downward $d = 0.9$.

$$P(S_2 = \$19.80) = P(Y_1 = 0.9) \cdot P(Y_2 = 1.1) + P(Y_1 = 1.1) \cdot P(Y_2 = 0.9)$$

$$= 2(1-p)p = \binom{2}{1} \cdot 0.4 \cdot 0.6 = 0.48$$

$$P(S_2 = \$16.20) = P(Y_1 = 0.9) \cdot P(Y_2 = 0.9)$$

$$= \binom{2}{0}(1-p)^2 = 0.4^2 = 0.16$$

We now have the complete probability space of the random variable S_2. One can see the connection between S_2 and X by the congruency of the probabilities of the individual outcomes, that is,

$P(S_2 = \$24.20) = P(X = 2)$
$P(S_2 = \$19.80) = P(X = 1)$
$P(S_2 = \$16.20) = P(X = 0)$

From this, we derive, again, the relationship

$$S_2 = S_0 \cdot 1.1^X \cdot 0.9^{2-X}$$

Thus, even though S_2, or, generally S_T, is not a distributed binomial itself, its probability distribution can be derived from the related binomial random variable X.[5]

Application to the Binomial Interest Rate Model

We next consider a binomial interest rate model of short rates, that is, one-period interest rates. Starting in $t = 0$, the short rate evolves over the subsequent two periods as depicted in Figure 9.4. In $t = 0$, we have $r_0 = 4\%$, which is the short rate for period 1. For the following period, period 2, the short rate is r_1 while finally, r_2 is valid for period 3, from $t = 2$ through $t = 3$. Both r_1 and r_2 are unknown in advance and assume values at random.

As we see, in each of the successive periods, the short rate either increases or decreases by 1% (i.e., 100 basis points). Each movement is assumed to occur with a probability of 50%. So, in period i, $i = 1, 2$, the change in interest rate, Δr_i, has $P(\Delta r_i = 1\%) = p = 0.5$ for an up-movement and $P(\Delta r_i = -1\%) = 1 - p = 0.5$ for a down-movement. For each period,

FIGURE 9.4 Binomial Interest Rate Model

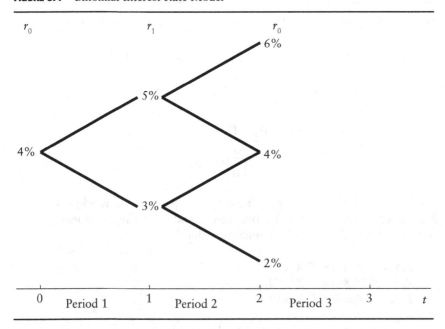

---
[5]Note that the successive prices S_1, ..., S_T depend on their respective predecessors. They are said to be *path-dependent*. Only the changes, or factors Y_{t+1}, for each period are independent.

we may model the interest rate change by some Bernoulli random variable where X_1 denotes the random change in period 1 and X_2 that of period 2. The $X_i = 1$ in case of an up-movement and $X_i = 0$ otherwise. The sum of both (i.e., $Y = X_1 + X_2$) is a binomially distributed random variable, precisely $Y \sim B(2, 0.5)$, thus, assuming values 0, 1, or 2.

To be able to interpret the outcome of Y in terms of interest rate changes, we perform the following transformations. A value of $X_i = 1$ yields $\Delta r_i = 1\%$ while $X_i = 0$ translates into $\Delta r_i = -1\%$. Hence, the relationship between Y and r_2 is such that when $Y = 0$, implying two down-movements in a row, $r_2 = r_0 - 2\% = 2\%$. When $Y = 1$, implying one up- and down-movement each, $r_2 = r_0 + 1\% - 1\% = 4\%$. And finally, $Y = 2$ corresponds to two up-movements such that $r_2 = r_0 + 2\% = 6\%$. So, we obtain the probability distribution:

r_2	$P(r_2)$
2%	$\binom{2}{0} 0.5^0 \cdot 0.5^2 = 0.25$
4%	$\binom{2}{1} 0.5^1 \cdot 0.5^1 = 0.5$
6%	$\binom{2}{2} 0.5^2 \cdot 0.5^0 = 0.25$

Application to the Binomial Default Distribution Model

Here we introduce a very simple version of the correlated binomial default distribution model. Suppose we have a portfolio of N assets that may default independently of each other within the next period. The portfolio is homogenous in the sense that all assets are identically likely to default. Hence, we have the common value p as each asset's probability to default. The assets are conceived as Bernoulli random variables X_i, $i = 1, 2, \ldots, N$. If asset i defaults, $X_i = 0$ and $X_i = 1$, otherwise. So, the total of defaulted assets is a binomial random variable Y with $Y = X_1 + X_2 + \ldots + X_N \sim B(N, p)$.

For now, we are unable to fully present the model since the notion of correlation and dependence between random variables will not be introduced until later in this book.

HYPERGEOMETRIC DISTRIBUTION

Recall that the prerequisites to obtain a binomial $B(n,p)$ random variable X is that we have n identically distributed random variables Y_i, all following the same Bernoulli law $B(p)$ of which the sum is the binomial random variable X. We referred to this type of random experiment as "drawing with replacement" so that for the sequence of individual drawings Y_i, we always have the same conditions.

Suppose instead that we do not "replace." Let's consider the distribution of "drawing without replacement." This is best illustrated with an urn containing N balls, K of which are black and $N - K$ are white. So, for the initial drawing, we have the chance of drawing a black ball equal to K/N, while we have the chance of drawing a white ball equal to $(N - K)/N$. Suppose, the first drawing yields a black ball. Since we do not replace it, the condition before the second drawing is such that we have $(K - 1)$ black balls and still $(N - K)$ white balls. Since the number of black balls has been reduced by one and the number of white balls is unchanged, the chance of drawing a black ball has been reduced compared to the chance of drawing a white ball; the total is also reduced by one. Hence, the condition is different from the first drawing. It would be similar if instead we had drawn a white ball in the first drawing, however, with the adverse effect on the chance to draw a white ball in the second drawing.

Now suppose in the second drawing another black ball is selected. The chances are increasingly adverse against drawing another black ball, in the third trial. This changing environment would be impossible in the binomial model of identical conditions in each trial.

Even if we had drawn first a black ball and then a white ball, the chances would not be the same as at the outset of the experiment before any balls were drawn because the total is now reduced to $N - 2$ balls. So, the chance of obtaining a black ball is now $(K - 1)/(N - 2)$, and that of obtaining a white ball is $(N - K - 1)/(N - 2)$. Mathematically, this is not the same as the original K/N and $(N - K)/(N)$. Hence, the conditions are altering from one drawing (or trial) to the next.

Suppose now that we are interested in the sum X of black balls drawn in a total of n trials. Let's look at this situation. We begin our reasoning with some illustration given specific values, that is,

$N = 10$
$K = 4$
$n = 5$
$k = 3$

FIGURE 9.5 Drawing $n = 5$ Balls without Replacement

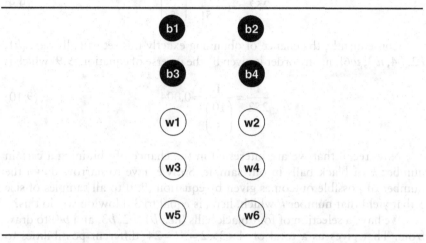

Note: $N = 10$, $K = 4$ (black), $n = 5$, and $k = 3$ (black).

The urn containing the black and white balls is depicted in Figure 9.5. Let's first compute the number of different outcomes we have to consider when we draw $n = 5$ out of $N = 10$ balls regardless of any color. We have 10 different options to draw the first ball; that is, $b1$ through $w6$ in Figure 9.5. After the first ball has been drawn, without replacement the second ball can be drawn from the urn consisting of the remaining nine balls. After that, the third ball is one out of the remaining eight, and so on until five balls have been successively removed. In total, we have

$$10 \times 9 \times 8 \times 7 \times 6 = 10!/5! = 30,240$$

alternative ways to withdraw the five balls.[6] For example, we may draw $b4$, $b2$, $b1$, $w3$, and $w6$. However, this is the same as $w6$, $w3$, $b4$, $b2$, and $b1$ or any other combination of these five balls. Since we do not care about the exact order of the balls drawn, we have to account for that in that we divide the total number of possibilities (i.e., 30,240) by the number of possible combinations of the very same balls drawn. The latter is equal to

$$5 \times 4 \times 3 \times 2 \times 1 = 5! = 120$$

Thus, we have 30,240/120 = 252 different nonredundant outcomes if we draw five out of 10 balls. Alternatively, this can be written as

---
[6] The factorial operator ! is introduced in Appendix A.

$$252 = \frac{10!}{5! \times 5!} = \binom{10}{5} \qquad (9.9)$$

Consequently, the chance of obtaining exactly this set of balls (i.e., $\{b1, b2, b4, w3, w6\}$) in any order is given by the inverse of equation (9.9) which is

$$\frac{1}{252} = \frac{1}{\binom{10}{5}} = 0.004 \qquad (9.10)$$

Now recall that we are interested in the chance of obtaining a certain number k of black balls in our sample. So, we have to narrow down the number of possible outcomes given by equation (9.9) to all samples of size 5 that yield that number k which, here, is equal to 3. How do we do this?

We have a selection of four black balls (i.e., $b1$, $b2$, $b3$, and $b4$) to draw from. That gives us a total of $4 \times 3 \times 2 = 4! = 24$ different possibilities to recover $k = 3$ black balls out of the urn consisting of four balls. Again, we do not care about the exact order in which we draw the black balls. To us, it is the same whether we select them, for example, in the order $b1 - b2 - b4$ or $b2 - b4 - b1$, as long as we obtain the set $\{b1, b2, b4\}$. So, we correct for this by dividing the total of 24 by the number of combinations to order these particular black balls; that is,

$$3 \times 2 \times 1 = 3! = 6$$

Hence, the number of combinations of drawing $k = 3$ black balls out of four is

$$24/6 = 4!/3! = 4$$

Next we need to consider the previous number of possibilities of drawing $k = 3$ black balls in combination with drawing $n - k = 2$ white balls. We apply the same reasoning as before to obtain two white balls from the collection of six (i.e., $\{w1, w2, w3, w4, w5, w6\}$). That gives us $6 \times 5/2 = 6!/2! = 15$ nonredundant options to recover two white balls, in our example.

In total, we have

$$4 \times 15 = \frac{4 \times 3 \times 2 \times 1}{3 \times 2 \times 1} \times \frac{6 \times 5 \times 4 \times 3 \times 2 \times 1}{2 \times 1} = \frac{4!}{3! \times 1!} \times \frac{6!}{2! \times 4!} = \binom{4}{3} \times \binom{6}{2} = 60$$

different possibilities to obtain three black and two white balls in a sample of five balls. All these 60 samples have the same implication for us (i.e., $k = 3$).

Combining these 60 possibilities with a probability of 0.004 as given by equation (9.10), we obtain as the probability for a sum of $k = 3$ black balls in a sample of $n = 5$

$$60/252 = 0.2381$$

Formally, we have

$$P(X=3) = \frac{\binom{4}{3}\binom{6}{2}}{\binom{10}{5}} = 0.2381$$

Then, for our example, the probability distribution of X is[7]

$$P(X=k) = \frac{\binom{4}{k}\binom{6}{n-k}}{\binom{10}{5}}, \quad k = 1,2,3,4 \qquad (9.11)$$

Let's advance from the special conditions of the example to the general case; that is, (1) at the beginning, some nonnegative integer N of black and white balls combined, (2) the overall number of black balls $0 \leq K \leq N$, (3) the sample size $0 \leq n \leq N$, and (4) the number $0 \leq k \leq n$ of black balls in the sample.

In equation (9.11), we have the probability of k black balls in the sample of $n = 5$ balls. We dissect equation (9.9) into three parts: the denominator and the two parts forming the product in the numerator. The denominator gives the number of possibilities to draw a sample of $n = 5$ balls out of $N = 10$ balls, no matter what the combination of black and white. In other words, we choose $n = 5$ out of $N = 10$. The resulting number is given by the binomial coefficient. We can extend this to choosing a general sample of n drawings out of a population of an arbitrary number of N balls. Analogous to equation (9.11), the resulting number of possible samples of length n (i.e., n drawings) is then given by

$$\binom{N}{n} \qquad (9.12)$$

Next, suppose we have k black balls in this sample. We have to consider that in equation (9.11), we chose k black balls from a population of $K = 4$

---

[7]Note that we cannot draw more than four black balls from $b1$, $b2$, $b3$, and $b4$.

yielding as the number of possibilities for this the binomial coefficient on the left-hand side in the numerator. Now we generalize this by replacing $K = 4$ by some general number of black balls ($K \leq N$) in the population. The resulting number of choices for choosing k out of the overall K black balls is then,

$$\binom{K}{k} \qquad (9.13)$$

And, finally, we have to draw the remaining $n - k$ balls, which have to be white, from the population of $N - K$ white balls. This gives us

$$\binom{N-K}{n-k} \qquad (9.14)$$

different nonredundant choices for choosing $n - k$ white balls out of $N - K$.

Finally, all we need to do is to combine equations (9.12), (9.13), and (9.14) in the same fashion as equation (9.11). By doing so, we obtain

$$P(X = k) = \frac{\binom{K}{k}\binom{N-K}{n-k}}{\binom{N}{n}}, \quad k = 1, 2, \ldots, n \qquad (9.15)$$

as the probability to obtain a total of $X = k$ black balls in the sample of length n *without replacement*.

Importantly, here, we start out with N balls of which K are black and, after each trial, we do not replace the ball drawn, so that the population is different for each trial. The resulting random variable is *hypergeometric* distributed with parameters (N,K,n); that is, $Hyp(N,K,n)$, and probability distribution equation (9.15).

The mean of a random variable X following a hypergeometric probability law is given by

$$E(X) = n \cdot \frac{K}{N}$$

and the variance of this $X \sim Hyp(N,K,n)$ is given by

$$Var(X) = \sigma^2 = n \cdot \frac{K}{N} \cdot \frac{N-K}{N} \cdot \frac{N-n}{N-1}$$

The hypergeometric and the binomial distributions are similar, though, not equivalent. However, if the population size N is large, the hypergeometric distribution is often approximated by the binomial distribution with equation (9.6) causing only little deviation from the true probabilities equation (9.15).

Application

Let's see how the hypergeometric distribution has been used applied in a Federal Reserve Bank of Cleveland study to assess whether U.S. exchange-rate intervention resulted in a desired depreciation of the dollar.[8]

Consider the following scenario. The U.S. dollar is appreciating against a certain foreign currency. This might hurt U.S. exports to the country whose sovereign issues the particular foreign currency. In response, the U.S. Federal Reserve might be inclined to intervene by purchasing that foreign currency to help depreciate the U.S. dollar through the increased demand for foreign currency relative to the dollar. This strategy, however, may not necessarily produce the desired effect. That is, the dollar might continue to appreciate relative to the foreign currency. Let's let an intervention by the Federal Reserve be defined as the purchase of that foreign currency. Suppose that we let the random variable X be number of interventions that lead to success (i.e., depreciation of the dollar). Given certain conditions beyond the scope of this book, the random variable X is approximately distributed hypergeometric.

This can be understood by the following slightly simplified presentation. Let the number of total observations be N days of which K is the number of days with a dollar depreciation (with or without intervention), and $N - K$ is the number of days where the dollar appreciated or remained unchanged. The number of days the Federal Reserve intervenes is given by n. Furthermore, let k equal the number of days the interventions are successful so that $n - k$ accounts for the unsuccessful interventions. The Federal Reserve could technically intervene on all N days that would yield a total of K successes and $N - K$ failures. However, the actual number of occasions n on which there are interventions might be smaller. The n interventions can be treated as a sample of length n taken from the total of N days without replacement.

The model can best be understood as follows. The observed dollar appreciations, persistence, or depreciations are given observations. The Federal Reserve can merely decide to intervene or not. Consequently, if it took action on a day with depreciation, it would be considered a success and the number of successes available for future attempts would, therefore,

---

[8]This application draws from Humpage (1998).

be diminished by one. If, on the other hand, the Federal Reserve decided to intervene on a day with appreciation or persistence, it would incur a failure that would reduce the number of available failures left by one. The $N - n$ days there are no interventions are treated as not belonging to the sample.

The randomness is in the selection of the days on which to intervene. The entire process can be illustrated by a chain with N tags attached to it containing either a + or − symbol. Each tag represents one day. A + corresponds to an appreciation or persistence of the dollar on the associated day, while a − to a depreciation. We assume that we do not know the symbol behind each tag at this point.

In total, we have K tags with a + and $N - n$ with a − tag. At random, we flip n of these tags, which is equivalent to the Federal Reserve taking action on the respective days. Upon turning the respective tag upside right, the contained symbol reveals immediately whether the associated intervention resulted in a success or not.

Suppose, we have $N = 3{,}072$ total observations of which $K = 1{,}546$ represents the number of days with a dollar depreciation, while on $N - K = 1{,}508$ days the dollar either became more valuable or remained steady relative to the foreign currency.

Again, let X be the hypergeometric random variable describing successful interventions. On $n = 138$ days, the Federal Reserve saw reason to intervene, that is, purchase foreign currency to help bring down the value of the dollar which was successful on $k = 51$ days and unsuccessful on the remaining $n - k = 87$ days. Concisely, the values are given by $N = 3{,}072$, $K = 1{,}546$, $N - K = 1{,}508$, $n = 138$, $k = 51$, and $n - k = 87$.

So, the probability for this particular outcome $k = 51$ for the number of successes X given $n = 138$ trials is

$$P(X = 51) = \frac{\binom{1546}{51}\binom{1508}{87}}{\binom{3072}{138}} = 0.00013429$$

which is an extremely small probability.

Suppose we state the simplifying hypothesis that the Federal Reserve is overall successful if most of the dollar depreciations have been the result of interventions (i.e., purchase of foreign currency). Then, this outcome with $n = 51$ successful interventions given a total of $N - K$ depreciations shows that the decline of the dollar relative to the foreign currency might be the result of something other than a Federal Reserve intervention. Hence, the

Federal Reserve intervention might be too vague a forecast of a downward movement of the dollar relative to the foreign currency.

MULTINOMIAL DISTRIBUTION

For our next distribution, the *multinomial distribution*, we return to the realm of drawing with replacement so that for each trial, there are exactly the same conditions. That is, we are dealing with independent and identically distributed random variables.[9] However, unlike the binomial distribution, let's change the population so that we have not only two different possible outcomes for one drawing, but a third or possibly more outcomes.

We extend the illustration where we used an urn containing black and white balls. In our extension, we have a total of N balls with three colors: K_w white balls, K_b black balls, and $K_r = N - K_w - K_b$ red balls. The probability of each of these colors is denoted by

$$P(Y = \text{white}) = p_w$$

$$P(Y = \text{black}) = p_b$$

$$P(Y = \text{red}) = p_r$$

with each of these probabilities representing the population share of the respective color: $p_i = K_i/N$, for i = white, black, and red. Since all shares combined have to account for all N, we set

$$p_r = 1 - p_b - p_w$$

For purposes of this illustration, let $p_w = p_b = 0.3$ and $p_r = 0.4$. Suppose that in a sample of $n = 10$ trials, we obtain the following result: $n_w = 3$ white, $n_b = 4$ black, and $n_r = n - n_w - n_b = 3$ red. Furthermore, suppose that the balls were drawn in the following order

Y_1	Y_2	Y_3	Y_4	Y_5	Y_6	Y_7	Y_8	Y_9	Y_{10}
r	w	b	b	w	r	r	b	w	b

where the random variable Y_i represents the outcome of the i-th trial.[10] This particular sample occurs with probability

---

[9] Once again we note that we are still short of a formal definition of independence in the context of probability theory. We use the term in the sense of "uninfluenced by."
[10] We denote w = white, b = black, and r = red.

$$P(Y_1 = r, Y_2 = w, \ldots, Y_{10} = b) = p_r \cdot p_w \cdot \ldots \cdot p_b$$
$$= p_r^3 \cdot p_w^3 \cdot p_b^4$$

The last equality indicates that the order of appearance of the individual values, once again, does not matter.

We introduce the random variable X representing the number of the individual colors occurring in the sample. That is, X consists of the three components X_w, X_b, and X_r or, alternatively, $X = (X_w, X_b, X_r)$. Analogous to the binomial case of two colors, we are not interested in the order of appearance, but only in the respective numbers of occurrences of the different colors (i.e., n_w, n_b, and n_r). Note that several different sample outcomes may lead to $X = (n_w, n_b, n_r)$. The total number of different nonredundant samples with n_w, n_b, and n_r is given by the multinomial coefficient introduced in Appendix C of this book, which here yields

$$\binom{n}{n_w \ n_b \ n_r} = \binom{10}{3 \ 3 \ 4} = 4{,}200$$

Hence, the probability for this value of $X = (k_w, k_b, k_r) = (3,4,3)$ is then

$$P(X = (3,4,3)) = \binom{10}{3 \ 3 \ 4} \cdot p_w^3 \cdot p_b^4 \cdot p_r^3$$
$$= 4{,}200 \cdot 0.3^3 \cdot 0.3^4 \cdot 0.4^3$$
$$= 0.0588$$

In general, the probability distribution of a multinomial random variable X with k components $X_1, X_2, \ldots, X_k$ is given by

$$P(X_1 = n_1, X_2 = n_2, \ldots, X_k = n_k)$$
$$= \binom{n}{n_1 \ n_2 \ \ldots \ n_k} \cdot p_1^{n_1} \cdot p_2^{n_2} \cdot \ldots \cdot p_k^{n_k} \quad (9.16)$$

where, for $j = 1, 2, \ldots, k$, n_j denotes the outcome of component j and the p_j the corresponding probability.

The means of the k components X_1 through X_k are given by

$$E(X_1) = p_1 \cdot n$$
$$\vdots$$
$$E(X_k) = p_k \cdot n$$

and their respective variances by

Discrete Probability Distributions

$$Var(X_1) = \sigma_1^2 = p_1 \cdot (1-p_1) \cdot n$$
$$\vdots$$
$$Var(X_k) = \sigma_k^2 = p_k \cdot (1-p_k) \cdot n$$

Multinomial Stock Price Model

We can use the multinomial distribution to extend the binomial stock price model described earlier. Suppose, we are given a stock with price S_0, in $t = 0$. In $t = 1$, the stock can have either price

$$S_1^{(u)} = S_0 \cdot u$$

$$S_1^{(l)} = S_0 \cdot l$$

$$S_1^{(d)} = S_0 \cdot d$$

Let the three possible outcomes be a 10% increase in price ($u = 1.1$), no change in price ($l = 1.0$), and a 10% decline in price ($d = 0.9$). That is, the price either goes up by some factor, remains steady, of drops by some factor. Therefore,

$$S_1^{(u)} = S_0 \cdot 1.1$$

$$S_1^{(l)} = S_0 \cdot 1.0$$

$$S_1^{(d)} = S_0 \cdot 0.9$$

Thus, we have three different outcomes of the price change in the first period. Suppose, the price change behaved the same in the second period, from $t = 1$ until $t = 2$. So, we have

$$S_2^{(u)} = S_1 \cdot 1.1$$

$$S_2^{(l)} = S_1 \cdot 1.0$$

$$S_2^{(d)} = S_1 \cdot 0.9$$

at time $t = 2$ depending on

$$S_1 \in \{S_1^{(u)}, S_1^{(l)}, S_1^{(d)}\}$$

Let's denote the random price change in the first period by Y_1 while the price change in the second period by the random variable Y_2. So, it is obvious that Y_1 and Y_2 independently assume some value in the set $\{u,l,d\} = \{1.1, 1.0, 0.9\}$. After two periods (i.e., in $t = 2$), the stock price is

$$S_2 = S_0 \cdot Y_1 \cdot Y_2 \in \left\{S_2^{(u)}, S_2^{(l)}, S_2^{(d)}\right\}$$

Note that the random variable S_2 is not multinomially distributed itself. However, as we will see, it is immediately linked to a multinomial random variable.

Since the initial stock price S_0 is given, the random variable of interest is the product $Y_1 \cdot Y_2$, which is in a one-to-one relationship with the multinomial random variable $X = (n_u, n_l, n_d)$ (i.e., the number of up-, zero-, and down-movements, respectively). The state space of $Y_1 \cdot Y_2$ is given by $\{uu, ul, ud, ll, ld, dd\}$. This corresponds to the state space of X, which is given by

$$\Omega' = \{(2,0,0), (0,2,0), (0,0,2), (1,1,0), (1,0,1), (0,1,1)\}$$

Note that since $Y_1 \cdot Y_2$ is a product, we do not consider, for example, $(Y_1 = u, Y_2 = d)$ and $(Y_1 = d, Y_2 = u)$ separately. With

$$P(Y_i = u) = p^u = 0.25$$

$$P(Y_i = l) = p^l = 0.50$$

$$P(Y_i = d) = p^d = 0.25$$

the corresponding probability distribution of X is given in the first two columns of Table 9.1. We use the multinomial coefficient

$$\binom{n}{n_u \; n_l \; n_d}$$

where

n = the number of periods
n_u = the number of up-movements
n_l = number of zero movements
n_d = number of down-movements

Now, if $S_0 = \$20$, then we obtain the probability distribution of the stock price in $t = 2$ as shown in columns 2 and 3 in Table 9.1. Note that

Discrete Probability Distributions

the probabilities of the values of S_2 are associated with the corresponding price changes X and, hence, listed on the same lines of Table 9.1. It is now possible to evaluate the probability of events such as, "a stock price S_2 of, at most, $22," from the σ-algebra $\mathbb{A}'$ of the multinomial probability space of X. This is given by

$$P(S_2 \leq \$22)$$
$$= P(S_2 = \$16.2) + P(S_2 = \$18) + P(S_2 = \$19.8) + P(S_2 = \$20) + P(S_2 = \$22)$$
$$= 0.25 + 0.125 + 0.25 + 0.25 + 0.0625$$
$$= 1 - P(S_2 = \$24.2)$$
$$= 0.9375$$

where the third line is the result of the fact that the sum of the probabilities of all disjoint events has to add up to one. That follows since any event and its complement account for the entire state space Ω'.

TABLE 9.1 Probability Distribution of the Two-Period Stock Price Model

$X = (n_u, n_l, n_d)$	$P(X = \cdot)$	$S_2 = \cdot$
(2,0,0)	$\binom{2}{2\ 0\ 0} p^u p^u = 0.0625$	$S_0 \cdot u^2 = 20 \cdot 1.1^2 = 24.2$
(1,1,0)	$\binom{2}{1\ 1\ 0} p^u p^l = 2 \cdot 0.25 \cdot 0.5 = 0.25$	$S_0 \cdot u \cdot l = 20 \cdot 1.1 \cdot 1.0 = 22$
(1,0,1)	$\binom{2}{1\ 0\ 1} p^u p^d = 2 \cdot 0.25^2 = 0.125$	$S_0 \cdot u \cdot d = 20 \cdot 1.1 \cdot 0.9 = 19.8$
(0,2,0)	$\binom{2}{0\ 2\ 0} p^l p^l = 0.5^2 = 0.25$	$S_0 \cdot l \cdot l = 20 \cdot 1.0^2 = 20$
(0,1,1)	$\binom{2}{0\ 1\ 1} p^l p^d = 2 \cdot 0.5 \cdot 0.25 = 0.25$	$S_0 \cdot l \cdot d = 20 \cdot 1.0 \cdot 0.9 = 18$
(0,0,2)	$\binom{2}{0\ 0\ 2} p^d p^d = 0.25^2 = 0.0625$	$S_0 \cdot d^2 = 20 \cdot 0.9^2 = 16.2$

In the first and second columns, we have the probability distribution of the two period stock price changes $X = Y_1 \cdot Y_2$ in the multinomial stock price model. In the third column, we have the probability distribution of the stock price S_2.

FIGURE 9.6 Multinomial Stock Price Model: Stock Price S_2, in $t = 2$

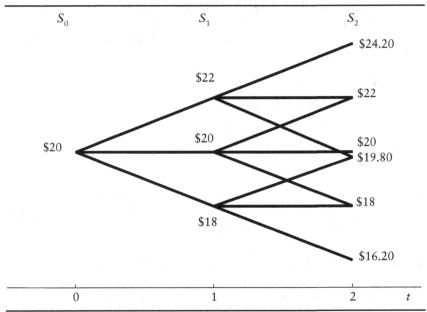

In Figure 9.6, we can see the evolution of the stock price along the different paths.

From equation (9.1), the expected stock price in $t = 2$ is computed as

$$E(S_2) = \sum_{s \in \Omega'} s \cdot P(S_2 = s) = \$24.2 \cdot 0.25 + \$18 \cdot 0.125 + \$19.8 \cdot 0.25$$
$$+ \$20 \cdot 0.25 + \$22 \cdot 0.0625 + \$24.2 \cdot 0.0625$$
$$= \$21.1375$$

So, on average, the stock price will evolve into $S_2 = \$21.14$ (rounded).

POISSON DISTRIBUTION

To introduce our next distribution, consider the following situation. A property and casualty insurer underwrites a particular type of risk, say, automotive damage. Overall, the insurer is interested in the total annual dollar amount of the claims from all policies underwritten. The total is the sum of the individual claims of different amounts. The insurer has to have enough equity as risk guarantee. In a simplified way, the sufficient amount is given by the number of casualties N times the average amount per claim.

In this situation, the insurer's interest is in the total number of claims N within one year. Note that there may be multiple claims per policy. This number N is random because the insurer does not know its exact value at the beginning of the year. The insurer knows, however, that the minimum number of casualties possible is zero. Theoretically, although it is unlikely, there may be infinitely many claims originating from the year of interest.

So far, we have considered the number of claims over the period of one year. It could be of interest to the insurer, however, to know the behavior of the random variable N over a period of different length, say five years, Or, even, the number of casualties related to one month could be of interest. It might be reasonable to assume that there will probably be fewer claims in one month than in one year or five years.

The number of claims, N, as a random variable should follow a probability law that accounts for the length of the period under analysis. In other words, the insurers want to assure that the probability distribution of N gives credit to N being proportional to the length of the period in the sense that if a period is n times as long as another, then the number of claims expected over the longer period should be n times as large as well.

As a candidate that satisfies these requirements, we introduce the **Poisson distribution** with parameter λ and formally expressed as $Poi(\lambda)$. We define that the parameter is a positive real number (i.e., $\lambda > 0$). A Poisson random variable N—that is, $X \sim Poi(\lambda)$—assumes nonnegative integer values. Formally, N is a function mapping the space of outcomes, Ω, into the state space

$$\Omega' = \{0, 1, 2, ...\}$$

which is the set $\mathbb{N}$ of the nonnegative integer numbers.

The probability measure of a Poisson random variable N for nonnegative integers $k = 0, 1, 2, ...$ is defined as

$$P(N = k) = \frac{\lambda^k}{k!} e^{-\lambda} \tag{9.17}$$

where $e = 2.7183$ is the Euler constant. Here, we have unit period length.

The mean of a Poisson random variable with parameter λ is

$$E(N) = \lambda$$

while its variance is given by

$$Var(N) = \sigma^2 = \lambda \tag{9.18}$$

So, both parameters, mean and variance, of $N \sim Poi(\lambda)$ are given by the parameter λ.

For a period of general length t, equation (9.17) becomes

$$P(N=k) = \frac{(\lambda t)^k}{k!} e^{-\lambda t} \qquad (9.19)$$

We can see that the new parameter is now λt, accounting for the time proportionality of the distribution of N, that is, $N = N(t)$ is the number of jumps of size 1 in the interval $(0,t)$. The mean changes to

$$EN(t) = \lambda t \qquad (9.20)$$

and analogous to the variance given by (9.18) is now

$$Var(N(t)) = \sigma^2(t) = \lambda t \qquad (9.21)$$

We can see by equation (9.20) that the average number of occurrences is the average per unit of time, λ, times the length of the period, t, in units of time. The same holds for the variance given by equation (9.21).

The Poisson distribution serves as an approximation of the hypergeometric distribution when certain conditions are met regarding sample size and probability distribution.[11]

Application to Credit Risk Modeling for a Bond Portfolio

The Poisson distribution is typically used in finance for credit risk modeling. For example, suppose we have a pool of 100 bonds issued by different corporations. By experience or empirical evidence, we may know that each quarter of a year the expected number to default is two; that is, $\lambda = 2$. Moreover, from prior research, we can approximate the distribution of N by the Poisson distribution, even though, theoretically, the Poisson distribution admits values k greater than 100. What is the number of bonds to default within the next year, on average? According to equation (9.3), since the mean is $E_{quarter}(N) = \lambda = 2$, per quarter, the mean per year ($t = 4$) is

$$E_{year}(N) = \lambda t = 2 \cdot 4 = 8$$

By equation (9.20), the variance is 8, from equation (9.19), the probability of, at most, 10 bonds to default is given by

---

[11]Suppose we are analyzing the random experiment of sampling n balls from a pool of K black and $N - K$ white balls, without replacement. Then, the conditions are: (1) a sample size $n \geq 30$; 2) a probability of black $K/N \leq 0.1$; and (3) a sample size ratio $n/N \leq 0.1$.

$$P(N \leq 10) = P(N=0) + P(N=1) + \ldots + P(N=10)$$
$$= e^{-2\times 4} \cdot \frac{(2\times 4)^0}{0!} + e^{-2\times 4} \cdot \frac{(2\times 4)^1}{1!} + \ldots + e^{-2\times 4} \cdot \frac{(2\times 4)^{10}}{10!}$$
$$= 0.8159$$

DISCRETE UNIFORM DISTRIBUTION

Consider a probability space $(\Omega', \mathbb{A}', P)$ where the state space is a finite set of, say n, outcomes, that is, $\Omega' = \{x_1, x_2, \ldots, x_n\}$. The σ-algebra $\mathbb{A}'$ is given by the power set of Ω'.

So far we have explained how drawings from this Ω' may be modeled by the multinomial distribution. In the multinomial distribution, the probability of each outcome may be different. However, suppose that the for our random variable X, we have a constant $P(X = x_j) = 1/n$, for all $j = 1, 2, \ldots, n$. Since all values x_j have the same probability (i.e., they are equally likely), the distribution is called the *discrete uniform distribution*. We denote this distribution by $X \sim DU_{\Omega'}$. We use the specification Ω' to indicate that X is a random variable on this particular state space.

The mean of a discrete, uniformly distributed random variable X on the state space $\Omega' = \{x_1, x_2, \ldots, x_n\}$ is given by

$$E(X) = \sum_{i=1}^{n} p_i \cdot x_i = \frac{1}{n} \sum_{i=1}^{n} x_i \qquad (9.22)$$

Note that equation (9.22) is equal to the arithmetic mean given by equation (3.2) in Chapter 3. The variance is

$$Var(X) = \sum_{i: x_i \in \Omega'} p_i \cdot (x_i - E(X))^2$$
$$= \frac{1}{n} \sum_{i: x_i \in \Omega'} (x_i - E(X))^2$$

with $E(X)$ from equation (9.22).

A special case of a discrete uniform probability space is given when $\Omega' = \{1, 2, \ldots, n\}$. The resulting mean, according to equation (9.22), is then,

$$E(X) = \sum_{i=1}^{n} p_i \cdot x_i = \frac{1}{n} \sum_{i=1}^{n} i = \frac{1}{n} \times \frac{n(n+1)}{2} = \frac{n+1}{2} \qquad (9.23)$$

For this special case of discrete uniform distribution of a random variable X, we use the notation $X \sim DU(n)$ with parameter n.

Let's once more consider the outcome of a toss of a dice. The random variable number of dots, X, assumes one of the numerical outcomes 1, 2, 3, 4, 5, 6 each with a probability of 1/6. Hence, we have a uniformly distributed discrete random variable X with the state space $\Omega' = \{1,2,3,4,5,6\}$. Consequently, we express this as $X \sim DU(6)$.

Next, we want to consider several independent trials, say $n = 10$, of throwing the dice. By $n_1, n_2, n_3, n_4, n_5,$ and n_6, we denote the number of occurrence of the values 1, 2, 3, 4, 5, and 6, respectively. With constant probability $p_1 = p_2 = \ldots = p_6 = 1/6$, we have a discrete uniform distribution, that is, $X \sim DU(6)$. Thus, the probability of obtaining $n_1 = 1, n_2 = 2, n_3 = 1, n_4 = 3, n_5 = 1,$ and $n_6 = 2$, for example, is

$$P(X_1 = 1, X_2 = 1, \ldots, X_6 = 2) = \binom{10}{1\ 2\ \ldots\ 2} \left(\frac{1}{6}\right)^{10}$$

$$= \frac{10!}{1! \times 2! \times \ldots \times 2!} \cdot \left(\frac{1}{6}\right)^{10}$$

$$= 151200 \cdot 0.00000016538$$

$$= 0.0025$$

Application to the Multinomial Stock Price Model

Let us resume the stock price model where in $t = 0$ we have a given stock price, say $S_0 = \$20$, where there are three possible outcomes at the end of the period. In the first period, the stock price either increases to

$$S_1^{(u)} = S_0 \cdot 1.1 = \$22$$

remains the same at

$$S_1^{(l)} = S_0 \cdot 1.0 = \$20$$

or decreases to

$$S_1^{(d)} = S_0 \cdot 0.9 = \$18$$

each with probability 1/3. Again, we introduce the random variable Y assuming the values $u = 1.1, l = 1.0,$ and $d = 0.9$ and, thus, representing the percentage change of the stock price between $t = 0$ and $t + 1 = 1$. The stock price in $t + 1 = 1$ is given by the random variable S_1 on the corresponding state space

$$\Omega_S = \left\{ S_1^{(u)}, S_1^{(l)}, S_1^{(d)} \right\}$$

Suppose, we have $n = 10$ successive periods in each of which the stock price changes by the factors u, l, or d. Let the multinomial random variable $X = (X_1, X_2, X_3)$ represent the total of up-, zero-, and down-movements, respectively. Suppose, after these n periods, we have $n_u = 3$ up-movements, $n_l = 3$ zero-movements, and $n_d = 4$ down-movements. According to equation (9.16), the corresponding probability is

$$P(X_1 = 3, X_2 = 3, X_3 = 4) = \binom{10}{3\ 3\ 4} \left(\frac{1}{3}\right)^{10}$$
$$= 4200 \cdot 0.00001935$$
$$= 0.0711$$

This probability corresponds to a stock price in $t = 10$ of

$$S_{10} = S_0 \cdot u^3 \cdot l^3 \cdot d^4 = \$20 \cdot 1.1^3 \cdot 1 \cdot 0.9^4 = \$17.47$$

This stock price is a random variable given by

$$S_{10} = S_0 \cdot Y_1 \cdot Y_2 \cdot \ldots \cdot Y_{10}$$

where the Y_i are the corresponding relative changes (i.e., factors) in the periods $i = 1, 2, \ldots, 10$. Note that S_{10} is not uniformly distributed even though it is a function of the random variables $Y_1, Y_2, \ldots, Y_{10}$ because its possible outcomes do not have identical probability.

CONCEPTS EXPLAINED IN THIS CHAPTER
(IN ORDER OF PRESENTATION)

Discrete random variables
Probability distribution
Probability law
Discrete law
Discrete cumulative distribution (cdf)
Variance
Standard deviation
Bernoulli distribution

Drawing with replacement
Binomial distribution
Binomial coefficient
Binomial tree
Hypergeometric distribution
Multinomial distribution
Poisson distribution
Discrete uniform distribution

APPENDIX List of Discrete Distributions

Name		Probability Law	Mean	Variance	Description
Bernoulli	$B(p)$	$P(X=1)=p$ $P(X=0)=1-p$	p	$p \cdot (1-p)$	One drawing from $\Omega' = \{1,0\}$
Binomial	$B(n,p)$	$P(X=k) = \binom{n}{k} \cdot p^k \cdot (1-p)^{n-k}$	$n \cdot p$	$n \cdot p \cdot (1-p)$	n drawings with replacement from $\Omega' = \{1,0\}$
Hypergeometric	$Hyp(N,K,n)$	$P(X=k) = \dfrac{\binom{K}{k}\binom{N-K}{n-k}}{\binom{N}{n}}$	$n \dfrac{K}{N}$	$n \dfrac{K}{N} \dfrac{N-K}{N} \dfrac{N-n}{N-1}$	n drawings without replacement from $\Omega' = \{1,0\}$
Multinomial		$P(X_1=n_1,\ldots,X_k=n_k) = \binom{n}{n_1 \ \cdots \ n_k} \cdot p_1^{n_1} \cdots p_k^{n_k}$	$p_1 \cdot n$ $\vdots$ $p_k \cdot n$	$p_1 \cdot (1-p_1) \cdot n$ $\vdots$ $\sigma_k^2 = p_k \cdot (1-p_k) \cdot n$	n drawings with replacement from $\Omega' = \{x_1,\ldots,x_k\}$
Poisson	$Poi(\lambda)$	$P(N=k) = \dfrac{\lambda^k}{k!} e^{-\lambda}$	λ	λ	One drawing from $\Omega' = \{0,1,2,\ldots\} = \mathbb{N}$. Unit period length
	$Poi(\lambda t)$	$P(N=k) = \dfrac{(\lambda t)^k}{k!} e^{-\lambda t}$	λt	λt	One drawing from $\Omega' = \{0,1,2,\ldots\} = \mathbb{N}$. Period length t
Discrete Uniform	DU_α	$P(X=x_1) = \ldots = P(X=x_n) = \dfrac{1}{n}$	(8.1)	(8.2)	One drawing from $\Omega' = \{x_1,\ldots,x_k\}$
	$DU(n)$	$P(X=1) = \ldots = P(X=n) = \dfrac{1}{n}$	$\dfrac{n+1}{2}$	(8.2)	One drawing from $\Omega' = \{1,\ldots,k\}$ Equal probability

Note: For the k components of a multinomial random variable, we have k means and variances.

$B(n,p)$, Binomial Probability Distribution

$$P(X=k) = \binom{n}{k} \cdot p^k \cdot (1-p)^{n-k} \text{ for } n = 5$$

p \ k	0.1	0.2	0.5	0.8	0.9
1	0.3281	0.4096	0.1563	0.0064	0.0005
2	0.0729	0.2048	0.3125	0.0512	0.0081
3	0.0081	0.0512	0.3125	0.2048	0.0729
4	0.0005	0.0064	0.1563	0.4096	0.3281
5	0	0.0003	0.0313	0.3277	0.5905

$B(n,p)$, Binomial Probability Distribution

$$P(X=k) = \binom{n}{k} \cdot p^k \cdot (1-p)^{n-k} \text{ for } n = 10$$

p \ k	0.1	0.2	0.5	0.8	0.9
1	0.3874	0.2684	0.0098	0	0
2	0.1937	0.3020	0.0439	0.0001	0
3	0.0574	0.2013	0.1172	0.0008	0
4	0.0112	0.0881	0.2051	0.0055	0.0001
5	0.0015	0.0264	0.2461	0.0264	0.0015
6	0.0001	0.0055	0.2051	0.0881	0.0112
7	0	0.0008	0.1172	0.2013	0.0574
8	0	0.0001	0.0439	0.3020	0.1937
9	0	0	0.0098	0.2684	0.3874
10	0	0	0.0010	0.1074	0.3487

Discrete Probability Distributions

$B(n,p)$, Binomial Probability Distribution

$$P(X=k) = \binom{n}{k} \cdot p^k \cdot (1-p)^{n-k} \text{ for } n = 50$$

p \ k	0.1	0.2	0.5	0.8	0.9
1	0.0286	0.0002	0	0	0
2	0.0779	0.0011	0	0	0
3	0.1386	0.0044	0	0	0
4	0.1809	0.0128	0	0	0
5	0.1849	0.0295	0	0	0
6	0.1541	0.0554	0	0	0
7	0.1076	0.0870	0	0	0
8	0.0643	0.1169	0	0	0
9	0.0333	0.1364	0	0	0
10	0.0152	0.1398	0	0	0
20	0	0.0006	0.0419	0	0
30	0	0	0.0419	0.0006	0
40	0	0	0	0.1398	0.0152
41	0	0	0	0.1364	0.0333
42	0	0	0	0.1169	0.0643
43	0	0	0	0.0870	0.1076
44	0	0	0	0.0554	0.1541
45	0	0	0	0.0295	0.1849
46	0	0	0	0.0128	0.1809
47	0	0	0	0.0044	0.1386
48	0	0	0	0.0011	0.0779
49	0	0	0	0.0002	0.0286
50	0	0	0	0	0.0052

$B(n,p)$, Binomial Probability Distribution

$$P(X=k) = \binom{n}{k} \cdot p^k \cdot (1-p)^{n-k} \text{ for } n = 100$$

k \ p	0.1	0.2	0.5	0.8	0.9
1	0.0003	0	0	0	0
2	0.0016	0	0	0	0
3	0.0059	0	0	0	0
4	0.0159	0	0	0	0
5	0.0339	0	0	0	0
6	0.0596	0.0001	0	0	0
7	0.0889	0.0002	0	0	0
8	0.1148	0.0006	0	0	0
9	0.1304	0.0015	0	0	0
10	0.1319	0.0034	0	0	0
20	0.0012	0.0993	0	0	0
30	0	0.0052	0	0	0
40	0	0	0.0108	0	0
50	0	0	0.0796	0	0
60	0	0	0.0108	0	0
70	0	0	0	0.0052	0
80	0	0	0	0.0993	0.0012
90	0	0	0	0.0034	0.1319
91	0	0	0	0.0015	0.1304
92	0	0	0	0.0006	0.1148
93	0	0	0	0.0002	0.0889
94	0	0	0	0.0001	0.0596
95	0	0	0	0	0.0339
96	0	0	0	0	0.0159
97	0	0	0	0	0.0059
98	0	0	0	0	0.0016
99	0	0	0	0	0.0003
100	0	0	0	0	0

Poi(λ), Poisson Probability Distribution

$$P(X=k) = \frac{\lambda^k \cdot e^{-\lambda}}{x!}$$ for Several Values of Parameter λ.

k \ λ	0.1	0.5	1	2	5	10
1	0.0905	0.3033	0.3679	0.2707	0.0337	0.0005
2	0.0045	0.0758	0.1839	0.2707	0.0842	0.0023
3	0.0002	0.0126	0.0613	0.1804	0.1404	0.0076
4	0	0.0016	0.0153	0.0902	0.1755	0.0189
5	0	0.0002	0.0031	0.0361	0.1755	0.0378
6	0	0	0.0005	0.0120	0.1462	0.0631
7	0	0	0.0001	0.0034	0.1044	0.0901
8	0	0	0	0.0009	0.0653	0.1126
9	0	0	0	0.0002	0.0363	0.1251
10	0	0	0	0	0.0181	0.1251
11	0	0	0	0	0.0082	0.1137
12	0	0	0	0	0.0034	0.0948
13	0	0	0	0	0.0013	0.0729
14	0	0	0	0	0.0005	0.0521
15	0	0	0	0	0.0002	0.0347
16	0	0	0	0	0	0.0217
17	0	0	0	0	0	0.0128
18	0	0	0	0	0	0.0071
19	0	0	0	0	0	0.0037
20	0	0	0	0	0	0.0019
50	0	0	0	0	0	0
100	0	0	0	0	0	0

CHAPTER 10
Continuous Probability Distributions

In this chapter, we introduce the concept of continuous probability distributions. We present the continuous distribution function with its corresponding density function, a function unique to continuous probability laws. In the chapter, parameters of location and scale such as the mean and higher moments—variance and skewness—are defined for the first time even though they will be discussed more thoroughly in Chapter 13.

The more commonly used distributions with appealing statistical properties that are used in finance will be presented in Chapter 11. In Chapter 12, we discuss the distributions that unlike the ones discussed in Chapter 11 are capable of dealing with extreme events.

CONTINUOUS PROBABILITY DISTRIBUTION DESCRIBED

Suppose we are interested in outcomes that are no longer countable. Examples of such outcomes in finance are daily logarithmic stock returns, bond yields, and exchange rates. Technically, without limitations caused by rounding to a certain number of digits, we could imagine that any real number could provide a feasible outcome for the daily logarithmic return of some stock. That is, the set of feasible values that the outcomes are drawn from (i.e., the space Ω) is uncountable. The events are described by *continuous intervals* such as, for example, (–0.05, 0.05], which, referring to our example with daily logarithmic returns, would represent the event that the return at a given observation is more than –5% and at most 5%.[1]

In the context of *continuous probability distributions*, we have the real numbers $\mathbb{R}$ as the uncountable space Ω. The set of events is given by the Borel σ-algebra $\mathbb{B}$, which, as we recall from Chapter 8, is based on the half-open intervals of the form $(-\infty, a]$, for any real a. The space $\mathbb{R}$ and the σ-algebra $\mathbb{B}$ form the *measurable space* $(\mathbb{R}, \mathbb{B})$, which we are to deal with, throughout this chapter.

---
[1]For those unfamiliar with continuity in a mathematical sense, we recommend reading Appendix A first.

DISTRIBUTION FUNCTION

To be able to assign a probability to an event in a unique way, in the context of continuous distributions we introduce as a device the *continuous distribution function* $F(a)$, which expresses the probability that some event of the sort $(-\infty,a]$ occurs (i.e., that a number is realized that is at most a).[2] As with discrete random variables, this function is also referred to as the *cumulative distribution function* (*cdf*) since it aggregates the probability up to a certain value.

To relate to our previous example of daily logarithmic returns, the distribution function evaluated at say 0.05, that is $F(0.05)$, states the probability of some return of at most 5%.[3]

For values x approaching $-\infty$, F tends to zero, while for values x approaching ∞, F goes to 1. In between, F is *monotonically increasing and right-continuous*.[4] More concisely, we list these properties below:

Property 1. $F(x) \xrightarrow{x \to -\infty} 0$
Property 2. $F(x) \xrightarrow{x \to \infty} 1$
Property 3. $F(b) - F(a) \geq 0$ for $b \geq a$
Property 4. $\lim_{x \downarrow a} F(x) = F(a)$

The behavior in the extremes—that is when x goes to either $-\infty$ or ∞—is provided by properties 1 and 2, respectively. Property 3 states that F should be monotonically increasing (i.e., never become less for increasing values). Finally, property 4 guarantees that F is right-continuous.

Let us consider in detail the case when $F(x)$ is a continuous distribution, that is, the distribution has no jumps. The continuous probability distribution function F is associated with the probability measure P through the relationship[5]

$$F(a) = P((-\infty, a])$$

that is, that values up to a occur, and

$$F(b) - F(a) = P((a,b]) \qquad (10.1)$$

---

[2] Formally, an outcome $\omega \in \Omega$ is realized that lies inside of the interval $(-\infty,a]$.
[3] The distribution function F is also referred to as the *cumulative probability distribution function* (often abbreviated *cdf*) expressing that the probability is given for the accumulation of all outcomes less than or equal to a certain value.
[4] For the definition of monotonically increasing, see Appendix A.
[5] For the requirements on a probability measure on an uncountable space, in particular, see the definition of the probability measure in Chapter 8.

Continuous Probability Distributions

Therefore, from equation (10.1) we can see that the probability of some event related to an interval is given by the difference between the value of F at the upper bound b of the interval minus the value of F at the lower bound a. That is, the entire probability that an outcome of at most a occurs is subtracted from the greater event that an outcome of at most b occurs. Using set operations, we can express this as

$$(a,b] = (-\infty, b] \setminus (-\infty, a]$$

For example as we have seen, the event of a daily return of more than -5% and, at most, 5% is given by $(-0.05, 0.05]$. So, the probability associated with this event is given by $P((-0.05, 0.05]) = F(0.05) - F(-0.05)$.

In contrast to a discrete probability distribution, a continuous probability distribution always assigns zero probability to countable events such as individual outcomes a_i or unions thereof such as

$$\bigcup_{i=1}^{\infty} a_i$$

That is,

$$P(\{a_i\}) = 0, \text{ for all } a_i$$
$$P\left(\bigcup_{i=1}^{\infty} a_i\right) = 0 \quad (10.2)$$

From equation (10.2), we can apply the left-hand side of equation (10.1) also to events of the form (a,b) to obtain

$$P((a,b)) = F(b) - F(a) \quad (10.3)$$

Thus, it is irrelevant whether we state the probability of the daily logarithmic return being more than -5% and at most 5%, or the probability of the logarithmic return being more than -5% and less than 5%. They are the same because the probability of achieving a return of exactly 5% is zero. With a space Ω consisting of uncountably many possible values such as the set of real numbers, for example, each individual outcome is unlikely to occur. So, from a probabilistic point of view, one should never bet on an exact return or, associated with it, one particular stock price.

Since countable sets produce zero probability from a continuous probability measure, they belong to the so called **P-null sets**. All events associated with P-null sets are unlikely events.

So, how do we assign probabilities to events in a continuous environment? The answer is given by equation (10.3). That, however, presumes knowledge of the distribution function F. The next task is to define the continuous distribution function F more specifically as explained next.[6]

DENSITY FUNCTION

The continuous distribution function F of a probability measure P on $(\mathbb{R}, \mathbb{B})$ is defined as follows

$$F(x) = \int_{-\infty}^{x} f(t) \, dt \tag{10.4}$$

where $f(t)$ is the *density function* of the probability measure P.

We interpret equation (10.4) as follows. Since, at any real value x the distribution function uniquely equals the probability that an outcome of at most x is realized, that is, $F(x) = P((-\infty, x])$, equation (10.4) states that this probability is obtained by integrating some function f over the interval from $-\infty$ up to the value x.

What is the interpretation of this function f? The function f is the marginal rate of growth of the distribution function F at some point x. We know that with continuous distribution functions, the probability of exactly a value of x occurring is zero. However, the probability of observing a value inside of the interval between x and some very small step to the right Δx (i.e., $[x, x + \Delta x)$) is not necessarily zero. Between x and $x + \Delta x$, the distribution function F increases by exactly this probability; that is, the increment is

$$F(x + \Delta x) - F(x) = P(X \in [x, x + \Delta x)) \tag{10.5}$$

Now, if we divide $F(x + \Delta x) - F(x)$ from equation (10.5) by the width of the interval, Δx, we obtain the average probability or average increment of F per unit step on this interval. If we reduce the step size Δx to an infinitesimally small step ∂x, this average approaches the *marginal rate of growth* of F at x, which we denote f; that is,[7]

$$\lim_{\Delta x \to 0} \frac{F(x + \Delta x) - F(x)}{\Delta x} = \frac{\partial F(x)}{\partial x} \equiv f(x) \tag{10.6}$$

---

[6] The concept of integration is explained in Appendix A.
[7] The expression $\partial F(x)$ is equivalent to the increment $F(x + \Delta x) - F(x)$ as Δx goes to zero.

Continuous Probability Distributions

At this point, let us recall the histogram with relative frequency density for class data as explained in Chapter 4. Over each class, the height of the histogram is given by the density of the class divided by the width of the corresponding class. Equation (10.6) is somewhat similar if we think of it this way. We divide the probability that some realization should be inside of the small interval. And, by letting the interval shrink to width zero, we obtain the marginal rate of growth or, equivalently, the *derivative* of F.[8] Hence, we call f the **probability density function** or simply the **density function**. Commonly, it is abbreviated as *pdf*.

Now, when we refocus on equation (10.4), we see that the probability of some occurrence of at most x is given by integration of the density function f over the interval $(-\infty, x]$. Again, there is an analogy to the histogram. The relative frequency of some class is given by the density multiplied by the corresponding class width. With continuous probability distributions, at each value t, we multiply the corresponding density $f(t)$ by the infinitesimally small interval width dt. Finally, we integrate all values of f (weighted by dt) up to x to obtain the probability for $(-\infty, x]$. This, again, is similar to histograms: in order to obtain the cumulative relative frequency at some value x, we compute the area covered by the histogram up to value x.[9]

In Figure 10.1, we compare the histogram and the probability density function. The histogram with density h is indicated by the dotted lines while the density function f is given by the solid line. We can now see how the probability $P((-\infty, x^*])$ is derived through integrating the marginal rate f over the interval $(-\infty, x^*]$ with respect to the values t. The resulting total probability is then given by the area A_1 of the example in Figure 10.1.[10] This is analogous to class data where we would tally the areas of the rectangles whose upper bounds are less than x^* and the part of the area of the rectangle containing x^* up to the dash-dotted vertical line.

Requirements on the Density Function

Given the uncountable space $\mathbb{R}$ (i.e., the real numbers) and the corresponding set of events given by the Borel σ-algebra $\mathbb{B}$, we can give a more rigorous formal definition of the density function. The *density function f* of probability measure P on the measurable space $(\mathbb{R}, \mathbb{B})$ with distribution function F is a Borel-measurable function f satisfying

---

[8] We assume that F is continuous and that the derivative of F exists. In Appendix A, we explain the principles of derivatives.
[9] See Chapter 4 for an explanation of the relationship between histograms and cumulative relative frequency distributions.
[10] Here we can see that integration can be intuitively thought of as summation of infinitely many values $f(t)$ multiplied by the infinitesimally small interval widths dt.

FIGURE 10.1 Comparison of Histogram and Density Function

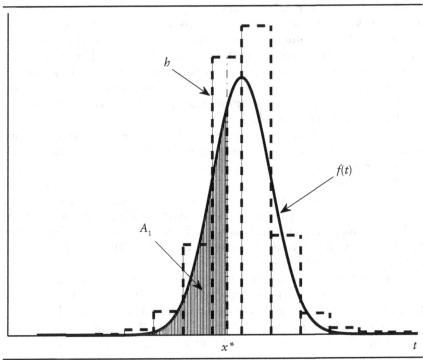

Area A_1 represents probability $P((-\infty, x^*])$ derived through integration of $f(t)$ with respect to t between $-\infty$ and x^*.

$$P((-\infty, x]) = F(x) = \int_{-\infty}^{x} f(t)\,dt \qquad (10.7)$$

with $f(t) \geq 0$, for all $t \in \mathbf{R}$ and

$$\int_{-\infty}^{\infty} f(t)\,dt = 1$$

Recall that by the requirement of Borel-measurability described in Chapter 8, we simply assure that the real-valued images generated by f have their origins in the Borel σ-algebra $\mathbb{B}$. Informally, for any value $y = f(t)$, we can trace the corresponding origin(s) t in $\mathbb{B}$ that is (are) mapped to y through the function f. Otherwise, we might incur problems computing the integral in equation (10.7) for reasons that are beyond the scope of this chapter.[11]

---
[11] The concept of an integral is discussed in Appendix A.

Continuous Probability Distributions

From definition of the density function given by equation (10.7), we see that it is reasonable that f be a function that exclusively assumes nonnegative values. Although we have not mentioned this so far, it is immediately intuitive since f is the marginal rate of growth of the continuous distribution function F. At each t, $f(t) \cdot dt$ represents the limit probability that a value inside of the interval $(t, t + dt]$ should occur, which can never be negative. Moreover, we require the integration of f over the entire domain from $-\infty$ to ∞ to yield 1, which is intuitively reasonable since this integral gives the probability that any real value occurs.

The requirement

$$\int_{-\infty}^{\infty} f(t)\,dt = 1$$

implies the graphical interpretation that the area enclosed between the graph of f over the entire interval $(-\infty, \infty)$ and the horizontal axis equals one. This is displayed in Figure 10.2 by the shaded area A. For example, to visualize graphically what is meant by

$$\int_{-\infty}^{x} f(t)\,dt$$

FIGURE 10.2 Graphical Interpretation of the Equality $A = \int_{-\infty}^{\infty} f(x)\,dx = 1$

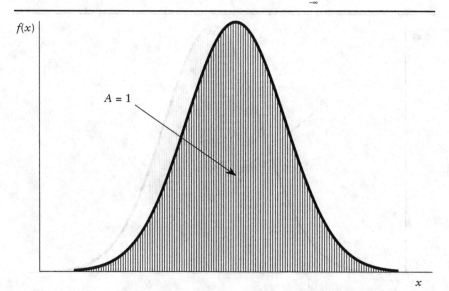

in equation (10.7), we can use Figure 10.1. Suppose the value x were located at the intersection of the vertical dash-dotted line and the horizontal axis (i.e., x^*). Then, the shaded area A_1 represents the value of the integral and, therefore, the probability of occurrence of a value of at most x. To interpret

$$\int_a^b f(t)\,dt$$

graphically, look at Figure 10.3. The area representing the value of the interval is indicated by A. So, the probability of some occurrence of at least a and at most b is given by A. Here again, the resemblance to the histogram becomes obvious in that we divide one area above some class, for example, by the total area, and this ratio equates the according relative frequency.

For the sake of completeness, it should be mentioned without indulging in the reasoning behind it that there are probability measures P on $(\mathbb{R}, \mathbb{B})$ even with continuous distribution functions that do not have density functions as defined in equation (10.7). But, in our context, we will only regard probability measures with continuous distribution functions with associated density functions so that the equalities of equation (10.7) are fulfilled.

FIGURE 10.3 Graphical Interpretation of $A = \int_a^b f(x)\,dx$

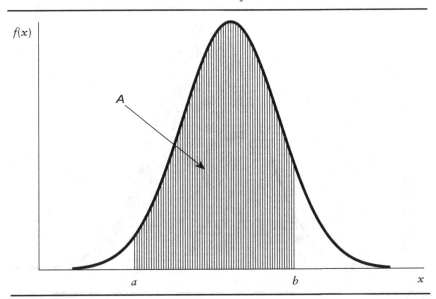

Sometimes, alternative representations equivalent to equation (10.7) are used. Typically, the following expressions are used

$$F(x) = \int_{\mathbb{R}} f(t) \cdot 1_{(-\infty, x]} \, dt \tag{10.8a}$$

$$F(x) = \int_{-\infty}^{\infty} f(t) \cdot 1_{(-\infty, x]} \, dt \tag{10.8b}$$

$$F(x) = \int_{(-\infty, x]} P(dt) \tag{10.8c}$$

$$F(x) = \int_{(-\infty, x]} dP(t) \tag{10.8d}$$

Note that in the first two equalities, (10.8a) and (10.8b), the indicator function $1_{(a,b]}$ is used. The indicator function is explained in Appendix A. The last two equalities, (10.8c) and (10.8d), can be used even if there is no density function and, therefore, are of a more general form. We will, however, predominantly apply the representation given by equation (10.7) and occasionally resort to the last two forms above.

We introduce the term *support* at this point to refer to the part of the real line where the density is truly positive, that is, all those x where $f(x) > 0$.

CONTINUOUS RANDOM VARIABLE

So far, we have only considered continuous probability distributions and densities. We yet have to introduce the quantity of greatest interest to us in this chapter, the *continuous random variable*. For example, stock returns, bond yields, and exchange rates are usually modeled as continuous random variables.

Informally stated, a continuous random variable assumes certain values governed by a probability law uniquely linked to a continuous distribution function F. Consequently, it has a density function associated with its distribution. Often, the random variable is merely described by its density function rather than the probability law or the distribution function.

By convention, let us indicate the random variables by capital letters. Recall from Chapter 8 that any random variable, and in particular a continuous random variable X, is a *measurable function*. Let us assume that X is a function from the probability space $\Omega = \mathbb{R}$ into the *state space* $\Omega' = \mathbb{R}$. That is, origin and image space coincide.[12] The corresponding σ-algebrae

---

[12] For a definition of these terms, see Chapter 8.

containing events of the elementary outcomes ω and the events in the image space $X(ω)$, respectively, are both given by the Borel σ-algebra $\mathbb{B}$. Now we can be more specific by requiring the continuous random variable X to be a $\mathbb{B} - \mathbb{B}$-measurable real-valued function. That implies, for example, that any event $X \in (a,b]$, which is in $\mathbb{B}$, has its origin $X^{-1}((a,b])$ in $\mathbb{B}$ as well. Measurability is important when we want to derive the probability of events in the state space such as $X \in (a,b]$ from original events in the probability space such as $X^{-1}((a,b])$. At this point, one should not be concerned that the theory is somewhat overwhelming. It will become easier to understand once we move to the examples.

COMPUTING PROBABILITIES FROM THE DENSITY FUNCTION

The relationship between the continuous random variable X and its density is given by the following.[13] Suppose X has density f, then the probability of some event $X \leq x$ or $X \in (a,b]$ is computed as

$$P(X \leq x) = \int_{-\infty}^{x} f(t)\,dt$$

$$P(X \in (a,b]) = \int_{a}^{b} f(t)\,dt \qquad (10.9)$$

which is equivalent to $F(x)$ and $F(b) - F(a)$ respectively, because of the one-to-one relationship between the density f and the distribution function F of X.

As explained earlier, using indicator functions, equations (10.9) could be alternatively written as

$$P(X \leq x) = \int_{-\infty}^{\infty} 1_{(-\infty,x]}(t) f(t)\,dt$$

$$P(X \in (a,b]) = \int_{-\infty}^{\infty} 1_{(a,b]}(t) f(t)\,dt$$

In the following, we will introduce parameters of location and spread such as the mean and the variance, for example.[14] In contrast to the data-dependent statistics, parameters of random variables never change. Some

---

[13] Sometimes the density of X is explicitly indexed f_X. We will not do so here, however, except where we believe by not doing so will lead to confusion. The same holds for its distribution function F.

[14] In Chapter 13, they will be treated more thoroughly.

probability distributions can be sufficiently described by their parameters. They are referred to as *parametric distributions*. For example, for the normal distribution we introduce shortly, it is sufficient to know the parameters mean and variance to completely determine the corresponding distribution function. That is, the shape of parametric distributions is governed only by the respective parameters.

LOCATION PARAMETERS

The most important location parameter is the *mean* that is also referred to as the *first moment*. It is the only location parameter presented in this chapter. Others will be introduced in Chapter 13.

Analogous to the discrete case, the mean can be thought of as an average value. It is the number that one would have to expect for some random variable X with given density function f. The mean is defined as follows: Let X be a real-valued random variable on the space $\Omega = \mathbb{R}$ with Borel σ-algebra $\mathbb{B}$. The *mean* is given by

$$E(X) = \int_{-\infty}^{\infty} x \cdot f(x) dx \qquad (10.10)$$

in case the integral on the right-hand side of equation (10.10) exists (i.e., is finite). Typically, the mean parameter is denoted as μ.

In equation (10.10) that defines the mean, we weight each possible value x that the random variable X might assume by the product of the density at this value, $f(x)$, and step size dx. Recall that the product $f(x) \cdot dx$ can be thought of as the limiting probability of attaining the value x. Finally, the mean is given as the integral over these weighted values. Thus, equation (10.10) is similarly understood as the definition of the mean of a discrete random variable where, instead of integrated, the probability-weighted values are summed.

DISPERSION PARAMETERS

We turn our focus towards measures of spread or, in other words, dispersion measures. Again, as with the previously introduced measures of location, in probability theory the dispersion measures are universally given parameters. Here, we introduce the moments of higher order, variance, standard deviation, and the skewness parameters.

Moments of Higher Order

It might sometimes be necessary to compute *moments of higher order*. As we already know from descriptive statistics, the mean is the moment of order one.[15] However, one might not be interested in the expected value of some quantity itself but of its square. If we treat this quantity as a continuous random variable, we compute what is the *second moment*.

Let X be a real-valued random variable on the space $\Omega = \mathbb{R}$ with Borel σ-algebra $\mathbb{B}$. The *moment of order k* is given by the expression

$$E\left(X^k\right) = \int_{-\infty}^{\infty} x^k \cdot f(x) \, dx \qquad (10.11)$$

in case the integral on the right-hand side of equation (10.11) exists (i.e., is finite).

From equation (10.11), we learn that higher-order moments are equivalent to simply computing the mean of X taken to the k-th power.

Variance

The variance involves computing the expected squared deviation from the mean $E(X) = \mu$ of some random variable X. For a continuous random variable X, the variance is defined as follows: Let X be a real-valued random variable on the space $\Omega = \mathbb{R}$ with Borel σ-algebra $\mathbb{B}$, then the *variance* is

$$Var(X) = \int_{-\infty}^{\infty} \left(x - E(X)\right)^2 \cdot f(x) \, dx = \int_{-\infty}^{\infty} (x - \mu)^2 \cdot f(x) \, dx \qquad (10.12)$$

in case the integral on the right-hand side of equation (10.12) exists (i.e., is finite). Often, the variance in equation (10.12) is denoted by the symbol σ^2.

In equation (10.12), at each value x, we square the deviation from the mean and weight it by the density at x times the step size dx. The latter product, again, can be viewed as the limiting probability of the random variable X assuming the value x. The square inflates large deviations even more compared to smaller ones. For some random variable to have a small variance, it is essential to have a quickly vanishing density in the parts where the deviations $(x - \mu)$ become large.

All distributions that we discuss in this chapter and the two that follow are parametric distributions. For some of them, it is enough to know the mean μ and variance σ^2 and consequently, we will resort to these two

---

[15]Alternatively, we often say the *first* moment. For the higher orders k, we consequently might refer to the *k-th* moment.

FIGURE 10.4 Two Density Functions Yielding Common Means, $\mu_1 = \mu_2$, but Different Variances, $\sigma_1^2 < \sigma_2^2$

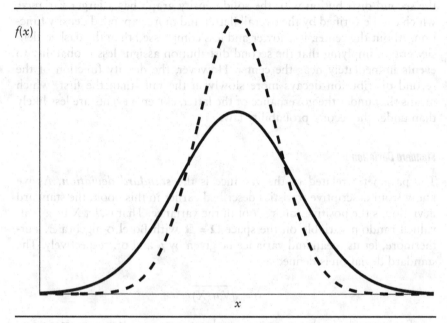

Note: Dashed graph: $\sigma_1^2 = 1$. Solid graph: $\sigma_2^2 = 1.5$.

parameters often. Historically, the variance has often been given the role of risk measure in context of portfolio theory. Suppose we have two random variables R_1 and R_2 representing the returns of two stocks, S_1 and S_2, with equal means μ_{R_1} and μ_{R_2}, respectively, so that $\mu_{R_1} = \mu_{R_2}$. Moreover, let R_1 and R_2 have variances $\sigma_{R_1}^2$ and $\sigma_{R_2}^2$, respectively, with $\sigma_{R_1}^2 < \sigma_{R_2}^2$. Then, omitting further theory, at this moment, we prefer S_1 to S_2 because of the S_1's smaller variance. We demonstrate this in Figure 10.4. The dashed line represents the graph of the first density function while the second one is depicted by the solid line. Both density functions yield the same mean (i.e., $\mu_1 = \mu_2$). However, the variance from the first density function, given by the dashed graph, is smaller than that of the solid graph (i.e., $\sigma_1^2 < \sigma_2^2$). Thus, using variance as the risk measure and resorting to density functions that can be sufficiently described by the mean and variance, we can state that density function for S_1 (dashed graph) is preferable. We can interpret the figure as follows.

Since the variance of the distribution with the dashed density graph is smaller, the probability mass is less dispersed over all x values. Hence, the density is more condensed about the center and more quickly vanishing in the extreme left and right ends, the so-called *tails*. On the other hand, the second distribution with the solid density graph has a larger variance, which can be verified by the overall flatter and more expanded density function. About the center, it is lower and less compressed than the dashed density graph, implying that the second distribution assigns less probability to events immediately near the center. However, the density function of the second distribution decays more slowly in the tails than the first, which means that under the governance of the latter, extreme events are less likely than under the second probability law.

Standard Deviation

The parameter related to the variance is the *standard deviation*. As we know from descriptive statistics described earlier in this book, the standard deviation is the positive square root of the variance. That is, let X be a real-valued random variable on the space $\Omega = \mathbb{R}$ with Borel σ-algebra $\mathbb{B}$. Furthermore, let its mean and variance be given by μ and σ^2, respectively. The standard deviation is defined as

$$\sigma = \sqrt{Var(X)}$$

For example, in the context of stock returns, one often expresses using the standard deviation the return's fluctuation around its mean. The standard deviation is often more appealing than the variance since the latter uses squares, which are a different scale from the original values of X. Even though mathematically not quite correct, the standard deviation, denoted by σ, is commonly interpreted as the average deviation from the mean.

Skewness

Consider the density function portrayed in Figure 10.5. The figure is obviously symmetric about some location parameter μ in the sense that $f(-x - \mu) = f(x - \mu)$.[16] Suppose instead that we encounter a density function f of some random variable X that is depicted in Figure 10.6. This figure is not symmetric about any location parameter. Consequently, some quantity stating the extent to which the density function is deviating from symmetry is needed. This is accomplished by a parameter referred to as *skewness*. This parameter measures the degree to which the density function leans to either one side, if at all.

---

[16] We will go further into detail on location parameters in Chapter 13.

Continuous Probability Distributions

FIGURE 10.5 Example of Some Symmetric Density Function $f(x)$

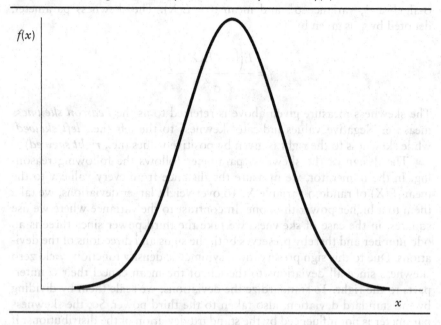

FIGURE 10.6 Example of Some Asymmetric Density Function $f(x)$

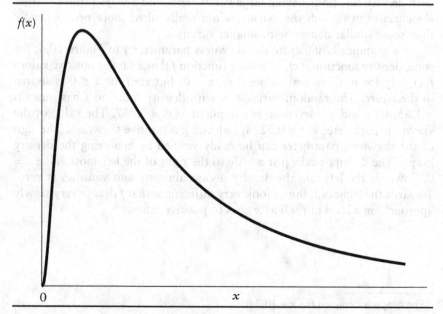

Let X be a real-valued random variable on the space $\Omega = \mathbb{R}$ with Borel σ-algebra $\mathbb{B}$, variance σ^2, and mean $\mu = E(X)$. The skewness parameter, denoted by γ, is given by

$$\gamma = \frac{E\left((x - E(X))^3\right)}{\sigma^{3/2}}$$

The skewness measure given above is referred to as the *Pearson skewness* measure. Negative values indicate skewness to the left (i.e., *left skewed*) while skewness to the right is given by positive values (i.e., *right skewed*).

The design of the skewness parameter follows the following reasoning. In the numerator, we measure the distance from every value x to the mean $E(X)$ of random variable X. To overweight larger deviations, we take them to a higher power than one. In contrast to the variance where we use squares, in the case of skewness we take the third power since three is an odd number and thereby preserves both the signs and directions of the deviations. Due to this sign preservation, symmetric density functions yield zero skewness since all deviations to the left of the mean cancel their counterparts to the right. To standardize the deviations, we scale them by dividing by the standard deviation, also taken to the third power. So, the skewness parameter is not influenced by the standard deviation of the distributions. If we did not scale the skewness parameter in this way, distribution functions with density functions having large variances would always produce larger skewness even though the density is not really tilted more pronouncedly than some similar density with smaller variance.

We graphically illustrate the skewness parameter γ in Figure 10.6, for some density function $f(x)$. A density function f that assumes positive values $f(x)$ only for positive real values (i.e., $x > 0$) but zero for $x \leq 0$ is shown in the figure. The random variable X with density function f has mean $\mu = 1.65$. Its standard deviation is computed as $\sigma = 0.957$. The value of the skewness parameter is $\gamma = 0.7224$, indicating a positive skewness. The sign of the skewness parameter can be easily verified by analyzing the density graph. The density peaks just a little to the right of the left-most value $x = 0$. Towards the left tail, the density decays abruptly and vanishes at zero. Towards the right tail, things look very different in that f decays very slowly approaching a level of $f = 0$ as x goes to positive infinity.[17]

---

[17] The graph is depicted for $x \in [0, 3.3]$.

CONCEPTS EXPLAINED IN THIS CHAPTER (IN ORDER OF PRESENTATION)

Continuous intervals
Continuous probability distributions
Measurable space
Continuous distribution function
Cumulative distribution function (cdf)
Monotonically
Monotonically increasing and right continuous
P-null sets
Density function
Marginal rate of growth
Derivative of F
Probability density function or density function (pdf)
Density function of f
Support
Continuous random variable
Measurable function
State space
Parametric distributions
Mean
First moment
Moments of higher order
Second moment
Moment of order k
Variance
Tails
Standard deviation
Skewness
Pearson skewness
Left skewed
Right skewed

CHAPTER 11

Continuous Probability Distributions with Appealing Statistical Properties

In the preceding chapter, we introduced the concept of continuous probability distributions. In this chapter, we discuss the more commonly used distributions with appealing statistical properties that are used in finance. The distributions discussed are the normal distribution, the chi-square distribution, the Student's *t*-distribution, the Fisher's *F*-distribution, the exponential distribution, the gamma distribution (including the special Erlang distribution), the beta distribution, and the log-normal distribution. Many of the distributions enjoy widespread attention in finance, or statistical applications in general, due to their well-known characteristics or mathematical simplicity. However, as we emphasize, the use of some of them might be ill-suited to replicate the real-world behavior of financial returns.

NORMAL DISTRIBUTION

The first distribution we discuss is the *normal distribution*. It is the distribution most commonly used in finance despite its many limitations. This distribution, also referred to as the *Gaussian distribution* (named after the mathematician and physicist C. F. Gauss), is characterized by the two parameters: mean (μ) and standard deviation (σ). The distribution is denoted by $N(\mu,\sigma^2)$. When $\mu = 0$ and $\sigma^2 = 1$, then we obtain the *standard normal distribution*.

For $x \in R$, the density function for the normal distribution is given by

$$f(x) = \frac{1}{\sqrt{2\pi}\sigma} \cdot e^{-\frac{(x-\mu)^2}{2\sigma^2}} \qquad (11.1)$$

The density in equation (11.1) is always positive. Hence, we have support (i.e., positive density) on the entire real line. Furthermore, the density

FIGURE 11.1 Normal Density Function for Various Parameter Values

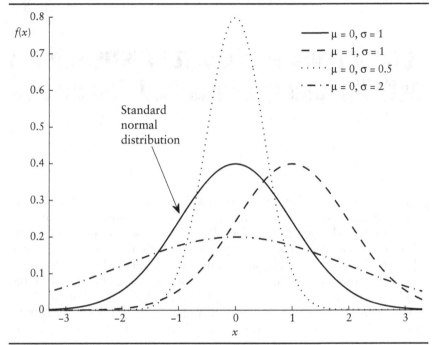

function is symmetric about μ. A plot of the density function for several parameter values is given in Figure 11.1. As can be seen, the value of μ results in a horizontal shift from 0 while σ inflates or deflates the graph. A characteristic of the normal distribution is that the densities are bell shaped.

A problem is that the distribution function cannot be solved for analytically and therefore has to be approximated numerically. In the particular case of the standard normal distribution, the values are tabulated. Standard statistical software provides the values for the standard normal distribution as well as most of the distributions presented in this chapter. The standard normal distribution is commonly denoted by the Greek letter Φ such that we have $\Phi(x) = F(x) = P(X \leq x)$, for some standard normal random variable X. In Figure 11.2, graphs of the distribution function are given for three different sets of parameters.

Properties of the Normal Distribution

The normal distribution provides one of the most important classes of probability distributions due to two appealing properties that generally are not shared by all distributions:

FIGURE 11.2 Normal Distribution Function for Various Parameter Values

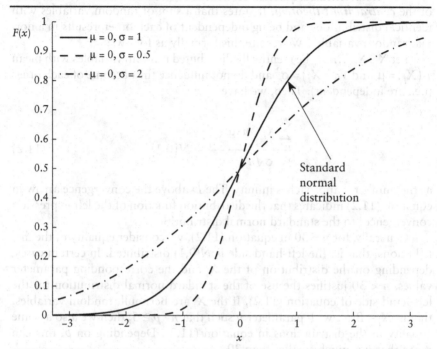

Property 1. The distribution is *location-scale invariant*.
Property 2. The distribution is *stable under summation*.

Property 1, the location-scale invariance property, guarantees that we may multiply X by b and add a where a and b are any real numbers. Then, the resulting $a + b \cdot X$ is, again, normally distributed, more precisely, $N(a+\mu, b\sigma)$. Consequently, a normal random variable will still be normally distributed if we change the units of measurement. The change into $a + b \cdot X$ can be interpreted as observing the same X, however, measured in a different scale. In particular, if a and b are such that the mean and variance of the resulting $a + b \cdot X$ are 0 and 1, respectively, then $a + b \cdot X$ is called the *standardization of X*.

Property 2, stability under summation, ensures that the sum of an arbitrary number n of normal random variables, $X_1, X_2, ..., X_n$ is, again, normally distributed provided that the random variables behave independently of each other. This is important for aggregating quantities.

These properties are illustrated later in the chapter.

Furthermore, the normal distribution is often mentioned in the context of the *central limit theorem*. It states that a sum of random variables with identical distributions and being independent of each other, results in a normal random variable.[1] We restate this formally as follows:

Let $X_1, X_2, ..., X_n$ be identically distributed random variables with mean $E(X_i) = \mu$ and $Var(X_i) = \sigma^2$ and do not influence the outcome of each other (i.e., are independent). Then, we have

$$\frac{\sum_{i=1}^{n} X_i - n \cdot \mu}{\sigma\sqrt{n}} \xrightarrow{D} N(0,1) \qquad (11.2)$$

as the number n approaches infinity. The D above the convergence arrow in equation (11.2) indicates that the distribution function of the left expression convergences to the standard normal distribution.

Generally, for $n = 30$ in equation (11.2), we consider equality of the distributions; that is, the left-hand side is $N(0,1)$ distributed. In certain cases, depending on the distribution of the X_i and the corresponding parameter values, $n < 30$ justifies the use of the standard normal distribution for the left-hand side of equation (11.2). If the X_i are Bernoulli random variables, that is, $X_i \sim B(p)$, with parameter p such that $n \cdot p \geq 5$, then we also assume equality in the distributions in equation (11.2). Depending on p, this can mean that n is much smaller than 30.

These properties make the normal distribution the most popular distribution in finance. But this popularity is somewhat contentious, however, for reasons that will be given as we progress in this and the following chapters.

The last property we will discuss of the normal distribution that is shared with some other distributions is the bell shape of the density function. This particular shape helps in roughly assessing the dispersion of the distribution due to a rule of thumb commonly referred to as the *empirical rule*. Due to this rule, we have

$$P(X \in [\mu \pm \sigma]) = F(\mu + \sigma) - F(\mu - \sigma) \approx 68\%$$

$$P(X \in [\mu \pm 2\sigma]) = F(\mu + 2\sigma) - F(\mu - 2\sigma) \approx 95\%$$

$$P(X \in [\mu \pm 3\sigma]) = F(\mu + 3\sigma) - F(\mu - 3\sigma) \approx 100\%$$

The above states that approximately 68% of the probability is given to values that lie in an interval one standard deviation σ about the mean

---

[1] There exist generalizations such that the distributions need no longer be identical. However, this is beyond the scope of this chapter.

Continuous Probability Distributions with Appealing Statistical Properties   251

μ. About 95% probability is given to values within 2σ to the mean, while nearly all probability is assigned to values within 3σ from the mean.

By comparison, the so called *Chebychev inequalities* valid for any type of distribution—so not necessarily bell-shaped—yield

$$P(X \in [\mu \pm \sigma]) \approx 0\%$$

$$P(X \in [\mu \pm 2\sigma]) \approx 75\%$$

$$P(X \in [\mu \pm 3\sigma]) \approx 89\%$$

which provides a much coarser assessment than the empirical rule as we can see, for example, by the assessed 0% of data contained inside of one standard deviation about the mean.

Applications to Stock Returns

Applying Properties 1 and 2 to Stock Returns

With respect to Property 1, consider an example of normally distributed stock returns r with mean μ. If μ is nonzero, this means that the returns are a combination of a constant μ and random behavior centered about zero. If we were only interested in the latter, we would subtract μ from the returns and thereby obtain a new random variable $\tilde{r} = r - \mu$, which is again normally distributed.

With respect to Property 2, we give two examples. First, let us present the effect of aggregation over time. We consider daily stock returns that, by our assumption, follow a normal law. By adding the returns from each trading day during a particular week, we obtain the week's return as r_w = $r_{Mo} + r_{Tu} + \ldots + r_{Fr}$ where $r_{Mo}, r_{Tu}, \ldots r_{Fr}$ are the returns from Monday through Friday. The weekly return r_w is normally distributed as well. The second example applies to portfolio returns. Consider a portfolio consisting of n different stocks, each with normally distributed returns. We denote the corresponding returns by R_1 through R_n. Furthermore, in the portfolio we weight each stock i with w_i, for $i = 1, 2, \ldots, n$. The resulting portfolio return $R_p = w_1 R_1 + w_2 R_2 + \ldots + w_n R_n$ is also a normal random variable.

Using the Normal Distribution to Approximate the Binomial Distribution

Again we consider the binomial stock price model from Chapter 9. At time t = 0, the stock price was S_0 = \$20. At time $t = 1$, the stock price was either up or down by 10% so that the resulting price was either S_0 = \$18 or S_0 = \$22.

Both up- and down-movement occurred with probability $P(\$18) = P(\$22) = 0.5$. Now we extend the model to an arbitrary number of n days. Suppose each day i, $i = 1, 2, \ldots, n$, the stock price developed in the same manner as on the first day. That is, the price is either up 10% with 50% probability or down 10% with the same probability. If on day i the price is up, we denote this by $X_i = 1$ and $X_i = 0$ if the price is down. The X_i are, hence, $B(0.5)$ random variables. After, say, 50 days, we have a total of $Y = X_1 + X_2 + \ldots + X_{50}$ up movements. Note that because of the assumed independence of the X_i, that Y is a $B(50, 0.5)$ random variable with mean $n \cdot p = 25$ and variance $n \cdot p \cdot (1-p) = 12.5$. Let us introduce

$$Z_{50} = \frac{Y - 25}{\sqrt{12.5}}$$

From the comments regarding equation (11.2), we can assume that Z_{50} is approximately $N(25, 12.5)$ distributed. So, the probability of at most 15 up-movements, for example, is given by $P(Y \leq 15) = \Phi((15-25)/\sqrt{12.5}) = 0.23\%$. By comparison, the probability of no more than five up-movements is equal to $P(Y \leq 5) = \Phi((5-25)/\sqrt{12.5}) \approx 0\%$.

Normal Distribution for Logarithmic Returns

As another example, let X be some random variable representing a quantitative daily market dynamic such as new information about the economy. A dynamic can be understood as some driving force governing the development of other variables. We assume that it is normally distributed with mean $E(X) = \mu = 0$ and variance $Var(X) = \sigma^2 = 0.2$. Formally, we would write $X \sim N(0, 0.2)$. So, on average, the value of the daily dynamic will be zero with a standard deviation of $\sqrt{0.2}$. In addition, we introduce a stock price S as a random variable, which is equal to S_0 at the beginning.

After one day, the stock price is modeled to depend on the dynamic X as follows

$$S_1 = S_0 \cdot e^X$$

where S_1 is the stock price after one day. The exponent X in this presentation is referred to as a *logarithmic return* in contrast to a *multiplicative return* R obtained from the formula $R = S_1/S_0 - 1$. So, for example, if $X = 0.01$, S_1 is equal to $e^{0.01} \cdot S_0$. That is almost equal to $1.01 \cdot S_0$, which corresponds to an increase of 1% relative to S_0.[2] The probability of X being, for instance, no greater than 0.01 after one day is given by[3]

$$P(X \leq 0.01) = \int_{-\infty}^{0.01} f(x)dx = \int_{-\infty}^{0.01} \frac{1}{\sqrt{2\pi}\sqrt{0.2}} e^{-\frac{x^2}{2 \cdot 0.2}} dx \approx 0.51$$

Consequently, after one day, the stock price increases, at most, by 1% with 51% probability, that is, $P(S_1 \leq 1.01 \cdot S_0) \approx 0.51$.

Next, suppose we are interested in a five-day outlook where the daily dynamics X_i, $i = 1, 2, \ldots, 5$ of each of the following consecutive five days are distributed identically as X and independent of each other. Since the dynamic is modeled to equal exactly the continuously compounded return—that is logarithmic returns—we refer to X as the return in this chapter. For the resulting five-day returns, we introduce the random variable $Y = X_1 + X_2 + \ldots + X_5$ as the linear combination of the five individual daily returns. We know that Y is normally distributed from Property 2. More precisely, $Y \sim N(0,1)$. So, on average, the return tends in neither direction, but the volatility measured by the standard deviation is now $\sqrt{5} \approx 2.24$ times that of the daily return X. Consequently, the probability of Y not exceeding a value of 0.01 is now,

$$P(Y \leq 0.01) = \int_{-\infty}^{0.01} \frac{1}{\sqrt{2\pi}\sqrt{1}} e^{-\frac{y^2}{2 \cdot 1}} dy \approx 0.50$$

We see that the fivefold variance results in a greater likelihood to exceed the threshold 0.01, that is,

$$P(Y > 0.01) = 1 - P(Y \leq 0.01) \approx 0.50 > 0.49 \approx P(X > 0.01)$$

We model the stock price after five days as

$$S_5 = S_0 \cdot e^Y = S_0 \cdot e^{X_1 + X_2 + \ldots + X_5}$$

So, after five days, the probability for the stock price to have increased by no more than 1% relative to S_0 is equal to

$$P(S_5 \leq e^{0.01} \cdot S_0) = P(S_5 \leq 1.01 \cdot S_0) \approx 0.50$$

---

[2] For values near 0, the logarithmic return X is virtually equal to the multiplicative return R. Rounding to two decimals, they are both equal to 0.01 here.
[3] For some computer software, the probability will be given as 0.5 due to rounding.

There are two reasons why in finance logarithmic returns are commonly used. First, logarithmic returns are often easier to handle than multiplicative returns. Second, if we consider returns that are attributed to ever shorter periods of time (e.g., from yearly to monthly to weekly to daily and so on), the resulting compounded return after some fixed amount of time can be expressed as a logarithmic return. The theory behind this can be obtained from any introductory book on calculus.

CHI-SQUARE DISTRIBUTION

Our next distribution is the *chi-square distribution*. Let Z be a standard normal random variable, in brief $Z \sim N(0,1)$, and let $X = Z^2$. Then X is distributed chi-square with one degree of freedom. We denote this as $X \sim \chi^2(1)$. The *degrees of freedom* indicate how many independently behaving standard normal random variables the resulting variable is composed of. Here X is just composed of one, namely Z, and therefore has one degree of freedom.

Because Z is squared, the chi-square distributed random variable assumes only nonnegative values; that is, the support is on the nonnegative real numbers. It has mean $E(X) = 1$ and variance $Var(X) = 2$.

In general, the chi-square distribution is characterized by the degrees of freedom n, which assume the values 1, 2, ... Let $X_1, X_2, ..., X_n$ be n $\chi^2(1)$ distributed random variables that are all independent of each other. Then their sum, S, is

$$S = \sum_{i=1}^{n} X_i \sim \chi^2(n) \qquad (11.3)$$

In words, the sum is again distributed chi-square but this time with n degrees of freedom. The corresponding mean is $E(X) = n$, and the variance equals $Var(X) = 2 \cdot n$. So, the mean and variance are directly related to the degrees of freedom.

From the relationship in equation (11.3), we see that the degrees of freedom equal the number of independent $\chi^2(1)$ distributed X_i in the sum. If we have $X_1 \sim \chi^2(n_1)$ and $X_2 \sim \chi^2(n_2)$, it follows that

$$X_1 + X_2 \sim \chi^2(n_1 + n_2) \qquad (11.4)$$

From property (11.4), we have that chi-square distributions have Property 2; that is, they are stable under summation in the sense that the sum

FIGURE 11.3 Density Functions of Chi-Square Distributions for Various Degrees of Freedom n

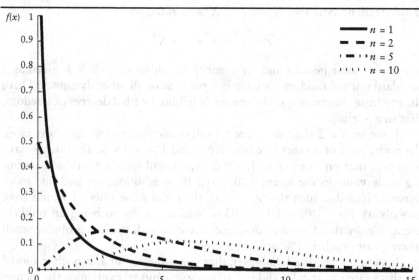

of any two chi-squared distributed random variables is itself chi-square distributed.

The chi-square density function with n degrees of freedom is given by

$$f(x) = \begin{cases} f(x) = \dfrac{1}{2^{n/2}\Gamma(n/2)} \cdot e^{-x/2} \cdot x^{n/2-1}, & x \geq 0 \\ 0 & x < 0 \end{cases}$$

for $n = 1, 2, \ldots$ where $\Gamma(\cdot)$ is the *gamma function.* The gamma function is explained in Appendix A. Figure 11.3 shows a few examples of the chi-square density function with varying degrees of freedom. As can be observed, the chi-square distribution is skewed to the right.

We will revisit this distribution in the context of estimation of the variance parameter σ^2 in Part Three of this book where we cover inductive statistics for estimating parameters and hypothesis testing.

Application to Modeling Short-Term Interest Rates

As an example of an application of the chi-square distribution, we present a simplified model of short-term interest rates, that is, so-called *short rates*.

The short rate given by r_t, at any time t, is assumed to be a nonnegative continuous random variable. Furthermore, we let the short rate be composed of d independent dynamics $X_1, X_2, \ldots, X_d$ according to

$$r_t = X_1^2 + X_2^2 + \ldots + X_d^2$$

where d is some positive integer number. In addition, each X_i is given as a standard normal random variable independent of all other dynamics. Then, the resulting short rate r_t is chi-square distributed with d degrees of freedom, that is, $r_t \sim \chi^2(d)$.

If we let $d = 2$ (i.e., there are two dynamics governing the short rate), the probability of a short rate between 0 and 1% is 0.5%. That is, we have to expect that on five out of 1,000 days, we will have a short rate assuming some value in the interval (0,0.01]. If, in addition, we had one more dynamic included such that $r_t \sim \chi^2(3)$, then, the same interval would have probability $P(r_t \in (0,0.01]) \approx 0.03\%$, which is close to being an unlikely event. We see that the more dynamics are involved, the less probable small interest rates such as 1% or less become.

It should be realized, however, that this is merely an approach to model the short rate statistically and not an economic model explaining the factors driving the short rate.

STUDENT'S *t*-DISTRIBUTION

An important continuous probability distribution when the population variance of a distribution is unknown is the *Student's t-distribution* (also referred to as the *t-distribution* and *Student's distribution*.[4]

The *t*-distribution is a mixture of the normal and chi-square distributions. To derive the distribution, let X be distributed standard normal, that is, $X \sim N(0,1)$, and S be chi-square distributed with n degrees of freedom, that is, $S \sim \chi^2(n)$. Furthermore, if X and Y are independent of each other (which is to be understood as not influencing the outcome of the other), then

$$Z = \frac{X}{\sqrt{S/n}} \sim t(n) \qquad (11.5)$$

In words, equation (11.5) states that the resulting random variable Z is Student's *t*-distributed with n degrees of freedom. The degrees of freedom are inherited from the chi-square distribution of S.

---

[4]This distribution was derived by William Gosset in 1908 who wrote under the pseudonym *Student* because his employer, Guinness Brewery in Dublin, did not allow him to publish under his own name.

How can we interpret equation (11.5)? Suppose we have a population of normally distributed values with zero mean. The corresponding normal random variable may be denoted as X. If one also knows the standard deviation of X,

$$\sigma = \sqrt{Var(X)}$$

with X/σ, we obtain a standard normal random variable.

However, if σ is not known, we have to use, for example,

$$\sqrt{S/n} = \sqrt{1/n \cdot (X_1^2 + X_2^2 + \ldots + X_n^2)}$$

instead where $X_1^2, X_2^2, \ldots, X_n^2$ are n random variables identically distributed as X. Moreover, $X_1, X_2, \ldots, X_n$ have to assume values independently of each other. Then, the distribution of

$$X/\sqrt{S/n}$$

is the t-distribution with n degrees of freedom, that is,

$$X/\sqrt{S/n} \sim t(n)$$

By dividing by σ or S/n, we generate rescaled random variables that follow a standardized distribution. Quantities similar to $X/\sqrt{S/n}$ play an important role in parameter estimation. (We encounter them in Chapters 17 through 19.)

The density function is defined as

$$f(x) = \frac{1}{\sqrt{n \cdot \pi}} \cdot \frac{\Gamma\left(\frac{n+1}{2}\right)}{\Gamma\left(\frac{n}{2}\right)} \cdot \left(1 + \frac{x^2}{n}\right)^{-\frac{n+1}{2}} \qquad (11.6)$$

where the gamma function Γ is incorporated again. The density function is symmetric and has support (i.e., is positive) on all R.

Basically, the Student's t-distribution has a similar shape to the normal distribution, but thicker tails. For large degrees of freedom n, the Student's t-distribution does not significantly differ from the standard normal distribution. As a matter of fact, for $n \geq 100$, it is practically indistinguishable from $N(0,1)$.

Figure 11.4 shows the Student's t-density function for various degrees of freedom plotted against the standard normal density function. The same is done for the distribution function in Figure 11.5.

FIGURE 11.4 Density Function of the *t*-Distribution for Various Degrees of Freedom *n* Compared to the Standard Normal Density Function ($N(0,1)$)

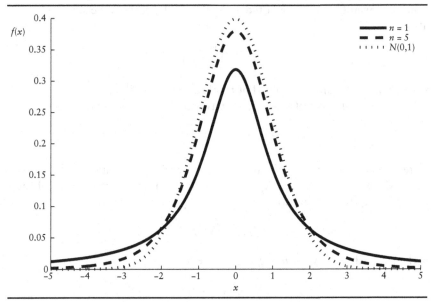

FIGURE 11.5 Distribution Function of the *t*-Distribution for Various Degrees of Freedom *n* Compared to the Standard Normal Density Function ($N(0,1)$)

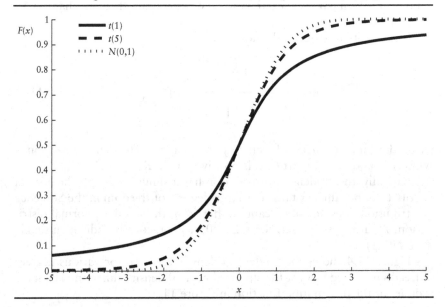

In general, the lower the degrees of freedom, the heavier the tails of the distribution, making extreme outcomes much more likely than for greater degrees of freedom or, in the limit, the normal distribution. This can be seen by the distribution function that we depicted in Figure 11.5 for $n = 1$ and $n = 5$ against the standard normal cumulative distribution function (cdf). For lower degrees of freedom such as $n = 1$, the solid curve starts to rise earlier and approach 1 later than for higher degrees of freedom such as $n = 5$ or the $N(0,1)$ case.

This can be understood as follows. When we rescale X by dividing by $\sqrt{S/n}$ as in equation (11.5), the resulting $X/\sqrt{S/n}$ obviously inherits randomness from both X and S. Now, when S is composed of few X_i, only, say $n = 3$, such that $X/\sqrt{S/n}$ has three degrees of freedom, there is a lot of dispersion from S relative to the standard normal distribution. By including more independent $N(0,1)$ random variables X_i such that the degrees of freedom increase, S becomes less dispersed. Thus, much uncertainty relative to the standard normal distribution stemming from the denominator in $X/\sqrt{S/n}$ vanishes. The share of randomness in $X/\sqrt{S/n}$ originating from X alone prevails such that the normal characteristics preponderate. Finally, as n goes to infinity, we have something that is nearly standard normally distributed.

The mean of the Student's t random variable is zero, that is $E(X) = 0$, while the variance is a function of the degrees of freedom n as follows

$$\sigma^2 = Var(X) = \frac{n}{n-2}$$

For $n = 1$ and 2, there is no finite variance. Distributions with such small degrees of freedom generate extreme movements quite frequently relative to higher degrees of freedom. Precisely for this reason, stock price returns are often found to be modeled quite well using distributions with small degrees of freedom, or alternatively, large variances.

Application to Stock Returns

Let us resume the example at the end of the presentation of the normal distribution. We consider, once again, the 5-day return Y with standard normal distribution. Suppose that now we do not know the variance. For this reason, at any point in time t, we rescale the observations of Y by

$$\sqrt{\frac{1}{5} \cdot \left(Y_{-1}^2 + Y_{-2}^2 + \ldots + Y_{-5}^2\right)}$$

where the $Y_{-1}^2, Y_{-2}^2, \ldots, Y_{-5}^2$ are the five independent weekly returns immediately prior to Y. The resulting rescaled weekly returns

$$Z = \frac{Y}{\sqrt{Y_{-1}^2 + Y_{-2}^2 + \ldots + Y_{-5}^2}}$$

then are $t(5)$ distributed. The probability of Y not exceeding a value of 0.01 is now

$$P(Y \leq 0.01) = F(0.01) = 0.5083$$

where F is the cumulative distribution function of the Student's t-distribution with five degrees of freedom. Under the $N(0,1)$, this probability was about the same.

Again, we model the stock price after five days as $S_5 = S_0 \cdot e^Y$ where S_0 is today's price. As we know, when $Y \leq 0.01$, then $S_5 \leq S_0 \cdot e^{0.01} = S_0 \cdot 1.01$. Again, it follows that the stock price increases by at most 1% with probability of about 0.51. So far there is not much difference here between the standard normal and the $t(5)$ distribution.

Let's analyze the stock of American International Group (AIG), in September 2008. During one week, that is, five trading days, the stock lost about 67% of its value. That corresponds to a value of the 5-day return of $Y = -1.0986$ because of $e^Y = e^{-1.0986} = 0.3333 = 1 - 0.6667$. In the $N(0,1)$ model, a decline of this magnitude or even worse would occur with probability

$$P(Y \leq -1.0986) = \Phi(-1.0986) = 13.6\%$$

while under the $t(5)$ assumption, we would obtain

$$P(Y \leq -1.0986) = F(-1.0986) = 16.1\%$$

This is 2.5% more likely in the $t(5)$ model. So, stock price returns exhibiting extreme movements such as that of the AIG stock price should not be modeled using the normal distribution.

F-DISTRIBUTION

Our next distribution is the **F-distribution**. It is defined as follows. Let $X \sim \chi^2(n_1)$ and $Y \sim \chi^2(n_2)$.

Furthermore, assuming X and Y to be independent, then the ratio

$$F(n_1, n_2) = \frac{X/n_1}{Y/n_2} \tag{11.7}$$

FIGURE 11.6 Density Function of the F-Distribution for Various Degrees of Freedom n_1 and n_2

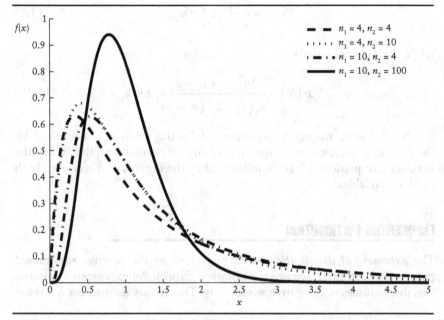

has an F-distribution with n_1 and n_2 degrees of freedom inherited from the underlying chi-square distributions of X and Y, respectively. We see that the random variable in equation (11.7) assumes nonnegative values only because neither X nor Y are ever negative. Hence, the support is on the nonnegative real numbers. Also like the chi-square distribution, the F-distribution is skewed to the right.

The F-distribution has a rather complicated looking density function of the form

$$f(x) = \begin{cases} \dfrac{F\left(\dfrac{n_1+n_2}{2}\right)}{F\left(\dfrac{n_1}{2}\right)F\left(\dfrac{n_2}{2}\right)} \cdot \left(\dfrac{n_1}{n_2}\right)^{n_1/2} \cdot \dfrac{x^{n_1/2-1}}{\left[1+x\cdot\dfrac{n_1}{2}\right]^{\frac{n_1+n_2}{2}}}, & x \geq 0 \\ 0 & x < 0 \end{cases} \qquad (11.8)$$

Figure 11.6 displays the density function (11.8) for various degrees of freedom. As the degrees of freedom n_1 and n_2 increase, the function graph becomes more peaked and less asymmetric while the tails lose mass.

The mean is given by

$$E(X) = \frac{n_2}{n_2 - 2}, \text{ for } n_2 > 2 \qquad (11.9)$$

while the variance equals

$$\sigma^2 = Var(X) = \frac{2n_2^2(n_1 + n_2 - 2)}{n_1(n_2 - 2)^2(n_2 - 4)}, \text{ for } n_2 > 4 \qquad (11.10)$$

Note that according to equation (11.9), the mean is not affected by the degrees of freedom n_1 of the first chi-square random variable, while the variance in equation (11.10) is influenced by the degrees of freedom of both random variables.

EXPONENTIAL DISTRIBUTION

The *exponential distribution* is characterized by the positive real-valued parameter λ. In brief, we use the notation Exp(λ). An exponential random variable assumes nonnegative values only. The density defined for $\lambda > 0$ by

$$f(x) = \begin{cases} \lambda \cdot e^{-\lambda x}, & x \geq 0 \\ 0 & x < 0 \end{cases}$$

is right skewed. Figure 11.7 presents the density function for various parameter values λ.

The distribution function is obtained by simple integration as

$$F(x) = 1 - e^{-\lambda x}$$

For identical parameter values as in Figure 11.7, we have plots of the exponential distribution function shown in Figure 11.8.

For this distribution, both the mean and variance are relatively simple functions of the parameter. That is, for the mean

$$E(X) = \frac{1}{\lambda}$$

and for the variance

$$Var(X) = \frac{1}{\lambda^2}$$

Continuous Probability Distributions with Appealing Statistical Properties

FIGURE 11.7 Exponential Density Function for Various Parameter Values λ

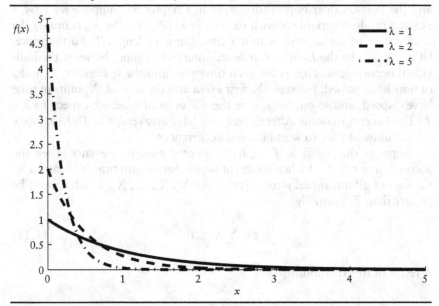

FIGURE 11.8 Distribution Function $F(x)$ of the Exponential Distribution for Various Parameter Values λ

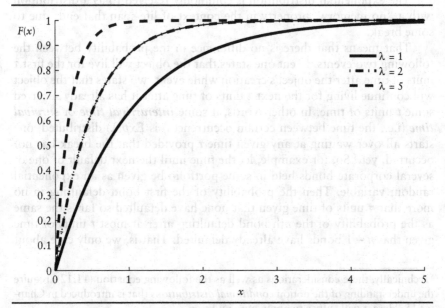

There is an inverse relationship between the exponential distribution and the Poisson distribution (discussed in Chapter 9). Suppose we have a Poisson random variable N with parameter λ, i.e., $N \sim Poi(\lambda)$, counting the occurrences of some event within a time frame of length T. Furthermore, let $X_1, X_2, \ldots$ be the $Exp(\lambda)$ distributed interarrival times between the individual occurrences. That is between time zero and the first event, X_1 units of time have passed, between the first event and the second, X_2 units of time have elapsed, and so on. Now, over these T units of time, we expect $T \cdot \lambda = T \cdot E(N)$ events to occur. Alternatively, we have an average of $T/(T \cdot \lambda) = 1/\lambda = E(X)$ units of time to wait between occurrences.

Suppose that by time T we have counted exactly n events. Then the accrued time τ elapsed when the event occurs for the nth time is obtained by the sum of all individual interarrival times $X_1, X_2, \ldots, X_n$, which cannot be greater than T. Formally

$$\tau = \sum_{i}^{n} X_i \leq T \tag{11.11}$$

A result of this relationship is

$$E(N) = \lambda = \frac{1}{E(X)}$$

The exponential distribution is commonly referred to as a distribution with a "no memory" property in the context of life-span that ends due to some break.

That means that there is no difference in the probability between the following two events. Event one states that the object will live for the first τ units of time after the object's creation while event two states that the object will continue living for the next τ units of time after it has already survived some t units of time. In other words, if some *interarrival time* or *survival time* (i.e., the time between certain occurrences) is $Exp(\lambda)$ distributed, one starts all over waiting at any given time t provided that the break has not occurred, yet.[5] So, for example, let the time until the next default of one of several corporate bonds held in some portfolio be given as an exponential random variable. Then the probability of the first bond defaulting in no more than t units of time given that none have defaulted so far is the same as the probability of the nth bond defaulting after at most t units of time given that $n - 1$ bonds have already defaulted. That is, we only care about

---

[5]Technically, these considerations as well as the following equation (11.12), require the understanding of the notion *conditional distributions* that is introduced in Chapter 15. Here it will suffice to apply pure intuition.

the probability distribution of the time of occurrence of the next default regardless of how many bonds have already defaulted.

Finally, an additional property of the exponential distribution is its relationship to the chi-square distribution. Let X be $Exp(\lambda)$. Then X is also chi-square distributed with two degrees of freedom, that is, $X \sim \chi^2(2)$.

Applications in Finance

In applications in finance, the parameter λ often has the meaning of a *default rate*, *default intensity*, or *hazard rate*. This can be understood by observing the ratio

$$\frac{P(X \in (t, t+dt])}{dt \cdot P(X > t)} \quad (11.12)$$

which expresses the probability of the event of interest such as default of some company occurring between time t and $t + dt$ given that it has not happened by time t, relative to the length of the horizon, dt. Now, let the length of the interval, dt, approach zero, and this ratio in equation (11.12) will have λ as its limit.

The exponential distribution is often used in credit risk models where the number of defaulting bonds or loans in some portfolio over some period of time is represented by a Poisson random variable and the random times between successive defaults by exponentially distributed random variables. In general, then, the time until the nth default is given by the sum in equation (11.11).

Consider, for example, a portfolio of bonds. Moreover, we consider the number of defaults in this portfolio in one year to be some Poisson random variable with parameter $\lambda = 5$, that is, we expect five defaults per year. The same parameter, then, represents the default intensity of the exponentially distributed time between two successive defaults, that is, $\tau \sim Exp(5)$, so that on average, we have to wait $E(\tau) = 1/5$ of a year or 2.4 months. For example, the probability of less than three months (i.e., 1/4 of a year) between two successive defaults is given by

$$P(\tau \leq 0.25) = 1 - e^{-5 \cdot 0.25} = 0.7135$$

or roughly 71%. Now, the probability of no default in any given year is then

$$P(\tau > 1) = e^{-5 \cdot 1} = 0.0067$$

or 0.67%.

RECTANGULAR DISTRIBUTION

The simplest continuous distribution we are going to introduce is the *rectangular distribution*. Often, it is used to generate simulations of random outcomes of experiments via transformation.[6] If a random variable X is rectangular distributed, we denote this by $X \sim Re(a,b)$ where a and b are the parameters of the distribution.

The support is on the real interval $[a,b]$. The density function is given by

$$f(x) = \begin{cases} \dfrac{1}{b-a}, & a \leq x \leq b \\ 0 & x \notin [a,b] \end{cases} \quad (11.13)$$

We see that this density function is always constant, either zero or between the bounds a and b, equal to the inverse of the interval width. Figure 11.9 displays the density function (11.13) for some general parameters a and b.

Through integration, the distribution function follows in the form

$$F(x) = \begin{cases} 0 & x < a \\ \dfrac{1}{b-a} & a \leq x \leq b \\ 1 & x > b \end{cases} \quad (11.14)$$

The mean is equal to

$$E(X) = \frac{a+b}{2}$$

and the variance is

$$Var(X) = \frac{(b-a)^2}{12}$$

In Figure 11.10, we have the distribution function given by equation (11.14) with some general parameters a and b. By analyzing the plot, we can see that the distribution function is not differentiable at a or b, since the derivatives of F do not exist for these values. At any other real value x, the derivative exists (being 0) and is continuous. We say in the latter case that f is *smooth* there.

---

[6] Any distribution function F can be treated as a rectangular random variable on $[0,1]$. Through the corresponding quantiles of F (to be introduced in Chapter 13), we obtain a realization of the distribution F.

Continuous Probability Distributions with Appealing Statistical Properties

FIGURE 11.9 Density Function of a $Re(a,b)$ Distribution

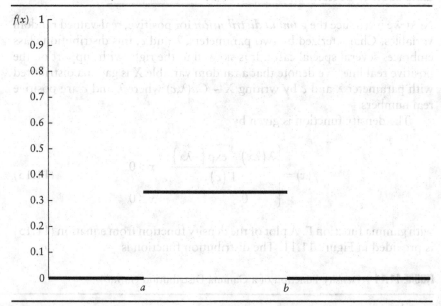

FIGURE 11.10 Distribution Function of a $Re(a,b)$ Distribution

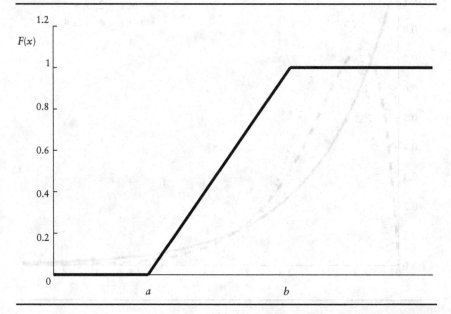

GAMMA DISTRIBUTION

Next we introduce the *gamma distribution* for positive, real-valued random variables. Characterized by two parameters, λ and c, this distribution class embraces several special cases. It is skewed to the right with support on the positive real line. We denote that a random variable X is gamma distributed with parameter λ and c by writing $X \sim Ga(\lambda,c)$ where λ and c are positive real numbers.

The density function is given by

$$f(x) = \begin{cases} \dfrac{\lambda(\lambda x)^{c-1} \exp\{-\lambda x\}}{\Gamma(c)}, & x > 0 \\ 0 & x \leq 0 \end{cases} \quad (11.15)$$

with gamma function Γ. A plot of the density function from equation (11.15) is provided in Figure 11.11. The distribution function is

FIGURE 11.11 Density Function of a Gamma Distribution $Ga(\lambda,b)$

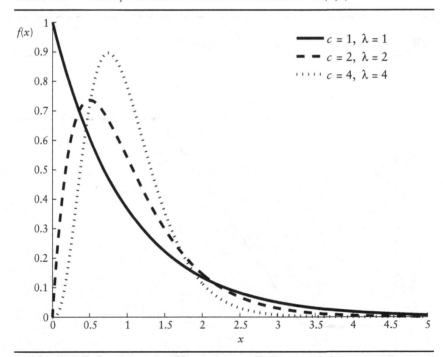

$$F(x) = \begin{cases} 0 & x < 0 \\ \dfrac{\int_0^{\lambda x} u^{c-1} e^{-u} du}{b^c \Gamma(c)}, & x \geq 0 \end{cases}$$

The mean is

$$E(X) = \frac{c}{\lambda}$$

with variance

$$Var(X) = \frac{c}{\lambda^2}$$

Erlang Distribution

A special case is the *Erlang distribution*, which arises for natural number values of the parameter c, that is, $c \in \mathbb{N}$. The intuition behind it is as follows. Suppose we have c exponential random variables with the same parameter λ, that is, $X_1, X_2, \ldots, X_c \sim Exp(\lambda)$ all being independent of each other. Then the sum of these

$$S = \sum_{i=1}^{c} X_i$$

is distributed $Ga(\lambda, c)$ such that the resulting distribution function is

$$F(s) = \begin{cases} 0 & s < 0 \\ 1 - e^{-\lambda s} \sum_{i=1}^{c-1} \dfrac{(\lambda s)^i}{i!}, & s \geq 0 \end{cases}$$

So, when we add the identically $Exp(\lambda)$ distributed inter-arrival times until the c th default, for example, the resulting combined waiting time is Erlang distributed with parameters c and λ.

BETA DISTRIBUTION

The *beta distribution* is characterized by the two parameters c and d that are any positive real numbers. We abbreviate this distribution by $Be(c,d)$. It has a density function with support on the interval [0,1], that is, only for $x \in [0,1]$ does the density function assume positive values. In the context of credit risk modeling, it commonly serves as an approximation for generating random defaults when the true underlying probabilities of default of certain companies are unknown.[7]

The density function is defined by

$$f(x) = \begin{cases} \dfrac{1}{B(c,d)} x^{c-1}(1-x)^{d-1}, & 0 \leq x \leq 1 \\ 0 & \text{else} \end{cases}$$

where $B(c,d)$ denotes the **beta function** with parameters c and d. It is introduced in Appendix A. The density function may assume various different shapes depending on c and d. For a few exemplary values, we present the plots in Figure 11.12. As we can see, for $c = d$, the density function is symmetric about $x = 0.5$.

FIGURE 11.12 Density Function of a Beta Distribution $Be(c,d)$

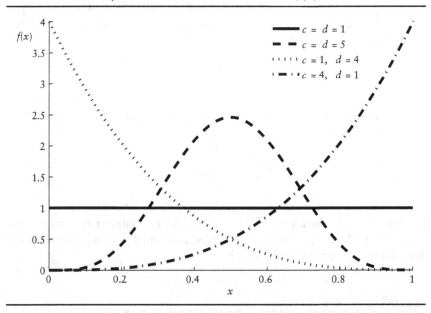

---

[7]We will introduce quantiles in Chapter 13. Quantiles are essential for generating realizations of probability distributions.

LOG-NORMAL DISTRIBUTION

Another important distribution in finance is the *log-normal distribution*. It is connected to the normal distribution via the following relationship. Let Y be a normal random variable with mean μ and variance σ^2. Then the random variable

$$X = e^Y$$

is log-normally distributed with parameters μ and σ^2. In brief, we denote this distribution by $X \sim Ln(\mu,\sigma^2)$.

Since the exponential function $e^Y = \exp(Y)$ only yields positive values, the support of the log-normal distribution is on the positive half of the real line, only, as will be seen by its density function given by

$$f(x) = \begin{cases} \dfrac{1}{x\sigma\sqrt{2\pi}} e^{\frac{(\ln x - \mu)^2}{2\sigma^2}}, & x > 0 \\ 0 & \text{else} \end{cases} \quad (11.16)$$

which looks strikingly similar to the normal density function given by (11.2). Figure 11.13 depicts the density function for several parameter values.

This density function results in the log-normal distribution function

$$F(x) = \Phi\left(\frac{\ln x - \mu}{\sigma}\right)$$

where $\Phi(\cdot)$ is the distribution function of the standard normal distribution.[8] A plot of the distribution function for different parameter values can be found in Figure 11.14.

Mean and variance of a log-normal random variable are

$$E(X) = e^{\left(\mu + \sigma^2/2\right)} \quad (11.17)$$

and

$$Var(X) = e^{\sigma^2}\left(e^{\sigma^2} - 1\right)e^{2\mu} \quad (11.18)$$

---

[8] This is the result of the one-to-one relationship between the values of a log-normal and a standard normal random variable.

FIGURE 11.13 Density Function of the Log-Normal Distribution for Various Values of μ and σ^2

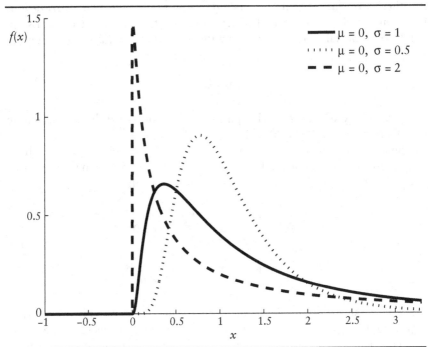

Application to Modeling Asset Returns

The reason for the popularity of the log-normal distribution is that logarithmic asset returns r have been historically modeled as normally distributed such that the related asset prices are modeled by a log-normal distribution. That is, let P_t denote today's asset price and, furthermore, let the daily return r be $N(\mu,\sigma^2)$. Then in a simplified fashion, tomorrow's price is given by $P_{t+1} = P_t \cdot e^r$ while the percentage change between the two prices, e^r, is log-normally distributed, that is, $Ln(\mu,\sigma^2)$.

The log-normal distribution is closed under special operations as well. If we let the n random variables $X_1, \ldots, X_n$ be log-normally distributed each with parameters μ and σ^2 and uninfluenced by each other, then multiplying all of these and taking the nth root we have that

$$\sqrt[n]{\prod_{i=1}^{n} X_i} \sim Ln(\mu,\sigma^2)$$

FIGURE 11.14 Distribution Function of the Log-Normal Distribution for Various Parameter Values μ and σ²

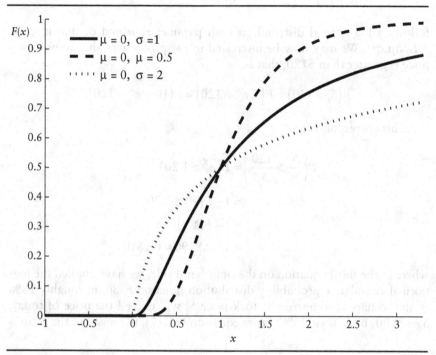

where the product sign is defined as[9]

$$\prod_{i=1}^{n} X_i \equiv X_1 \times X_2 \times \ldots \times X_n$$

As an example, we consider a very simplified stock price model. Let $S_0 = \$100$ be today's stock price of some company. We model tomorrow's price S_1 as driven by the 1-day dynamic X from the previous example of the normal distribution. In particular, the model is

$$S_1 = S_0 \cdot e^X$$

By some slight manipulation of the above equation, we see that the ratio of tomorrow's price over today's price

---

[9]If one multiplies some number x several times by itself, say n times, to obtain the nth power of x, that is, x^n, one applies the nth root to retrieve x again, that is $\sqrt[n]{x^n} = x$.

$$\frac{S_1}{S_0} = e^X$$

follows a log-normal distribution with parameters μ and σ, that is, $S_1/S_0 \sim LN(\mu,\sigma^2)$. We may now be interested in the probability that tomorrow's price is greater than \$120; that is,

$$P(S_1 > 120) = P(S_0 e^X > 120) = P(100 \cdot e^X > 120)$$

This corresponds to

$$P\left(\frac{S_1}{S_0} > \frac{120}{S_0}\right) = P(e^X > 1.20)$$
$$= 1 - P(e^X \leq 1.20)$$
$$= 1 - F(1.2)$$
$$= 1 - 0.8190 = 0.1810$$

where in the third equation on the right-hand side, we have applied the log-normal cumulative probability distribution function F. So, in roughly 18% of the scenarios, tomorrow's stock price S_1 will exceed the price of today, $S_0 = \$100$, by at least 20%. From equation (11.17), the mean of the ratio is

$$E\left(\frac{S_1}{S_0}\right) = \mu_{S_1/S_0} = e^{0+\frac{0.2}{2}} = 1.1052$$

implying that we have to expect tomorrow's stock price to be roughly 10% greater than today, even though the dynamic X itself has an expected value of 0. Finally, equation (11.18) yields the variance

$$Var\left(\frac{S_1}{S_0}\right) = \sigma^2_{S_1/S_0} = e^{0.2}(e^{0.2}-1) = 0.2704$$

which is only slightly larger than that of the dynamic X itself.

The statistical concepts learned to this point can be used for pricing certain types of derivative instruments. In Appendix D, we present an explicit computation for the price formula for a certain derivative instrument when stock prices are assumed to be log-normally distributed. More specifically, we present the price of a European call option and the link to an important pricing model in finance known as the Black-Scholes option pricing model.

CONCEPTS EXPLAINED IN THIS CHAPTER (IN ORDER OF PRESENTATION)

Normal distribution
Gaussian distribution
Standard normal distribution
Location-scale invariant
Stable under summation
Standardization of X
Central limit theorem
Empirical rule
Chebychev inequalities
Logarithmic return
Multiplicative return
Chi-square distribution
Degrees of freedom
Gamma function
Short rates
Student's t-distribution (t-distribution, Student's distribution)
F-distribution
Exponential distribution
Inter-arrival time
Survival time
Default rate
Default intensity
Hazard rate
Rectangular distribution
Gamma distribution
Erlang distribution
Beta distribution
Beta function
Log-normal distribution

CHAPTER 12

Continuous Probability Distributions Dealing with Extreme Events

In this chapter, we present a collection of continuous probability distributions that are used in finance in the context of modeling extreme events. While in Chapter 11 the distributions discussed were appealing in nature because of their mathematical simplicity, the ones introduced here are sometimes rather complicated, using parameters that are not necessarily intuitive. However, due to the observed behavior of many quantities in finance, there is a need for more flexible distributions compared to keeping models mathematically simple.

While the Student's t-distribution discussed in Chapter 11 is able to mimic some behavior inherent in financial data such as so-called *heavy tails* (which means that a lot of the probability mass is attributed to extreme values), it fails to capture other observed behavior such as skewness. Hence, we decided not to include it in this chapter.

In this chapter, we will present the generalized extreme value distribution, the generalized Pareto distribution, the normal inverse Gaussian distribution, and the α-stable distribution together with their parameters of location and spread. The presentation of each distribution is accompanied by some illustration to help render the theory more appealing.

GENERALIZED EXTREME VALUE DISTRIBUTION

Sometimes it is of interest to analyze the probability distribution of extreme values of some random variable rather than the entire distribution. This occurs in risk management (including operational risk, credit risk, and market risk) and risk control in portfolio management. For example, a portfolio manager may be interested in the maximum loss a portfolio might incur with a certain probability. For this purpose, *generalized extreme value* (GEV) *distributions* are designed. They are characterized by the real-valued parameter ξ. Thus, the abbreviated appellation for this distribution is $GEV(\xi)$.

Technically, one considers a series of identically distributed random variables $X_1, X_2, \ldots, X_n$, which are independent of each other so that each one's value is unaffected by the others' outcomes. Now, the GEV distributions become relevant if we let the length of the series n become ever larger and consider its largest value, that is, the maximum.

The distribution is not applied to the data immediately but, instead, to the so-called *standardized data*. Basically, when standardizing data x, one reduces the data by some constant real parameter a and divides it by some positive parameter b so that one obtains the quantity $(x - a)/b$.[1] The parameters are usually chosen such that this standardized quantity has zero mean and unit variance. We have to point out that neither variance nor mean have to exist for all probability distributions.

Extreme value theory, a branch of statistics that focuses solely on the extremes (tails) of a distribution, distinguishes between three different types of generalized extreme value distributions: Gumbel distribution, Fréchet distribution, and Weibull distribution.[2] The three types are related in that we obtain one type from another by simply varying the value of the parameter ξ. This makes GEV distributions extremely pleasant for handling financial data.

For the *Gumbel distribution*, the general parameter is zero (i.e., $\xi = 0$) and its density function is

$$f(x) = e^{-x} \exp\{-e^{-x}\}$$

A plot of this density is given by the dashed graph in Figure 12.1 that corresponds to $\xi = 0$. The distribution function of the Gumbel distribution is then

$$F(x) = \exp\{-e^{-x}\}$$

Again, for $\xi = 0$, we have the distribution function displayed by the dashed graph in Figure 12.2.

The second $GEV(\xi)$ distribution is the *Fréchet distribution*, which is given for $\xi > 0$ and has density

$$f(x) = (1 + \xi x)^{-\xi - 1} \cdot \exp\{-x^{-\xi}\}$$

---

[1] Standardization is a linear transform of the random variable such that its location parameter becomes zero and its scale one.
[2] In the extreme value theory literature, these distributions are referred to respectively as Type I, Type II, and Type III.

Continuous Probability Distributions Dealing with Extreme Events

FIGURE 12.1 $GEV(\xi)$ Density Function for Various Parameter Values

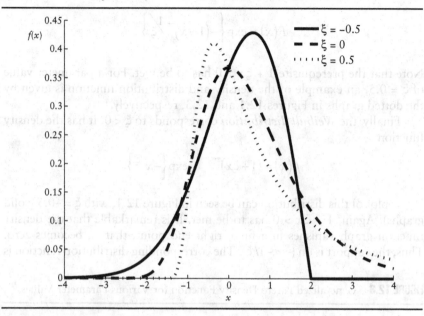

FIGURE 12.2 $GEV(\xi)$ Distribution Function for Various Parameter Values

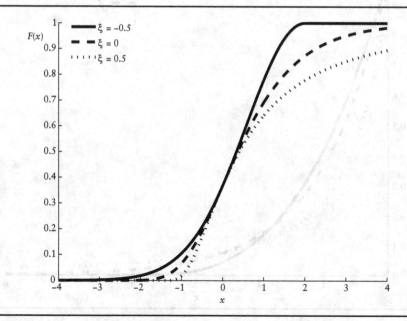

with corresponding distribution function

$$F(x) = \exp\left\{-(1+x)^{-1/\xi}\right\}$$

Note that the prerequisite $1 + \xi x > 0$ has to be met. For a parameter value of $\xi = 0.5$, an example of the density and distribution function is given by the dotted graphs in Figures 12.1 and 12.3, respectively.

Finally, the **Weibull distribution** corresponds to $\xi < 0$. It has the density function

$$f(x) = (1 + \xi x)^{-\xi - 1} \cdot \exp\left\{-x^{-\xi}\right\}$$

A plot of this distribution can be seen in Figure 12.1, with $\xi = -0.5$ (solid graphs). Again, $1 + \xi x > 0$ has to be met. It is remarkable that the density function graph vanishes in a finite right end point, that is, becomes zero. Thus, the support is on $(-\infty, -1/\xi)$. The corresponding distribution function is

FIGURE 12.3 Generalized Pareto Density Function for Various Parameter Values

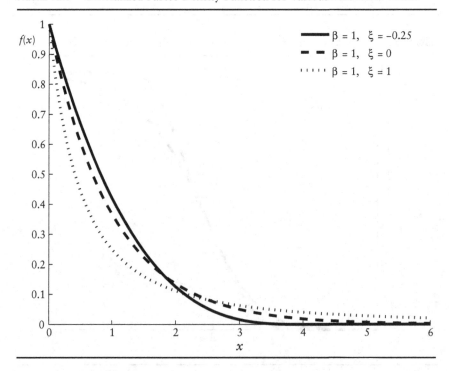

Continuous Probability Distributions Dealing with Extreme Events

$$F(x) = \exp\left\{-(1+x)^{-1/\xi}\right\}$$

a graph of which is depicted in Figure 11.2 for $\xi = -0.5$ (solid line).

Notice that the extreme parts of the density function (i.e., the tails) of the Fréchet distribution vanish more slowly than that of the Gumbel distribution. Consequently, a Fréchet type distribution should be applied when dealing with scenarios of extremes.

GENERALIZED PARETO DISTRIBUTION

A distribution often employed to model large values, such as price changes well beyond the typical change, is the *generalized Pareto distribution* or, as we will often refer to it here, simply *Pareto distribution*. This distribution serves as the distribution of the so called "peaks over thresholds," which are values exceeding certain benchmarks or loss severity.

For example, consider n random variables $X_1, X_2, \ldots, X_n$ that are all identically distributed and independent of each other. Slightly idealized, they might represent the returns of some stock on n different observation days. As the number of observations n increases, suppose that their maximum observed return follows the distribution law of a GEV distribution with parameter ξ. Furthermore, let u be some sufficiently large threshold return. Suppose that on day i, the return exceeded this threshold. Then, given the exceedance, the amount return X_i surpassed u by, that is, $X_i - u$, is a generalized Pareto distributed random variable.

The following density function characterizes the Pareto distribution

$$f(x) = \begin{cases} \dfrac{1}{\beta}\left(1+\xi\dfrac{x}{\beta}\right)^{-1-1/\xi}, & x \geq 0 \\ 0 & \text{else} \end{cases}$$

with $\beta > 0$ and $1 + (\xi x)/\beta > 0$. Hence, the distribution is right skewed since the support is only on the positive real line. The corresponding distribution function is given by

$$F(x) = \frac{1}{\beta}\left(1+\xi\frac{x}{\beta}\right)^{-1-1/\xi}, x \geq 0$$

As we can see, the Pareto distribution is characterized by two parameters β and ξ. In brief, the distribution is denoted by $Pa(\beta,\xi)$. The parameter β serves as a scale parameter while the parameter ξ is responsible for the overall shape as becomes obvious by the density plots in Figure 12.3. The distribution function is displayed, in Figure 12.4, for a selection of parameter values.

For $\beta < 1$, the mean is

$$E(X) = \beta/1 - \xi$$

When β becomes very small approaching zero, then the distribution results in the exponential distribution with parameter $\lambda = 1/\beta$.

The Pareto distribution is commonly used to represent the tails of other distributions. For example, while in neighborhoods about the mean, the normal distribution might serve well to model financial returns, for the tails (i.e., the end parts of the density curve), however, one might be better advised to apply the Pareto distribution. The reason is that the normal distribution may not assign sufficient probability to more pronounced price

FIGURE 12.4 Generalized Pareto Distribution Function for Various Parameter Values

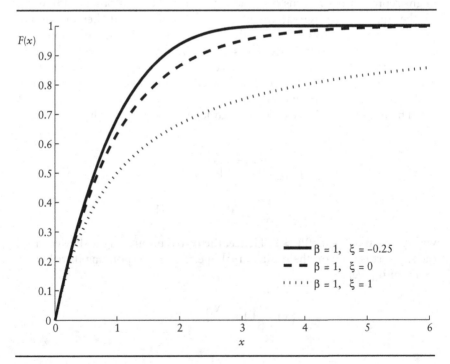

changes measured in log-returns. On the other hand, if one wishes to model behavior that attributes less probability to extreme values than the normal distribution would suggest this could be accomplished by the Pareto distribution as well. The reason why the class of the Pareto distributions provides a prime candidate for these tasks is due to the fact that it allows for a great variety of different shapes one can smoothly obtain by altering the parameter values.

NORMAL INVERSE GAUSSIAN DISTRIBUTION

Another candidate for the modeling of financial returns is the *normal inverse Gaussian distribution*. It is considered suitable since it assigns a large amount of probability mass to the tails. This reflects the inherent risks in financial returns that are neglected by the normal distribution since it models asset returns behaving more moderately. But in recent history, we have experienced more extreme shocks than the normal distribution would have suggested with reasonable probability.

The distribution is characterized by four parameters, a, b, c, and d. In brief, the distribution is denoted by $NIG(a,b,\mu,\delta)$. For real values x, the density function is given by

$$f(x) = \frac{a \cdot \delta}{\pi} \exp\left\{\delta\sqrt{a^2 - b^2} + b(x-\mu)\right\} \frac{K_1\left(a\sqrt{\delta^2 + (x-\mu)^2}\right)}{\sqrt{\delta^2 + (x-\mu)^2}}$$

where K_1 is the so-called *Bessel function of the third kind*, which is described in Appendix A. In Figure 12.5, we display the density function for a selection of parameter values.

The distribution function is, as in the normal distribution case, not analytically presentable. It has to be determined with the help of numerical methods. We display the distribution function for a selection of parameter values, in Figure 12.6.

The parameters have the following interpretation. Parameter a determines the overall shape of the density while b controls skewness. The location or position of the density function is governed via parameter μ and δ is responsible for scaling. These parameters have values according to the following

$a > 0$
$0 \leq b < a$

FIGURE 12.5 Normal Inverse Gaussian Density Function for Various Parameter Values

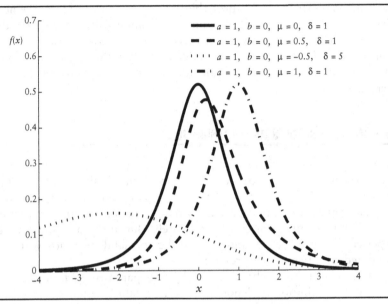

FIGURE 12.6 Normal Inverse Gaussian Distribution Function for Various Parameter Values

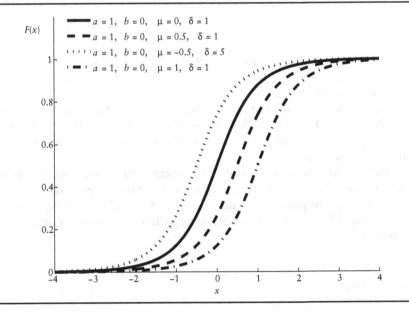

Continuous Probability Distributions Dealing with Extreme Events

$\mu \in R$
$\delta > 0$

The mean of a *NIG* random variable is

$$E(X) = \mu + \frac{\delta \cdot b}{\sqrt{a^2 - b^2}}$$

and the variance is

$$Var(X) = \delta \frac{a^2}{\left(\sqrt{a^2 - b^2}\right)^3}$$

Normal Distribution versus Normal Inverse Gaussian Distribution

Due to a relationship to the normal distribution that is beyond the scope here, there are some common features between the normal and NIG distributions.

The *scaling property* of the *NIG* distribution guarantees that any *NIG* random variable multiplied by some real constant is again a *NIG* random variable. Formally, for some $k \in R$ and $X \sim NIG(a,b,\mu,\delta)$, we have that

$$k \cdot X \sim NIG\left(\frac{a}{k}, \frac{b}{k}, k \cdot \mu, k \cdot \delta\right) \quad (12.1)$$

Amongst others, the result in equation (12.1) implies that the factor k shifts the density function by the k-fold of the original position. Moreover, we can reduce skewness in that we inflate X by some factor k.

Also, the *NIG* distribution is summation stable such that, under certain prerequisites concerning the parameters, independent *NIG* random variables are again *NIG*. More precisely, if we have the random variables $X_1 \sim NIG(a,b,\mu_1,\delta_1)$ and $X_2 \sim NIG(a,b,\mu_2,\delta_2)$, the sum is $X_1 + X_2 \sim NIG(a,b,\mu_1 + \mu_2, \delta_1 + \delta_2)$. So, we see that only location and scale are affected by summation.

α-STABLE DISTRIBUTION

The final distribution we introduce is the class of *α-stable distributions*. Often, these distributions are simply referred to as *stable distributions*. While many models in finance have been modeled historically using the normal distribution based on its pleasant tractability, concerns have been raised that it underestimates the danger of downturns of extreme magnitude inherent in stock markets. The sudden declines of stock prices experienced during several crises since the late 1980s—October 19, 1987 ("Black Monday"), July 1997 ("Asian currency crisis"), 1998–1999 ("Russian ruble crisis"), 2001 ("Dot-Com Bubble"), and July 2007 and following ("Subprime mortgage crisis")—are examples that call for distributional alternatives accounting for extreme price shocks more adequately than the normal distribution. This may be even more necessary considering that financial crashes with serious price movements might become even more frequent in time given the major events that transpired throughout the global financial markets in 2008. The immense threat radiating from heavy tails in stock return distributions made industry professionals aware of the urgency to take them serious and reflect them in their models.

Many distributional alternatives providing more realistic chances to severe price movements have been presented earlier, such as the Student's t in Chapter 11 or *GEV* distributions earlier in this chapter, for example. In the early 1960s, Benoit Mandelbrot suggested as a distribution for commodity price changes the class of stable distributions. The reason is that, through their particular parameterization, they are capable of modeling moderate scenarios as supported by the normal distribution as well as extreme ones beyond the scope of most of the distributions that we have presented in this chapter.

The stable distribution is characterized by the four parameters α, β, σ, and μ. In brief, we denote the α-stable distribution by $S(\alpha,\beta,\sigma,\mu)$. Parameter α is the so called **tail index** or **characteristic exponent**. It determines how much probability is assigned around the center and the tails of the distribution. The lower the value α, the more pointed about the center is the density and the heavier are the tails. These two features are referred to as **excess kurtosis** relative to the normal distribution. This can be visualized graphically as we have done in Figure 12.7 where we compare the normal density to an α-stable density with a low $\alpha = 1.5$.[3] The density graphs are obtained from fitting the distributions to the same sample data of arbitrarily generated numbers. The parameter α is related to the parameter ξ of the Pareto distribution resulting in the tails of the density functions of α-stable random variables to vanish at a rate proportional to the Pareto tail.

The tails of the Pareto as well as the α-stable distribution decay at a rate with fixed power α, $x^{-\alpha}$ (i.e., *power law*), which is in contrast to the normal distribution whose tails decay at an exponential rate (i.e., roughly $e^{-x^2/2}$). We illustrate the effect focusing on the probability of exceeding some value x

---

[3]The parameters for the normal distribution are $\mu = 0.14$ and $\sigma = 4.23$. The parameters for the stable distribution are $\alpha = 1.5$, $\beta = 0$, $\sigma = 1$, and $\mu = 0$. Note that symbols common to both distributions have different meanings.

FIGURE 12.7 Comparison of the Normal (Dash-Dotted) and α-Stable (Solid) Density Functions

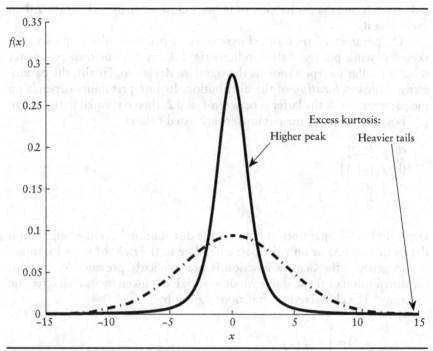

somewhere in the upper tail, say $x = 3$. Moreover, we choose the parameter of stability to be $\alpha = 1.5$. Under the normal law, the probability of exceedance is roughly $e^{-3^2/2} = 0.011$ while under the power law it is about $3^{-1.5} = 0.1925$. Next, we let the benchmark x become gradually larger. Then the probability of assuming a value at least twice or four times as large (i.e., $2x$ or $4x$) is roughly

$$e^{-\frac{(2\times 3)^2}{2}} \approx 0$$

or

$$e^{-\frac{(4\times 3)^2}{2}} \approx 0$$

for the normal distribution. In contrast, under the power law, the same exceedance probabilities would be $(2 \times 3)^{-1.5} = 0.068$ or $(4 \times 3)^{-1.5} \approx 0.024$. This is a much slower rate than under the normal distribution. Note that the value of $x = 3$ plays no role for the power tails while the exceedance probability of the normal distribution decays the faster the further out we are in the tails (i.e., the larger is x). The same reasoning applies to the lower tails considering the probability of falling below a benchmark x rather than exceeding it.

The parameter β indicates *skewness* where negative values represent left skewness while positive values indicate right skewness. The *scale* parameter σ has a similar interpretation as the standard deviation. Finally, the parameter μ indicates *location* of the distribution. Its interpretability depends on the parameter α. If the latter is between 1 and 2, then μ is equal to the mean.

Possible values of the parameters are listed below:

α (0,2]
β [−1,1]
σ (0,∞)
μ $\mathbb{R}$

Depending on the parameters α and β, the distribution has either support on the entire real line or only the part extending to the right of some location.

In general, the density function is not explicitly presentable. Instead, the distribution of the α-stable random variable is given by its characteristic function.[4] The characteristic function is given by

$$\varphi(t) = \int_{-\infty}^{\infty} e^{itx} f(x) dx$$

$$= \begin{cases} \exp\left\{-\sigma^{\alpha}|t|^{\alpha}\left[1 - i\beta\mathrm{sign}(t)\tan\frac{\pi\alpha}{2}\right] + i\mu t\right\} & \alpha \neq 1 \\ \exp\left\{-\sigma|t|\left[1 - i\beta\frac{2}{\pi}\mathrm{sign}(t)\ln(t)\right] + i\mu t\right\} & \alpha = 1 \end{cases} \quad (12.2)$$

The density, then, has to be retrieved by an inverse transform to the characteristic function. Numerical procedures are employed for this task to approximate the necessary computations. The characteristic function (12.2) is presented here more for the sake of completeness rather than necessity. So, one should not be discouraged if it appears overwhelmingly complex.

In Figures 12.8 and 12.9, we present the density function for varying parameter β and α, respectively. Note in Figure 12.9 that for a $\beta = 1$, the density is positive only on a half-line towards the right as α approaches its upper-bound value of 2.

---

[4]See Appendix A for an introduction of the characteristic function.

FIGURE 12.8 Stable Density Function for Various Values of β

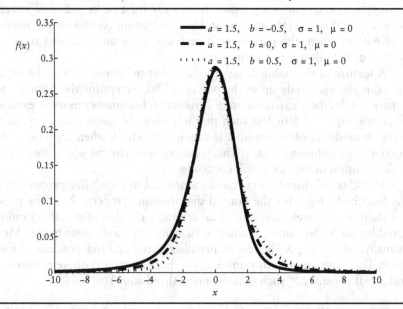

FIGURE 12.9 Stable Density Function (totally right-skewed) for Various Values of α

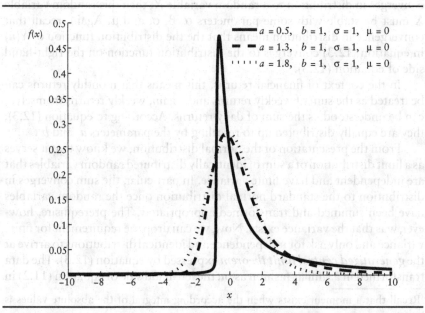

Only in the case of an α of 0.5, 1, or 2, can the functional form of the density be stated. For our purpose here, only the case α = 2 is of interest Because for this special case, the stable distribution represents the normal distribution. Then, the parameter β ceases to have any meaning since the normal distribution is not asymmetric.

A feature of the stable distributions is that moments such as the mean, for example, exist only up to the power α.[5] So, except for the normal case (where α = 2), there exists no finite variance. It becomes even more extreme when α is equal to 1 or less such that not even the mean exists any more. The non-existence of the variance is a major drawback when applying stable distributions to financial data. This is one reason why the use of this family of distribution in finance is still contended.

This class of distributions owes its name to the *stability property* that we described earlier for the normal distribution (Property 2) in the previous chapter: The weighted sum of an arbitrary number of α-stable random variables with the same parameters is, again, α-stable distributed. More formally, let $X_1, ..., X_n$ be identically distributed and independent of each other. Then, assume that for any $n \in \mathbf{N}$, there exists a positive constant a_n and a real constant b_n such that the normalized sum $Y(n)$

$$Y(n) = a_n \left(X_1 + X_2 + ... + X_n \right) + b_n \sim S(\alpha, \beta, \sigma, \mu) \qquad (12.3)$$

converges in distribution to a random variable X, then this random variable X must be stable with some parameters α, β, σ, and μ. Again, recall that convergence in distribution means that the the distribution function of $Y(n)$ in equation (12.3) converges to the distribution function on the right-hand side of equation (12.3).

In the context of financial returns, this means that monthly returns can be treated as the sum of weekly returns and, again, weekly returns themselves can be understood as the sum of daily returns. According to equation (12.3), they are equally distributed up to rescaling by the parameters a_n and b_n.

From the presentation of the normal distribution, we know that it serves as a limit distribution of a sum of identically distributed random variables that are independent and have finite variance. In particular, the sum converges in distribution to the standard normal distribution once the random variables have been summed and transformed appropriately. The prerequisite, however, was that the variance exists. Now, we can drop the requirement for finite variance and only ask for independence and identical distributions to arrive at the *generalized central limit theorem* expressed by equation (12.3). The data transformed in a similar fashion as on the left-hand side of equation (11.2) in

---

[5] Recall that a moment exists when the according integral of the absolute values is finite.

Chapter 11 will have a distribution that follows a stable distribution law as the number n becomes very large. Thus, the class of α-stable distributions provides a greater set of limit distributions than the normal distribution containing the latter as a special case. Theoretically, this justifies the use of α-stable distributions as the choice for modeling asset returns when we consider the returns to be the resulting sum of many independent shocks.

Let us resume the previous example with the random dynamic and the related stock price evolution. Suppose, now, that the 10-day dynamic was S_α distributed. We denote the according random variable by V_{10}. We select a fairly moderate stable parameter of $\alpha = 1.8$. A value in this vicinity is commonly estimated for daily and even weekly stock returns. The skewness and location parameters are both set to zero, that is, $\beta = \mu = 0$. The scale is $\sigma = 1$, so that if the distribution was normal, that is, $\alpha = 2$, the variance would be 2 and, hence, consistent with the previous distributions. Note, however, that for $\alpha = 1.8$, the variance does not exist. Here the probability of the dynamic's exceedance of the lower threshold of 1 is

$$P(V_{10} > 1) = 0.2413 \qquad (12.4)$$

compared to 0.2398 and 0.1870 in the normal and Student's t cases, respectively. Again, the probability in (12.4) corresponds to the event that in 10 days, the stock price will be greater than \$271. So, it is more likely than in the normal and Student's t model.

For the higher threshold of 3.5, we obtain

$$P(V_{10} > 3.5) = 0.0181$$

compared to 0.0067 and 0.0124 from the normal and Student's t cases, respectively. This event corresponds to a stock price beyond \$3,312, which is an immense increase. Under the normal distribution assumption, this event is virtually unlikely. It would happen in less than 1% of the 10-day periods. However, under the stable as well as the Student's t assumption, this could happen in 1.81% or 1.24% of the scenarios, which is three times or double the probability, respectively. Just for comparison, let us assume $\alpha = 1.6$, which is more common during a rough market climate. The dynamic would now exceed the threshold of 1 with probability

$$P(V_{10} > 1) = 0.2428$$

which fits in with the other distribution. For 3.5, we have

$$P(V_{10} > 3.5) = 0.0315 \qquad (12.5)$$

which is equal to five times the probability under the normal distribution and almost three times the probability under the Student-t-distribution assumption. For this threshold, the same probability as in equation (12.5) could only be achieved with a variance of $\sigma^2 = 4$, which would give the overall distribution a different shape. In the Student's t case, the degree of freedom parameter would have to be less than 3 such that now the variance would not exist any longer.

For the stable parameters chosen, the same results are obtained when the sign of the returns is negative and losses are considered. For example, $P(V_{10} < -3.5) = 0.0315$ corresponds to the probability of obtaining a stock price of \$3 or less. This scenario would only be given 0.67% probability in a normal distribution model. With respect to large portfolios such as those managed by large banks, negative returns deserve much more attention since losses of great magnitude result in wide-spread damages to industries beyond the financial industry.

As another example, let's look at what happened to the stock price of American International Group (AIG) in September 2008. On one single day, the stock lost 60% of its value. That corresponds to a return of about -0.94.[6] If we choose a normal distribution with $\mu = 0$ and $\sigma^2 = 0.0012$ for the daily returns, a drop in price of this magnitude or less has near zero probability. The distributional parameters were chosen to best mimic the behavior of the AIG returns. By comparison, if we take an α-stable distribution with $\alpha = 1.6$, $\beta = 0$, $\mu = 0$, and $\sigma = 0.001$ where these parameters were selected to fit the AIG returns, we obtain the probability for a decline of at least this size of 0.00003, that is, 0.003%. So even with this distribution, an event of this impact is almost negligible. As a consequence, we have to chose a lower parameter α for the stable distribution. That brings to light the immense risk inherent in the return distributions when they are truly α-stable.

CONCEPTS EXPLAINED IN THIS CHAPTER (IN ORDER OF PRESENTATION)

Heavy tails
Generalized extreme value distributions
Standardized data
Extreme value theory
Gumbel distribution
Fréchet distribution
Weibull distribution

---

[6]One has to keep in mind that we are analyzing logarithmic returns.

Generalized Pareto distribution
Normal inverse Gaussian distribution
Bessel function of the third kind
Scaling property
α-stable distributions
Stable distributions
Tail index
Characteristic exponent
Excess kurtosis
Power law
Skewness
Scale
Location
Stability property
Generalized central limit theorem

CHAPTER 13

Parameters of Location and Scale of Random Variables

In the previous four chapters, we presented discrete and continuous probability distributions. It is common to summarize distributions by various measures. The most important of these measures are the parameters of location and scale. While some of these parameters have been mentioned in the context of certain probability distributions in the previous chapters, we introduce them here as well as additional ones.

In this chapter, we present as parameters of location quantiles, the mode, and the mean. The mean is introduced in the context of the moments of a distribution. Quantiles help in assessing where some random variable assumes values with a specified probability. In particular, such quantiles are given by the lower and upper quartiles as well as the median. In the context of portfolio risk, the so-called value-at-risk measure is used. As we will see, this measure is defined as the minimum loss some portfolio incurs with specified probability.

As parameters of scale, we introduce moments of higher order: variance together with the standard deviation, skewness, and kurtosis. The variance is the so-called *second central moment* and the related standard deviation are the most commonly used risk measures in the context of portfolio returns. However, their use can sometimes be misleading because, as was noted in Chapter 11, some distributions particularly suited to model financial asset returns have no finite variance. The skewness will be introduced as a parameter of scale that helps in determining whether some probability distribution is asymmetric. As we will see, most financial asset returns are skewed. The kurtosis offers insight into the assignment of probability mass in the tails of the distribution as well as about the mode. Excess kurtosis, which describes non-normality of the probability distribution in the sense of heavier tails as well as a more accentuated density function about the mode relative to the normal distribution, is presented since it is commonly used in describing financial asset returns. In the appendix to this chapter, we list all parameters introduced here for the continuous distribution functions introduced in Chapters 11 and 12.

PARAMETERS OF LOCATION

In general, location parameters give information on the positioning of the distribution on the real line, $\Omega = \mathbf{R}$. For example, the information can be about the smallest or the largest values possible or the value that is expected before the drawing for a certain random variable.

Quantiles

Quantiles are parameters of location that account for a certain quantity of probability. An alternative term for quantile is *percentile*. Usually, quantiles are given with respect to a value α being some real number between 0 and 1 and denoted by α-quantile. The α indicates the proportion of probability that is assigned to the values that are equal to the α-quantile or less. The α-quantile denoted as q_α is equal to the value x where the distribution function $F(x)$ assumes or exceeds the value α for the first time.

Formally, we define a quantile as

$$q_\alpha = \inf\{x : F(x) \geq \alpha\} \tag{13.1}$$

where *inf* is short for *infimum* meaning lowest bound.[1]

In this sense of equation (13.1), the quantiles can be interpreted as obtained through an inverse function F^{-1} to the cumulative distribution function F. That is, for any level of probability α, the inverse F^{-1} yields the α-quantile through $q_\alpha = F^{-1}(\alpha)$.

Note that when X is a discrete random variable, it can be that the distribution function is exactly equal to α (i.e., $F(x) = \alpha$) for several values of x. So, according to the definition of a quantile given by equation (13.1), q_α should be equal to the smallest such x. However, by similar reasoning as in the discussion of the quantile statistic presented in Chapter 3, the quantile q_α is sometimes defined as any value of the interval between this smallest x and the smallest value for which $F(x) > \alpha$.

For particular values of α, we have special names for the corresponding quantiles. For $\alpha = 0.25$, $q_{0.25}$ is called the *lower quartile*; for $\alpha = 0.5$, $q_{0.5}$ is the *median;* and for $\alpha = 0.75$, $q_{0.75}$ is called the *upper quartile*. The median,

---

[1]Technically, in contrast to the minimum, which is the smallest value that can be assumed, the infimum is some limiting value that need not be assumed. Here, we will interpret the notation inf{x: $F(x) \geq \alpha$} as *the smallest value x that satisfies $F(x) \geq \alpha$.*

for example, is the value for which, theoretically, half the realizations will not exceed that value while the other half will not fall below that very same value.

Computation of Quantiles for Various Distributions

We know from the definition in equation (13.1) that, for example, the 0.2-quantile of some distribution is the value $q_{0.2}$ where for the first time, the distribution function F is no longer smaller than 0.2, that is, $F(q_{0.2}) \geq 0.2$ and $F(x) < 0.2$ for all $x < q_{0.2}$. We illustrate this in Figure 13.1, where we depict the cumulative distribution function of the discrete binomial $B(5,0.5)$ distribution introduced in Chapter 9. The level of $\alpha = 0.2$ is given by the horizontal dashed line. Now, the first time $F(x)$ assumes a value on or above this line is at $x = 2$ with $F(2) = 0.5$. Hence, the 0.2-quantile is $q_{0.2} = 2$ as indicated by the vertical line extending from $F(2) = 0.5$ down to the horizontal axis.

If for this $B(5,0.5)$ distribution we endeavor to compute the median, we find that for all values in the interval [2,3], the distribution is $F(x) = 0.5$. By the definition for a quantile given by equation (13.1), the median is uniquely given as $q_{0.5} = 2 = \inf\{x : F(x) \geq 0.5\}$. However, by the alternative defini-

FIGURE 13.1 Determining the 0.2-Quantile $q_{0.2}$ of the $B(5,0.5)$ Distribution as the x-Value 2 where the Distribution Function F Exceeds the 0.2 Level (dashed line) for the First Time

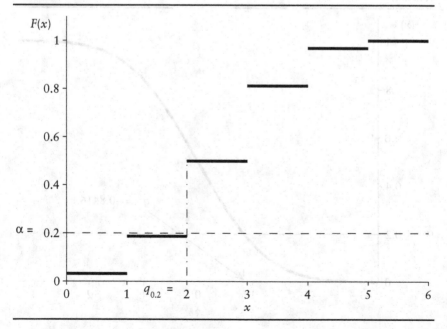

tion, we have that all values in the interval [2,3] are the median, including $x = 3$ where $F(x) > 0.5$ for the first time. Consequently, the median of the $B(5,0.5)$ distribution is not unique when the alternative definition is used. There are other discrete probability distributions for which this definition yields more than one median.

Next, we have a look at the standard normal distribution that is a continuous distribution described in Chapter 11. In Figure 13.2, we display the $N(0,1)$ cumulative distribution function. Again, we obtain the 0.2-quantile as the value x where the distribution function $F(x)$ intersects the dashed horizontal line at level $\alpha = 0.2$. Alternatively, we can determine the 0.2-quantile by the area under the probability density function $f(x)$ in Figure 13.3 as well. At $q_{0.2} = 2$, this area is exactly equal to 0.2.

Value-at-Risk

Let's look at how quantiles are related to an important risk measure used by financial institutions called *value-at-risk* (VaR). Consider a portfolio consisting of financial assets. Suppose the return of this portfolio is given by r_p. Denoting today's portfolio value by P_0, the value of the portfolio tomorrow is assumed to follow

FIGURE 13.2 Determining the 0.2-Quantile Using the Cumulative Distribution Function

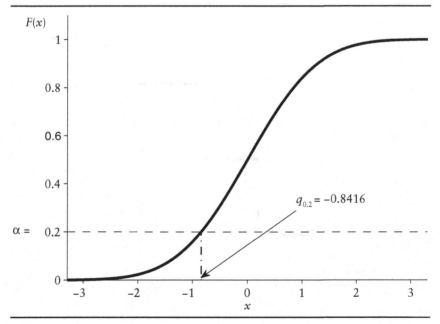

FIGURE 13.3 Determining the 0.2-Quantile of the Standard Normal Distribution with the Probability Density Function

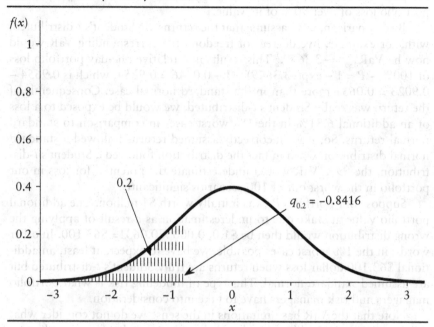

$$P_1 = P_0 \cdot e^{r_p}$$

As is the case for quantiles, in general, VaR is associated with some level α. Then, VaR_α states that with probability $1 - \alpha$, the portfolio manager incurs a loss of VaR_α or more.

Computing the VaR of Various Distributions

Let us set $\alpha = 0.99$. Then with probability 0.01, the return of the portfolio will be equal to its 0.01-quantile or less. Formally, this is stated as

$$P(r_p \leq \text{VaR}_{0.99}) = 0.01$$

What does this figure mean in units of currency? Let's use some numbers to make this concept clearer. Let the daily portfolio return follow the standard normal distribution (i.e., $N(0,1)$), which we introduced in Chapter 11. The 1% quantile for this distribution is equal to -2.3263. Then, we have $\text{VaR}_{0.99} = q_{0.01} = -2.3263$, which translates into a relative one-day portfolio loss of $100\% - P = 1 - \exp(-2.3263) = 1 - 0.0977 = 0.9023$ where P is the next day's portfolio value as a percentage of the original value. So, when the portfolio return is standard normally distributed, the $\text{VaR}_{0.99}$ yields a portfolio loss of over 90% of its value.

By comparison, if we assume that the returns are Student's t-distributed, with, for example, five degrees of freedom, the corresponding VaR would now be $\text{VaR}_{0.99}^{t} = -3.3649$.[2] This results in a relative one-day portfolio loss of $100\% - P = 1 - \exp(-3.3649) = 1 - 0.0346 = 0.9654$, which is $0.9654 - 0.9023 = 0.0631$ more than in the standard normal case. Consequently, if the return was really Student's t-distributed, we would be exposed to a loss of an additional 6.31% in the 1% worst cases in comparison to standard normal returns. So, if we incorrectly assumed returns followed a standard normal distribution when in fact the distribution followed a Student's t-distribution, the 99% VaR would underestimate the potential for loss in our portfolio in the worst out of 100 scenarios significantly.

Suppose our portfolio P_0 was initially worth $1 million. The additional portfolio value at stake due to underestimation as a result of applying the wrong distribution would then be $1,000,000 × 0.0631 = $63,100. In other words, in the 1% worst cases possible, we have to expect, at least, an additional $63,100 dollar loss when returns are truly Student's t-distributed but are assumed standard normal. This type of modeling risk is what portfolio managers and risk managers have to take into consideration.

Note that the VaR has limitations in the sense we do not consider what the distribution looks like for values below q_α. Consider, for example, two hypothetical portfolios with weekly returns r_p^1 and r_p^2, respectively. Suppose, $r_p^1 \sim N(0,1)$ while $r_p^2 \sim N(0.965, 2)$. So the return of portfolio 2 has a greater volatility as indicated by the larger standard deviation (i.e., $\sigma_2 = \sqrt{2} > \sigma_1 = 1$), while its return has a positive expected value, $\mu_2 = 0.9635$ compared to an expected value of 0 for the return of portfolio 1.

But, even though the distributions are different, their VaRs are the same, that is, $\text{VaR}_{0.99}^{(1)} = \text{VaR}_{0.99}^{(2)} = -2.3263$ where $\text{VaR}_{0.99}^{(1)}$ is the 99% VaR of portfolio 1 and $\text{VaR}_{0.99}^{(2)}$ of portfolio 2. We illustrate this in Figure 13.4, where we depict both portfolio return density functions for values of $x \in [-3.5, -2]$. As we can see by the solid line representing the cumulative distribution of the returns of portfolio 2, it is more likely for portfolio 2 to achieve returns strictly less than -2.3263 than for portfolio 1, even though the probability for each portfolio to obtain a return of -2.3263 or less is identical, namely 1%. This additional risk, however, in the returns of portfolio 2 is not captured in the $\text{VaR}_{0.99}^{(2)}$. One solution to this problem is provided by the *expected shortfall*, which is defined as the expected loss given that the loss is worse than a certain benchmark. It builds on the concept of conditional probability discussed in Chapter 15.

---

[2]The Student's t-distribution is covered in Chapter 11.

FIGURE 13.4 Comparison of Left Tails of Two Normal Distributions with Identical VaR$_{0.99}$

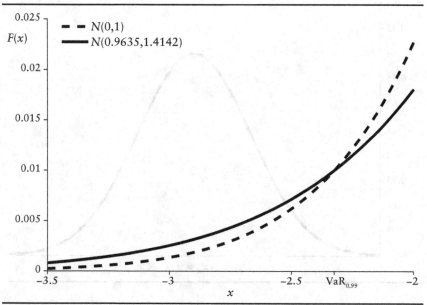

Mode

Let the random variable X follow some probability law, discrete or continuous. Suppose one is interested in the value that occurs with the highest probability if X is discrete or highest density function value for a continuous X. The location parameter having this attribute is called the *mode*.

When there is only one mode for a distribution, we refer to the probability distribution as *unimodal*. This is illustrated for the standard normal distributon in Figure 13.5, which depicts the standard normal density function. Note that in general probability distributions with bell-shaped density functions such as the normal distribution are unimodal.

As an example, let us consider standard normally distributed portfolio returns. The return value with the highest density value is $r_p = 0$ with $f(0) = 0.3989$. For all other values, the density function is lower.

The mode does not need to be unique, as we will see in the following example. Let X be some $B(5,0.5)$ random variable. In the table below, we give the corresponding probability distribution

	$x=0$	$x=1$	$x=2$	$x=3$	$x=4$	$x=5$	$x=6,7,\ldots$
$P(X=x)$	0.0313	0.1562	0.3125	0.3125	0.1562	0.0313	0

FIGURE 13.5 Mode of the Standard Normal Distribution

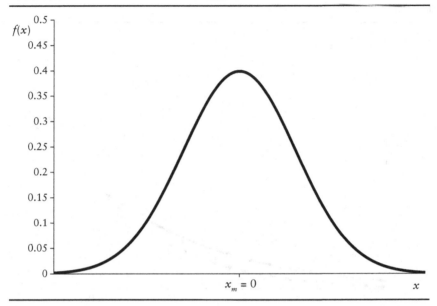

The modes are $x_m = 2$ and $x_m = 3$ because we have $P(2) = P(3) = 0.3125$ and $P(x) < 0.3125$, for $x = 0, 1, 4,$ and 5.

Mean (First Moment)

Of all information given by location parameters about some distribution, the *mean* reveals the most important insight. Although we already mentioned this location parameter when we presented discrete and continuous distributions here we provide a more formal presentation within the context of location parameters.

Mathematically, the mean of some distribution is defined as

$$E(X) = \sum_{x \in \Omega'} x \cdot P(X = x) \qquad (13.2)$$

if the corresponding random variable is discrete, and

$$E(X) = \int_{-\infty}^{\infty} x \cdot f(x) dx \qquad (13.3)$$

if the random variable is continuous with density function $f(x)$. In equation (13.2), we summed over all possible values in the state space Ω' that random variable X can assume. In equation (13.3), we integrate over all possible values weighted by the respective density values.

The mean can thus be interpreted as the *expected value*. However, we have to make sure that the mean actually exists in the sense that equation (13.2) or equation (13.3) are finite for absolute values. That is, we have to check that

$$\sum_{x \in \Omega'} |x| \cdot P(X = x) < \infty \qquad (13.4)$$

or

$$\int_{-\infty}^{\infty} |x| \cdot f(x) dx < \infty \qquad (13.5)$$

is satisfied, depending on whether the random variable is discrete or continuous. The requirements equations (13.4) and (13.5) are referred to as *absolute convergence*. This is an important criterion when modeling asset returns because for some distributions used in finance, a mean may not exist. For example, in Chapter 12, we discussed the α-stable distribution as a distribution that models well extreme behavior. Numerous empirical studies have found that this distribution better characterizes the return distribution of assets than the normal distribution. For α-stable distributions with $\alpha \leq 1$, however, the mean does not exist.

Whether the random variable is discrete or continuous, the mean represents the value we can expect, on average. Such extreme heavy-tailed distributions can occur when considering loss distributions in financial operational risk.

Mean of the Binomial Distribution

For example, consider again the binomial stock price model that we introduced in Chapter 9 where the up- and down-movements were given by the binomial random variable X with parameters $n = 5$ and $p = 0.5$, that is, $X \sim B(5, 0.5)$. Then by equation (13.2), the mean is computed as

$$E(X) = 0 \cdot \binom{5}{0} 0.5^0 \cdot 0.5^5 + 1 \cdot \binom{5}{1} 0.5^1 \cdot 0.5^4 + \ldots + 5 \cdot \binom{5}{5} 0.5^5 \cdot 0.5^0$$
$$= 5 \cdot 0.5 = 2.5$$

which is the result of the shortcut $E(X) = n \cdot p$ as we know from equation (9.7) in Chapter 9. So after five days, we have to expect 2.5 up-movements, on average.

Mean of the Poisson Distribution

As another example, suppose we are managing an insurance company that deals with director and officer liability coverage. By experience, we might know that the number of claims within a given period of time, say one year, may be given by a Poisson random variable N that we covered in Chapter 9. The parameter of the Poisson distribution λ is assumed here to be equal to 100. Then, the mean of the number of claims per year is given to be

$$E(X) = 0 \cdot \frac{100^0}{0!} \cdot e^{-100} + 1 \cdot \frac{100^1}{1!} \cdot e^{-100} + 2 \cdot \frac{100^2}{2!} \cdot e^{-100} + \ldots$$

$$= e^{-100} \cdot \left[1 \cdot \frac{100^1}{1!} + 2 \cdot \frac{100^2}{2!} + \ldots \right]$$

$$= 100 \cdot e^{-100} \cdot \left[1 + \frac{100^1}{1!} + \frac{100^2}{2!} + \ldots \right]$$

$$= 100 \cdot 1 = 100 = \lambda$$

where we used the fact that the last term in the third equality is exactly the probability $P(N \leq \infty)$ that naturally, equals one. This mean was already introduced in the coverage of the Poisson distribution in Chapter 9. Thus, our insurance company will have to expect to receive 100 claims per year.

Mean of the Exponential Distribution

As an example of a continuous random variable, let us consider the inter-arrival time τ between two consecutive claims in the previous example. The number N of claims inside of one year was previously modeled as a Poisson random variable with parameter $\lambda = 100$. We use this very same parameter for the distribution of the inter-arrival time to express the connectivity between the distribution of the claims and the time we have to wait between their arrival at the insurance company. Thus, we model τ as an exponential random variable with parameter $\lambda = 100$, that is, $\tau \sim Exp(100)$. So, according to equation (13.3), the mean is computed as

$$E(\tau) = \int_{-\infty}^{\infty} t \cdot f(t) \, dt = \int_{0}^{\infty} t \cdot 100 \cdot e^{-100 \cdot t} \, dt$$

$$= \left[-t \cdot e^{-100 \cdot t} \right]_{0}^{\infty} + \int_{0}^{\infty} e^{-100 \cdot t} \, dt$$

$$= \left(-\frac{1}{100} \right) \cdot \left[e^{-100 \cdot t} \right]_{0}^{\infty} = \frac{1}{100}$$

which is exactly $1/\lambda$ as we know already from our coverage of the exponential distribution in Chapter 11.[3] So, on average, we have to wait one hundreds of one year, or roughly 3.5 days, between the arrival of two successive claims.

The time between successive defaults in a portfolio of bonds is also often modeled as an exponential random variable. The parameter λ is interpreted as the default intensity, that is, the marginal probability of default within a vanishingly small period of time. This procedure provides an approach in credit risk management to model prices of structured credit portfolios called collateralized debt obligations.

Mean of the Normal Distribution

Let's compute the expected value of a stock price at time $t = 1$. From the perspective of time $t = 0$, the stock price at time $t = 1$ is given by a function of the random return r_1 and the initial stock price S_0 according to

$$S_1 = S_0 \cdot e^{r_1}$$

Moreover, the return may be given as a normal random variable with mean μ and variance σ^2, that is, $r_1 \sim N(0,02)$. We know from Chapter 11 that if a normally distributed random variable X enters as the exponent of an exponential function, the resulting random variable Y is log-normally distributed with the same parameters as X. Consequently, we have the ratio

$$\frac{S_1}{S_0} = e^{r_1}$$

as our log-normally distributed random variable. More precisely,

$$\frac{S_1}{S_0} \sim Ln(0, 0.2)$$

Thus by equation (13.3) and the presentation of the log-normal mean from Chapter 11, the expected value of the ratio is[4]

---

[3] We solved this integral in the first equality by the so-called method of *partial integration* that is explained in any introductory calculus textbook.
[4] The actual computation of the integral in the first equality requires knowledge beyond the scope assumed in this book.

$$E\left(\frac{S_1}{S_0}\right) = \int_0^\infty y \cdot \frac{1}{y\sqrt{2\cdot\pi\cdot 0.2}} \cdot e^{-\frac{(\ln y - 0)^2}{2\cdot 0.2}} dy$$
$$= e^{0.5 \cdot 0.2} = e^{0.1}$$
$$= 1.1052$$

So, on average, the stock price in $t = 1$ will be 10.52% greater than in $t = 0$. Suppose the initial stock price was \$100. This translates into an expected stock price at $t = 1$ of

$$E(S_1) = S_0 \cdot 1.1052 = \$110.52$$

In Figure 13.6 we display the mean of this $N(0,2)$ distributed return relative to its density function. We also present the quartiles as well as the median for this return in the figure.

PARAMETERS OF SCALE

While location parameters reveal information about the position of parts of the distribution, *scale parameters* indicate how dispersed the probability

FIGURE 13.6 Mean, Median, Mode, and Quartiles of the $N(0,02)$ Distributed One-Day Stock Price Return

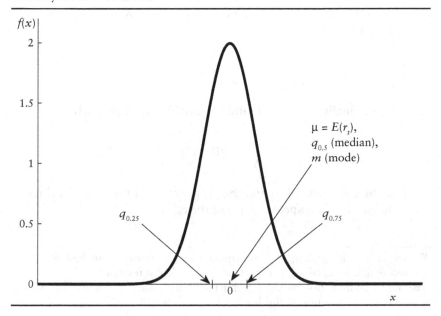

mass is relative to certain values or location parameters and whether it is distributed asymmetrically about certain location parameters.

Moments of Higher Order

In the previous section, we introduced the mean as the first moment. Here we introduce a more general version, the *k-th moment* or *moment of order k* measuring the dispersion relative to zero. It is defined as

$$E(X) \quad \sum_{\omega \in \Omega'} x^k \cdot P(X = x) \tag{13.6}$$

for discrete distributions, that is, on countable sets, and as

$$E(X) = \int_{-\infty}^{\infty} x^k \cdot f(x) dx \tag{13.7}$$

for continuous distributions. As with the mean, we have to guarantee first that equations (13.6) and (13.7) converge absolutely. That is, we need to check whether

$$\sum_{x \in \Omega'} |x|^k \cdot P(X = x) < \infty \tag{13.8}$$

or, alternatively,

$$\int_{-\infty}^{\infty} |x|^k \cdot f(x) dx < \infty \tag{13.9}$$

are met.

Moments of higher order than the mean are helpful in assessing the influence of extreme values on the distribution. For example, if $k = 2$, that is, the so-called *second moment*, not only are the signs of the values ignored, but all values are squared as a particular measure of scale. This has the effect that, relative to small values such as between -1 and 1, large values become even more pronounced.

Second Moment of the Poisson Distribution

As an example, consider the number N of firms whose issues are held in a bond portfolio defaulting within one year. Since we do not know the exact number at the beginning of the year, we model it as a Poisson distributed random variable. Note that the Poisson distribution is only an approximation since, technically, any non-negative integer number is feasible as an outcome under the Poisson law, while we only have a finite number of firms in our portfolio.

We decide to set the parameter equal to $\lambda = 5$. So, on average, we expect five companies to default within one year. With absolute convergence in equation (13.8) is fulfilled for $k = 2$, the second moment can be computed according to equation (13.6) as[5]

$$E(N^2) = \sum_{k=0}^{\infty} k^2 \cdot \frac{5^k}{k!} \cdot e^{-5}$$
$$= 5 + 5^2 = 30$$

So, 30 is the expected number of square defaults per year.

We can derive from the second equality that the second moment of a Poisson random variable with parameter λ is equal to $E(N^2) = \lambda + \lambda^2$. While this by itself has no interpretable meaning, it will be helpful, as we will see, in the computation of the variance, which we are going to introduce as a scale parameter shortly.

Second Moment of the Log-Normal Distribution

As another illustration, let us again consider the stock price model presented in the illustration of the mean earlier in this chapter. The stock price return r_1 from $t = 0$ to $t = 1$ was given by a normal random variable with mean $\mu = 0$ and variance $\sigma^2 = 0.2$. Then, the ratio of tomorrow's stock price to today's (i.e., S_1/S_0) is log-normally distributed with the same parameters as r_1. The second moment of S_1/S_0 is[6]

$$E\left(\left(S_1/S_0\right)^2\right) = \int_0^{\infty} y^2 \cdot \frac{1}{y\sqrt{2 \cdot \pi \cdot 0.2}} \cdot e^{-\frac{(\ln y - 0)^2}{2 \cdot 0.2}} dy$$
$$= e^{2(0 - 0.2^2)} = 0.9231$$

We see by the second equality that the second moment of a log-normal random variable Y with parameters μ and σ^2 is given by $E(Y^2) = \exp(\mu - \sigma^2)$.

Nonexistence of the Second Moment of the α-Stable Distribution

Now suppose that the one-day return r_1 is not normally distributed but, instead, distributed α-stable with parameters $\alpha = 1.8, \beta = 0, \sigma = 1$, and $\mu = 0$.[7]

---

[5]The mathematical steps to arrive at the second equality are too advanced for this text and, hence, not necessary to be presented here.
[6]The explicit computation of this integral is beyond the scope of this text.
[7]The α-stable distribution is explained in Chapter 12.

For r_1, the condition of absolute convergence (13.9) with $k = 2$ is not fulfilled. Thus, the return does not have a second moment. The same is true for the ratio S_1/S_0. We see that, under the assumption of an α-stable distribution, extreme movements are considered so common that higher moments explode while for the normal distribution, moments of all orders exist. This demonstrates the importance of the right choice of distribution when modeling stock returns. In particular daily or, even more so, intra-daily returns, exhibit many large movements that need to be taken care of by the right choice of a distribution.

Variance

In our discussion of the variance, skewness, and kurtosis, we will make use of the following definition of the *central moment of order k*.

Moments of some random variable X of any order that are centered about the mean μ rather than zero are referred to as *central moments*, and computed as

$$E\left((X-\mu)^k\right) \qquad (13.10)$$

provided that this expression is finite.

Earlier the second moment was presented as a particular measure of spread about zero. Here, we introduce a slightly different measure which is directly related to the second moment: the variance, abbreviated Var.[8] In contrast to the second moment, however, it is computed relative to the mean. It is referred to as the *second central moment* with k equal to 2 in definition (13.10). Formally, the variance is defined as the expected value of the squared deviations from the mean.

In the case of a discrete random variable X with mean $E(X) = \mu$, we have

$$Var(X) = \sum_{x \in \Omega}(x-\mu)^2 \cdot P(X = x) \qquad (13.11)$$

while when X is a continuous random variable, the definition is given by

$$Var(X) = \int_{-\infty}^{\infty}(x-\mu)^2 f(x)dx \qquad (13.12)$$

One can show that both equations (13.11) and (13.12) can be represented in a version using the second moment and the mean that in general, simplifies computations. This alternative representation is

---

[8] Do not confuse *Var* for variance with VaR, which stands for value-at-risk.

$$Var(X) = E(X^2) - (E(X))^2 \qquad (13.13)$$

which is valid for both discrete and continuous random variables. Sometimes, in particular if the mean and second moment are already known, it is more convenient to use version (13.13) rather than having to explicitly compute equations (13.11) or (13.12).

Variance of the Poisson Distribution

As an illustration, we consider the random number N of firms whose bonds are held in a portfolio defaulting within one year. Again, we model the distribution of N using the Poisson distribution with parameter $\lambda = 5$ such that we expect five firms to default within one average year. However, in each individual year, the actual number of defaults may vary significantly from five. This we measure using the variance. According to equation (13.11), it is computed as[9]

$$\begin{aligned} Var(N) &= \sum_{k=0}^{\infty} (k-5)^2 \cdot \frac{5^k}{k!} \cdot e^{-5} \\ &= \sum_{k=0}^{\infty} (k^2 - 2 \cdot k \cdot 5 + 25) \cdot \frac{5^k}{k!} \cdot e^{-5} \\ &= \sum_{k=0}^{\infty} k^2 \cdot \frac{5^k}{k!} \cdot e^{-5} - 10 \underbrace{\sum_{k=0}^{\infty} k \frac{5^k}{k!} \cdot e^{-5}}_{E(N)=5} + 25 \underbrace{\sum_{k=0}^{\infty} \frac{5^k}{k!} \cdot e^{-5}}_{1} \\ &= 5 \cdot \sum_{k=0}^{\infty} (k-1+1) \cdot \frac{5^{k-1}}{(k-1)!} \cdot e^{-5} - 25 \\ &= 5 \cdot \underbrace{\sum_{k=1}^{\infty} (k-1) \cdot \frac{5^{k-1}}{(k-1)!} \cdot e^{-5}}_{E(N)=5} + 5 \cdot \underbrace{\sum_{k=1}^{\infty} \frac{5^{k-1}}{(k-1)!} \cdot e^{-5}}_{1} - 25 \\ &= 5 = \lambda \end{aligned}$$

We see that the variance of a Poisson distribution is, as well as the mean, equal to parameter λ. With $E(N^2) = 25$ and $E(N) = 5$, we can conveniently compute the variance as $Var(N) = E(N^2) - (E(N))^2 = 30 - 25 = 5$, according to equation (13.13).

---

[9] For the interested reader, we have included some of the intermediate calculations. They are, however, not essential to understand the theory

Variance of the Exponential Distribution

In our next illustration, let's return to the insurance company example and consider the exponentially distributed time τ between two successive claims reported. Recall that τ was distributed $\tau \sim Exp(100)$; that is, we have to expect to wait 1/100 of a year, or roughly 3.5 days, between one and the next claim. Equation (13.12) yields the variance as

$$Var(X) = \int_0^\infty \left(x - \frac{1}{100}\right)^2 \cdot 100 \cdot e^{-100 \cdot x} dx$$

$$= \int_0^\infty \left(x^2 - \frac{2x}{100} + \frac{1}{100^2}\right) \cdot 100 \cdot e^{-100 \cdot x} dx$$

$$= \int_0^\infty x^2 \cdot 100 \cdot e^{-100 \cdot x} dx - \frac{2}{100} \underbrace{\int_0^\infty x \cdot 100 \cdot e^{-100 \cdot x} dx}_{E(\tau) = 1/100} + \frac{1}{100^2} \underbrace{\int_0^\infty 100 \cdot e^{-100 \cdot x} dx}_{1}$$

$$= \int_0^\infty x^2 \cdot 100 \cdot e^{-100 \cdot x} dx - \frac{2}{100} \underbrace{\int_0^\infty x \cdot 100 \cdot e^{-100 \cdot x} dx}_{E(\tau) = 1/100} + \frac{1}{100^2} \underbrace{\int_0^\infty 100 \cdot e^{-100 \cdot x} dx}_{1}$$

$$= \left[-x^2 \cdot e^{-100 \cdot x}\right]_0^\infty + \frac{2}{100} \underbrace{\int_0^\infty x \cdot 100 \cdot e^{-100 \cdot x} dx}_{E(\tau) = 1/100} - \frac{2}{100^2} + \frac{1}{100^2}$$

$$= 0 + \frac{1}{100^2} = \frac{1}{100^2}$$

We can see that the variance of an exponential random variable is equal to $1/\lambda^2$.[10]

Variance of the Normal Distribution and α-Stable Distribution

Now, let us recall the normally distributed one-day stock return r_1 from previous examples. The mean of the distribution is given by $\mu = 0$ while its variance equals $\sigma^2 = 0.2$. So, we are done since the variance is already provided by one of the parameters. The variance of the corresponding log-normally distributed ratio of tomorrow's stock price to today's (i.e., S_1/S_0) is[11]

---

[10] In the second last equation of the above calculations, we used partial integration. Note that this is only addressed to the reader interested in the calculation steps and not essential for the further theory.

[11] The explicit computation is beyond the scope of this text. Note that the formula of the variance is stated in Chapter 12, in the presentation of the log-normal distribution.

$$Var(S_1/S_0) = e^{2 \cdot 0 + 0 \cdot ?}\left(e^{0.2} - 1\right) = 0.2704$$

If, on the other hand, the return was α-stable distributed with α = 1.8, β = 0, σ = 1, and μ = 0, the variance does neither exist for the return nor the stock price ratio since the respective second moments are infinite, as was mentioned already.

Standard Deviation

Instead of the variance, one often uses the ***standard deviation***, often abbreviated as *std. dev.* and in equations denoted by σ, which is simply the square root of the variance, that is,

$$\sigma = \sqrt{Var(X)}$$

In many cases, the standard deviation is more appealing than the variance since the latter uses squares and, hence, yields a value that is differently scaled than the original data. By taking the square root, the resulting quantity (i.e., the standard deviation) is in alignment with the scale of the data.

At this point, let us recall the Chebychev inequality, discussed in Chapter 11, that states that the probability of any random variable X deviating by more than c units from its mean μ is less than or equal to the ratio of its variance to c^2. From this inequality, we have that

$P(|X - \mu \leq \sigma|) > 0$ Roughly 0% of the data are within one standard deviation about the mean.

$P(|X - \mu \leq 2 \cdot \sigma|) > \dfrac{3}{4}$ More than 75% of the data are within two standard deviations about the mean.

$P(|X - \mu \leq 3 \cdot \sigma|) > \dfrac{8}{9}$ More than 89% of the data are within three standard deviations about the mean.

This, however, is a very coarse estimate since it has to apply for any type of distribution. A more refined guidance derived from the normal distribution can be given if the distribution is symmetric about the median and unimodal. Then, by the following empirical rule of thumb, we have

$P(|X - \mu \leq \sigma|) \approx 0.683$ Roughly 68% of the data are within one standard deviation about the mean.

$P(|X - \mu \leq 2 \cdot \sigma|) \approx 0.955$ More than 96% of the data are within two standard deviations about the mean.

$$P(|X-\mu| \leq 3 \cdot \sigma) \approx 0.997$$ About all of the data are within three standard deviations about the mean.

With our previous one-day stock price return, which we assumed to follow the $N(0,0.2)$ law, we have to expect that on 68 out of 100 days we have a return between $-\sqrt{0.2} = -0.4472$ and $\sqrt{0.2} = 0.4472$. Furthermore, in 96 out of 100 days, we have to expect the return to be inside of the interval

$$\left[-2 \cdot \sqrt{0.2}, 2 \cdot \sqrt{0.2}\right] = \left[-0.8944, 0.8944\right]$$

and virtually no day will experience a return below $-3 \cdot \sqrt{0.2} = -1.3416$ or above $3 \cdot \sqrt{0.2} = 1.3416$. We demonstrate this in Figure 13.7.

This translates to the following bounds for the stock price ratio S_1/S_0

[0.6394, 1.5639]   in 68%,
[0.4088, 2.4459]   in 95%, and
[0.2614, 3.8253]   in 100% of the cases,

which, setting $S_0 = \$100$, corresponds to bounds for tomorrow's stock price of

[\$63.94, \$156.39]   in 68%,
[\$40.88, \$244.59]   in 95%, and
[\$26.14, \$382.53]   in 100% of the cases.

FIGURE 13.7 Rule of Thumb for $N(0,0.2)$ One-Day Return r_t

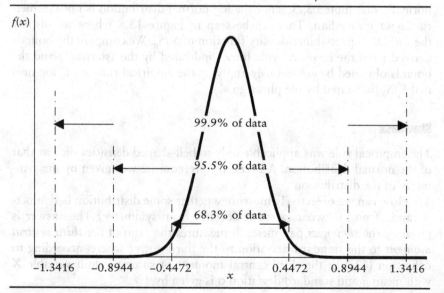

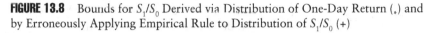

FIGURE 13.8 Bounds for S_1/S_0 Derived via Distribution of One-Day Return (∗) and by Erroneously Applying Empirical Rule to Distribution of S_1/S_0 (+)

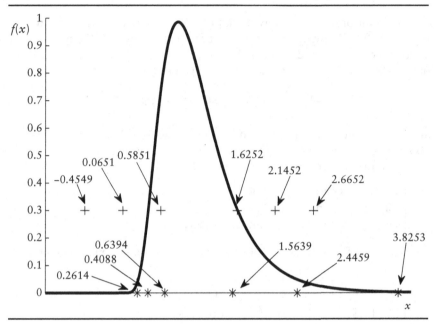

Note that we cannot apply the rule of thumb immediately to the log-normally distributed S_1/S_0 since the log-normal distribution is not symmetric about its median. This can be seen in Figure 13.8 where we display the $Ln(0,0.2)$ probability density function of S_1/S_0. We compare the bounds derived from the empirical rule for r_1 (indicated by the asterisk ∗) and the bounds obtained by erroneously applying the empirical rule to the log-normal S_1/S_0 (indicated by the plus sign +).

Skewness

The empirical rule was applicable only to bell-shaped densities such as that of the normal distribution. A necessary prerequisite was given by the symmetry of the distribution.

How can we objectively measure whether some distribution is symmetric and, if not, to what extent it deviates from symmetry? The answer is given by the *skewness* parameter. It measures the ratio of the *third central moment* to the standard deviation to the third power where, according to definition (13.10), the third central moment of some random variable X with mean μ and standard deviation σ is given by

Parameters of Location and Scale of Random Variables

$$E\left((X-\mu)^3\right)$$

such that the skewness is formally defined as

$$\sigma_3 = \frac{E\left((X-\mu)^3\right)}{\sigma^3} \qquad (13.14)$$

Since, in contrast to the standard deviation, the third central moment is sensitive to the signs of the deviations $(x - \mu)$, the skewness parameter σ_3 can assume any real value. We distinguish between three different sets of values for σ_3: less than zero, zero, and greater than zero. The particular kind of skewness that some distribution displays is given below:

$\sigma_3 < 0$ *left-skewed*
$\sigma_3 = 0$ *not skewed*
$\sigma_3 > 0$ *right-skewed*

Skewness of Normal and Exponential Distributions

Since the density function of the normal distribution is symmetric, its skewness is zero.

As another example, consider the exponentially distributed time τ between two successive failures of firms in our bond portfolio illustration where we assumed parameter $\lambda = 5$, that is, $\tau \sim Exp(5)$. The computation of the skewness yields $\sigma_3 = 2 > 0$ such that the exponential distribution is right-skewed. Note that the mode of τ is $m = 0$, the median is $q_{0.5} = 0.1386$, and the mean equals $E(\tau) = 0.2$. Hence, the numeric order of these three location parameters is $m < q_{0.5} < E(\tau)$. We display this in Figure 13.9. This is always the case for right-skewed distributions with one mode, only (i.e., unimodal distributions). The exact opposite numerical order is true for left-skewed distributions with one mode.[12]

Skewness of GE Daily Returns

As another illustration of the skewness in stock returns, let's look at the daily returns of the General Electric (GE) stock. It is commonly found that daily financial data exhibit skewness, in particular, negative skewness. We will verify this using the daily 7,300 observations from April 24, 1980 until

---

[12]Note that distributions can have more than one mode, such as the rectangle distribution.

FIGURE 13.9 Location Parameters of Right-Skewed $Exp(5)$ Distribution

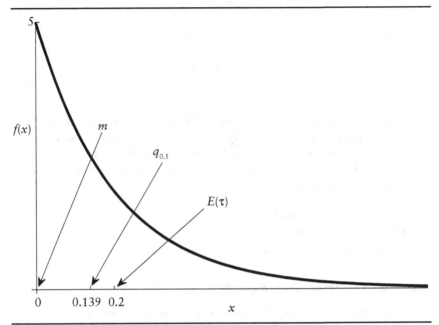

March 30, 2009. We display the returns in Figure 13.10. As we can see, the returns roughly assume values between −0.20 and 0.20 in that period.

To compute the skewness as given by equation (13.14), we need knowledge of the probability distribution. Since due to its symmetry the normal distribution cannot capture skewness in any direction, we have to resort to some probability distribution whose parameters allow for asymmetry. Let us assume that the daily GE returns are normal inverse Gaussian (NIG) distributed. From Chapter 12 where we discuss this distribution, we know that the *NIG* distribution has four parameters, (a,b,μ,δ). For the GE data, we have $a = 31.52$, $b = -0.7743$, $\mu = 0.0006$, and $\delta = 0.0097$ obtained from estimation.

Now, the skewness in equation (13.14) for a *NIG* random variable can be specified as

$$\sigma_3 = 3 \cdot \frac{b}{a} \cdot \frac{1}{\sqrt{d \cdot \sqrt{a^2 - b^2}}}$$

Thus with the parameter values given, this skewness is approximately equal to −0.1330. Hence, we observe left-skewed returns.

FIGURE 13.10 GE Daily Returns Between April 24, 1980 and March 30, 2009

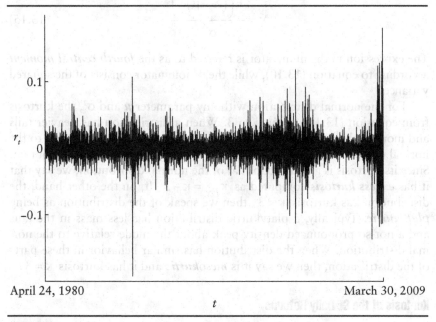

Note: Data obtained from finance.yahoo.com

Kurtosis

Many probability distributions appear fairly similar to the normal distribution, particularly if we look at the probability density function. However, they are not the same. They might distinguish themselves from the normal distribution in very significant aspects.

Just as the existence of positive or negative skewness found in some data is a reason to look for alternatives to the normal distribution, there may also be other characteristics inherent in the data that render the normal distribution inappropriate. The reason could be that there is too much mass in the tails of the distribution of the data relative to the normal distribution, referred to as *heavy tails*, as well as an extremely accentuated peak of the density function at the mode. That is, we have to look at the *kurtosis* of the distribution of the data.

These parts of the distribution just mentioned (i.e., the mass in the tails) as well as the shape of the distribution near its mode, are governed by the kurtosis parameter. For a random variable X with mean μ and variance σ^2, the kurtosis parameter, denoted by κ, is defined as

$$\kappa = \frac{E\left((X-\mu)^4\right)}{\sigma^4} \qquad (13.15)$$

The expression in the numerator is referred to as the *fourth central moment* according to equation (13.10), while the denominator consists of the squared variance.

For the normal distribution with any parameter μ and σ^2, the kurtosis from equation (13.15) is equal to 3. When a distribution has heavier tails and more probability is assigned to the area about the mode relative to the normal distribution, then $\kappa > 3$ and we speak of a *leptokurtic* distribution. Since its kurtosis is greater than that of the normal distribution, we say that it has *excess kurtosis* computed as $\kappa_{Ex} = \kappa - 3$. If, on the other hand, the distribution has kurtosis $\kappa < 3$, then we speak of the distribution as being *platykurtic*. Typically, a platykurtic distribution has less mass in the tails and a not-so-pronounced density peak about the mode relative to the normal distribution. When the distribution has similar behavior in these parts of the distribution, then we say it is *mesokurtic* and it has kurtosis $\kappa = 3$.

Kurtosis of the GE Daily Returns

As an illustration, let us continue with the analysis of the 7,300 GE daily returns between April 24, 1900 and March 30, 2009. For the *NIG* distribution, the kurtosis from equation (13.15) can be specified as

$$\kappa = 3 + 3 \cdot \left(1 + 4\left(\frac{\beta}{\alpha}\right)^2\right) \cdot \frac{1}{\delta\left(\sqrt{\alpha^2 - \beta^2}\right)}$$

With the parameters given previously when we illustrated skewness calculated for the GE daily returns, the kurtosis is equal to $\kappa = 12.8003$, which is well above 3. Thus, using the *NIG* distribution, the parameter values suggest that the GE daily returns are leptokurtic; that is, they are heavy tailed with a high peak at the mode.

For comparison, we also choose the normal distribution as an alternative for the returns. When the GE daily returns are modeled as a normal random variable, the parameters are given as $\mu = 0.0003$ and $\sigma^2 = 0.0003$. We display both, the NIG (solid) and normal (dashed) density functions in Figure 13.11.[13] Note how the leptokurtosis of the *NIG* is distinctly higher

---

[13] For the computation of the NIG density function, we greatly appreciate the open-source MATLAB code provided by Dr. Ralf Werner, Allianz, Group Risk Controlling, Risk Methodology, Koeniginstr. 28, D-80802 Muenchen.

FIGURE 13.11 Modeling GE Daily Returns with the Normal and the NIG Distribution

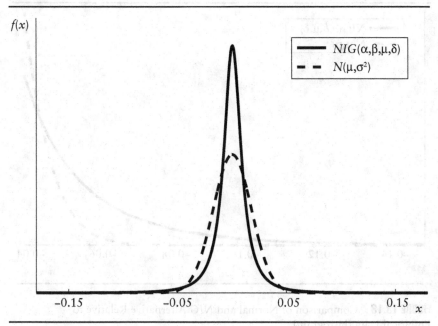

than that for the normal distribution. In addition, the heaviness of the NIG tails relative to the normal distribution can be seen in Figure 13.12 where we compare the respective left, that is, negative, tails.

Finally, the two distributional alternatives are compared with the empirical data given by the histogram (gray) in Figure 13.13.[14] As we can see, the NIG density curve (solid) reflects the features of the histogram better than the normal density (dashed).

---
[14]The histogram of empirical data is explained in Chapter 4.

FIGURE 13.12 Comparison of the Left Tails of the GE Daily Returns: Normal Distribution versus *NIG* Distribution

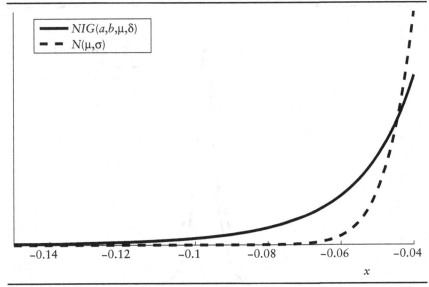

FIGURE 13.13 Comparison of Normal and *NIG* Alternative Relative to Empirical Data (histogram)

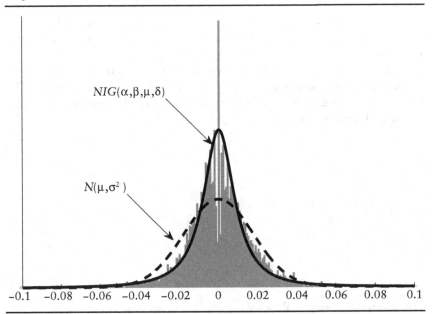

CONCEPTS EXPLAINED IN THIS CHAPTER (IN ORDER OF PRESENTATION)

Quantiles
Percentile
Infimum (inf)
Lower quartile
Upper quartile
Value-at-risk (VaR)
Expected shortfall
Mode
Unimodal
Mean
Expected Value
Absolute convergence
Scale parameters
k-th moment (moment of order k)
Second moment
Central moment of order k
Central moments
Second central moment
Standard deviation
Skewness
Third central moment
Heavy tails
Kurtosis
Fourth central moment
Leptokurtic
Excess kurtosis
Platykurtic
Mesokurtic

APPENDIX: PARAMETERS FOR VARIOUS DISTRIBUTION FUNCTIONS

Distribution	Mean	Variance	Skewness	Kurtosis
Normal	μ	σ^2	0	3
Chi-square	f	$2f$	$\sqrt{\dfrac{8}{f}}$	$3+\dfrac{12}{f}$
Student's t	0, if $n \geq 2$	$\dfrac{n}{n-2}$, if $n \geq 3$	0, if $n \geq 4$	$3+\dfrac{6}{n-4}$, if $n \geq 5$
F	$\dfrac{n_2}{n_2-2}$, if $n_2 > 2$	$\dfrac{2n_2^2(n_1+n_2-2)}{n_1(n_2-2)^2(n_2-4)}$, if $n_2 > 4$	$\dfrac{2n_1+n_2-2}{n_2-6}\sqrt{\dfrac{8(f_2-4)}{f_1(f_1+f_2-2)}} > 0$, if $n_2 > 6$	$3+\dfrac{12\left[(f_2-2)^2(f_2-4)+f_1(f_1+f_2-2)(5f_2-22)\right]}{f_1(f_2-6)(f_2-8)(f_1+f_2-2)} > 3$, if $n_2 > 8$
Exponential	$\dfrac{1}{\lambda}$	$\dfrac{1}{\lambda^2}$	2	9
Weibull	$a+b\Gamma\left(1+\dfrac{1}{c}\right)$	$b^2\left[\Gamma\left(1+\dfrac{2}{c}\right)-\Gamma^2\left(1+\dfrac{1}{c}\right)\right]$	> 0, if $c \leq 3.6$; ≈ 0, if $c \approx 3.6$; < 0, if $c \geq 3.6$	Does not exist
Rectangular	$\dfrac{a+b}{2}$	$\dfrac{(a-b)^2}{12}$	0	1.8
Gamma	$\dfrac{c}{\lambda}$	$\dfrac{c}{\lambda^2}$	$\dfrac{2}{\sqrt{c}}$	$3+\dfrac{6}{c}$

Distribution		Mean	Variance	Skewness	Kurtosis
Beta		$\dfrac{c}{c+d}$	$\dfrac{c \cdot d}{(c+d)^2(c+d+1)}$	> 0, if $d < c$ $= 0$, if $c \approx 3.6$ < 0, if $d > c$	$\dfrac{3(c+d+1)\left[c^2(d+2)+d^2(c+2)-2cd\right]}{cd(c+d+2)(c+d+3)}$
Log-normal		$e^{\mu+\frac{\sigma^2}{2}}$	$e^{\sigma^2}\cdot\left(e^{\sigma^2}-1\right)\cdot e^{2\mu}$	$\left(e^{\sigma^2}+2\right)\sqrt{e^{\sigma^2}-1}$	$\left(e^{\sigma^2}\right)^4 + 2\left(e^{\sigma^2}\right)^3 + 3\left(e^{\sigma^2}\right)^2 - 3$
Generalized Error	I[a]	0.5772	$1.6449 \cdot b$	1.1396	≈ 5.4
	II[b]				
	III[c]				
Pareto		$\dfrac{\beta}{\xi(\xi-1)}$ if $1 > \xi$	$\dfrac{\beta^2}{(1-\xi)^2(1-2\xi)}$ if $1 > 2\xi$	$\dfrac{2(\xi+1)}{(1-3\xi)}\sqrt{1-2\xi}$, if $1 > 3\xi$	$3 + \dfrac{6\left(1+\xi+6\xi^2-2\xi^3\right)}{(1-3\xi)(1-4\xi)}$ if $1 > 4\xi$
Normal Inverse Gaussian		$\mu+\delta\dfrac{b}{\sqrt{a^2-b^2}}$	$\delta\dfrac{a^2}{\left(\sqrt{a^2-b^2}\right)^3}$	$3\cdot\dfrac{b}{a}\cdot\dfrac{b}{\left(\delta\cdot\sqrt{a^2-b^2}\right)^{1/2}}$	$3\left(1+4\left(\dfrac{b}{a}\right)^2\right)\dfrac{b}{\delta\cdot\sqrt{a^2-b^2}}$
α-Stable		μ, if $\alpha > 1$	$2\cdot\sigma^2$, if $\alpha = 2$	It exists only if $\alpha = 2$	It exists only if $\alpha = 2$

[a] We give the GEV distributions for standardized random variables. The parameter b is the scale of the non-standardized random variables.
[b] Due to the complexity of the formulae not listed, here.
[b] Due to the complexity of the formulae not listed, here.

CHAPTER 14
Joint Probability Distributions

In previous chapters, we explained the properties of a probability distribution of a single random variable; that is, the properties of a univariate distribution. Univariate distributions allow us to analyze the random behavior of individual assets, for example. In this chapter, we move from the probability distribution of a single random variable (univariate distribution) to the probability law of two or more random variables, which we call a *multivariate distribution*. Understanding multivariate distributions is important in financial applications such as portfolio selection theory, factor modeling, and credit risk modeling, where the random behavior of more than one quantity needs to be modeled simultaneously. For example, Markowitz portfolio theory builds on multivariate randomness of assets in a portfolio.

We begin this chapter by introducing the concept of joint events and joint probability distributions with the joint density function. For the latter, we present the contour lines of constant density. From there, we proceed to the marginal probability distribution, followed by the immensely important definition of stochastic dependence. As common measures of joint random behavior, we will give the covariance and correlation parameters along with their corresponding matrices.

In particular, we present as continuous distributions the multivariate normal and multivariate Student's t-distribution, as well the elliptical distributions. All of these distributions play an important role in financial modeling. In Chapter 16 we will learn that the dependence structure of a distribution function is given by a concept called a copula and introduce the tail dependence that provides an important device in assessing the probability of joint extreme movements of different stock returns, for example. Wherever necessary in this chapter, we will make a distinction between discrete and continuous cases.

HIGHER DIMENSIONAL RANDOM VARIABLES

In the theory of finance, it is often necessary to analyze several random variables simultaneously. For example, if we only invest in the stock of one particular company, it will be sufficient to model the probability distribution of that stock's return exclusively. However, as soon as we compose a portfolio of stocks of several companies, it will very likely not be enough to analyze the return distribution of each stock individually. The reason is that the return of a stock might influence the return of another stock, a feature that is neglected if we consider stock returns in isolation. So, when managing a portfolio, we need to orchestrate the component stock returns jointly.

Discrete Case

Let's begin by introducing discrete probability distributions. Let X be a discrete random variable consisting of, say, k components so that we can write X as

$$X = (X_1, X_2, \ldots, X_k) \tag{14.1}$$

where in equation (14.1) each of the components $X_1, X_2, \ldots, X_k$ can be treated as a univariate random variable. Because of its multidimensionality, the random variable X is also referred to as **random vector**. The random vector assumes values which we denote by the k-tuple $(x_1, x_2, \ldots, x_k)$.[1]

For discrete random vectors, the space Ω can be viewed as the countable set of all outcomes $(x_1, x_2, \ldots, x_k)$ where the $x_1, x_2, \ldots, x_k$ are real numbers.[2] The σ-algebra $\mathbb{A}(\Omega)$ introduced in Chapter 8 then consists of sets formed by the outcome k-tuples $(x_1, x_2, \ldots, x_k)$.

For example, suppose we have two assets. Let us assume that asset one, denoted by A_1, can only take on one of the following three values next period: $90, $100, or $110. Suppose asset two, denoted by A_2, can take on only one of the following two values next period: $80 or $120. So, the space containing all outcome pairs (x_1, x_2) is given by

$$\Omega = \{(90,80),(100,80),(110,80),(90,120),(100,120),(110,120)\}$$

The first component, x_1, of each pair represents the value of A_1 next period, while the second component, x_2, represents the value of A_2 next period. The corresponding σ-algebra with the combined events is then

---

[1] If $k = 2$, a tuple is referred to as a *pair* while for $k = 3$, a tuple is called a *triple*.
[2] More generally, we could write $(\omega_1, \ldots, \omega_k)$ to indicate elementary events or, equivalently, outcomes. But since we are only dealing with real numbers as outcomes, we use the notation $(x_1, \ldots, x_k)$, which are the values that the random vector X can assume.

formed from these elements of Ω. For example, such a combined event B could be that A_1 is either equal to $90 or $100 and $A_2 = $120. This event B is given by the element $\{(90,120), (100,120)\}$ of the σ-algebra.

Continuous Case

In contrast to the discrete case just described, we now assume the components of the random vector X to be random variables with values in uncountable sets. That is, X is now continuous. Consequently, the k-dimensional random vector X is associated with outcomes in an uncountable set Ω of k-dimensional elements. The values given by the k-tuple $(x_1, x_2, ..., x_k)$ are now such that each of the k components x_i is a real number.

The corresponding σ-algebra $\mathbb{A}(\Omega)$ is slightly difficult to set up. We can no longer simply take all possible outcome tuples and combine them arbitrarily. As discussed in Chapter 8 as a main example of Ω being an uncountable set, we can take the k-dimensional σ-algebra of the Borel sets, $\mathbb{B}^k$, instead, which contains, for example, all real numbers, all sorts of open and half-open intervals of real numbers, and all unions and intersections of them.

For example, let us now model the two previous assets A_1 and A_2 as continuous random variables. That is, both A_1 and A_2 assume values in uncountable sets such that the resulting combination, that is, the pair (A_1, A_2), takes on values (x_1, x_2) in an uncountable set Ω itself. Suppose the possible values for A_1 were in the range [90,110]; that is A_1 could have any price between $90 and $110. For A_2, let us take the range [80,120], that is, A_2 has any value between $80 and $120. Thus, the space containing all outcomes for the random vector (A_1, A_2) is now formally given by[3]

$$\Omega = \{(x_1, x_2) | x_1 \in [90,110] \text{ and } x_2 \in [80,120]\}$$

As the set of events, we have the σ-algebra of the 2-dimensional Borel sets, this is, $\mathbb{B}^k$. For example, "A_1 between $93.50 and $100 and A_2 less than $112.50" corresponds to

$$\{(x_1, x_2) | x_1 \in [93.50, 100] \text{ and } x_2 \in (-\infty, 112.50]\}$$

or "A_1 equal to $90 and A_2 at least $60" corresponds to

$$\{(90, x_2) | x_2 \in [60, \infty)\}$$

could be such events from that σ-algebra.

---
[3] A set of the form $\{(x_1, x_2) | x_1 \in A \text{ and } x_2 \in B\}$ is to be understood as any pair where the first component is in set A and the second component in set B.

JOINT PROBABILITY DISTRIBUTION

Now let's introduce the probability distribution for random vectors such as the previously introduced X. Since the randomness of each component of X has to be captured in relation to all other components, we end up treating the components' random behavior jointly with the *multivariate* or *joint probability distribution*.

Discrete Case

We begin with the analysis of joint probability distributions in the discrete case. Suppose Ω was our countable set composed of the k-dimensional outcomes $(x_1, \ldots, x_k)$. Each particular outcome is assigned a probability between 0 and 1 such that we can express by

$$P(X = (x_1, \ldots, x_k))$$

the probability that random vector X is equal to $(x_1, \ldots, x_k)$. In the discrete case, it is easy to give the probability that X assumes a value in set B where B is in the σ-algebra. Since B consists of outcomes, the probability of B, that is, $P(B) = P(X \in B)$, is computed by simply collecting all outcomes that are contained in B. Since outcomes are mutually exclusive, we then add the probabilities of the respective outcomes and obtain the probability of B. More formally, we express this probability in the form

$$P(B) = \sum_{(x_1, \ldots, x_k) \in B} P(X = (x_1, \ldots, x_k))$$

Suppose we are interested in the joint event that each component X_i of random vector X was less than or equal to some respective value a_i. This can be expressed by the event

$$B = \{(x_1, \ldots, x_k) | x_1 \leq a_1, \ldots, x_k \leq a_k\} \quad (14.2)$$

We know from previous chapters that the probability of some univariate random variable Y assuming values less than or equal to a, $P(Y \leq a)$, is given by the distribution function of Y evaluated at a, that is, $F_Y(a)$. A similar approach is feasible for events of the form (14.2) in that we introduce the joint distribution function for discrete random variables. Formally, the *joint cumulative distribution function* (cdf) is given by[4]

---

[4] In words, the notation $(x_1, \ldots, x_k): x_1 \leq a_1, \ldots, x_k \leq a_k$ using the colon means "all k-tuples $(x_1, \ldots, x_k)$ such that $x_1 \leq a_1, \ldots, x_k \leq a_k$."

$$F_X((a_1,\ldots,a_k)) = \sum_{(x_1,\ldots,x_k):x_1\leq a_1,\ldots,x_k\leq a_k} P(X=(x_1,\ldots,x_k)) \qquad (14.3)$$

Since the form in equation (14.3) is rather complicated, we introduce shorthand notation. The outcomes are abbreviated by $x = (x_1, \ldots, x_k)$. Then, event B from (14.2) can be rewritten in the form $B = \{x \leq a\}$, where a represents the tuple $(a_1, \ldots, a_k)$. So, we can write the cdf from equation (14.3) in the more appealing form

$$F_X(a) = \sum_{x\leq a} P(X=x) \qquad (14.4)$$

To illustrate the importance of understanding joint probability distributions, we use as a very simple example the two assets A_1 and A_2 from our earlier example. For convenience, below we relist the values each asset can assume:

A_1	$90	$100	$110

and

A_2	$80	$120

As pointed out earlier, if we formed a portfolio out of these two assets, the mere knowledge of the values each asset can separately assume is not enough. We need to know their joint behavior. For example, certain values of A_1 might never occur with certain other values of A_2. As an extreme case, consider the joint probability distribution given by the following 2-by-3 table:

A_2 \ A_1	$90	$100	$110
$80		0.05	0.45
$120	0.5		

That is, we have only three different possible scenarios. With 5% probability, A_1 will have a value of $100 while simultaneously A_2 will have a value of $80. In another 45% of the cases, the value of A_1 will be equal to $110 while, at the same time, that of A_2 will be $80 again. And, finally, with 50% probability, we have a value for A_1 of $90 while, at the same time, A_2 is worth $120. Besides these three combinations, there are no value pairs for A_1 and A_2 with positive probability.

Continuous Case

As in the univariate case, when the probability distribution is continuous, a single value for $x = (x_1, x_2, ..., x_n)$ occurs with probability zero. Only events corresponding to sets of the form

$$A = \{(x_1,...,x_k) | a_1 \leq x_1 \leq b_1, a_2 \leq x_2 \leq b_2, ..., a_k \leq x_k \leq b_k\} \quad (14.5)$$

can have positive probability. The set A in equation (14.5) comprises all values whose first component lies between a_1 and b_1, whose second component is between a_2 and b_2, and so forth. Set A, thus, generates a k-dimensional *volume* with edge lengths $b_1 - a_1$ through $b_k - a_k$ that all need to be greater than zero for the volume to have positive probability.[5]

If a random variable X has a continuous probability law, we define the probability of an event represented by the form of A as

$$P(A) = \int_{a_1 \leq x_1 \leq b_1,...,a_k \leq x_k \leq b_k} \cdots \int f_X(x_1, x_2, ..., x_k) \cdot dx_1 \cdot dx_2 \cdots dx_k \quad (14.6)$$

That is, we obtain $P(A)$ by integrating the *joint* or *multivariate probability density function* (pdf) f_X of X over the values in A.[6] Analogous to the univariate case presented in Chapter 10, the joint probability density function represents the rate of probability increment at some value $(x_1, x_2, ..., x_k)$ by taking infinitesimally small positive steps dx_1 through dx_k in the respective components.

A brief comment on notation is in order. In equation (14.6) we should have actually written k integrals explicitly because we are integrating with respect to k random components, x_1 through x_k. For simplicity, however, we omit the explicit presentation and merely indicate them by the "$\cdots$" symbols. Moreover, the multiple integral in equation (14.6) can also be written in the short version

$$P(A) = \int_{a \leq x \leq b} f(x) dx$$

where we represent the k-tuples $(x_1, ..., x_k)$ in the abbreviated notation x.

For the probability of the particular event that all components are less than or equal to some respective bound, we introduce the joint cdf of multivariate continuous random variables. Let B denote such an event, which can be formally expressed as

---

[5] In the case of $k = 2$, the volume is an area.
[6] Instead of probability density function, we will often simply refer to it as *density function*.

$$B = \{(x_1,\ldots,x_k) | x_1 \leq b_1, x_2 \leq b_2, \ldots, x_k \leq b_k\} \tag{14.7}$$

or, in brief, $B = \{x \leq b\}$. Then, we can state the probability of this event B as

$$\begin{aligned} P(B) &= \int_{x_1 \leq b_1, x_2 \leq b_2, \ldots, x_k \leq b_k} \cdots \int f_X(x_1, x_2, \ldots, x_k) \cdot dx_1 \cdot dx_2 \cdot \ldots \cdot dx_k \\ &= F_X(b_1, b_2, \ldots, b_k) \end{aligned} \tag{14.8}$$

where in the first equality of equation (14.8), we used the representation given by equation (14.6). The second equality in equation (14.8) accounts for the fact that the probability of this event is given by the cumulative distribution function F_X evaluated at $(b_1, b_2, \ldots, b_k)$. So, equation (14.8) can be regarded as a definition of the joint (or multivariate) cumulative distribution function of some continuous random variable X.

To illustrate, consider the two-dimensional case (i.e., $k = 2$). Suppose we have two stocks, S_1 and S_2, whose daily returns can be modeled jointly by a two-dimensional random variable $r = (r_1, r_2)$ where component r_1 denotes the return of S_1 and r_2 that of S_2. The space of r may be given by the two-dimensional real numbers (i.e., $\Omega = \mathbb{R}^2$) such that the events for r are contained in the σ-algebra $\mathbb{A}(\Omega) = \mathbb{B}^2$. Suppose next that we are interested in the probability that r_1 assumes some value between 0 and 0.1 while r_2 is between −0.5 and 0.5. This event corresponds to the set

$$A = \{r_1, r_2 | 0 \leq r_1 \leq 1 \text{ and } -0.5 \leq r_2 \leq 0.5\} \tag{14.9}$$

According to equation (14.6), we obtain as probability of A from equation (14.9) the integral

$$P(A) = \int_0^1 \int_{-0.5}^{0.5} f_r(r_1, r_2) dr_1 dr_2 \tag{14.10}$$

where f_r is the joint probability density function of the random return vector r. We demonstrate this graphically in Figure 14.1. With the grid surface (mesh), we outline the two-dimensional joint pdf between −0.5 and 1 on the r_1-axis and between −1 and 1 on the r_2-axis. The volume representing $P(A)$ generated by the integral (14.10) is encompassed by the dash-dotted lines on the $r_1 - r_2$-plane, the vertical dashed lines, and the $f_r(r_1, r_2)$ values, indicated by the shaded surface for r_1 values between 0 and 1 and r_2 values between −0.5 and 0.5. For the return density function, not presented here in detail, the probability in integral (14.10) is 0.1887.[7]

---

[7] The probability distribution of this return vector is given by the two-dimensional normal distribution to be introduced later in the chapter.

FIGURE 14.1 Extract of the Two-Dimensional pdf f_r of the Returns Vector r

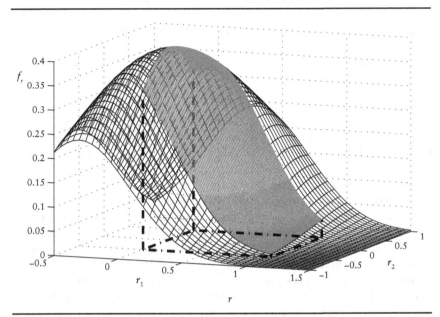

Note: The indicated volume represents the probability $P(A)$ from in integral (14.10).

Note that the volume under the entire joint density function is equal to 1 since this represents the probability

$$F(\infty,\infty) = \int_{\infty}^{\infty} \int_{-\infty}^{\infty} f_r(r_1,r_2)\,dr_1 dr_2 = 1$$

that the component random returns of r assume any value. This equals the probability of the event B of the form (14.7) with b_i going to infinity.

Contour Lines

With multivariate probability distributions it is sometimes useful to compare the so-called *contour lines* or *isolines* of their respective density functions. A contour line is the set of all outcomes $(x_1, ..., x_k)$ such that the density function f_X equals some constant; that is,

$$\left\{(x_1, x_2, ..., x_k) \mid f_X(x_1, x_2, ..., x_k) = c\right\} \qquad (14.11)$$

FIGURE 14.2 Probability Density Function f_r of Random Return Vector r Displayed on 2-Dimensional Subset $[-3,3]$ of Space $\Omega = \mathbb{R}^2$

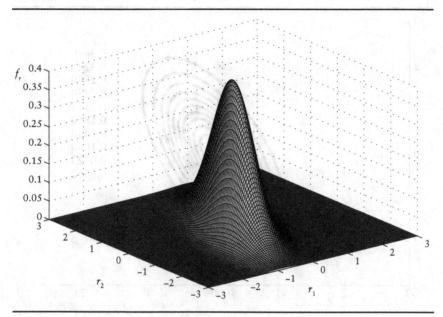

With the density function f_r of the multivariate random return vector r, depicted in Figure 14.2, we illustrate the meaning of equation (14.11) for a selection of contour lines in Figure 14.3. The contour lines correspond to the 10 different values for c given in the legend. Beginning with the inner lines which represent higher pdf values for c, we obtain lower levels for c of the pdf as we gradually proceed to the outer contour lines. The maximum level of c (0.3979) is assumed at the singular joint return of $r = (0,0)$, which is not depicted in the figure.

MARGINAL DISTRIBUTIONS

In the previous two sections, we analyzed the joint random behavior of the components of some random vector, which was either discrete or continuous. Here we will consider how to retrieve the probability distributions of the individual components from the joint probability distribution. That is, we determine the *marginal probability distribution* in order to treat each component random variable isolated from the others.

FIGURE 14.3 Contour Lines of Joint Probability Density Function of Random Return Vector r

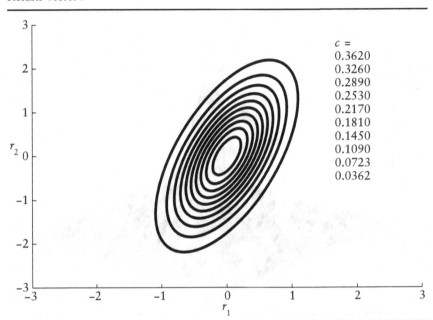

Note: Legend: values $f_r(r_1, r_2) = c$ in descending order represent the contour lines from inner to outer.

Discrete Case

In the case of a discrete random vector, the marginal probability of a certain component random variable X_i assuming some value x_i is obtained by summing up the probabilities of all values of x where in component i the value x_i occurs. For example, in order to obtain the probability that component i of some random vector $X = (X_1, X_2, ..., X_k)$ is equal to 0, we compute the marginal probability of this event according to

$$P(X_i = 0) = \sum_{x \in \mathbb{R}^k : x_i = 0} P(x_1, x_2, ..., x_k) \qquad (14.12)$$

where in equation (14.12) we sum over all those outcomes $x = (x_1, x_2, ..., x_k)$ whose i-th component is equal to 0.

As an illustration, suppose we know the joint distribution of the returns of a collection of assets such as A_1 and A_2 from our earlier example, and we are also interested in the individual probability distributions of each asset's return. That is, we endeavor to obtain the marginal probability distributions of A_1 and A_2. For convenience, we reproduce the previous distribution table here

	A_1 $90	$100	$110
A_2			
$80		0.05	0.45
$120	0.5		

Let us begin with the marginal distribution of A_1. Computing the marginal probabilities of A_1 assuming the values $90, $100, and $110 analogously to equation (14.12), we obtain[8]

$$P(A_1 = \$90) = P(A_1 = \$90, A_2 = \$120) = 0.50$$
$$P(A_1 = \$100) = P(A_1 = \$100, A_2 = \$80) = 0.05$$
$$P(A_1 = \$110) = P(A_1 = \$110, A_2 = \$80) = 0.45$$
$$\text{Total} = 1.00$$

We see that the sum of the marginal probabilities of all possible values of A_1 is equal to 1.

We see that in 50% of all scenarios, A_1 has a value of $90. Further, it has a value of $100 in 5% of the scenarios, while in 45% of all scenarios, it is worth $90. The mean of A_1 is simply computed as $E(A_1) = 0.5 \times \$90 + 0.05 \times \$100 + 0.45 \times \$110 = \$99.50$. Similarly, for A_2, we have

$$P(A_2 = \$80) = P(A_1 = \$100, A_2 = \$80) + P(A_1 = \$110, A_2 = \$80)$$
$$= 0.05 + 0.45 = 0.50$$
$$P(A_2 = \$120) = P(A_1 = \$90, A_2 = \$120) = 0.50$$
$$\text{Total} = 1.00$$

So, in half the scenarios, A_2 is worth $80 while its value is $120, in the other 50% of all scenarios. The mean is $E(A_2) = 0.5 \times \$80 + 0.5 \times \$120 = \$100$.

Note that we always have to check that for each random variable the marginal probabilities add up to one, which is the case here.

The change in perspective becomes obvious now when we compare joint and marginal distributions. Suppose a portfolio manager wanted to know the probability that portfolio consisting of 10,000 shares of each of A_1 and A_2 being worth at least $2,200,000. Individually, A_1 has a value

---
[8] Analogous to equation (14.12), we mean that we respectively sum the probabilities of all events where instead of using $x_1 = 0$ in equation (14.12), $A_1 = \$90, \$100$, or $\$110$.

greater than or equal to $100 in 50% of all cases while A_2 is worth $120 in 50% of the cases.

Does that mean that the portfolio is worth at least 10,000 × $100 + 10,000 × $120 = $2,200,000 in 50% of the cases? To correctly answer this question, we need to consider all possible scenarios with their respective probability as given by the joint probability distribution and not the marginal distributions. There are three different scenarios, namely scenario 1: A_1 = $100, A_2 = $80, scenario 2: A_1 = $110, A_2 = $80, and scenario 3: A_1 = $90, A_2 = $120. So, with a 5% chance, we have a portfolio value of $1,800,000, with a 45% chance a portfolio value of $1,900,000, and with a 50% chance the portfolio value is equal to $2,100,000. We see that the portfolio is never worth $2,200,000 or more. This is due to the particular joint probability distribution that assigns much probability to value pairs where when one stock is priced high while the other stock assumes a lower value. On average, the portfolio will be worth 0.05 × $1,800,000 + 0.5 × $2,100,000 = $1,995,000. This is exactly 10,000 times E(A_1) + E(A_2), which is an immediate result from the fact that the mean of a sum is equal to the sum of the means.

Continuous Case

For continuous random vectors, the concept behind the marginal distribution is similar to the discrete case. Let the i-th component be the one whose marginal distribution we endeavor to compute at, say, 0. Then, as in the discrete case, we need to aggregate all outcomes $x = (x_1, x_2, ..., x_k)$ where the i-th component is held constant at 0. However, remember that we do not have positive probability for individual outcomes. Hence, we cannot simply add up the probability as we have done with discrete random variables as in equation (14.12). Instead, we need to resort to the joint density function. Since the components of a continuous random vector are continuous themselves, the marginal distributions will be defined by the respective marginal density functions.

The *marginal density function* f_{X_i} of component X_i at some value a is obtained by integrating the joint density function f_X over all values x with the i-th component equal to a. That is,

$$f_{X_i}(a) = \int_{-\infty}^{\infty} ... \int_{-\infty}^{\infty} f_X(x_1, x_2, ..., a, ..., x_k) dx_1 \cdot ... \cdot dx_{i-1} \cdot dx_{i+1} \cdot ... dx_k \quad (14.13)$$

In definition (14.13) of the marginal density function, we have $k - 1$ integrals since we do not integrate with respect to the i-th component. Definition (14.13) can be alternatively written in a brief representation as

Joint Probability Distributions

$$f_{X_i}(a) = \int_{x:x_i=a} f_X(x)dx$$

We illustrate the concept of the marginal density function f_{X_1} of the first component X_1 graphically for some two-dimensional random vector $X = (X_1, X_2)$ in Figure 14.4. In the figure, the gray shaded area equals the marginal density function evaluated at $x_1 = 0$, that is, $f_{X_1}(0)$. The shaded area can be thought of as being the cut obtained by cutting through the volume under the joint density surface along the line whose x_1 value is kept constant at $x_1 = 0$.

As an example, recall the random return vector $r = (r_1, r_2)$ from an earlier illustration in this chapter. Suppose we wanted to analyze the second asset's return, r_2, separately from the other. From the joint distribution—which we need not know in detail here—we obtain from equation (14.13) the marginal density of r_2

$$f_{r_2}(a) = \int_{-\infty}^{\infty} f_r(r_1, a) \cdot dr_1 \tag{14.14}$$

FIGURE 14.4 Marginal Density Function f_{X_1} of Component Random Variable X_1, Evaluated at $x_1 = 0$

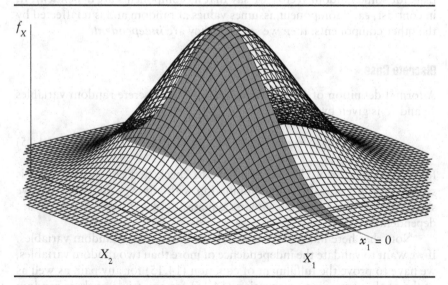

Note: The gray shaded area is equal to $f_{X_1}(0)$.

What is the probability then that r_2 is negative? We simply have to integrate the marginal density function in equation (14.14) over r_2 from $-\infty$ to 0. That is, we compute

$$F_{r_2}(0) = \int_{-\infty}^{0} f_{r_2}(r_2) \cdot dr_2$$

which is equal to the cumulative distribution function of r_2, evaluated at 0. With the particular multivariate probability distribution of r, this happens to be the standard normal cumulative distribution function at 0, that is, $\Phi(0) = 0.5$. So in 50% of the scenarios, r_2 will be negative.

DEPENDENCE

Let us consider the k-dimensional random vector $X = (X_1, X_2, ..., X_k)$. Then each of the components of X is a random variable assuming real numbered values. The joint probability distribution may have the following influence on the component random variables. Although individually the components may assume any value, jointly certain combinations of values may never occur. Furthermore, certain values of some particular component, say j, may only occur with certain values of the other combinations. If this should be true, for some random vector, we say that its components are *dependent*. If, in contrast, each component assumes values at random and is unaffected by the other components, then we say that they are *independent*.

Discrete Case

A formal definition of *independence* of any two discrete random variables X_i and X_j is given by

$$P(X_i = a, X_j = b) = P(X_i = a) \cdot P(X_j = b) \qquad (14.15)$$

for all values $a, b \in \mathbb{R}$. That is, we have to check that equation (14.15) is satisfied for any X_i value a in combination with any X_j value b. In case we should find a pair (a,b) that violates equation (14.15), then X_i and X_j are dependent.

Note that here independence is only defined for two random variables. If we want to validate the independence of more than two random variables, we have to prove the fulfillment of equation (14.15) for any pair, as well as of the analog extension of equation (14.15) for any triple of these random variables, any set of four, and so on. That is, we have to validate

$$P(X_{i_1} = a, X_{i_2} = b, X_{i_3} = c)$$
$$= P(X_{i_1} = a) \cdot P(X_{i_2} = b) \cdot P(X_{i_3} = c)$$
$$P(X_{i_1} = a, X_{i_2} = b, X_{i_3} = c, X_{i_4} = d)$$
$$= P(X_{i_1} = a) \cdot P(X_{i_2} = b) \cdot P(X_{i_3} = c) \cdot P(X_{i_4} = d)$$
$$\ldots$$

Recall that in the example with the discrete asset random vector $A = (A_1, A_2)$ that a portfolio value of \$2,100,000 could not be achieved. This was the case even though the marginal distributions permitted the isolated values $A_1 = \$110$ and $A_2 = \$120$ such that \$2,100,000 might be obtained or even exceeded. However, these combinations of A_1 and A_2 to realize \$2,100,000, such as (\$110,\$120), for example, were assigned zero probability by the joint distribution such that they never happen. Let us validate the condition (14.15) for our two assets taking the value pair (\$110,\$120).[9] The probability of this event is $P(\$110,\$120) = 0$ according to the joint probability table given earlier in this chapter. The respective marginal probabilities are $P(A_1 = \$110) = 0.45$ and $P(A_2 = \$120) = 0.5$. Thus, we have $P(\$110,\$120) = 0 \neq 0.45 \cdot 0.5 = P(A_1 = \$110) \cdot P(A_2 = \$120)$, and, consequently the asset values are dependent.

Continuous Case

For the analysis of dependence of two continuous random variables X_i and X_j, we perform a similar task as we have done in equation (14.15) in the discrete case. However, since the random variables are continuous, the probability of single values is always zero. So, we have to use the joint and marginal density functions. In the continuous case, independence is defined as

$$f_{X_i, X_j}(a,b) = f_{X_i}(a) \cdot f_{X_j}(b) \qquad (14.16)$$

where f_{X_i, X_j} is the joint density function of X_i and X_j.[10] Again, if we find a pair (a,b) such that condition (14.16) is not met, we know that X_i and X_j are dependent.

---

[9]The first component of the pair is the A_1 value and the second the A_2 value.
[10]The joint density function f_{X_i, X_j} can be retrieved from the k-dimensional joint density f_X by integration over all components from $-\infty$ to ∞.

We can extend the condition (14.16) to a k-dimensional generalization that, when $X_1, X_2, \ldots, X_k$ are independent, their joint density function has the following appearance

$$f_X(a_1, a_2, \ldots, a_k) = f_{X_1}(a_1) \cdot f_{X_2}(a_2) \cdot \ldots \cdot f_{X_k}(a_k) \qquad (14.17)$$

So if we can write the joint density function as the product of the marginal density functions, as done on the right side of equation (14.17), we conclude that the k random variables are independent.

At this point, let's introduce the concept of *convolution*. Technically, the convolution of two functions f and g is the integral

$$h(z) = \int_{-\infty}^{\infty} f(x) \cdot g(z-x) \, dx$$

for any real number z. That is, it is the product of two functions f and g integrated over all real numbers such that for each value x, f is evaluated at x and, simultaneously, g is evaluated at the difference $z - x$. Now, let's think of f and g as the density functions of some independent continuous random variables X and Y, for example. Then, the convolution integral h yields the density of the sum $Z = X + Y$. If we want to add a third independent continuous random variable, say U, with density function p, then the resulting sum $U + X + Y = U + Z$ is itself a continuous random variable with density function

$$k(u) = \int_{-\infty}^{\infty} p(z) \cdot h(u-z) \, dz$$

In this fashion, we can keep adding additional independent continuous random variables and always obtain the density function of the resulting sum through the convolution of the respective marginal density functions.

Thus, let $X_1, X_2, \ldots, X_n$ be n independent and identically distributed continuous random variables such that their joint density function is of the form (14.17). Then, the distribution of the sum $X_1 + X_2 + \ldots + X_n$ is itself continuous and its density function is also obtained through convolution of the marginal density functions from equation (14.17).

Let us turn to our previous illustration of stock returns. Recall the random vector r consisting of the two random returns r_1 and r_2.[11] Without going into details, we state here that the two marginal distributions are normal with parameters $\mu_1 = 0, \sigma_1^2 = 0.25$ and $\mu_2 = 0, \sigma_2^2 = 1$, respectively. That is, we have $r_1 \sim N(0, 0.25)$ with density function

---

[11] A two-dimensional random vector is referred to as a *bivariate* random variable.

Joint Probability Distributions

$$f_{r_1}(a) = \frac{1}{\sqrt{2\cdot\pi\cdot 0.25}} \cdot e^{-\frac{a^2}{2\cdot 0.25}}$$

and $r_2 \sim N(0,1)$ with density function

$$f_{r_2}(b) = \frac{1}{\sqrt{2\cdot\pi}} \cdot e^{-\frac{b^2}{2}}$$

The joint density of the returns is

$$f_r(a,b) = \frac{1}{2\cdot\pi\cdot 0.32} \cdot e^{-\frac{4\cdot a^2 - 1.44\cdot a\cdot b + b^2}{1.28}} \quad (14.18)$$

Let us validate condition (14.16) at the joint value (0,0), that is, both returns are equal to zero. Multiplication of the two marginal densities evaluated at zero ($a = 0$ and $b = 0$) yields

$$\frac{1}{2\cdot\pi\cdot 0.25} \cdot e^{-\frac{4\cdot 0^2 + 0^2}{2}} = 0.6366$$

while the joint density in equation (14.18), evaluated at (0,0) yields 0.4974. Since $0.6366 \neq 0.4974$, we conclude that the two returns are dependent. This dependence can be detected when analyzing the contour lines in Figure 14.3. The lines form ellipses revealing the dependence in the direction of the longer radius. Had they been perfectly circular, the returns would have been independent.

COVARIANCE AND CORRELATION

In the previous section, we learned that dependence of two random variables can be validated using either equation (14.15) for discrete random variables or equation (14.16) for continuous random variables. There is an alternative method—one of the most commonly used techniques for this purpose—we introduce here. It builds on the reasoning of the previous section and includes the expected joint deviation from the respective means by the two random variables. This measure is called the *covariance*.

Let us denote the covariance of the two random variables X and Y as $Cov(X,Y)$. If the value of the covariance is different from zero, then we know that the two random variables are dependent. If it is equal to zero, on the other hand, we cannot state that they are independent. However, we can draw a slightly weaker conclusion, namely, that the two random variables are *uncorrelated*. The opposite of this would be that the random variables are *correlated* that, in turn, corresponds to a covariance different from zero.

We summarize this concept below:

	X,Y
$Cov(X,Y) = 0$	Uncorrelated
$Cov(X,Y) \neq 0$	Correlated, dependent

Discrete Case

When the joint probability distribution is discrete, the covariance is defined as

$$Cov(X,Y) = E\left[(X - E(X)) \cdot (Y - E(Y))\right]$$
$$= \sum_x \sum_y (x - E(X)) \cdot (y - E(Y)) \cdot P(X = x, Y = y) \quad (14.19)$$

where $E(X)$ is the mean of X and $E(Y)$ is the mean of Y. The double summation in equation (14.19) indicates that we compute the sum of joint deviations multiplied by their respective probability over all combinations of the values x and y. For the sake of formal correctness, we have to mention that the covariance is only defined if the expression on the right side of equation (14.19) is finite.

For example, using our previous example of two assets whose values A_1 and A_2 were found to be dependent, we want to validate this using the covariance measure given by equation (14.19). We know that the mean of A_1, $E(A_1)$, is \$99.5 and that of A_2, $E(A_2)$, is \$100. Then, we obtain as the covariance[12]

$$Cov(X,Y) = (90 - 99.5)(120 - 100) \times 0.5 + (100 - 99.5)(80 - 100) \times 0.05$$
$$+ (110 - 99.5)(80 - 100) \times 0.45$$
$$= -190$$

which is different from zero and therefore supports our previous finding of dependence between the two random variables.

Continuous Case

In the continuous case, the covariance is also a measure of joint deviation from the respective means. In contrast to the definition in equation (14.19), we use the double integral of the joint density $f_{X,Y}$ rather than the double

---
[12]For simplicity, we omit the dollar sign in the computation.

summation of probabilities. So, the covariance of two continuous random variables is defined as

$$Cov(X,Y) = E\big[(X - E(X))\cdot(Y - E(Y))\big]$$
$$= \int_{-\infty}^{\infty} \int_{-\infty}^{\infty} (x - E(X))\cdot(y - E(Y))\cdot f_{X,Y}(x,y)\,dx\cdot dy \quad (14.20)$$

in that we integrate the joint $f_{X,Y}$ over x and y from $-\infty$ to ∞.

In the discrete as well as the continuous case, the right sides of equations (14.19) and (14.20), respectively, have to be finite for the covariance to exist.

We will illustrate equation (14.20) using the example of the random return vector r. The covariance is computed as

$$Cov(X,Y) = \int_{-\infty}^{\infty} \int_{-\infty}^{\infty} x\cdot y \cdot \frac{1}{2\cdot\pi\cdot 0.32}\cdot e^{-\frac{4\cdot x^2 - 1.44\cdot x\cdot y + y^2}{1.28}}\,dx\cdot dy$$
$$= 0.3$$

One should not worry if these integrals seem too difficult to compute. Most statistical software packages provide the computation of the covariance as a standard task.

Aspects of the Covariance and Covariance Matrix

Now we explain what happens to the covariance if we change the random variables to a different scale. Also, we will introduce the covariance matrix containing the covariances of any two components of a random vector.

Covariance of Transformed Random Variables

Note that if one computes the covariance of a random variable X with itself, we obtain the variance of X, that is, $Cov(X,X) = Var(X)$. Moreover, the covariance is symmetric in the sense that a permutation of X and Y has no effect, that is, $Cov(Y,X) = Cov(X,Y)$.

If we add a constant to either X or Y, or both, the resulting covariance is not affected by this, that is, $Cov(X + a, Y + b) = Cov(X,Y)$. For this reason, we can say that the covariance is invariant with respect to linear shifts. The multiplication of either one of the two random variables is reflected in the covariance, however. That is by multiplying X by some constant a and Y by some constant b, the covariance changes to $Cov(aX,bY) = a\cdot b\cdot Cov(X,Y)$.

For example, consider a portfolio P consisting of several risky assets. When we add a position in a risk-free asset C to our portfolio, the resulting portfolio return, R^P, is consequently composed of some risky return R and the risk-free interest rate R_f (that is, R_f is a constant). Now, in case we want to analyze the joint random behavior of the overall return of this portfolio, $R^P = R + R_f$, with the return of some other portfolio, R_2^P, we compute their covariance, which is equal to

$$Cov(R^P, R_2^P) = Cov(R + R_f, R_2^P) = Cov(R, R_2^P)$$

Due to the linear shift invariance of the covariance, the risk-free part of the portfolio plays no role in the computation of the covariance of these two portfolios.

If, instead of adding a risk-free position, we doubled our risky position, the resulting portfolio return $R^P = 2R$ and the other portfolio R_2^P would have covariance given by $Cov(R^P, R_2^P) = Cov(2R, R_2^P)$. So, by doubling our risky position in the first portfolio, we have doubled the covariance as well.

Covariance Matrix

Suppose we have a vector X consisting of the k components X_1, X_2, ..., X_k. Now, any component X_i with any other component has a covariance, which we denoted by $Cov(X_i, X_j)$. The *covariance matrix*, usually denoted by the symbol Σ, contains the covariances of all possible pairs of components as shown below:

$$\Sigma = \begin{pmatrix} Var(X_1) & Cov(X_2, X_1) & & Cov(X_k, X_1) \\ Cov(X_1, X_2) & Var(X_2) & & \vdots \\ \vdots & & \ddots & Cov(X_k, X_{k-1}) \\ Cov(X_1, X_k) & \cdots & & Var(X_k) \end{pmatrix} \quad (14.21)$$

The presentation of the covariance matrix given by form (14.21) is somewhat tedious. Instead of $Cov(X_i, X_j)$ and $Var(X_i)$, we use the parameter notation with σ_{ij} and σ_i^2, respectively. Thus, we can alternatively present the covariance matrix in the form

$$\Sigma = \begin{pmatrix} \sigma_1^2 & \sigma_{21} & & \sigma_{k1} \\ \sigma_{12} & \sigma_2^2 & & \vdots \\ \vdots & & \ddots & \sigma_{k1} \\ \sigma_{1k} & \cdots & & \sigma_k^2 \end{pmatrix}$$

Due to the symmetry of the covariance, that is, $\sigma_{ij} = \sigma_{ji}$, the covariance matrix is symmetric. This means that transposing the covariance matrix yields itself. That attribute of the covariance matrix will facilitate the estimation of the variances and covariances. Rather than having to estimate $k \times k = k^2$ parameters, only $k \times (k+1)/2$ parameters need be estimated.

In our previous example with the two-dimensional returns, we have the following covariance matrix

$$\Sigma = \begin{pmatrix} 0.25 & 0.30 \\ 0.30 & 1.00 \end{pmatrix}$$

Correlation

As we just explained, the covariance is sensitive to changes in the units of measurement of the random variables. As a matter of fact, we could increase the covariance a-fold by transforming X into $a \cdot X$. Hence, the covariance is not bounded by some value. This is not satisfactory since this scale dependence makes it difficult to compare covariances of pairs of random variables measured in different units or scales. For example, if we change from daily to say weekly returns, this change of units will be noticed in the covariance.

For this reason, we need to scale the covariance somehow so that effects such as multiplication of a random variable by some constant do not affect the measure used to quantify dependence. This scaling is accomplished by dividing the covariance by the standard deviations of both random variables. That is, the *correlation coefficient*[13] of two random variables X and Y, denoted by $\rho_{X,Y}$ is defined as

$$\rho_{X,Y} = \frac{Cov(X,Y)}{\sqrt{\sigma_X^2} \cdot \sqrt{\sigma_Y^2}} \tag{14.22}$$

We expressed the standard deviations as the square roots of the respective variances σ_X^2 and σ_Y^2. Note that the correlation coefficient is equal to one, that is, $\rho_{X,X} = 1$, for the correlation between the random variable X with itself. This can be seen from (14.22) by inserting σ_X^2 for the covariance in the numerator, and having σ_X^2, in the denominator. Moreover, the correlation coefficient is symmetric. This is due to definition (14.22) and the fact that the covariance is symmetric.

---

[13]More specifically, this correlation coefficient is referred to as the Pearson product moment correlation coefficient in honor of Karl Pearson, the mathematician/statistician who derived the formula.

The correlation coefficient given by (14.22) can take on real values in the range of −1 and 1 only. When its value is negative, we say that the random variables X and Y are *negatively correlated*, while they are *positively correlated* in the case of a positive correlation coefficient. When the correlation is zero, due to a zero covariance, we refer to X and Y as *uncorrelated*. We summarize this below:

$-1 \leq \rho_{X,Y} \leq 1$

$-1 \leq \rho_{X,Y} < 0$ X and Y negatively correlated

$\rho_{X,Y} = 0$ X and Y uncorrelated

$0 < \rho_{X,Y} \leq 1$ X and Y positively correlated

As with the covariances of a k-dimensional random vector, we list the correlation coefficients of all pairwise combinations of the k components in a k–by-k matrix

$$\Gamma = \begin{pmatrix} 1 & \rho_{21} & & \rho_{k1} \\ \rho_{12} & 1 & & \vdots \\ \vdots & & \ddots & \rho_{k1} \\ \rho_{1k} & \cdots & & 1 \end{pmatrix}$$

This matrix, referred to as the *correlation coefficient matrix* and denoted by Γ, is also symmetric since the correlation coefficients are symmetric.

For example, suppose we have a portfolio consisting of two assets whose prices are denominated in different currencies, say asset A in U.S. dollars ($) and asset B in euros (€). Furthermore, suppose the exchange rate was constant at $1.30 per €1. Consequently, asset B always moves 1.3 times as much when translated into the equivalent amount of dollars then when measured in euros. The covariance of A and B when expressed in their respective local currencies is $Cov(A,B)$. When translating B into the dollar equivalent, we obtain $Cov(A, 1.3 \cdot B) = 1.3 \times Cov(A,B)$. So, instead we should compute the correlation of the two. Suppose the variance of A in dollars is σ_A^2 and that of B in euros is σ_B^2. According to the transformation rules of the variance that we covered in Chapter 3 and which also apply to the variance parameter of probability distributions, the variance of B can be expressed as $(1.3)^2 \times \sigma_B^2$ when asset B is given in dollar units. Consequently, the correlation coefficient of A and B measured in dollars is computed as

$$\rho_{A,B} = \frac{Cov(A, 1.3 \times B)}{\sqrt{\sigma_A^2} \cdot \sqrt{(1.3)^2 \times \sigma_B^2}} = \frac{1.3 \times Cov(A,B)}{1.3 \times \sqrt{\sigma_A^2} \cdot \sqrt{\sigma_B^2}} = \frac{Cov(A,B)}{\sqrt{\sigma_A^2} \cdot \sqrt{\sigma_B^2}}$$

which is the same as when B is denominated in euros.

As another example, let's continue our previous analysis of the continuous return vector r. With the respective standard deviations of the component returns and their covariance given as $\sigma_1 = 0.5$, $\sigma_2 = 1$, and $\sigma_{1,2} = 0.3$, respectively, we obtain the correlation coefficient as

$$\rho_{1,2} = \frac{0.3}{0.5 \times 1} = 0.6$$

which indicates a clear positive correlation between r_1 and r_2.

Criticism of the Correlation and Covariance as a Measure of Joint Randomness

As useful and important a measure that correlation and covariance are, they are not free from criticism as a sufficient measure of joint randomness. Very often, the pairwise correlations or covariances are parameters of the multivariate distribution characterizing part of the joint random behavior of the components. However, the covariance is unable to capture certain aspects. For example, financial returns reveal dependencies with respect to joint extreme movements even though their covariances may be zero. This is particularly dangerous for a portfolio manager who only focuses on the covariance and ignores the other forms of dependence. There exist, however, devices shedding light on these aspects neglected by the covariance.

One such device will be introduced in Chapter 16 as a measure of *tail-dependence*. The positive left-tail-dependence between the random returns of two assets expresses the probability of one return performing very badly given that the other already performs poorly. Despite the fact that the correlation can be very close to 1 (but not 1, say, 0.9), extreme losses or returns can be practically independent. In other words, the correlation is not a meaningful measure if we are interested in the dependence between extreme losses or returns.

SELECTION OF MULTIVARIATE DISTRIBUTIONS

In this section, we introduce a few of the most common multivariate distributions used in finance.

Multivariate Normal Distribution

In finance, it is common to assume that the random variables are normally distributed. The joint distribution is then referred to as a ***multivariate nor-***

mal distribution. To get a first impression of multivariate distributions, Figures 14.2 and 14.4 show the surfaces and Figure 14.3 a contour plot of the bivariate (2-dimensional) normal probability density functions with standard normal marginals. We are going to explain how such a distribution can be constructed from the univariate normal distribution and how an explicit expression for the density function can be obtained.

Mathematically, the joint distribution of a random vector $X = (X_1, X_2, ..., X_k)$ is said to be a multivariate normal distribution if, for any real numbered weights $a_1, a_2, ..., a_k$, the linear combination of the form $a_1 X_1 + a_2 X_2 + ... + a_k X_k$ of its components is also a normally distributed random variable.

This is very important in the field of portfolio optimization. Let the X_1, X_2, ..., and X_k represent assets contained in the portfolio. Then, the resulting portfolio, as a linear combination of them, is itself normally distributed.

Properties of Multivariate Normal Distribution

There are important properties of the multivariate normal distribution. To explain them, we will discuss the special case where there are two random variables. This case is referred to as a ***bivariate normal distribution***. The two properties are:

Property 1: If X and Y have a bivariate normal distribution, then for some constants $a, b \in \mathbb{R}$, $Z = aX + bY$ is normally distributed with an expected value equal to

$$\mu_Z = a\mu_X + b\mu_Y \tag{14.23}$$

where μ_X and μ_Y are the expected values of X and Y. The variance of Z is

$$\sigma_Z^2 = a^2 \sigma_X^2 + b^2 \sigma_Y^2 + 2ab\sigma_{X,Y}$$

with σ_X^2 and σ_Y^2 being the respective variances.

Hence, the resulting variance is not simply the weighted sum of the marginal variances but of the covariance as well. So, if the latter is greater than zero, the variance of Z, σ_Z^2, becomes larger than $a^2 \sigma_X^2 + b^2 \sigma_Y^2$ by exactly $2ab\sigma_{X,Y}$. If, on the other hand, the covariance is negative, then σ_Z^2 is less than $a^2 \sigma_X^2 + b^2 \sigma_Y^2$ by $2ab\sigma_{X,Y}$. These effects call for the strategy of ***diversification*** in a portfolio in order to reduce σ_Z^2 and, consequently, are immensely important for a portfolio manager to understand.

Property 2: If X and Y have a bivariate (2-dimensional) normal distributions and the covariance between the two variables is zero, then the two variables are independent.

Note that, from our prior discussion, the covariance is always zero when the random variables are independent; however, in general, the converse does not hold. So, Property 2 of the multivariate normal distribution is an immensely powerful statement. It is another reason for the widespread popularity of the normal distribution in finance.

Density Function of a General Multivariate Normal Distribution

If we want to characterize the density function of a univariate normal distribution, we have to specify the mean μ and the variance σ^2. In the bivariate setting, in addition to both means (μ_X and μ_Y) and variances (σ_X^2 and σ_Y^2), we need the correlation parameter $\rho_{X,Y}$ which determines the dependence structure.

The density function of a general multivariate normal distribution of the random vector $X = (X_1, X_2, ..., X_k)$ is defined by

$$f_X(x_1, x_2, ..., x_k) = \frac{1}{\sqrt{|\Sigma|(2\pi)^d}} \cdot e^{-\frac{1}{2}(x-\mu)\Sigma^{-1}(x-\mu)^T}$$

where Σ is the covariance matrix of X, $|\Sigma|$ its determinant, $\mu = (\mu_1, \mu_2, ..., \mu_k)$ is the vector of all k means, and $(x - \mu)^T$ denotes the transpose of the vector $(x - \mu)$.[14] It is necessary that $|\Sigma| > 0$, which also ensures that the inverse matrix Σ^{-1} exists.

Now, if X is a multivariate normal random vector, we state that in brief as

$$X \sim N(\mu, \Sigma)$$

Note that if $|\Sigma| = 0$, the density function f_X does not exist but we can still state the probability distribution through its characteristic function. Recall from Chapter 12 and Appendix A that the characteristic function of the probability law P of some random variable X evaluated at any real number t is defined as

$$\varphi_X(t) = E(e^{itX})$$

where $E(\cdot)$ denotes the expected value with respect to P. The number e is roughly equal to 2.7183 while i is the so-called *imaginary number* defined

---
[14] For a discussion vector transpose and determinant, see Appendix B.

as $i = \sqrt{-1}$ such that $i^2 = -1$. In other words, the characteristic function is the expected value of the random variable e^{itX}.[15] Any and, in particular, any continuous probability distribution has a unique characteristic function even if the probability density function does not exist. The characteristic function of the multivariate normal distribution is given by

$$\varphi_X(t) = e^{\left(i\mu^T t - \frac{1}{2} t^T \Sigma t\right)}$$

where μ^T and t^T denote the vector transpose of μ and t, respectively.

As a first illustration, refer to Figures 14.2 and 14.4. In Figure 14.2, we found the joint density function of a bivariate normal distribution with mean vector $\mu = (0,0)$ and covariance matrix

$$\Sigma = \begin{pmatrix} 0.25 & 0.30 \\ 0.30 & 1.00 \end{pmatrix}$$

This was the distribution of the return vector r. In Figure 14.4, we display the density function of a random vector X with

$$X \sim N(\mu, \Sigma)$$

with $\mu = (0,0)$, again, and

$$\Sigma = \begin{pmatrix} 1 & 0 \\ 0 & 1 \end{pmatrix}$$

which corresponds to independent univariate standard normal component random variables X_1 and X_2. In Figure 14.5, we depict the corresponding cumulative distribution function.

Application to Portfolio Selection

For our next illustration, we turn to portfolio optimization. Suppose a portfolio manager manages a portfolio consisting of n stocks whose daily returns $R_1, R_2, ..., R_n$ are jointly normally distributed with corresponding mean vector $\mu = (\mu_1, \mu_2, ..., \mu_n)$ and covariance matrix Σ. Here we define the daily return of stock i as the relative change in price between today ($t = 0$) and tomorrow ($t = 1$), that is,

---

[15]The random variable e^{itX} assumes *complex* values. Recall from Appendix A that complex numbers that contain real numbers are always of the form $a + i \cdot b$ where a and b are real numbers and i is the imaginary number.

FIGURE 14.5 Bivariate Normal Distribution with Mean Vector $\mu = (0,0)$ and Covariance Matrix
$$\Sigma = \begin{pmatrix} 1 & 0 \\ 0 & 1 \end{pmatrix}$$

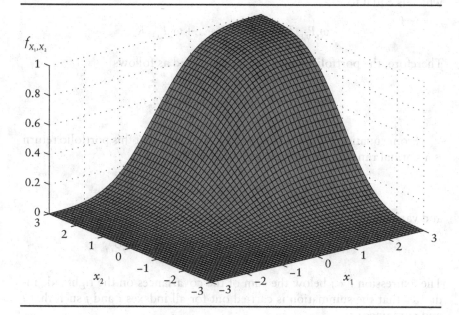

$$R_i = \frac{P_{1,i}}{P_{0,i}}$$

For each stock i, we denote the number of shares by N_i such that, today, the portfolio is worth

$$P_{0,PF} = N_1 P_{0,1} + N_2 P_{0,2} \ldots + N_n P_{0,n}$$

Let $(\omega_1, \omega_2, \ldots, \omega_n)$ denote the relative weights of the n stocks in the portfolio, then the contribution of stock i to the portfolio value is

$$\omega_i \cdot P_{0,PF} = N_i \cdot P_{0,i}, \text{ for } i = 1, 2, \ldots, n$$

So, we can represent the portfolio value of today as

$$P_{0,PF} = \omega_1 P_{0,PF} + \omega_2 P_{0,PF} \ldots + \omega_n P_{0,PF}$$

Tomorrow, the same portfolio will be worth the random value

$$P_{1,PF} = \omega_1 P_{1,PF} + \omega_2 P_{1,PF} \ldots + \omega_n P_{1,PF}$$

which is equal to

$$\omega_1 P_{0,PF} R_1 + \omega_2 P_{0,PF} R_2 + \ldots + \omega_n P_{0,PF} R_n$$

Therefore, the portfolio return can be computed as follows

$$R_{PF} = \frac{P_{1,PF}}{P_{0,PF}} = \omega_1 R_{1,1} + \omega_2 R_{1,2} \ldots + \omega_n R_{1,n}$$

From equation (14.23) of *Property 1*, we know that this portfolio return is a normal random variable as well, with mean

$$\mu_{PF} = \omega_1 \mu_1 + \omega_2 \mu_2 \ldots + \omega_n \mu_n$$

and variance

$$\sigma_{PF}^2 = \omega_1^2 \sigma_1^2 + \omega_2^2 \sigma_2^2 + \ldots + \omega_n^2 \sigma_n^2 + 2 \sum_{i \neq j} \omega_i \omega_j \sigma_{i,j}$$

The expression $i \neq j$ below the sum of the covariances on the right side indicates that the summation is carried out for all indexes i and j such that i and j are unequal.

Now, it may be the objective of the portfolio manager to maximize the return of the portfolio given that the variance remains at a particular level. In other words, the portfolio manager seeks to find the optimal weights (ω_1, ω_2, ..., ω_k) such that μ_{PF} is maximized for a given level of risk as measured by the portfolio variance. We will denote the risk level by C, that is, $\sigma_{PF}^2 = C$. Conversely, the objective could be to achieve a given portfolio return, that is, $\mu_{PF} = E$, with the minimum variance possible. Again, the portfolio manager accomplishes this task by finding the optimal portfolio weights suitable for this problem.

The set of pairs of the respective minimum variances for any portfolio return level μ_{PF} for some bivariate return vector $r = (r_1, r_2)$, where now $E(r_1) \neq E(r_2)$, is depicted in Figure 14.6.[16] The covariance matrix is still,

$$\Sigma = \begin{pmatrix} 0.25 & 0.30 \\ 0.30 & 1.00 \end{pmatrix}$$

---

[16] Here, we chose the fictitious expected asset returns $\mu_1 = 0.10$ and $\mu_2 = 1.60$.

FIGURE 14.6 Global Minimum-Variance Portfolio, Set of Feasible Portfolios, and Efficient Frontier for the Bivariate Asset Returns Case

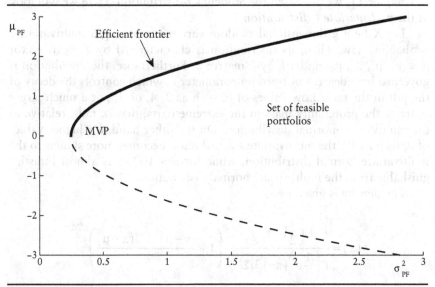

as we know it already from some previous example. The region bounded by the parabola is the set of all possible portfolios that can be created and is referred to as the set of *feasible portfolios*. The set of minimum-variance portfolios lies on the parabola composed of the solid and dashed curves. The solid curve is referred to as *efficient frontier*. The point (i.e., portfolio) where the portfolio variance is the lowest is called the *global minimum variance portfolio* and denoted by *MVP*. In our example the *MVP is* located here roughly at $\left(\sigma_{PF}^2, \mu_{PF}\right) = (0.25, -0.02)$. Inside of the parabola, we find all feasible portfolios that yield the desired given expected portfolio return but fail to accomplish this with the least variance possible. That is, for the set of feasible portfolios there are certain portfolios that dominate other portfolios. This means either that for a given portfolio variance, the expected portfolio return is greater or that for a given expected portfolio return, the portfolio variance is smaller. Note that we had to alter the expected return vector $E(r) = \mu$ since, with $(\mu_1, \mu_2) = (0,0)$, we would not have been able to achieve any expected portfolio return other than $\mu_{PF} = 0$.

This framework was first introduced in 1952 by Harry Markowitz and is popularly known as *Markowitz portfolio selection* or *mean-variance portfolio optimization*.[17]

---

[17]The original theory was presented in Markowitz (1952) and extended in Markowtiz (1959).

Multivariate *t*-Distribution

In Chapter 11, we discussed the Student's *t*-distribution. Here we will look at the *multivariate t-distribution*.

Let X be a k dimensional random variable following a multivariate t probability law. Then, its distribution is characterized by a mean vector $\mu = (\mu_1, \mu_2, ..., \mu_k)$ and a k-by-k matrix Σ. Furthermore, the distribution is governed by a degrees of freedom parameter v, which controls the decay of the pdf in the tails. Low values of v, such as 3, 4, or 5, put a much larger share of the probability mass in the extreme parts, that is, tails, relative to the multivariate normal distribution. On the other hand, for higher values of v, such as 10, the multivariate *t*-distribution becomes more similar to the multivariate normal distribution, while for $v > 100$, it is almost indistinguishable from the multivariate normal distribution.

The density is given by

$$f(x) = \frac{\Gamma\left(\frac{1}{2}(v+k)\right)}{\Gamma\left(\frac{1}{2}v\right)(\pi v)^{k/2}|\Sigma|^{1/2}} \cdot \left(1 + \frac{(x-\mu)^T \Sigma^{-1}(x-\mu)}{v}\right)^{-\frac{v+k}{2}}$$

where

$\Gamma(\cdot)$ denotes the gamma function.[18]
$|\Sigma|$ is the determinant of Σ.
Σ^{-1} is the inverse of Σ.[19]

Again, it is necessary that $|\Sigma| > 0$, which also ensures that the inverse matrix Σ^{-1} exists.

Note that, here, Σ is not exactly the covariance matrix. To obtain it, assuming v is greater than 2, we have to multiply Σ by $v/v-2$. However, Σ exhibits the same correlation structure as the covariance matrix since it is only changed by some constant factor.

For a bivariate example, with $\mu = (0,0)$ and

$$\Sigma = \begin{pmatrix} 0.20 & 0.24 \\ 0.24 & 0.80 \end{pmatrix}$$

which corresponds to the covariance parameters in a prior illustration on the bivariate normal distribution, we display the corresponding joint density

---
[18] We explain the gamma function in Appendix A.
[19] The inverse A^{-1} of some matrix A is defined by $A \cdot A^{-1} = I_{k \times k}$, that is, their product yields the k-dimensional identity matrix. In other words, the inverse and the original matrix have the exact inverse effect.

Joint Probability Distributions

function as well as a contour plot in Figures 14.7 and 14.8, respectively. The degrees of freedom are given by $v = 10$ so that the distribution is still sufficiently different from the bivariate normal distribution.

Focusing on Figure 14.7, we observe that the peak height of 0.4974 in the center is much higher than that of the bivariate normal alternative whose height of 0.3979 is generated by the same means and covariance matrix depicted in Figure 14.2. Moreover, we see that the density decays more slowly in the tails, in diagonal direction, than for the normal case.

These two findings are further emphasized by the contour lines in Figure 14.8. Note that while their shape is very similar to that of the normal distribution depicted in Figure 14.3, they are characterized by much higher values in the center, as well as a slower decent in the outer parts. We see that the multivariate t-distribution is more capable of modeling random vectors whose components are characterized by more risk of extreme joint movements. The implication for a portfolio manager or a risk manager is that the appropriateness of the multivariate normal distribution should be validated for the assets under management. If it seems unfit, a portfolio manager or risk manager should seek for an alternative distribution such as the multivariate t-distribution.

FIGURE 14.7 Density Plot of the Bivariate Student's t-Distribution on $[-3,3] \times [-3,3]$

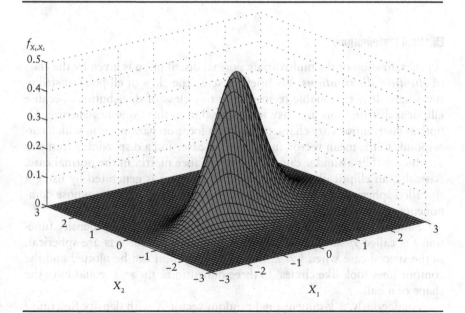

FIGURE 14.8 Contour Lines of of the Bivariate Student's t Density Function

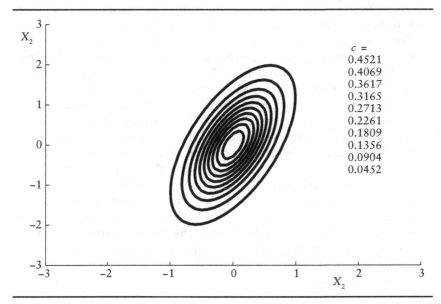

Note: Levels c in descending order correspond to contour lines going from inside to outside.

Elliptical Distributions

A generalization of the multivariate normal distribution is given by the class of *elliptical distributions*. We briefly discuss the class of elliptical distributions here. It is reasonable to introduce this class of distributions because elliptical distributions are easy to handle due to their simple structure. Elliptical distributions are characterized by a location parameter μ, which corresponds to the mean vector, in the normal case, and a dispersion parameter Σ, which fulfills a similar duty as the covariance matrix in the normal case. Basically, an elliptically distributed random vector is generated by using a so-called spherical random vector, which might be shifted and whose components are individually rescaled.

Simply speaking, a k-dimensional random vector X with density function f is called *spherically distributed* if all the contour sets are spherical. In the special case when $k = 2$, the density function can be plotted and the contour lines look like circles. In three dimensions, the sets would have the shape of a ball.

Analogously, a k-dimensional random vector X with density function f is said to be elliptically distributed if all contours are ellipsoids. This results from the individual rescaling of the components of the spherical random vector whereas the center of the ellipsoid is determined by the shift.

When $k = 2$, the ellipsoids appear as ellipses. As an example, we can look at the contour plots in Figures 14.3 and 14.8. Representatives of elliptical distributions include all multivariate normal distributions, multivariate t-distributions, logistic distributions, Laplace distributions, and a part of the multivariate stable distributions. Because of their complexity, we have not described the last three distributions.

Properties of the Elliptical Class

Here we state some of the properties of the elliptical class without mathematical rigor.[20]

Let us begin with the first property of *closure under linear transformations*. If X is a d-dimensional elliptical random variable and we shift it by some constant vector b and then multiply it by some matrix B with l rows and d columns, then the resulting random variable $Y = b + BX$ is elliptically distributed. Moreover, parameters of this distribution are easily derived from those of X. This property allows us to construct any elliptical random variable from another one.

The second property guarantees closure of the class of elliptical distributions with respect to dimension. If once again X denotes a d-dimensional elliptical random vector, then any of the d components X_i are elliptical as well with location parameters given by the components of the location parameter of X. Moreover, their scale parameters are given by the diagonal elements of the covariance matrix Σ of X.

The third property, which may be considered the most essential one of those presented here for purposes of application to portfolio theory, is the *closure under convolution*. Namely, let X and Y be two independent elliptical random variables, then their sum will also be elliptical. Hence, taking several assets whose returns are independent and elliptically distributed, any portfolio comprised of those assets has elliptically distributed returns as well. As another application, consider the observations of some series of asset returns. On each observation date, the return may be modeled as elliptical random variables. Over some period of time then, the aggregated returns will also be elliptical.

In general, elliptical distributions provide a rich class of distributions that can display important features such as heavy-tails of marginal distributions, for example, as well as tail-dependence both of which have been observed to be exhibited by asset returns in real-world financial markets.

---

[20] A complete listing of these properties can be found in a standard textbook on multivariate risk modeling.

However, due to their simple structure, a common criticism is that elliptical distributions are confined to symmetric distributions and whose dependence structure depends on the correlation or covariance matrix.

As we will see in Chapter 16 where we describe the concept of a copula, by using a copula one is able to account for tail-dependence and asymmetry, which one fails to achieve if one models dependence with the use of only the covariance matrix.

CONCEPTS EXPLAINED IN THIS CHAPTER (IN ORDER OF PRESENTATION)

Multivariate distribution
Random vector
Multivariate probability distribution
Joint probability distribution
Joint cumulative distribution function
Volume
Joint probability density function
Multivariate probability density function
Contour lines
Isolines
Marginal probability distribution
Marginal density function
Dependent
Independent
Independence
Convolution
Covariance
Uncorrelated
Corrleated
Covariance matrix
Correlation coefficient
Negatively correlated
Positively correlated
Correlation coefficient matrix
Tail dependence
Multivariate normal distribution
Bivariate normal distribution
Diversification
Imaginary number
Feasible portfolios

Efficient frontier
Global minimum variance portfolio
Markowitz portfolio selection
Mean-variance portfolio optimization
Multivariate t distribution
Elliptical distributions
Spherically distributed
Closure under linear transformations
Closure under convolution

CHAPTER 15

Conditional Probability and Bayes' Rule

In Chapter 8 we explained that one interpretation of the probability of an event is that it is the relative frequency of that event when an experiment is repeated or observed a very large number of times. Here are three examples: (1) a mortgage company observed that over the past 10 years, 8% of borrowers are delinquent in making their monthly mortgage payments; (2) the risk manager of a bank observed that over the past 12 years, 5% of corporate loans defaulted; and (3) an asset management firm has observed that over the past eight years, in 20% of the months the managers of its stock portfolios underperformed a client's benchmark by more than 50 basis points. For these examples, suppose that the following information is available: (1) during recessionary periods, the mortgage service company observes that 15% of borrowers are delinquent in making their monthly mortgage payment, (2) the risk manager of the bank observes that during recessionary periods, 11% of corporate loans defaulted, and (3) during periods of a declining stock market, the asset management firm observes that in 30% of the months its stock portfolio managers under-performed a client's benchmark by more than 50 basis points.

These three hypothetical examples suggest that taking into account the available additional knowledge (economic recession in our first two examples and a declining stock market in the third example) could result in revised (and more accurate) probabilities about the events of interest to us. We call these revised probabilities *conditional probabilities* and discuss them in more detail in this chapter.

This chapter is coauthored with Biliana Bagasheva.

CONDITIONAL PROBABILITY

Let's consider the experiment of tossing a fair dice. We are now able to compute easily probabilities such as the probability of observing an odd number (1/3) and the probability of obtaining a number greater than 3 (1/2). These "stand-alone" probabilities are called *unconditional probabilities* or *marginal probabilities* since to determine them we do not consider any other information apart from the experiment "tossing a die."

Suppose now we know that an even number came up at a particular throw of the die. Given this knowledge, what is the probability of obtaining the number 2 at that throw? While to most the answer would come in an almost automatic fashion, spelling out the reasoning behind it is instructive in understanding the mechanism of computing conditional probabilities. If an even number has come up on a die, that number must be either 2 or 4 or 6. Since the die is fair and the desired number 2 is one of these possibilities, it must be true that its probability is 1/3.

Let's define the following events:

Event A = {The number 2 shows up}.

Event B = {An even number shows up} = {2,4,6}.

Let's denote the sample space (the collection of all possible outcomes) by Ω and write it as

$$\Omega = \{1; 2; 3; 4; 5; 6\}$$

The Venn diagram in Figure 15.1 further helps to illustrate the experiment and the events in question.

As we know from Chapter 8, the *unconditional* probability of A is given by

$$P(A) = \frac{P(2)}{P(1) + P(2) + P(3) + \cdots + P(6)}$$

Substituting in the values for those probabilities, we obtain

$$P(A) = \frac{1/6}{6 \times 1/6} = 1/6$$

The knowledge that event B occurred serves to restrict the sample space, Ω, to only the three outcomes representing even numbers. We can "forget" about the remaining three outcomes representing odd numbers. The *conditional* probability of A given B is written as $P(A|B)$ and computed as

FIGURE 15.1 Venn Diagram for the Experiment "Tossing a Dice"

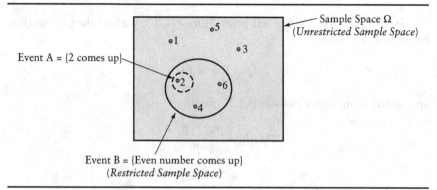

$$P(A|B) = \frac{P(2)}{P(2) + P(4) + P(6)}$$
$$= \frac{1/6}{3 \times 1/6} = 1/3$$

Notice that A's probability is updated substantially (from 1/6 to 1/3) when the information that B occurred is taken into account.

Although in simple experiments we can always use tools such as Venn diagrams to help us determine the conditional probability of an event, in more abstract problems that is not possible and we need a general formula for conditional probability.

Formula for Conditional Probability

The conditional probability that event A occurs given that event B occurs is computed as the ratio of the probability that both A and B occurred to the probability of B,

$$P(A|B) = \frac{P(A \cap B)}{P(B)} \qquad (15.1)$$

Of course, we assume that B can occur, that is, $P(B) \neq 0$. Recall that $A \cap B$ denotes the intersection between A and B (the event that A and B occur together).

Consider again Figure 15.1. The event A in fact also represents the intersection between A and B.

To illustrate using the formula in equation (15.1), let's consider a new event, C, and define it as

C = {A number greater than 3 shows up}.

To compute the conditional probability $P(C|B)$, we find the intersection between B and C as

$$C \cap B = \{4,6\}$$

and substituting into equation (15.1) obtain

$$P(C|B) = \frac{P(C \cap B)}{P(B)}$$
$$= \frac{2/6}{1/2} = 2/3$$

Illustration: Computing the Conditional Probability

"Which stocks are cheap?" is a question upon which anyone with an interest in stock market investing has pondered. One of the indicators most commonly employed to value stocks is the price/earnings (P/E) ratio. The potential out-(under)performance of a stock may be assessed on the basis of its P/E ratio relative to the average industry (or broad market) P/E ratio. Consider an example in which over a period of five years data were collected for the relative performance and P/E ratios of companies in a specific industry.

Consider the data in Table 15.1. The entries in the central part of the table (excluding the last row and rightmost column) represent joint probabilities. For example, the probability that a company in the industry has a P/E ratio lower than the industry average and at the same time outperforms the industry stock index is 6%.[1] Based on the hypothetical data in the Table 15.1, if a company is a candidate for an investment portfolio, what is the probability that it will outperform the industry index given that its P/E ratio

TABLE 15.1 P/E Ratios and Stocks' Relative Performance

Performance Relative to Industry Index	P/E Ratio Relative to Industry Average			
	Low	Average	High	Total
Underperforming	6%	11%	8%	25%
Equally performing	11%	19%	5%	35%
Outperforming	21%	15%	4%	40%
Total	38%	45%	17%	100%

---

[1] Table 15.1 is an example of a contingency table.

is the same as the industry average? To answer this question, let's denote by A the event that a company's P/E ratio is equal to the industry average and by B the event that a company's stock outperforms the industry index. The probability of A can be found as

$$P(A) = 0.11 + 0.19 + 0.15 = 0.45 \text{ or } 45\%$$

Companies with average P/E ratios and performance better than that of the index make up 15% of the companies in the industry according to the data in the Table 15.1. That is,

$$P(A \cap B) = 0.15 \text{ or } 15\%$$

Using these values and applying the formula in equation (15.1), we can compute the conditional probability that a company outperforms the industry index given that its average P/E ratio is equal to the average industry P/E ratio:

$$P(B|A) = \frac{P(A \cap B)}{P(A)} = \frac{0.15}{0.45} = 0.33 \text{ or } 33\%$$

Notice that the unconditional probability that a company outperforms the industry index is 40%. Taking into account the additional information about the company's P/E ratio modified the probability that its stock outperforms the index to 33%.

We can observe that the restricted sample space in this illustration is represented by the middle column of Table 15.1.

Suppose that an investment analyst contemplates whether an exceptionally good cotton harvest in South Asia would have an effect on the probability of outperformance of a stock in the U.S. high-tech industry. In all likelihood, the state of the cotton crop would not have an effect on the probability of the high tech stock performance. Such events are called "independent events." We provide a formal way to check for statistical independence of events in the next section.

INDEPENDENT EVENTS

Two events, A and B, are called *independent events* if the occurrence of one of them does not affect the probability that the other one occurs. We express this in terms of the conditional probabilities in the following ways:

$$P(A|B) = P(A) \text{ (if } P(B) > 0) \quad \text{or} \quad P(B|A) = P(B) \text{ (if } P(A) > 0) \quad (15.2)$$

Using equation (15.2), together with a simple manipulation of the expression in equation (15.1), provides an alternative definition of events' independence. A and B are independent if the product of their (unconditional) probabilities is equal to their joint probability,

$$P(A)P(B) = P(A \cap B) \qquad (15.3)$$

Note that equation (15.3) is true even if $P(A) = 0$, or/and $P(B) = 0$. From now on, we will assume that both $P(A) > 0$ and $P(B) > 0$, so then all three definitions in equations (15.2) and (15.3) are equivalent.

Events for which the expressions above do not hold are called ***dependent events***. Consequently, to check whether two events are independent or not, we simply need to see if the conditional probabilities are equal to the unconditional probabilities. It is sufficient to perform the check for one of the events.

Consider for example the experiment of throwing the same fair coin twice and define the events A = {Head on first toss} and B = {Head on second toss}. Clearly,

$$P(A) = P(B) = 1/2$$

It is also obvious that

$$P(B|A) = P(\text{Head on second toss} \mid \text{Head on first toss}) = 1/2$$

Since we have that $P(B) = P(B|A)$, events A and B are independent.

As another example, suppose that there are two black balls and a white ball in an urn and we draw two balls at random, without replacement.[2] Define A = {White ball on first draw} and B = {White ball on second draw}. The probability of A is $P(A) = 1/3$. The probability of B (in the absence of any condition) is also 1/3 since all balls have an equal chance of being selected. To find the conditional probability of B given A, consider the number and color of balls left in the urn after a white one is selected: two black balls remain and it is impossible to draw a white one. Therefore, we have

$$P(B|A) = 0$$

Since $P(B)$ is not equal to $P(B|A)$, we can conclude that events A and B are dependent.

---

[2] Recall also our discussion of the hypergeometric distribution in Chapter 9, applicable to situations of sampling without replacement.

MULTIPLICATIVE RULE OF PROBABILITY

We can use the discussion of conditional probability above to derive a rule for the probability of an intersection of two events, the so-called *joint probability*.

Recall that the formula for the probability of an event A conditional on event B is given by

$$P(A|B) = \frac{P(A \cap B)}{P(B)}$$

Multiplying both sides by $P(B)$, we obtain a formula for the probability of the intersection of A and B, called the *Multiplicative Rule of Probability*,

$$P(A \cap B) = P(A|B)P(B) \tag{15.4}$$

Equivalently, we can write

$$P(A \cap B) = P(B|A)P(A)$$

Let's consider an example for applying the Multiplicative Rule of Probability.

Illustration: The Multiplicative Rule of Probability

The importance of electronic trading on the foreign-exchange (FX) markets has increased dramatically since the first electronic markets (electronic communication networks or ECNs) appeared in the late 1990s. In 2006, more than half of the total traded FX volume was executed through electronic trading.[3] Some advantages that ECNs provide to market participants are increased speed of trade execution, lower transaction costs, access to a greater number and variety of market players, opportunity for clients to observe the whole order book, etc. The growth of algorithmic trading strategies is related to the expansion of electronic trading. An algorithmic trading strategy, in general, relies on a computer program to execute trades based on a set of rules determined in advance, in order to minimize transaction costs. Depending on the complexity of those rules, such computer programs are capable of firing and executing multiple trade orders per second. Even at that speed, it is not unlikely that by the time a trade order reaches the ECN, the price of the financial instrument has changed, so that a market order

---

[3]See, for example, the survey report of Greenwich Associates at http://www.e-forex.net/Files/surveyreportsPDFs/Greenwich.pdf.

either gets filled at an unintended price or fails altogether. This phenomenon is referred to as "execution slippage." Naturally, execution slippage entails a cost that both market participants and ECNs are eager to minimize.

Suppose we have estimated that, for a given FX market participant and a given ECN, there is a 50% chance that the trade execution time is 500 milliseconds. Given the execution time of 500 milliseconds, there is a 30% chance of execution slippage.

On a given day, the execution of a trade for the FX market participant is observed at random. What is the probability that the execution time was 500 milliseconds and that slippage occurred?

Let's define the following events of interest:

Event A = {Trade execution time is 500 milliseconds}.

Event B = {Execution slippage occurs}.

The following two probabilities are provided in the information above:

$$P(A) = 0.5 \quad \text{and} \quad P(B|A) = 0.3$$

Now, applying the Multiplicative Rule of Probability, we can compute that the probability that the execution time was 500 milliseconds and slippage occurred is

$$P(A \cap B) = P(A)P(B|A)$$
$$= 0.5 \times 0.3 = 0.15 \text{ or } 15\%$$

Multiplicative Rule of Probability for Independent Events

Recall that for two independent events A and B it is true that $P(A|B) = P(A)$. The Multiplicative Rule of Probability is modified in the following way for independent events:

$$P(A \cap B) = P(A)P(B) \qquad (15.5)$$

The modification of the Multiplicative Rule of Probability gives us another way to check for the independence of two events: the events are independent if their joint probability is equal to the product of their unconditional probabilities.

Recall that in Chapter 8 we discussed the case where two events could not occur together and called those events "mutually exclusive." The probability of the intersection of mutually exclusive events is 0. What is the relationship between mutually exclusive events and independent events?

Let A and B be two mutually exclusive events. They are generally not independent events if A occurs, the probability of B occurring is zero, so that $P(A \cap B) \neq P(A)P(B)$. The exception is when $P(A) = 0$ and/or $P(B) = 0$. Then, $P(A \cap B) = 0$ (this can be easily seen by rearranging equation (15.1)) and

$$P(A \cap B) = P(A)P(B)$$

so that A and B are independent.

Law of Total Probability

In Chapter 8, we expressed the probability of an event as the following sum:

$$P(A) = P(A \cap B) + P(A \cap B^c)$$

where B^c denotes the complement of event B.

Using the Multiplicative Rule of Probability, we can rewrite the probability of A as

$$P(A) = P(A|B)P(B) + P(A|B^c)P(B^c) \qquad (15.6)$$

The expression above is known as the *Law of Total Probability*. Let's see how it is applied in the following illustration.

Illustration: The Law of Total Probability

Typically, corporate bonds are assigned a credit rating based on their credit risk. These ratings are assigned by specialized companies called rating agencies and the three major ones include Moody's Investors Service, Standard & Poor's, and Fitch Ratings. Rating agencies assign a letter to a bond issue to describe its credit risk. For example, the letter classification used by Standard & Poor's and Fitch is AAA, AA, A, BBB, BB, B, CCC, and D. Moody's uses the classification Aaa, Aa, A, Baa, Ba, B, Caa, Ca, C. In both classifications, credit risk increases from lowest to highest. The letters D and C mean that the bond issue is in payment default. Bonds with ratings AAA to BBB (Aaa to Baa) are considered investment-grade bonds. Bonds with lower ratings are speculative-grade bonds, also commonly referred to as high-yield bonds or junk bonds.

TABLE 15.2 Credit Migration Table

	In 5th Year		
At Issuance	Investment Grade	Speculative Grade	In Default
Investment Grade	94.7%	5%	0.3%
Speculative Grade	1.2%	87.5%	11.3%
In Default	0%	0%	0%

Credit ratings can change during the lives of bond issues. Credit risk specialists use the so-called "credit migration tables" to describe the probabilities that bond issues' ratings change in a given period.

Suppose that a team of corporate bond analysts at a large asset management firm follow 1,000 corporate names and have compiled the following information for those corporate names. At the time of issuance, 335 of them were assigned speculative-grade rating and 665 of them were assigned investment-grade rating. Over a five-year period,

- A company with an investment-grade rating at the time of issuance has a 5% chance of downgrade to speculative-grade rating and a 0.03% chance of default (i.e., failure to pay coupons and/or principal) on its bond issue.
- A company with a speculative-grade rating at the time of issuance has an 11.3% chance of default and a 1.2% chance of upgrade to investment-grade rating.

A simplified credit migration table summarizing that information is provided in Table 15.2. Notice that the entries in the table represent conditional probabilities, so that, for example,

$$P(\text{Default}|\text{Investment grade at issuance}) = 0.003$$

$$P(\text{Upgrade}|\text{Speculative grade at issuance}) = 0.012$$

and so on.[4] What is the probability that a company defaults within a five-year period?

Let's define the events:

Event A = {The company has speculative-grade rating at time of issuance}.

---

[4] In practice, credit migration tables contain conditional probabilities for all possible transitions among all credit ratings over a much shorter period of time, usually a year.

Event B = {The company defaults within a five-year period}.

The credit migration table in Table 15.2 provides the conditional probabilities

$$P(B|A) = 0.113 \quad \text{and} \quad P(B|A^c) = 0.003$$

The unconditional probability that a company is assigned a speculative-grade rating at time of issuance is given by

$$P(A) = \frac{335}{1,000} = 0.335$$

while the unconditional probability that a company is assigned a nonspeculative-grade rating (that is, investment-grade rating in our example) at time of issuance is computed as

$$P(A^c) = \frac{665}{1,000} = 0.665$$

Now, we can readily substitute into the expression for the Law of Total Probability and find that the chance that a company among the 1,000 followed by the team of bond analysts defaults within a five-year period.

$$\begin{aligned}P(B) &= P(B|A)P(A) + P(B|A^c)P(A^c) \\ &= 0.113 \times 0.335 + 0.003 \times 0.665 \\ &= 0.04 \quad \text{or} \quad 4\%\end{aligned}$$

The Law of Total Probability for More than Two Events

The expression for the Law of Total Probability in equation (15.6) can easily be generalized to the case of K events:

$$\begin{aligned}P(A) &= P(A \cap B_1) + P(A \cap B_2) + \cdots + P(A \cap B_K) \\ &= P(A|B_1)P(B_1) + P(A|B_2)P(B_2) + \cdots + P(A|B_K)P(B_K)\end{aligned}$$

The only "catch" is that we need to take care that the events $B_1, B_2, \ldots, B_K$ exhaust the sample space (their probabilities sum up to 1) and are mutually exclusive. (In the two-event case, B and B^c clearly fulfil these requirements.)

BAYES' RULE

Bayes' rule, named after the eighteenth-century British mathematician Thomas Bayes, provides a method for expressing an unknown conditional probability $P(B|A)$ with the help of the known conditional probability $P(A|B)$. Bayes' rule, for events A and B, is given by the following expression

$$P(B|A) = \frac{P(A|B)P(B)}{P(A)}$$

Another formulation of Bayes' rule is by using the Law of Total Probability in the denominator in place of $P(A)$. Doing so, we obtain

$$P(B|A) = \frac{P(A|B)P(B)}{P(A|B)P(B) + P(A|B^c)P(B^c)}$$

Generalized to K events, Bayes' rule is written as

$$P(B_i|A) = \frac{P(A|B_i)P(B_i)}{P(A)}$$

where the subscript i denotes the i-th event and $i = 1, 2, \ldots, K$. Using the Law of Total Probability, we have

$$P(B_i|A) = \frac{P(A|B_i)P(B_i)}{P(A|B_1)P(B_1) + P(A|B_2)P(B_2) + \cdots + P(A|B_K)P(B_K)}$$

Illustration: Application of Bayes' Rule

The hedge fund industry has experienced exceptional growth, with assets under management more than doubling between 2005 and 2007. According to HedgeFund.net, the total assets under management of hedge funds (excluding funds of hedge funds), as of the second half of 2007, are estimated to be about $2.7 trillion.[5] Hedge funds vary by the types of investment strategies (styles) they employ. Some of the strategies with greatest proportion of assets under management allocated to them are long/short equity, arbitrage, global macro, and event driven.

Consider a manager of an event-driven hedge fund. This type of hedge fund strategy is focused on investing in financial instruments of companies

---

[5] "The 2008 Hedge Fund Asset Flows & Trends Report," HedgeFund.net.

Conditional Probability and Bayes' Rule

undergoing (expected to undergo) some event, for example, bankruptcy, merger, etc. Suppose that the goal of the manager is to identify companies that may become acquisition targets for other companies. The manager is aware of some empirical evidence that companies with a very high value of a particular indicator—the ratio of stock price to free cash flow per share (PFCF)—are likely to become acquisition targets and wants to test this hypothesis.

The hedge fund manager gathers the following data about the companies of interest

- The probability that a company becomes an acquisition target during the course of an year is 40%.
- 75% of the companies that became acquisition targets had values of PFCF more than three times the industry average.
- Only 35% of the companies that were not targeted for acquisition had PFCF higher than three times the industry average.

Suppose that a given company has a PFCF higher than three times the industry average. What is the chance that this company becomes a target for acquisition during the course of the year? Let's define the following events:

Event A = {The company becomes an acquisition target during the year}.

Event B = {The company's PFCF is more than three times the industry average}.

From the information provided in the problem, we can determine the following probabilities:

$$P(A) = 0.4$$

$$P(B|A) = 0.75$$

$$P(B|A^c) = 0.35$$

We need to find the conditional probability that a company becomes an acquisition target given that it has a PFCF exceeding the industry average by more than three times, that is, $P(A|B)$.

Since the conditional probabilities $P(B|A)$ and $P(B|A^c)$ are known, we can find $P(A|B)$ by applying Bayes' rule,

$$P(A|B) = \frac{P(B|A)P(A)}{P(B|A)P(A) + P(B|A^c)P(A^c)}$$

$$= \frac{0.75 \times 0.4}{0.75 \times 0.4 + 0.35 \times 0.6}$$

$$= 0.59 \text{ or } 59\%$$

In the second line above, 0.6 is the probability that a company does not become an acquisition target in a given year, which we find by subtracting $P(A)$ from 1. When we incorporate the evidence that a company has a high PFCF, we obtain a higher probability of acquisition than when that evidence is not taken into account. The probability of acquisition is thus "updated." The updated probability is more accurate since it is based on a larger base of available knowledge.

CONDITIONAL PARAMETERS

Let us consider again the three scenarios from the beginning of this chapter. We may ask the following additional questions about them, respectively:

- During recessionary periods, what is the expected number of clients of the mortgage company that are delinquent in making their monthly mortgage payments?
- During recessionary periods, what is the variability in the number of corporate loans that default?
- During periods of declining stock market, what is the expected number of managers of stock portfolios that underperform their clients' benchmarks by more than 50 basis points?

The questions point to the definition of parameters given the realization of some random variable. We call those parameters *conditional parameters* and in this section we focus on three of them: the *conditional expectation*, the conditional variance, and the conditional value-at-risk.

Conditional Expectation

From Chapter 13, we are already familiar with the concept of expectation of a random variable. All the following variables have unconditional expectations:

- State of the economy.

Conditional Probability and Bayes' Rule

- State of the stock market.
- Number of clients who are delinquent in making their monthly payments.
- Number of corporate loans that default.
- Number of portfolio managers that underperform clients' benchmarks.

These are the expected states of the economy or the stock market and the expected numbers of delinquent clients or defaulted corporate loans or underperforming managers.

In the three questions above, we are in fact considering scenarios based on the realizations of the state of the economy variable and the state of the stock market variable. If the state of the economy is "recession," for example, we do not know what the number of delinquent clients or the number of defaulted corporate loans are. Since they are random variables on the restricted space of "recession," we can compute their expected values, that is, their conditional expectations, conditional on the realized scenario.

In the discrete setting (for instance, the three examples above), where both the conditioning variable and the variable whose expectation we are interested in are discrete, the conditional expectation can be computed using the definition of conditional probability. In particular, the conditional expectation of a random variable X, given the event B, is equal to the unconditional expectation of the variable X set to zero outside of B and divided by the probability of B:

$$E[X|B] = \frac{E[I_B(X)]}{P(B)}$$

The term $I_B(X)$ is the indicator function of the set B. It is equal to 1 whenever X is in set B and 0 when X is in the complementary set B^c. That is, the indicator function helps isolate those values of X that correspond to the realization of B. As an example, consider again the first of the three questions in the beginning of the section. We have

$$E[\text{Number of delinquent clients}|\text{State of economy} = \text{recession}]$$
$$= \frac{E[I_{(\text{State of economy} = \text{recession})}(\text{Number of delinquent clients})]}{P(\text{State of economy} = \text{recession})}$$

As another example, suppose we consider the return on a stock portfolio ABC tomorrow and the value of the S&P 500 equity index tomorrow. We may be interested in computing the expected return on portfolio ABC tomorrow if we know S&P 500's value tomorrow. Let us assume that the returns of the S&P 500 are discrete variables so that the probability that returns assume a given value are finite. We would then have:

$$E[\text{Portfolio ABC's return}|\text{S\&P 500 value} = s]$$

$$= \frac{E[I_{(\text{S\&P 500 value} = s)}(\text{Portfolio ABC's return})]}{P(\text{S\&P 500 value} = s)}$$

Note that if the returns of the S&P 500 were continuous variables, the denominator in the previous expression would be zero. In order to define conditional expectations for continuous variables, we need a more general and abstract definition of conditional expectation as outlined here. Formally, suppose that the random variable X is defined on the probability space (Ω, A, P).[6] Further, suppose that G is a subset of A and is a sub-σ-algebra of A. G fulfills all the requirements on a σ-algebra we are already familiar with. It corresponds, for instance, to the subspaces defined by the variables "state of the economy" or "state of the stock market" in the examples above. Then, the conditional expectation of X given G, denoted by $E(X|G)$, is defined as any random variable *measurable with respect to* G such that its expectation on any set of G is equal to the expectation of X on the same set. The conditional expectation states that under the condition that an event in G has occurred we would expect X to take a value $E(X|G)$.

Conditional Variance

The *conditional variance* can be understood in a way completely analogous to the conditional expectation. The condition (for example, realization of a scenario) only serves to restrict the sample space. The conditional variance is a function that assigns, for each realization of the condition, the variance of the random variable on the restricted space. That is, for X defined on the probability space (Ω, A, P), the conditional variance is a random variable measurable with respect to the sub-σ-algebra, G, and denoted as var$[X|G]$ such that its variance on each set of G is equal to the variance of X on the same set. For example, with the question "During recessionary periods, what is the variability in the number of corporate loans that default?," we are looking to compute

var[Number of default corporate loans | State of the economy = Recession]

Using the definition of variance from Chapter 13, we could equivalently express var$[X|G]$ in terms of conditional expectations,

$$\text{var}[X|G] = E[X^2|G] + E[X|G]^2$$

---

[6]We defined probability space in Chapter 8.

Expected Tail Loss

In Chapter 13, we explained that the value-at-risk (VaR) is one risk measure employed in financial risk measurement and management. VaR is a feature of the unconditional distribution of financial losses (negative returns). Recall that the 99% VaR of the loss distribution, for instance, is the loss value such that, with a 1% chance, the financial asset will have a loss bigger than 99% VaR over the given period.

The VaR risk measure provides us with only a threshold. What loss could we expect if the 99% VaR level is broken? To answer this question, we must compute the expectation of losses conditional on the 99% VaR being exceeded:

$$E[-R_t \mid -R_t > 99\% \text{ VaR}]$$

where R_t denotes the return on a financial asset at time t and $-R_t$—the loss at time t. In the general case of $\text{VaR}_{(1-\alpha)100\%}$, the conditional expectation above takes the form

$$E[-R_t \mid -R_t > \text{VaR}_{(1-\alpha)100\%}]$$

and is known as the $(1 - \alpha)100\%$ *expected tail loss* (ETL) or *conditional value-at-risk* (CVaR) if the return R_t has density.

CONCEPTS EXPLAINED IN THIS CHAPTER (IN ORDER OF PRESENTATION)

Conditional probabilities
Unconditional probabilities
Marginal probabilities
Independent events
Dependent events
Joint probability
Multiplicative Rule of Probability
Law of Total Probability
Bayes' rule
Conditional parameters
Conditional expectations
Conditional variance
Expected tail loss
Conditional value-at-risk

CHAPTER 16
Copula and Dependence Measures

In previous chapters of this book, we introduced multivariate distributions that had distribution functions that could be presented as functions of their parameters and the values x of the state space; in other words, they could be given in closed form.[1] In particular, we learned about the multivariate normal and multivariate t-distributions. What these two distributions have in common is that their dependence structure is characterized by the covariance matrix that only considers the linear dependence of the components of the random vectors. However, this may be too inflexible for practical applications in finance that have to deal with all the features of joint behavior exhibited by financial returns.

Portfolio managers and risk managers have found that not only do asset returns exhibit heavy tails and a tendency to simultaneously assume extreme values, but assets exhibit complicated dependence structures beyond anything that could be handled by the distributions described in previous chapters of this book. For example, a portfolio manager may know for a portfolio consisting of bonds and loans the constituents' marginal behavior such as probability of default of the individual holdings. However, their aggregate risk structure may be unknown to the portfolio manager.

In this chapter, in response to the problems just mentioned, we introduce the *copula* as an alternative approach to multivariate distributions that takes us beyond the strict structure imposed by the distribution functions that analytically have a closed form. As we will see, the copula provides access to a much richer class of multivariate distributions.

For a random vector of general dimension d, we will show how a copula is constructed and list the most important specifications and properties of a copula. In particular, these will be the copula density, increments of a copula, copula bounds, and invariance of the copula under strictly monotonically increasing transformations. Furthermore, we introduce the simu-

---

[1]An exception is the class of α–stable distributions introduced in Chapter 12 for which we could neither present the density nor distribution function, in general.

lation of financial return data using the copula. We will then focus on the special case of two dimensions, repeating all theory we introduced for $d = 2$. Additionally, we will introduce alternative dependence measures. These will include the rank correlation measures such as Spearman's rho as well as tail dependence. For all the theoretical concepts we present, we will provide illustrations to aid in understanding the concepts.

As we stated elsewhere in this book, we may not always apply the strictest of mathematical rigor as we present the theoretical concepts in this chapter, but instead prefer an intuitive approach occasionally making a statement that may be mathematically ambiguous.

COPULA

We will begin by presenting the definition of the *copula* including its most important properties. We first introduce the general set-up of d dimensions and then focus on the particular case of $d = 2$.

Let $X = (X_1, X_2, \ldots, X_d)$ be a random vector such as a cross-sectional collection of asset returns where each of the components X_i has a marginal distribution function denoted $F_i(x_i)$. Now, since the X_i are random variables, the $F_i(X_i)$ are also random depending on the outcome of X_i. Recall from Chapter 13 that the α-quantile, q_α, of some probability distribution is the value that the random variable does not exceed with probability α. With the quantiles determined through the inverse of the distribution function, F_i^{-1}, such that $q_\alpha = F_i^{-1}(\alpha)$, there is a relationship between a random variable and its distribution function.[2]

The dependence structure between a finite number of random variables is determined by their joint distribution function. Moreover, the joint distribution function determines the marginal distributions as well. Thus the joint distribution function explains the entire probability structure of a vector of random variables. Next we will show that the joint distribution function is uniquely determined by the so-called copula and the marginal distribution functions.

Construction of the Copula

For a random vector $X = (X_1, X_2, \ldots, X_d)$, the joint distribution function F can be expressed as

---

[2] This relationship is unique at least for continuous random variables. However, as we know from Chapter 13, there might be several quantiles for a certain level α for a discrete random variable.

$$F(x_1, x_2, \ldots, x_d) = P(X_1 \leq x_1, X_2 \leq x_2, \ldots, X_d \leq x_d)$$
$$= C(P(X_1 \leq x_1), P(X_2 \leq x_2), \ldots, P(X_d \leq x_d))$$
$$= C(F_1(x_1), F_2(x_2), \ldots, F_d(x_d))$$

where the F_i are the not necessarily identical marginal distribution functions of the components X_i and C is a function on d-dimensional unit cube, and is called the copula of the random vector X, or copula of the joint distribution function F. It can be shown that the random variables $U_i = F_i(X_i)$ are uniformly distributed on the unit interval [0,1] at least for strictly increasing continuous distribution functions F_i.

The following theorem, known as **Sklar's Theorem**, establishes the relationship between (i) the joint multivariate distribution of a random vector X, and (ii) the copula function C and set of d-univariate marginal distribution functions of X.

Let F be the joint distribution function of the random vector $X = (X_1, X_2, \ldots, X_d)$. Moreover, let F_i denote the distribution functions of the components X_i. Then, through a copula function

$$C : [0,1]^d \to [0,1]$$

the joint distribution function F can be represented as

$$F(x_1, x_2, \ldots, x_d) = C(F_1(x_1), F_2(x_2), \ldots, F_d(x_d)) \quad (16.1)$$

for $-\infty \leq x_i \leq \infty$, $i = 1, 2, \ldots, d$.

So by equation (16.1), we see that any multivariate distribution function can be written as a function C that maps the set of d-dimensional real numbers between 0 and 1 into the interval [0,1]. That is, the joint distribution of the random vector X is determined by the marginal distribution functions of its components—assuming values between 0 and 1—coupled through C. Thus, any joint distribution can be decomposed into marginal probability laws and its dependence structure.

Conversely, Sklar's Theorem guarantees that with any copula C and arbitrary marginal distribution functions F_i, we always obtain some joint distribution function F of a random vector $X = (X_1, X_2, \ldots, X_d)$.

Specifications of the Copula

Now let's look at what functions qualify as a copula. We specify the copula by the following definition.

The copula is a function

$$C:[0,1]^d \to [0,1] \quad (16.2)$$

such that, for any d-tuple $u=(u_1,u_2,\ldots,u_d)\in[0,1]^d$, the following three requirements hold

Requirement 1: $C(u_1, u_2, \ldots, u_d)$ is increasing in each component u_i.
Requirement 2: $C(1, \ldots, 1, u_i, 1, \ldots, 1) = u_i$ for all $i = 1, 2, \ldots, d$.
Requirement 3: For any two points $u=(u_1,u_2,\ldots,u_d)\in[0,1]^d$ and $v=(v_1,v_2,\ldots,v_d)\in[0,1]^d$, such that v is at least as great as u in each component (i.e., $u_i \le v_i$, for $i = 1, 2, \ldots, d$), we have

$$\sum_{i_1=1}^{2}\cdots\sum_{i_d=1}^{2}(-1)^{i_1+i_2+\ldots+i_d}C\left(z_{i_1,1},z_{i_2,2},\ldots,z_{i_d,d}\right)\ge 0$$

where the j-th component of z is either $z_{1,j}=u_j$ or $z_{2,j}=v_j$ depending on the index i_j.

Let's discuss the meaning of these three requirements. Requirement 1 is a consequence of the fact that an increase in the value u_i of the marginal distribution F_i should not lead to a reduction of the joint distribution C. This is because an increase in u_i corresponds to an increase of the set of values that X_i can assume that, by the general definition of a probability measure explained in Chapter 8, leads to no reduction of the joint distribution of $X = (X_1, X_2, \ldots, X_d)$. For requirement 2, because by construction the $U_i \in [0,1]$ are uniformly distributed, the probability of any of the U_i assuming a value between 0 and u_i is exactly u_i. And, since $C(1,\ldots,1,u_i,1,\ldots,1)$ represents the probability that $U_i \in [0,u_i]$ while all remaining components assume any feasible value (i.e., in [0,1]), the joint probability is equal to u_i. Requirement 3 is a little more complicated. Basically, it guarantees that the copula as a distribution function never assigns a negative probability to any event. Combining requirements 1, 2, and 3, we see that a copula function can be any distribution function whose marginal distribution functions are uniform on [0,1]. We demonstrate this in more detail for the two-dimensional (i.e., $d = 2$) case a little later in this chapter.

The copula is not always unique, however. Only when the marginal distributions are continuous is it unique. In contrast, if some marginal distributions of the random vector are discrete, then a unique copula of the joint distribution does not exist. Consequently, several functions fulfilling the three requirements above qualify as the function C on the right-hand side of equation (16.1).

Properties of the Copula

The four essential properties of a copula are copula density, increments of the copula, bounds of the copula, and invariance under strictly monotonically increasing transformations. We describe these properties below. Some of these properties, as we will point out, only apply to continuous random vectors.

Copula Density

Just like for a multivariate distribution function of a random vector with continuous components described in Chapter 14, we can compute the density function of a copula. It is defined as

$$c(u_1, u_2, \ldots, u_d) = \frac{\partial C(u_1, u_2, \ldots, u_d)}{\partial u_1 \cdot \partial u_2 \cdot \ldots \cdot \partial u_d}$$

The copula density function may not always exist. However, for the commonly used *Gaussian copula* and *t-copula* that we discuss later in this chapter, they do exist.

Increments of the Copula

If we compute the derivative of a copula with respect to component u_i at any point such that $u_1, \ldots, u_{i-1}, u_{i+1}, \ldots, u_d \in [0,1]$ and $u_i \in (0,1)$, we would obtain

$$0 \leq \frac{\partial C(u_1, u_2, \ldots, u_d)}{\partial u_i}$$

This reveals that the copula never decreases if we increment any arbitrary component. This property is consistent with requirement 3 above.

Bounds of the Copula

We know that the copula assumes values between 0 and 1 to represent a joint distribution function. However, it is not totally unrestrained. There exists a set of bounds that any copula is always restricted by. These bounds are called the **Fréchet lower bound** and **Fréchet upper bound**. So, let C be any copula and $u = (u_1, u_2, \ldots, u_d) \in [0,1]^d$, then we have the restrictions[3]

---

[3] The function max $\{x,y\}$ is equal to x if $x \geq y$ and equal to y in any other case. The function min $\{x,y\}$ is equal to x if $y \geq x$ and equal to y in any other case.

$$\underbrace{\max\left\{\sum_{i=1}^{d} u_i - (d-1), 0\right\}}_{lower\ bound} \leq C(u_1, u_2, \ldots, u_d) \leq \underbrace{\min\{u_1, u_2, \ldots, u_d\}}_{upper\ bound}$$

While the Fréchet lower bound is a little more complicated to understand and just taken as given here, the reasoning behind the Fréchet upper bound is as follows. By requirement 2 of the definition of the copula, the copula is never greater than u_i when all other components equal 1. Now, letting the other components assume values strictly less than 1, then the copula can definitely never exceed u_i. Moreover, since this reasoning applies to all u_i, $i = 1, 2, \ldots, d$, simultaneously, the copula can never be greater than the minimum of these components (i.e., min $\{u_i\}$).

Invariance under Strictly Monotonically Increasing Transformations

A very important property whose use will often be essential in the examples presented later is that the copula is invariant under a strictly monotonically increasing transformation that we may denote as T where $T(y) > T(x)$ whenever $y > x$.[4] Invariance means that the copula is still the same whether we consider the random variables $X_1, X_2, \ldots, X_d$ themselves or strictly monotonically increasing transformations $T(X_1), T(X_2), \ldots, T(X_d)$ of them because their dependence structures are the same. And since the dependence structure remains unaffected by this type of transformation, it is immaterial for the copula whether we use the original random variables X_i or standardized versions of them such as

$$Z_i = (X_i - E(X_i))/\sqrt{Var(X_i)}$$

This is helpful because it is often easier to work with the joint distribution of standardized random variables.

Independence Copula Our first example of a copula is a very simple one. It is the copula that represents the joint distribution function of some random vector $Y = (Y_1, Y_2, \ldots, Y_d)$ with independent (continuous) components. Because of this characteristic, it is called an *independence copula*. Its construction is very simple. For $u = (u_1, u_2, \ldots, u_d) \in [0,1]$, it is defined as

---

[4]Monotonically increasing functions were introduced in Appendix A. Strictly monotonically increasing functions are such that they never remain at a level but always have positive slope. The horizontal line, for example, is monotonically increasing, however, it is not strictly monotonically increasing.

$$C_\theta^I(u_1, u_2, \ldots, u_d) = \prod_{i=1}^{d} u_i$$
$$= u_1 \cdot u_2 \cdot \ldots \cdot u_d$$
$$= P(Y_1 \leq F_1^{-1}(u_1), Y_2 \leq F_2^{-1}(u_2), \ldots, Y_d \leq F_d^{-1}(u_d))$$

where F_i^{-1} denotes the inverse of the marginal distribution function F of random variable Y_i, for each $i = 1, 2, \ldots, d$.

Next, we give two examples of copulas for which the exact multivariate distribution is known.

Gaussian Copula In this illustration, we present the copula of multivariate normal random vectors. Let the random vector $Y = (Y_1, Y_2, \ldots, Y_d)$ have the joint distribution $N(\mu, \Sigma)$ where μ denotes the vector of means and Σ the covariance matrix of the components of Y. Moreover, let $X = (X_1, X_2, \ldots, X_d) \sim N(0, \Gamma)$ be a vector of multivariate normally distributed random variables, as well, but with zero mean and Γ denoting the correlation matrix of Y. Then, because each component of X is a standardized transform of the corresponding component of Y, that is,

$$X_i = T(Y_i) = (Y_i - E(Y_i))/\sqrt{Var(Y_i)}$$

the copulae of X and Y coincide.[5] Hence,

$$C_\Gamma^N(u_1, u_2, \ldots, u_d) = P(\Phi(X_1) \leq u_1, \Phi(X_2) \leq u_2, \cdots, \Phi(X_d) \leq u_d)$$
$$= P(X_1 \leq \Phi^{-1}(u_1), X_2 \leq \Phi^{-1}(u_2), \cdots, X_d \leq \Phi^{-1}(u_d))$$
$$= \Phi_\Gamma(\Phi^{-1}(u_1), \Phi^{-1}(u_2), \cdots, \Phi^{-1}(u_d))$$
$$= F(F_1^{-1}(u_1), F_1^{-1}(u_2), \cdots, F_1^{-1}(u_d))$$
$$= P(F_1(Y_1) \leq u_1, F_2(Y_2) \leq u_2, \cdots, F_d(Y_d) \leq u_d) \quad (16.3)$$

which shows that C_Γ^N is the copula of both $N(\mu, \Sigma)$ and $N(0, \Gamma)$.

Here, we used F_i to denote the normal marginal distribution functions of the Y_i, while Φ denotes the standard normal distribution functions of the X_i. Accordingly, $F_i^{-1}(u_i)$ is the u_i-quantile of the corresponding normal distribution of Y_i just as $\Phi^{-1}(u_i)$ represents the standard normal u_i-quantile of X_i. Finally, Φ_Γ denotes the joint distribution function of the $N(0, \Gamma)$ distribution. From the third line in (16.3), we conclude that the Gaussian copula

---

[5] For more than one copula, we use *copulae* even though in the literature, one often encounters the use of *copulas*.

C_Γ^N is completely specified by the correlation matrix Γ. Graphical illustrations will follow later in this chapter dedicated to the case where $d = 2$.

t-Copula Here we establish the copula of the d-dimensional multivariate t-distribution function. We use $t_d(\nu,\mu,\Sigma)$ for this distribution where ν are the degrees of freedom, μ the vector of means, and Σ the dispersion matrix. Let $Y = (Y_1, Y_2, ..., Y_d)$ be the $t_d(\nu,\mu,\Sigma)$ distributed random vector, then the parameters indicating the degrees of freedom are identical for all d components of the random vector Y.

Due to the invariance of the copula with respect to transformation, it makes no difference whether we analyze the dependence structure of a $t_d(\nu,\mu,\Sigma)$ distributed random vector or a $t_d(\nu,0,P)$ one (i.e., a standardized t random vector that has zero mean and dispersion matrix P that is the correlation matrix associated with Σ). The corresponding copula, *t-copula*, is given by

$$C_{P,\nu}^t(u_1,u_2,...,u_d)$$
$$= P(t_\nu(X_1) \leq u_1, t_\nu(X_2) \leq u_2,...,t_\nu(X_d) \leq u_d)$$
$$= P(X_1 \leq t_\nu^{-1}(u_1), X_2 \leq t_\nu^{-1}(u_2),...,X_d \leq t_\nu^{-1}(u_d))$$
$$= F(t_\nu^{-1}(u_1), t_\nu^{-1}(u_2),...,t_\nu^{-1}(u_d))$$
$$= \frac{\Gamma\left(\frac{\nu+d}{2}\right)}{\Gamma\left(\frac{\nu}{2}\right)\sqrt{|P|(\nu\pi)^d}} \int_{-\infty}^{t_\nu^{-1}(u_1)} \cdots \int_{-\infty}^{t_\nu^{-1}(u_d)} \left(1+\frac{y'P^{-1}y}{\nu}\right)^{-\frac{\nu+d}{2}} dy_1...dy_d \quad (16.4)$$

where

$t_\nu^{-1}(u_i)$ denotes the respective u_i-quantiles of the marginal Student's t-distributions (denoted by $t(\nu)$) with distribution functions t_ν.
F is the distribution function of $t_d(\nu,0,P)$.
|P| is the determinant of the correlation matrix P.
$y = (y_1, y_2, ..., y_d)^T$ is the vector of integration variables.[6]

We will present illustrations for the special case where $d = 2$ later in the chapter.

Simulation of Financial Returns Using the Copula

Copulae and in particular the converse of Sklar's theorem (i.e., deriving a multivariate distribution function from some given copula and univariate

---
[6] See Chapter 11 for the univariate Student's t-distribution.

(marginal) distribution functions) enjoy intense use in the modeling of financial risk where the dependence structure in the form of the copula is known but the joint distribution may not be known.

With this procedure that we now introduce, we can simulate behavior of aggregated risk such as the returns of a portfolio of stocks.

Let $Y = (Y_1, Y_2, \ldots, Y_d)$ be the random vector of interest such as a set of some financial returns with copula C. In the first step, we generate the realizations $u_i \in [0,1]$ of the random variables U_i representing the marginal distribution functions of the d returns Y_i from the given copula C. This can be done by using the realizations x_i of d random variables X_i with marginal distribution functions F_i and joint distribution function F specified by the very same copula C and transforming them through $u_i = F_i(x_i)$.[7]

In the second step, we produce the realizations y_i for the Y_i. We denote the respective marginal distribution function of Y_i by $F_{Y,i}$ with its inverse $F_{Y,i}^{-1}(U_i)$ yielding the U_i-quantile. The y_i are now easily obtained by simply entering the realizations u_i from the first step into the transformation

$$y_1 = F_{Y,1}^{-1}(u_1), y_2 = F_{Y,2}^{-1}(u_2), \ldots, y_d = F_{Y,d}^{-1}(u_d)$$

and we are done.

We illustrate this procedure for the two-dimensional case as an example later.

The Copula for Two Dimensions

We now examine the particular case where $d = 2$ in order to get a better understanding of the theory presented thus far. The reduction to only two dimensions often appeals more to intuition than the general case.

In the case where $d = 2$, the function in equation (16.2) becomes

$$C:[0,1]^2 \to [0,1]$$

showing that the two-dimensional copula maps the values of two marginal distribution functions into the state space between 0 and 1.

Suppose we have two random variables, say X_1 and X_2, then we can give their bivariate distribution function F by

$$F(x_1, x_2) = C(F_1(x_1), F_2(x_2))$$

for any pair of real numbers (x_1, x_2).

From the definition of the copula, we have in the particular case of $d = 2$ that

---
[7] Of course, this approach works only if the marginal distributions are known.

1. For any value $u \in [0,1]$, that is, any value that a distribution function may assume, the following hold:

$$C(0,u) = C(u,0) = 0 \tag{16.5}$$

and

$$C(1,u) = C(u,1) = u \tag{16.6}$$

Equation (16.5) can be interpreted as the probability that one component assumes some value below the lowest value that it can possibly assume, which is unlikely, while the other has a value of, at most, its corresponding u-quantile, q_u.[8] Equation (16.6) is the probability of one component assuming any value at all, while the other is less than or equal to its u-quantile, q_u. So, the only component relevant for the computation of the probability in equation (16.6) is the one that is less than or equal to q_u.

2. For any two points $u = (u_1, u_2)$ and $v = (v_1, v_2)$ in $[0,1]^2$ such that the first is smaller than the second in both components (i.e., $u_1 \leq v_1$ and $u_2 \leq v_2$), we have

$$C(v_1, v_2) - C(v_1, u_2) - C(u_1, v_2) + C(u_1, u_2) \geq 0 \tag{16.7}$$

That is, we require of the copula that for it to be a bivariate distribution function, it needs to produce nonnegative probability for any rectangular event bounded by the points u and v.

We illustrate the second property in Figure 16.1. Throughout the following, we have to keep in mind that the marginal distribution functions F_1 and F_2 of the random variables X_1 and X_2 are treated as random variables, say U and V, with joint distribution given by the copula $C(U,V)$. Moreover, we will make extensive use of the relationship between these latter random variables and the corresponding quantiles $q_U = F_1^{-1}(U)$ and $q_V = F_2^{-1}(V)$ of X_1 and X_2, respectively.

As we can see from Figure 16.1, the rectangular event—we call it E—is given by the shaded area that lies inside of the four points (u_1, u_2), (u_1, v_2), (v_1, u_2), and (v_1, v_2). Now, E represents the event that random variable U assumes some values between u_1 and v_1 while V is between u_2 and v_2. To compute the probability of E, we proceed in three steps. First, we compute $C(v_1, v_2)$ representing the probability of U (i.e., the first random component) being less than or equal to v_1 and simultaneously V (i.e., the second random

---

[8] For continuous random variables, equation (16.5) is the case of one component assuming a value below its support. The term "support" was introduced in Chapter 10.

FIGURE 16.1 Event of Rectangular Shape (shaded area) Formed by the Four Points (u_1, u_2), (u_1, v_2), (v_1, u_2), and (v_1, v_2)

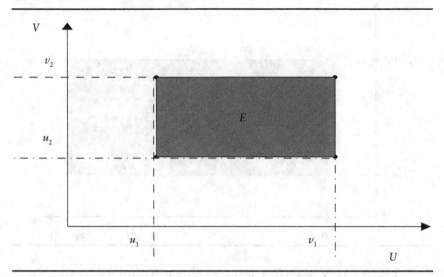

component) being less than or equal to v_2, which corresponds to the probability $P\left(X_1 \leq F_1^{-1}(v_1), X_2 \leq F_2^{-1}(v_2)\right)$.

Since we are considering more than E, in the second step, we subtract the probability $C(u_1, v_2)$ of the event of U being at most u_1 and V being no more than v_2, which is indicated in Figure 16.1 by the area bounded by the dashed lines with top-right corner (u_1, v_2) and open to the left and beneath. This is equivalent to subtracting the probability $P\left(X_1 \leq F_1^{-1}(u_1), X_2 \leq F_2^{-1}(v_2)\right)$ from $P\left(X_1 \leq F_1^{-1}(v_1), X_2 \leq F_2^{-1}(v_2)\right)$. Also, we subtract $C(v_1, u_2)$, that is, the probability of the event that U and V have values no greater than v_1 and u_2, respectively, indicated by the area with dash-dotted bounds, top-right corner (v_1, u_2), and open to the left and beneath. This amounts to reducing the probability $P\left(X_1 \leq F_1^{-1}(v_1), X_2 \leq F_2^{-1}(v_2)\right)$ minus $P\left(X_1 \leq F_1^{-1}(u_1), X_2 \leq F_2^{-1}(v_2)\right)$ additionally by $P\left(X_1 \leq F_1^{-1}(v_1), X_2 \leq F_2^{-1}(u_2)\right)$.

As a result, we have twice reduced the probability from step 1 by $C(u_1, u_2)$, which is the probability of the event of U and V being less than or equal to u_1 and u_2, respectively. In Figure 16.1, this event is the area with top-right corner (u_1, u_2), dashed upper bound, dash-dotted right bound, and open to the left and below. So, in the third step, we need to add $C(u_1, u_2)$ corresponding to the probability $P\left(X_1 \leq F_1^{-1}(u_1), X_2 \leq F_2^{-1}(u_2)\right)$ to obtain the probability of the event E. To see that this is nonnegative and consequently a probability, we need to realize that we have simultaneously computed the probability

FIGURE 16.2 Event (shaded area) Corresponding to the Event E from Figure 16.1

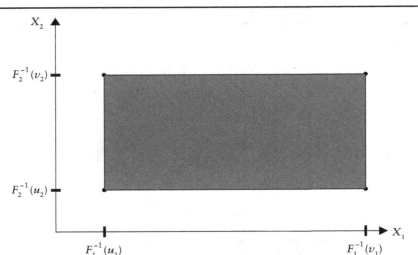

$$P\left(F_1^{-1}(u_1) \leq X_1 \leq F_1^{-1}(v_1), F_2^{-1}(u_2) \leq X_2 \leq F_2^{-1}(v_2)\right)$$
$$= P\left(X_1 \leq F_1^{-1}(v_1), X_2 \leq F_2^{-1}(v_2)\right) - P\left(X_1 \leq F_1^{-1}(u_1), X_2 \leq F_2^{-1}(v_2)\right)$$
$$- P\left(X_1 \leq F_1^{-1}(v_1), X_2 \leq F_2^{-1}(u_2)\right) + P\left(X_1 \leq F_1^{-1}(u_1), X_2 \leq F_2^{-1}(u_2)\right)$$
$$= F_X\left(F_1^{-1}(v_1), F_2^{-1}(v_2)\right) - F_X\left(F_1^{-1}(u_1), F_2^{-1}(v_2)\right)$$
$$- F_X\left(F_1^{-1}(v_1), F_2^{-1}(u_2)\right) + F_X\left(F_1^{-1}(u_1), F_2^{-1}(u_2)\right)$$

of the shaded event in Figure 16.2 that, by definition of a probability measure, can never become negative. Consequently, postulation (16.7) follows for the two-dimensional case. Here, F_X represents the joint distribution function of the random variables X_1 and X_2.

The reason we concentrated on the rectangular events in requirement (16.7) is that one can approximate any kind of event for bivariate distributions from rectangles.

In the following, we present the two-dimensional cases of the previous examples.

Gaussian Copula (d = 2)

In the particular case of $d = 2$, we concretize the Gaussian copula from equation (16.3) as

Copula and Dependence Measures

FIGURE 16.3 Gaussian Copula ($d = 2$) for $\rho = 0$

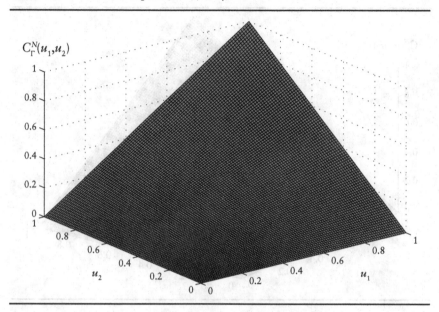

$$C_\Gamma^N(u_1, u_2) = \int_{-\infty}^{\Phi^{-1}(u_1)} \int_{-\infty}^{\Phi^{-1}(u_2)} \frac{1}{2\pi\sqrt{1-\rho^2}} e^{-\frac{x_1^2 - 2\rho x_1 x_2 + x_2^2}{2(1-\rho^2)}} dx_1 dx_2$$

Note that for the computation of the copula, we integrate the bivariate normal density function with correlation matrix

$$\Gamma = \begin{pmatrix} 1 & \rho \\ \rho & 1 \end{pmatrix}$$

where ρ denotes the correlation coefficient between X_1 and X_2.

We illustrate this copula for the cases where $\rho = 0$ (Figure 16.3), $\rho = 0.5$ (Figure 16.4), and $\rho = 0.95$ (Figure 16.5). By doing so, we increase the relationship between the two random variables from independence to near perfect dependence. At first glance, all three figures appear virtually identical. However, careful examination indicates that in Figure 16.3, the copula is a very smooth surface slightly protruding above the diagonal from $u = (0,0)$ to $u = (1,1)$. In contrast, in Figure 16.5, the copula looks like two adjoining flanks of a pyramid with a sharp edge where they meet. Figure 16.4 provides something in between the two shapes.

FIGURE 16.4 Gaussian Copula ($d = 2$) for $\rho = 0.5$

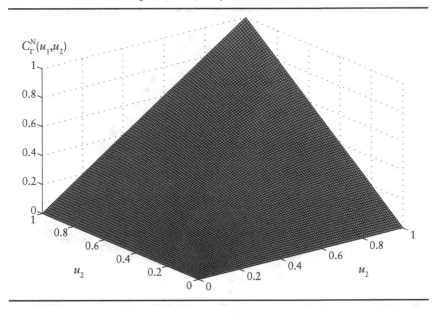

FIGURE 16.5 Gaussian Copula ($d = 2$) for $\rho = 0.95$

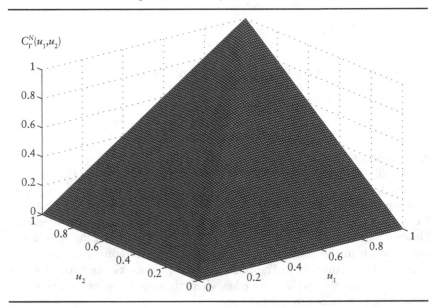

FIGURE 16.6 Contour Plots of the Gaussian Copula ($d = 2$) for $\rho = 0$

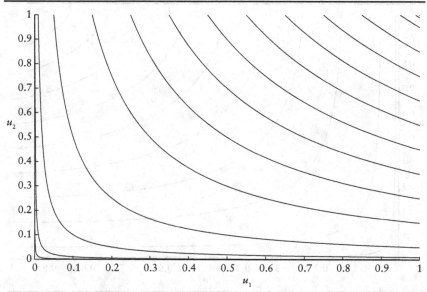

Now, if we look at the contour lines of the copula depicted in Figures 16.6 through 16.8 for the same cases of correlation, the difference becomes easily visible. While in the case of independence ($\rho = 0$), the lines change from circular to straight for increasing levels c of the copula (i.e., going from bottom-left to top-right), the graphic in Figure 16.8 representing almost perfect dependence looks much different. The contour lines there are almost everywhere straight with accentuated edges along the diagonal from $u = (0,0)$ to $u = (1,1)$. In the case of mediocre dependence illustrated in Figure 16.7, we have more pronounced curves than in Figure 16.6, but not as extreme as in Figure 16.8.

t-Copula (d = 2)

Let's establish the two-dimensional illustration of the t copula from equation (16.4). We concretize the copula as

$$C^t_{\Gamma,\nu}(u_1,u_2) = \frac{\Gamma\left(\frac{\nu}{2}+1\right)}{\Gamma\left(\frac{\nu}{2}\right)\sqrt{|P|}(\nu\pi)^2} \int_{-\infty}^{t_\nu^{-1}(u_1)} \int_{-\infty}^{t_\nu^{-1}(u_2)} \left(1+\frac{y'P^{-1}y}{\nu}\right)^{-\left(\frac{\nu}{2}+1\right)} dy_1 dy_2$$

FIGURE 16.7 Contour Plots of the Gaussian Copula ($d = 2$) for $\rho = 0.5$

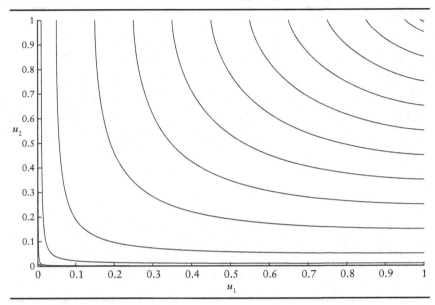

FIGURE 16.8 Contour Plots of the Gaussian Copula ($d = 2$) for $\rho = 0.95$

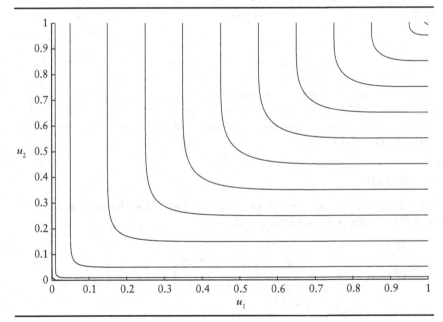

FIGURE 16.9 The t-Copula ($d = 2$) for $\rho = 0$ and $\nu = 5$

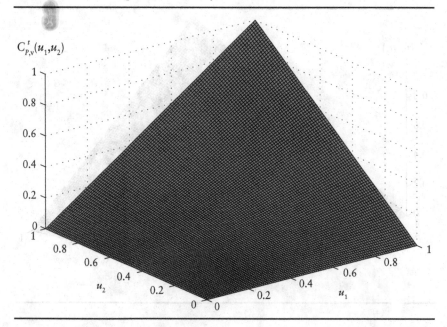

with the correlation matrix

$$P = \begin{pmatrix} 1 & \rho \\ \rho & 1 \end{pmatrix}$$

for the correlation coefficients $\rho = 0$, $\rho = 0.5$, and $\rho = 0.95$ as in the Gaussian copula example.[9] Moreover, we choose as degrees of freedom $\nu = 5$ to obtain a t-distribution much heavier (or thicker) tailed than the normal distribution.[10] The copula plots are depicted in Figures 16.9 through 16.11. We see that the situation is similar to the Gaussian copula case with the copula gradually turning into the shape of two adjoining pyramid flanks as the correlation increases.

For a more detailed analysis of the behavior of the t-copula, let's look at the contour plots displayed in Figures 16.12 through 16.14 where we plot the lines of the same levels c as in the prior example. At first glance, they appear to replicate the structure of the contour lines of the Gaussian copula. However, there is a noticeable difference that is most pronounced for the

---

[9]Note that we used P instead of Γ since here the latter is used for the gamma function introduced in Appendix A.
[10]For a discussion on tail behavior of a distribution, see Chapter 12.

FIGURE 18.10 The t-Copula ($d = 2$) for $\rho = 0.5$ and $\nu = 5$

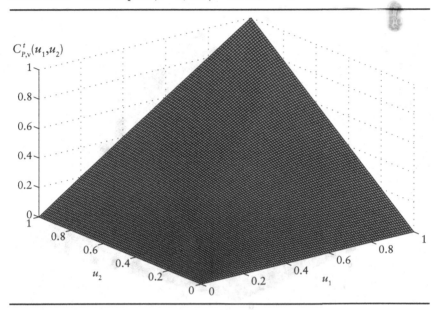

FIGURE 18.11 The t-Copula ($d = 2$) for $\rho = 0.95$ and $\nu = 5$

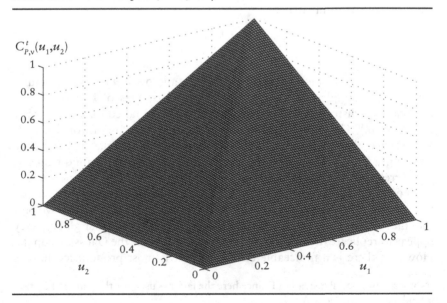

FIGURE 16.12 Contour Plots of the t-Copula ($d = 2$) for $\rho = 0$ and $\nu = 5$

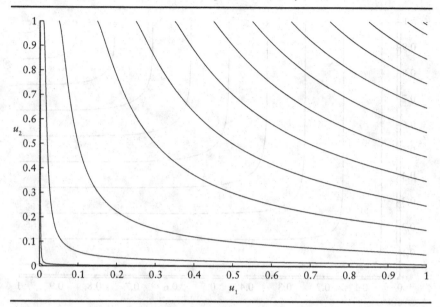

FIGURE 16.13 Contour Plots of the t-Copula ($d = 2$) for $\rho = 0.5$ and $\nu = 5$

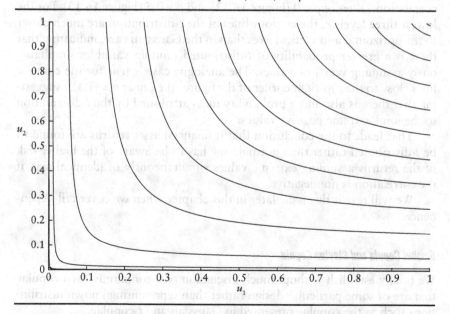

FIGURE 16.14 Contour Plots of the t-Copula ($d = 2$) for $\rho = 0.95$ and $\nu = 5$

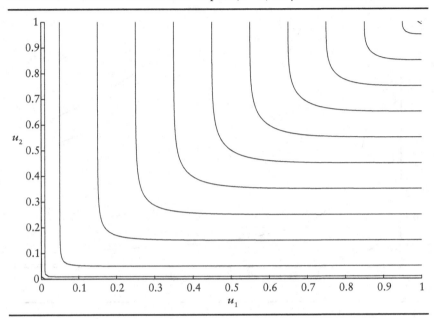

correlation values of $\rho = 0$ (Figure 16.12) and $\rho = 0.5$ (Figure 16.13). For the lowest three levels c, the contour lines of the t-distribution are much closer to the horizontal and vertical axes than in the Gaussian case, indicating that there is a greater probability of t-distributed random variables simultaneously assuming very low values. The analogue case is true for the contour lines close to the top-right corner of the figures (i.e., near $u = (1,1)$), suggesting that there is also more probability mass attributed by the t-distribution to the joint extreme positive values.

That leads to the conclusion that if financial asset returns are found to be t-distributed rather than normal, one has to be aware of the higher risk of the returns assuming extreme values simultaneously of identical sign if the correlation is nonnegative.

We will revisit this issue later in this chapter when we cover tail-dependence.

Gumbel Copula and Clayton Copula

We briefly establish without much discussion two commonly used copulae that are of some particular design rather than representing known distributions such as the copulae presented in Gaussian and t-copulae.

The so-called **Gumbel copula**, named in honor of Emil Julius Gumbel (1891–1966), a German mathematician, is characterized by the parameter θ. With parameter values $1 \leq \theta < \infty$, it is defined as

$$C_\theta^G(u_1, u_2) = e^{-\left((-\ln u_1)^\theta + (-\ln u_2)^\theta\right)^{1/\theta}}$$

When θ = 1, then we obtain the independence copula.

The **Clayton copula**, named after David Clayton, a British statistician, is defined as

$$C_\theta^C(u_1, u_2) = \left(u_1^{-\theta} + u_2^{-\theta} - 1\right)^{-1/\theta}$$

For its parameter θ, we have the values $0 < \theta < \infty$.

The Gumbel and Clayton copulae belong to the class of *Archimedean copulae* meeting certain requirements that we will not discuss.

Simulation with the Gaussian Copula ($d = 2$)

Let's illustrate the simulation procedure for the case $d = 2$ using the Gaussian copula C_Γ^N from equation (16.3) and two alternative sets of marginal distributions for Y_1 and Y_2. The first set is given by identical normal distributions with zero mean and variance 2 for both components, that is, $Y_i \sim N(0,2)$, $i = 1, 2$. The second set is given by identical Student's t-distributions, also with zero mean and variance 2, and with ν = 4 degrees of freedom. For both marginal distributions, we set the correlation matrix equal to

$$\Gamma = \begin{pmatrix} 1 & 0.5 \\ 0.5 & 1 \end{pmatrix}$$

In step one, we generate $n = 1{,}000$ realizations $u^{(i)} = (u_1^{(i)}, u_2^{(i)}) \in [0,1]^2$ of the random vector $U = (U_1, U_2)$ with copula C_Γ^N.[11] This can be achieved by generating 1,000 drawings $x^{(i)} = (x_1^{(i)}, x_2^{(i)})$ of a multivariate normally distributed random vector $X = (X_1, X_2)$ with identical correlation matrix Γ, and then transforming them by $u_1^{(i)} = \Phi(x_1^{(i)})$ and $u_2^{(i)} = \Phi(x_2^{(i)})$.[12] The realizations $u^{(1)}, u^{(2)}, \ldots, u^{(n)}$ are displayed in Figure 16.15.

In step two, we transform these $u^{(i)}$ into realizations $y^{(i)} = (y_1^{(i)}, y_2^{(i)})$ through the relationships $y_1^{(i)} = F_{Y,1}^{-1}(u_1^{(i)})$ and $y_2^{(i)} = F_{Y,2}^{-1}(u_2^{(i)})$ using the respective normal or, alternatively, Student's t inverses. Since the two marginal distribution func-

---

[11] The superscript (i) indicates the i-th drawing.
[12] The symbol Φ denotes the standard normal distribution function of the random components.

FIGURE 16.15 1,000 Generated Values $u^{(i)} = \left(u_1^{(i)}, u_2^{(i)}\right) \in [0,1]^2$ from the Gaussian Copula with $\rho = 0.5$

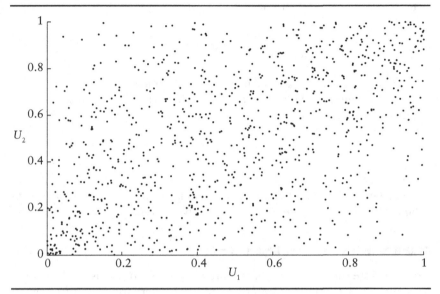

tions of Y_1 and Y_2 are modeled identically for both distributional choices, respectively, we have $F_{Y,1}^{-1} = F_{Y,2}^{-1}$, which we denote by F_Y^{-1} for simplicity.

First, we use the $N(0,2)$ as marginal distribution functions. We display these values $y^{(i)}$ in Figure 16.16. Next, we obtain the marginal Student's t-distributed realizations that we depict in Figure 16.17. In connection with both types of marginal distributions, we use the same Gaussian copula and identical correlations.

Comparing Figures 16.16 and 16.17, we see that the Gaussian copula generates data that appear to be dispersed about an ascending diagonal, for the marginal normal as well as Student's t-distributions. However, if we compare the quantiles of the marginal distributions, we see that the Student's t-distributions with $\nu = 4$ generate values that are scattered over a larger area than in the case of the normal distribution.

It should be clear that it is essential to consider both components of the joint distribution (i.e., the dependence structure as well as the marginal distributions) since variation in either one can lead to completely different results.

Simulating GE and IBM Data Through Estimated Copulae

In this illustration, we will present the simulation of stock returns. As data, we choose the daily returns of the stocks of the companies General Electric

FIGURE 16.16 1,000 Generated N(0,2) Observations $y^{(i)} = \left(y_1^{(i)}, y_2^{(i)}\right)$ from Gaussian Copula with $\rho = 0.5$

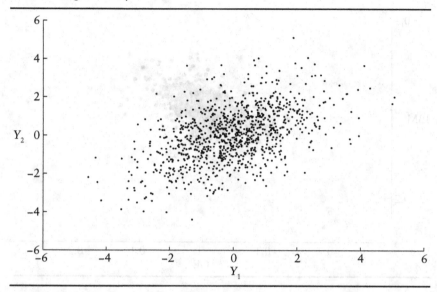

FIGURE 16.17 1,000 Generated Student's t Observations $y^{(i)} = \left(y_1^{(i)}, y_2^{(i)}\right)$ with Four Degrees of Freedom from Gaussian Copula with $\rho = 0.5$

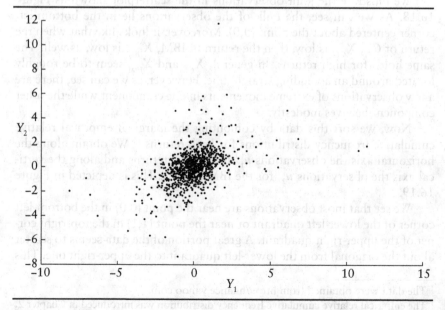

FIGURE 16.18 Bivariate Observations of the Daily Returns of GE and IBM

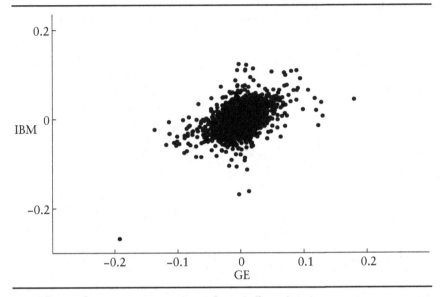

(GE) as well as International Business Machines (IBM) between April 24, 1980 and March 30, 2009. This accounts for 7,300 observations.[13]

We illustrate the joint observations in the scatterplot shown as Figure 16.18. As we can see, the bulk of the observations lie in the bottom-left corner centered about the point (0,0). Moreover, it looks like that when the return of GE, X_{GE}, is low, then the return of IBM, X_{IBM}, is low, as well. The same holds for high returns. In general, X_{GE} and X_{IBM} seem to be roughly located around an ascending straight line. However, as we can see, there are a few observations of extreme movements in one component while the other component behaves modestly.

Now, we sort this data by computing the marginal empirical relative cumulative frequency distributions for both returns.[14] We obtain along the horizontal axis the observations $u_1^{(i)}$ for the GE returns and along the vertical axis the observations $u_2^{(i)}$ for the IBM returns. This is depicted in Figure 16.19.

We see that most observations are near the point (0,0) in the bottom-left corner of the lower-left quadrant or near the point (1,1) in the top-right corner of the upper-right quadrant. A great portion of the data seems to scatter about the diagonal from the lower-left quadrant to the upper-right one. This

---

[13]The data were obtained from http://finance.yahoo.com.
[14]The empirical relative cumulative frequency distribution was introduced in Chapter 2.

FIGURE 16.19 Joint Observations $(u_1^{(i)}, u_2^{(i)})$ of the Empirical Cumulative Relative Frequency Distributions of the Daily Returns of GE and IBM

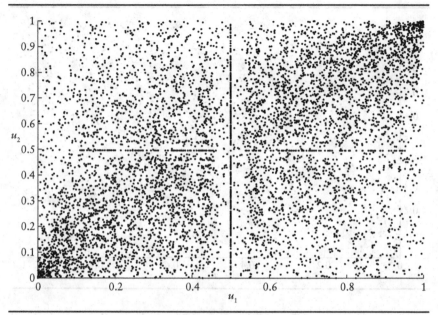

is in line with the observed simultaneous movements predominantly in the same direction (i.e., either up or down) of the returns.[15]

From these observations $\left(u_1^{(i)}, u_2^{(i)}\right)$ of the empirical cumulative relative frequency distribution, we estimate the respective copula. However, we will not pay any attention to the estimation techniques here and simply take the results as given. We fit the data to both the Gaussian copula and t-copula.

From both estimations, we obtain roughly the correlation matrix

$$\Gamma = \begin{pmatrix} 1 & 0.5 \\ 0.5 & 1 \end{pmatrix}$$

Moreover, for the t-copula, we compute an estimate of the degrees of freedom of slightly less than 4. For simplicity, however, we set this parameter $\nu = 4$.

With these two estimated copulae, C_Γ^N and $C_{\Gamma,\nu}^t$, we next generate 1,000 observations (u_1, u_2) each for the normal and Student's t-distribution

---

[15]Note the cross dissecting the plane into quadrants going through the points (0,0.5), (0.5,0), (1,0.5), and (0.5,1). The many observations (u_1, u_2) forming this cross-shaped array are associated with the many observed return pairs virtually zero in one—that is, either GE or IBM—or both components.

FIGURE 16.20 1,000 Generated Values of the Marginal Normal Distribution Functions of GE and IBM Daily Returns through the Gaussian Copula

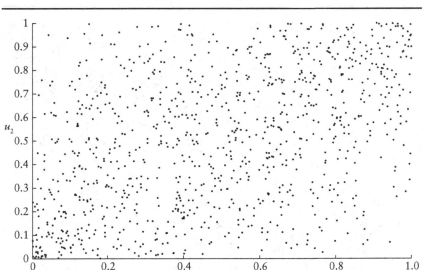

functions of both the GE and IBM returns. These observations are depicted in Figures 16.20 and 16.21, respectively.[16]

In the final step, we transform these (u_1, u_2) into the respective distribution's quantiles. That is, for the Gaussian copula, we compute returns for GE and IBM via $x_{GE} = F_{GE}^{-1}(u_1)$ and $x_{IBM} = F_{IBM}^{-1}(u_2)$, respectively. Here, F_{GE}^{-1} denotes the inverse of the estimated normal distribution of the GE returns while F_{IBM}^{-1} is that of the IBM returns. The generated Gaussian returns are shown in Figure 16.22. For the t-copula with marginal Student's t-distributions with four degrees of freedom, we obtain the returns analogously by $x_{GE} = F_{GE}^{-1}(u_2)$ and $x_{IBM} = F_{IBM}^{-1}(u_2)$, where here both F_{GE}^{-1} and F_{IBM}^{-1} represent the inverse function of the $t(4)$ distribution. The generated Student's t returns are displayed in Figure 16.23.

While both Figures 16.22 and 16.23 seem to copy the behavior of the center part of the scatterplot in Figure 16.18 quite well, only the t-copula with Student's t marginal returns seems to capture the behavior in the extreme parts as can be seen by the one observation with a return of -0.2 for GE.

---

[16] For the generated Student's t data, we scaled the values slightly to match the range of the observed returns. With only four degrees of freedom, very extreme values become too frequent compared to the original data.

Copula and Dependence Measures

FIGURE 16.21 1,000 Generated Values of the Marginal Student's t-Distribution Functions of GE and IBM Daily Returns through the t-Copula

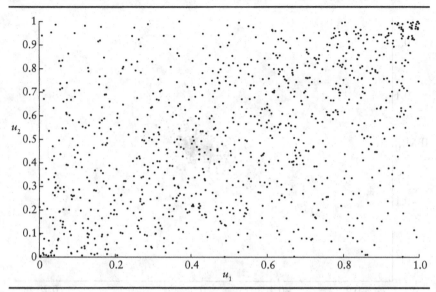

FIGURE 16.22 1,000 Generated Returns of GE and IBM Daily Returns through the Gaussian Copula with Marginal Normal Distribution Functions

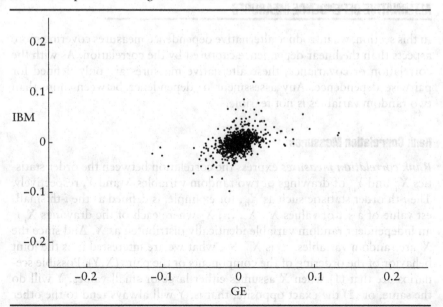

FIGURE 16.23 1,000 Generated Returns of GE and IBM Daily Returns through the *t*-Copula with Marginal Student's *t*-Distribution Functions

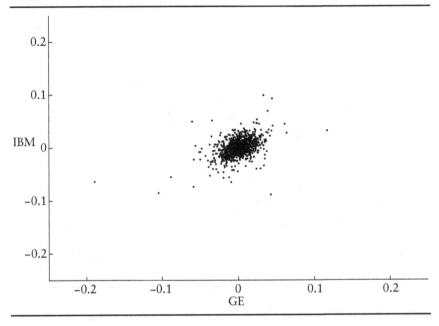

ALTERNATIVE DEPENDENCE MEASURES

In this section, we introduce alternative dependence measures covering more aspects than the linear dependence captured by the correlation. As with the correlation or covariance, these alternative measures are only defined for pairwise dependence. Any assessment of dependence between more than two random variables is not feasible.

Rank Correlation Measures

Rank correlation measures express the correlation between the order statistics $X_{(i)}$ and $Y_{(i)}$ of drawings of two random variables X and Y, respectively. The i-th order statistic such as $X_{(i)}$, for example, is defined as the i-th smallest value of a set of values $X_1, X_2, ..., X_n$ where each of the drawings X_i is an independent random variable identically distributed as X. And since the X_i are random variables, so is $X_{(i)}$. So, what we are interested in is the joint behavior of the ordering of the components of the pairs (X,Y). Possible scenarios are that (1) when X assumes either large or small values, Y will do the same, or (2) the exact opposite, that is, Y will always tend to the other extreme as X, or (3) Y will assume any values completely uninfluenced by the values of X.

Suppose that C is the copula of the joint distribution of X and Y, then the rank correlation only depends on this copula while the linear correlation measured by the coefficient ρ is also influenced by the exact specifications of the marginal distributions of X and Y. The reason is that for the rank correlation, we do not need the exact values of the $X_1, X_2, ..., X_n$ and $Y_1, Y_2, ..., Y_n$ but only the relationship between the ranks of the components X_i and Y_i of the pairs (X_i, Y_i).

For example, let X_i be the fifth smallest value of the n drawings $X_1, X_2, ..., X_n$ while Y_i is the smallest of all drawings $Y_1, Y_2, ..., Y_n$.[17] Then, the pair (X_i, Y_i) turns into $(5,1)$. In other words, we have rank 5 in the first component and rank 1 in the second component. That is, we first rank the values of each component and then replace the values by their ranks.

The rank coefficients are helpful to find a suitable copula for given two-dimensional data and specify the copula parameters.

Spearman's Rho

The rank correlation *Spearman's rho* is defined as follows. Let X and Y be two random variables with respective marginal distribution functions F_X and F_Y. Then, Spearman's rho is the rank correlation of X and Y given by the linear correlation of F_X and F_Y and denoted by $\rho^S(X,Y)$. Formally, this is

$$\rho^S(X,Y) = \rho(F_X(X), F_Y(Y)) \qquad (16.8)$$

where ρ denotes the linear correlation coefficient defined in Chapter 14.

The rank correlation measure is symmetric, that is, we have $\rho^S(X,Y) = \rho^S(Y,X)$. Finally, as with the correlation coefficient, $\rho^S(X,Y) = 0$ when the random variables X and Y are independent with the converse relationship not holding in general. Later we will give an example of the Spearman's rho.

Of the most commonly used rank correlation measures, we only introduced this one. However, we encourage the interested reader to refer to books on the copula to become acquainted with other ones such as **Kendall's tau**.

Tail Dependence

In finance, particularly risk management, it is important to have a measure of dependence that only focuses on the behavior of two random variables with respect to the extreme parts of the state space, that is, either very small

---
[17]Using order statistics notation, we write $X_{(5)}$ and $Y_{(1)}$.

or very large values for both random variables. Any such measure is referred to as a *measure of tail dependence*. We begin by defining a lower-tail and upper-tail dependence measure.

The *lower-tail dependence* is defined as follows. Let X and Y be two random variables with marginal distribution functions F_X and F_Y, respectively. Moreover, let F_X^{-1} and F_Y^{-1} denote the corresponding inverse functions. Then the lower tail dependence is measured by

$$\tau_l(X,Y) = \lim_{u \downarrow 0} P\left(Y < F_Y^{-1}(u) \middle| X < F_X^{-1}(u)\right) \quad (16.9)$$

where $u \downarrow 0$ indicates that u approaches 0 from above.[18]

We can alternatively replace the roles of X and Y in equation (16.9) to obtain

$$\tau_l(X,Y) = \lim_{u \downarrow 0} P\left(X < F_X^{-1}(u) \middle| Y < F_Y^{-1}(u)\right)$$

as an equivalent expression.

The lower-tail dependence expresses the probability of one component assuming very small values given that the other component is already very low. For example, the measure of lower-tail dependence is helpful in assessing the probability of a bond held in a portfolio defaulting given that another bond in the portfolio has already defaulted.

The other important measure of tail dependence is given by the following definition. The *upper-tail dependence* is defined as follows. Let X and Y be two random variables with marginal distribution functions F_X and F_Y, respectively. Moreover, let F_X^{-1} and F_Y^{-1} denote the corresponding inverse functions. Then the upper tail dependence is measured by

$$\tau_u(X,Y) = \lim_{u \uparrow 1} P\left(Y > F_Y^{-1}(u) \middle| X > F_X^{-1}(u)\right) \quad (16.10)$$

where $u \uparrow 1$ indicates that u approaches 1 from below.

As with the lower tail dependence, equation (16.10) could have alternatively been written as

$$\tau_u(X,Y) = \lim_{u \uparrow 1} P\left(X > F_X^{-1}(u) \middle| Y > F_Y^{-1}(u)\right)$$

The upper tail dependence is the analog of the lower-tail dependence. That is, by equation (16.10) we measure to what extent it is probable that a random variable assumes very high values given that some other random variable already is very large. For example, a portfolio manager may bet on

---

[18]The probability P(A|B) is called the *conditional probability of event A given the event B*. This concept is explained in Chapter 15.

a bull stock market and only add those stocks into his portfolio that have high upper tail dependence. So when one stock increases in price, the others in the portfolio are very likely to do so as well, yielding a much higher portfolio return than a well-diversified portfolio or even a portfolio consisting of positively correlated stocks.

Tail dependence is a feature that only some multivariate distributions can exhibit as we will see in the following two examples.

Tail Dependence of a Gaussian Copula

In this illustration, we examine the joint tail behavior of a bivariate distribution function with Gaussian copula. Furthermore, let we the marginal distributions be standard normal such that the components of the random vector are $X \sim N(0,1)$ and $Y \sim N(0,1)$.[19] Leaving out several difficult intermediate steps, we state that the asymptotic tail dependence of the Gaussian copula can be computed as

$$\begin{aligned} \tau_l^N(X,Y) &= 2 \cdot \lim_{u \downarrow 0} P\left(Y \leq \Phi^{-1}(u) \middle| X = \Phi^{-1}(u)\right) \\ &= 2 \cdot \lim_{y \to -\infty} P(Y \leq y | X = y) \\ &= 2 \cdot \lim_{y \to -\infty} \Phi\left(y\sqrt{1-\rho}\middle|\sqrt{1+\rho}\right) \\ &= 0 \end{aligned} \qquad (16.11)$$

which is true only for $0 \leq \rho < 1$.[20] The result in equation (16.11) does not hold when X and Y are perfectly positively correlated (i.e., $\rho = 1$). In that case, the random variables are obviously tail dependent because one component always perfectly mimics the other one.

For reason of symmetry of the jointly normally distributed random variables, the upper coefficient of tail dependence coincides with that in equation (16.11), that is, $\tau_u^N(X,Y) = 0$.

So, we see that jointly normally distributed random variables have no tail dependence unless they are perfectly correlated. Since empirically, tail dependence is commonly observed in stock returns, the Gaussian copula with normal marginal distributions seems somewhat dangerous to use because it may neglect the potential joint movements of large stock price changes, particularly since the correlation between stocks is commonly found to be less than 1.

---

[19] Note that because of the invariance of the copula with respect to standardization, the Gaussian copula is identical for any choice of parameter values of the normal marginal distributions.
[20] Here ρ denotes the correlation coefficient of X and Y, Φ the standard normal distribution function, and Φ^{-1} its inverse function.

Tail Dependence of t-Copula

In this illustration, we will establish the joint tail behavior of two random variables X and Y with marginal $t(\nu)$ distributions (i.e., Student's t with ν degrees of freedom) and whose joint distribution is governed by a t-copula. Furthermore, as in the Gaussian copula example, we let ρ denote the correlation coefficient of X and Y. Then, omitting the difficult intermediate steps, we obtain for the coefficient of lower tail dependence

$$\tau_l^t(X,Y) = 2 \cdot t_{\nu+1}\left(-\sqrt{\frac{(\nu+1)(1-\rho)}{1+\rho}}\right)$$

which equals $\tau_l^t = 0$ if $\rho = -1$, $\tau_l^t = 1$ if $\rho = 1$, and $\tau_l^t > 0$ if $\rho \in (-1,1)$.[21] In other words, if the random variables are perfectly negatively correlated, they are not lower-tail dependent. The reason is that, in that case, the two random variables can never move in the same direction. However, if the two random variables are perfectly positively correlated, their coefficient of lower-tail dependence is 1. That is identical to the behavior of the Gaussian copula. Finally, if the two random variables are somewhat correlated but not perfectly, they have some lower-tail dependence.

Without going into detail, we state that the lower-tail dependence increases for smaller degrees of freedom (i.e., the more the multivariate t differs from the normal distribution). Moreover, the lower-tail dependence increases with increasing linear correlation.

Since a joint distribution with the t-copula is symmetric, upper and lower coefficients of tail dependence coincide (i.e., $\tau_l^t = \tau_u^t$).

To illustrate the difference between the normal and t-distributions, we compare 1,000 generated bivariate Gaussian data in Figure 16.24 with the same number of generated bivariate t data in Figure 16.25. The Gaussian as well as the t data sets are almost perfectly correlated ($\rho = 0.99$). We see that while the observations of both sets of data are scattered about lines with approximately identical slopes, the bivariate t data spread further into the extreme parts of the state space. To see this, we have to be aware that the Gaussian data cover only the denser middle part of the t data; this is about the range from −3.3 to 3.3 in both components. The t data set has most of its observations located in the same area. However, as we can see, there are some observations located on the more extreme extension of the imaginary diagonal. For example, the point in the lower-left corner of Figure 16.25 represents the observation (−7.5, −8.5), which is well below −3.3 in both components.

---

[21]Here, $t_{\nu+1}(x)$ denotes the Student's t-distribution function evaluated at x.

FIGURE 16.24 Tail Dependence of Two Normally Distributed Random Variables with $\rho = 0.99$

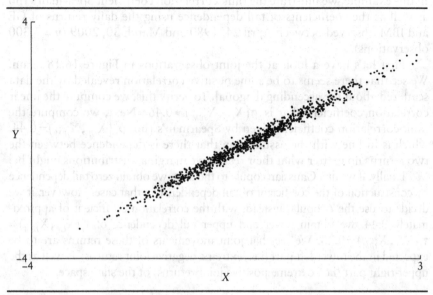

FIGURE 16.25 Tail Dependence of Two Student's t-Distributed Random Variables with $\rho = 0.99$

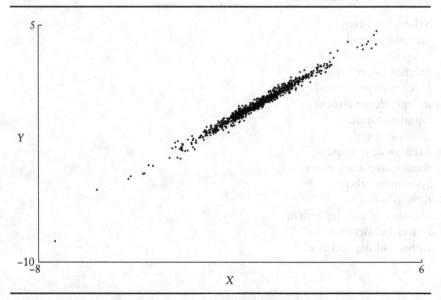

Spearman's Rho and Tail Dependence of GE and IBM Data

In this example, we illustrate the rank correlation coefficient Spearman's rho as well as the coefficients of tail dependence using the daily returns of GE and IBM observed between April 24, 1980 and March 30, 2009 ($n = 7,300$ observations).

First let's have a look at the joint observations in Figure 16.18 again. We see that there seems to be some positive correlation revealed by the data scattered about the ascending diagonal. To verify this, we compute the linear correlation coefficient that is $\rho(X_{GE}, X_{IBM}) \approx 0.46$. Next, we compute the rank correlation coefficient given by Spearman's rho, $\rho^S(X_{GE}, X_{IBM}) = 0.44$, which is in line with the assumption that there is dependence between the two returns no matter what their respective marginal distributions might be.

Finally, if we fit a Gaussian copula to the data, we obtain zero tail dependence by construction of the coefficient of tail dependence in that case. However, if we decide to use the t-copula, instead, with the correlation coefficient of approximately 0.44, we obtain lower and upper tail dependence of $\tau_l^t(X_{GE}, X_{IBM}) = \tau_l^t(X_{GE}, X_{IBM}) = 0.23$. We see that joint movements of these returns are to be expected in the lower-left part (i.e., extreme negative joint returns) as well as the upper-right part (i.e., extreme positive joint returns) of the state space.

CONCEPTS EXPLAINED IN THIS CHAPTER (IN ORDER OF PRESENTATION)

Sklar's Theorem
Gaussian copula
t-copula
Fréchet lower bound
Fréchet upper bound
Independence copula
Gumbel copula
Clayton copula
Archimedean copulae
Rank correlation measures
Spearman's rho
Kendall's tau
Measure of tail dependence
Lower-tail dependence
Upper-tail dependence

PART Three

Inductive Statistics

CHAPTER 17
Point Estimators

Information shapes a portfolio manager's or trader's perception of the true state of the environment such as, for example, the distribution of the portfolio return and its volatility or the probability of default of a bond issue held in a portfolio. A manager needs to gain information on population parameters to make well-founded decisions.

Since it is generally infeasible or simply too involved to analyze the entire population in order to obtain full certainty as to the true environment—for example, we cannot observe a portfolio for an infinite number of years to find out about the expected value of its return—we need to rely on a small sample to retrieve information about the population parameters. To obtain insight about the true but unknown parameter value, we draw a sample from which we compute statistics or estimates for the parameter.

In this chapter, we will learn about samples, statistics, and estimators. Some of these concepts we already covered in Chapter 3. In particular, we present the linear estimator, explain quality criteria (such as the bias, mean-square error, and standard error) and the large-sample criteria. In the context of large-sample criteria, we present the idea behind consistency, for which we need the definition of convergence in probability and the law of large numbers. As another large-sample criterion, we introduce the unbiased efficiency, explaining the best linear unbiased estimator or, alternatively, the minimum variance linear unbiased estimator. We then discuss the maximum likelihood estimation technique, one of the most powerful tools in the context of parameter estimation. The Cramér-Rao lower bounds for the estimator variance will be introduced. We conclude the chapter with a discussion of the exponential family of distributions and sufficient statistics.

SAMPLE, STATISTIC, AND ESTIMATOR

The probability distributions that we introduced so far in this book all depend on one or more parameters. In this chapter, we will refer to simply the parameter θ, which will have one or several components such as the parameter θ = (μ,σ^2) of the normal distribution, for example. The set of parameters is given by Θ, which will be called the *parameter space*.

The general problem that we will address in this chapter is the process of gaining information on the true population parameter such as, for example, the mean of some portfolio returns. Since we do not actually know the true value of θ, we merely are aware of the fact that is has to be in Θ. For example, the normal distribution has the parameter θ = (μ,σ^2) where the first component, the mean μ, can technically be any real number between minus and plus infinity (i.e., $\mathbb{R}$). The second component, the variance σ^2, is any positive real number (i.e., $\mathbb{R}^{++}$). And since the values of the two parameter components can be combined arbitrarily, which we express by the × symbol, we finally write the parameter space in the form $\Theta = \mathbb{R} \times \mathbb{R}^{++}$.

Sample

Let Y be some random variable with a probability distribution that is characterized by parameter θ. To obtain the information about this population parameter, we draw a *sample* from the population of Y. A sample is the total of n drawings $X_1, X_2, \ldots, X_n$ from the entire population. Note that until the drawings from the population have been made, the X_i are still random. The actually observed values (i.e., realizations) of the n drawings are denoted by $x_1, x_2, \ldots, x_n$. Whenever no ambiguity will arise, we denote the vectors $(X_1, X_2, \ldots, X_n)$ and $(x_1, x_2, \ldots, x_n)$ by the short hand notation X and x, respectively.

To facilitate the reasoning behind this, let us consider the value of the Dow Jones Industrial Average (DJIA) as some random variable. To obtain a sample of the DJIA, we will "draw" two values. More specifically, we plan to observe its closing value on two days in the future, say June 12, 2009 and January 8, 2010. Prior to these two dates, say on January 2, 2009, we are still uncertain as to value of the DJIA on June 12, 2009 and January 8, 2010. So, the value on each of these two future dates is random. Then, on June 12, 2009, we observe that the DJIA's closing value is 8,799.26, while on January 8, 2010 it is 10,618.19. Now, after January 8, 2010, these two values are *realizations* of the DJIA and not random any more.

Let us return to the theory. Once we have realizations of the sample, any further decision will then be based solely on the sample. However, we have to bear in mind that a sample provides only incomplete information since it will be impractical or impossible to analyze the entire population. This process of deriving a conclusion concerning information about a population's parameters from a sample is referred to as *statistical inference* or, simply, *inference*.

Formally, we denote the set of all possible sample values for samples of given length n (which is also called the *sample size*) by **X**. The sample space is similar to the space Ω containing all outcomes which we introduced in Chapter 8.

Sampling Techniques

There are two types of sampling methods: with replacement and without replacement. For example, the binomial distribution described in Chapter 9 is the distribution of random sampling with replacement of a random variable that is either 0 or 1. The analogue case without replacement is represented by the hypergeometric distribution that we discussed in Chapter 9, as well. We prefer sampling with replacement since this corresponds to independent draws such that the X_i are independent and identically distributed (i.i.d.).

Throughout this chapter, we will assume that individual draws are performed independently and under identical conditions (i.e., the $X_1, X_2, ..., X_n$ are i.i.d.). Then, the corresponding sample space is

$$\mathbf{X} = \Omega \times \Omega \times ... \times \Omega = \Omega^n$$

which is simply the space of the n-fold repetition of independent drawings from the space Ω. For instance, if we draw each of the n individual X_i from the real numbers (i.e., $\Omega = \mathbb{R}$), then the entire sample is drawn from $\mathbf{X} = \mathbb{R} \times \mathbb{R} \times ... \times \mathbb{R} = \mathbb{R}^n$.

As explained in Chapter 14, we know that the joint probability distribution of independent random variables is obtained by multiplication of the marginal distributions. That is, if the X_i, $i = 1, 2, ..., n$, are discrete random variables, the joint probability distribution will look like

$$P(X_1 = x_1, ..., X_n = x_n) = P(X_1 = x_1) \cdot ... \cdot P(X_n = x_n) \quad (17.1)$$

whereas, if the X_i, $i = 1, 2, ..., n$, are continuous, the joint density will be given as

$$f_{X_1, ..., X_n}(x_1, ..., x_n) = f_Y(x_1) \cdot ... \cdot f_Y(x_n) \quad (17.2)$$

with f_Y denoting the identical marginal density functions of the X_i since they are all distributed as Y.

Illustrations of Drawing with Replacement

As an illustration, consider the situation faced by a property and casualty insurance company for claims involving losses due to fires. Suppose the number of claims against the company has been modeled as a Poisson random variable introduced in Chapter 9. The company management may be uncertain as to the true value of the parameter λ. For this reason, a sample $(X_1, X_2, ..., X_{10})$ of the number of claims of the last 10 years is taken. The observations are given below:

x_1	x_2	x_3	x_4	x_5
517,500	523,000	505,500	521,500	519,000

x_6	x_7	x_8	x_9	x_{10}
519,500	526,000	511,000	524,000	530,500

For any general value of the parameter λ, the probability of this sample is given by

$$P(X_1 = x_1, X_2 = x_2, ..., X_{10} = x_{10}) = \prod_{i=1}^{10} \frac{\lambda^{x_i}}{x_i!} e^{-\lambda}$$

$$= e^{-\lambda} \cdot \left[\frac{\lambda^{517,500}}{517,500!} \cdot \frac{\lambda^{523,000}}{523,000!} \cdot \ldots \cdot \frac{\lambda^{530,500}}{530,500!} \right]$$

As another example, consider the daily stock returns of General Electric (GE) modeled by the continuous random variable X.[1] The returns on 10 different days $(X_1, X_2, ..., X_{10})$ can be considered a sample of i.i.d. draws. In reality, however, stock returns are seldom independent. If, on the other hand, the observations are not made on 10 consecutive days but with larger gaps between them, it is fairly reasonable to assume independence. Furthermore, the stock returns are modeled as normal (or Gaussian) random variables. We know from Chapter 11 that the normal distribution is characterized by the parameter (μ, σ^2). Now we may have observed for GE the sample from Table 17.1. The joint density, according to equation (17.2) and the density of equation (11.1) in Chapter 11, follows as

---

[1] As we have previously done, the stock returns we refer to are actually logarithmic, or log-returns, obtained from continuous compounding. They are computed as $X_t = \ln(P_t) - \ln(P_{t-1})$ where the P_t and P_{t-1} are the stock prices of today and yesterday, respectively, and ln denotes the natural logarithms or logarithm to the base $e = 2.7183$.

TABLE 17.1 Eleven Observations of the GE Stock Price from the NYSE Producing 10 Observations of the Return.

Date	Observed Stock Price		Observed Return	
Jan. 16, 2009	P_0	$13.96		
Jan. 23, 2009	P_1	$12.03	X_1	−0.1488
Jan. 30, 2009	P_2	$12.13	X_2	0.0083
Feb. 6, 2009	P_3	$11.10	X_3	−0.0887
Feb. 13, 2009	P_4	$11.44	X_4	0.0302
Feb. 20, 2009	P_5	$9.38	X_5	−0.1985
Feb. 27, 2009	P_6	$8.51	X_6	−0.0973
Mar. 6, 2009	P_7	$7.06	X_7	−0.1868
Mar. 13, 2009	P_8	$9.62	X_8	0.3094
Mar. 20, 2009	P_9	$9.54	X_9	−0.0084
Mar. 27, 2009	P_{10}	$10.78	X_{10}	0.1222

Source: Data obtained from finance.yahoo.

$$f_{X_1,\ldots,X_n}(-0.1488, 0.0083, \ldots, 0.1222)$$

$$= \frac{1}{\sqrt{2\pi}} e^{-\frac{(-0.1488-\mu)^2}{2\sigma^2}} \cdot \frac{1}{\sqrt{2\pi}} e^{-\frac{(0.0083-\mu)^2}{2\sigma^2}} \cdot \ldots \cdot \frac{1}{\sqrt{2\pi}} e^{-\frac{(0.1222-\mu)^2}{2\sigma^2}}$$

$$= \left(\frac{1}{\sqrt{2\pi}}\right)^{10} \cdot e^{-\frac{\left[(-0.1488-\mu)^2 + (0.0083-\mu)^2 + \ldots + (0.1222-\mu)^2\right]}{2\sigma^2}}$$

for general values of μ and σ^2.

Statistic

In our discussion here, certain terms and concepts we repeat from Chapter 3. In particular, we will point out the distinction between *statistic* and *population parameter*. In the context of estimation, the population parameter is inferred with the aid of the statistic. As we know, a statistic assumes some value That holds for a specific sample only, while the parameter prevails in the entire population.

The statistic, in most cases, provides a single number as an estimate of the population parameter generated from the n-dimensional sample. If the true but unknown parameter consists of, say, k components, the statistic will provide at least k numbers, that is at least one for each component.[2] We need to be aware of the fact that the statistic will most likely not equal the population parameter due to the random nature of the sample from which its value originates

Technically, the statistic is a function of the sample $(X_1, X_2, ..., X_n)$. We denote this function by t. Since the sample is random, so is t and, consequently, any quantity that is derived from it. As a random variable, t is a function from the set of sample values **X** into the k-dimensional space of real numbers:

$$t : \mathbf{X} \to \mathbb{R}^k \qquad (17.3)$$

where in most cases, $k = 1$ as just mentioned.

Even more technically, for the random variable t, we demand that it is measurable with respect to the two sets of possible events: one in the origin space and the other in the state space. That is, t has to be $A(\mathbf{X}) - \mathbb{B}^k$-measurable where $A(\mathbf{X})$ is the σ-algebra of the sample space **X** (origin) and $\mathbb{B}^k$ is the k-dimensional Borel σ-algebra. This simply means that any k-dimensional real value of t has its origin in the set of events $A(\mathbf{X})$.[3] We need to postulate measurability so that we can assign a probability to any values of the function $t(x_1, x_2, ..., x_n)$. For the remainder of this chapter, we assume that the statistics introduced will rely on functions t that all meet this requirement and, thus, we will avoid any further discussion on measurability.

Whenever it is necessary to express the dependence of statistic t on the outcome of the sample (x), we write the statistic as the function $t(x)$. Otherwise, we simply refer to the function t without explicit argument.

The statistic t as a random variable inherits its theoretical distribution from the underlying random variables (i.e., the random draws $X_1, X_2, ..., X_n$). If we vary the sample size n, the distribution of the statistics will, in most cases, change as well. This distribution expressing in particular the dependence on n is called the ***sampling distribution*** of t. Naturally, the sampling distribution exhibits features of the underlying population distribution of the random variable.

We will provide an illustration of a sampling distribution in the next section.

Estimator

The easiest way to obtain a number for the population parameter would be to simply guess. But this method lacks any foundation since it is based on

---

[2]In most (if not in all) cases we will encounter, the statistics will then provide exactly k numbers, that is, one number for each parameter component.
[3]For a discussion on measurabilty, see Chapter 8.

nothing but luck; in the best case, a guess might be justified by some experience. However, this approach is hardly analytical. Instead, we should use the information obtained from the sample, or better, the statistic.

When we are interested in the estimation of a particular parameter θ, we typically do not refer to the estimation function as statistic but rather as *estimator* and denote it by

$$\hat{\theta}: X \to \Theta \qquad (17.4)$$

As we see, in equation (17.4) the estimator is a function from the sample space X mapping into the parameter space Θ. We use the set Θ as state space rather than the more general set of the k-dimensional real numbers. We make this distinction between Θ and $\mathbb{R}^k$ in order to emphasize that the estimator particularly provides values for the parameter of interest, θ, even if the parameter space is all k-dimensional real numbers (i.e., $\Theta = \mathbb{R}^k$).

The estimator can be understood as some instruction of how to process the sample to obtain a valid representative of the parameter θ. The exact structure of the estimator is predetermined before the sample is realized. After the estimator has been defined, we simply need to enter the sample values accordingly.

Due to the estimator's dependence on the random sample, the estimator is itself random. A particular value of the estimator based on the realization of some sample is called an *estimate*. For example, if we realize 1,000 samples of given length n, we obtain 1,000 individual estimates $\hat{\theta}_i$, $i = 1, 2, \ldots, 1,000$. Sorting them by value—and possibly arranging them into classes—we can compute the distribution function of these realizations, which is similar to the empirical cumulative distribution function introduced in Chapter 2. Technically, this distribution function is not the same as the theoretical sampling distribution for this estimator for given sample length n introduced earlier. For increasing n, however, the distribution of the realized estimates will gradually become more and more similar in appearance to the sampling distribution.

Sometimes the distinction between statistics and estimators is non-existent. We admit that the presentation of much of this chapter's content is feasible without it. However, the treatment of sufficient statistics and exponential families explained later in this chapter will be facilitated when we are a little more rigorous here.

Estimator for the Mean

As an illustration, we consider normally distributed returns Y with parameters μ and σ^2. To obtain an estimate for the parameter component μ, we compute the familiar sample mean

$$\bar{X} = \frac{1}{n}\sum_{i=1}^{n} X_i$$

from the n independent draws X_i. Here let $n = 10$. Now, since $Y \sim N(\mu, \sigma^2)$, the sample mean is theoretically distributed

$$\bar{X} \sim N(\mu, \sigma^2/\sqrt{10})$$

by Properties 1 and 2 described in Chapter 11. The corresponding density function graph is displayed in Figure 17.1 by the solid line. It follows that it is tighter than the density function of the original random variable Y given by the dashed line due to the fact that the standard deviation is shrunk by the factor $\sqrt{10} = 3.1623$ compared to that of Y. Note that the values -1, 0, and 1 are indicated in Figure 17.1, for a better orientation, even though any numbers could be used since the parameter (μ, σ^2) is supposed to be unknown.

Now, from the same normal distribution as Y, we generate 730 samples of size $n = 10$ and compute for each the sample mean to obtain $\bar{x}_1, \bar{x}_2, \ldots, \bar{x}_{730}$.

FIGURE 17.1 Sampling Distribution Density Function of the Sample Mean $\bar{X} = 1/10 \sum_{i=1}^{10} X_i$ (solid) and of the Normal Random Variable Y (dashed)

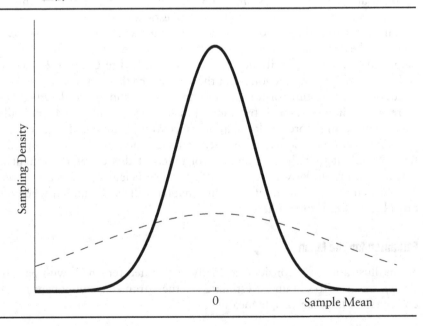

FIGURE 17.2 Histogram of 730 Generated Sample Means $\bar{x}$ for Normally Distributed Samples of Size $n = 10$

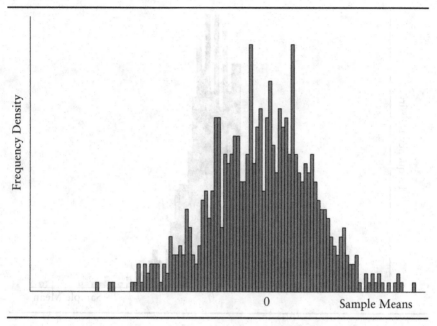

For display purposes, the sample means are gathered in 100 equidistant classes. The resulting histogram is given in Figure 17.2. We see that the distribution of the sample means copy quite well the appearance of the theoretical sample distribution density function in Figure 17.1. This should not be too surprising since we artificially generated the data.

We also sampled from real data. We observe the daily stock returns of GE between April 24, 1980 and March 30, 2009. This yields 7,300 observed returns. We partitioned the entire data series into 730 samples of size 10 and for each we computed the sample mean. As with the artificially generated sample means, the resulting sample mean values of the observed return data are classified into 100 equidistant classes, which are displayed in Figure 17.3. If GE daily returns are truly normally distributed, then this plot would be a consequence of the sampling distribution of the Gaussian sample means. Even though we are not exactly sure what the exact distribution of the GE returns is like, it seems that the sample means follow a Gaussian law. This appears to be the result of the Central Limit Theorem discussed in Chapter 11.

FIGURE 17.3 Histogram of Sample Means $\bar{x} = 1/10 \sum_{i=1}^{10} x_i$ of *GE* Daily Return Data Observed Between April 24, 1980 and March 30, 2009

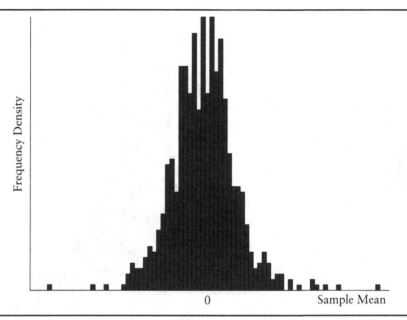

Source: Data obtained from finance.yahoo.

Linear Estimators

We turn to a special type of estimator, the *linear estimator*. Suppose we have a sample of size n such that $X = (X_1, X_2, ..., X_n)$. The linear estimator then has the following form

$$\hat{\theta} = \sum_{i=1}^{n} a_i X_i$$

where each draw X_i is weighted by some real a_i, for $i = 1, 2, ..., n$. By construction, the linear estimator weights each draw X_i by some weight a_i. The usual constraints on the a_i is that they sum to 1, that is,

$$\sum_{i=1}^{n} a_i = 1$$

A particular version of the linear estimator is the sample mean where all $a_i = 1/n$.

Let's look at a particular distribution of the X_i, the normal distribution. As we know from Chapter 11, this distribution can be expressed in closed form under linear affine transformation by Properties 1 and 2. That is, by adding several X_i and multiplying the resulting sum by some constant, we once again obtain a normal random variable. Thus, any linear estimator will be normal. This is an extremely attractive feature of the linear estimator.

Even if the underlying distribution is not the normal distribution, according to the Central Limit Theorem as explained in Chapter 11, the sample mean (i.e., when $a_i = 1/n$) will be approximately normally distributed as the sample size increases. This result facilitates parameter estimation for most distributions.

What sample size n is sufficient? If the population distribution is symmetric, it will not require a large sample size, often less than 10. If, however, the population distribution is not symmetric, we will need larger samples. In general, an n between 25 and 30 suffices. One is on the safe side when n exceeds 30.

The Central Limit Theorem requires that certain conditions on the population distribution are met such as finiteness of the variance. If the variance or even mean do not exist, another theorem, the so-called *Generalized Central Limit Theorem*, can be applied under certain conditions. (See equation (12.3) in Chapter 12.) However, these conditions are beyond the scope of this book. But we will give one example. The class of α-stable distributions provides such a *limiting distribution*, that is, one that certain estimators of the form

$$\sum_{i=1}^{n} a_i X_i$$

will approximately follow as n increases. We note this distribution because it is one that has been suggested by financial economists as a more general alternative to the Gaussian distribution to describe returns on financial assets.

Estimating the Parameter *p* of the Bernoulli Distribution

As our first illustration of the linear estimator, we consider the binomial stock price model, which appeared in several examples in earlier chapters. As we know, tomorrow's stock price is either up by some constant proportion with probability p, or down by some other constant proportion with probability $(1 - p)$. The relative change in price between today and tomorrow is related to a Bernoulli random variable Y, which is equal to 1 when the stock price moves up and 0 when the stock goes down. To obtain an estimate of the parameter p, which is equivalent to the expected value $p = E(Y)$, we use the sample mean, that is,

$$\hat{\theta} = \bar{X} = 1/n \sum_{i=1}^{n} X_i$$

FIGURE 17.4 Histogram of the Sample Means of a Bernoulli Population Contrasted with the Limiting Normal Distribution Indicated by the Density Function (Dashed)

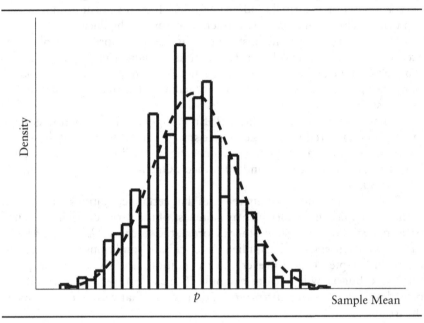

Now, if we have several samples each yielding an observation of the sample mean, we know that as the sample size increases, the means will approximately follow a normal distribution with parameter (μ,σ^2). We can take advantage of the fact that eventually $\mu = p$ because the expected values of the sample means, $E(\bar{X}) = p$, and the mean μ of the limiting normal distribution have to coincide. So, even though the exact value of (μ,σ^2) is unknown to us, we can simply observe the distribution of the sample means and try to find the location of the center. Another justification for this approach is given by the so-called Law of Large Numbers that is presented a little later in this chapter.

We demonstrate this by drawing 1,000 Bernoulli samples of size $n = 1,000$ with parameter $p = 0.5$. The corresponding sample means are classified into 30 equidistant classes. The histogram of the distribution is depicted in Figure 17.4. The dashed line in the figure represents the limiting normal density function with $\mu = 0.5$. Note that this would have worked with any other value of p, which accounts for the generalized label of the horizontal axis in the figure.

Estimating the Parameter λ of Poisson Distribution

The next illustration follows the idea of some previous examples where the number of claims of some property and casualty insurance company is modeled as a Poisson random variable N. Suppose that we were ignorant of the exact value of the parameter λ. We know from Chapter 9 that the expected value of N is given by $E(N) = λ$. As estimator of this parameter we, once again, take the sample mean,

$$\hat{\theta} = \bar{X} = 1/n \sum_{i=1}^{n} X_i$$

whose expected value is

$$E(\bar{X}) = E\left(1/n\left[X_1 + X_2 + \ldots + X_n\right]\right) = \lambda$$

as well, where the individual draws X_i are also Poisson distributed with parameter λ. We illustrate the distribution of 1,000 sample means of the Poisson samples of size $n = 1,000$ in Figure 17.5. We see that the distribution of the sample means fits the limiting normal distribution with $\mu = \lambda$ quite well. Note that the exact sample distribution of the sample means for finite n is not yet normal. More precisely, the sum

FIGURE 17.5 Histogram of the Sample Means of a Poisson Population Contrasted with the Limiting Normal Distribution Indicated by the Density Function (dashed)

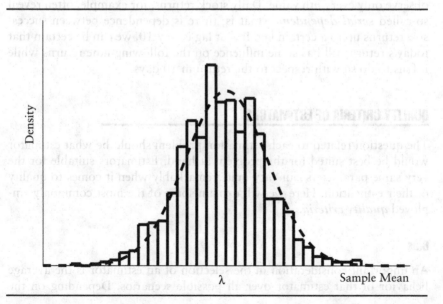

$$\sum_{i=1}^{n} X_i$$

is Poisson distributed with parameter $n\lambda$. The sample mean

$$\bar{X} = 1/n \sum_{i=1}^{n} X_i$$

is no longer Poisson distributed because of the division by n, but it is not truly Gaussian either.

Linear Estimator with Lags

As our third illustration, we present a linear estimator whose coefficients a_i are not equal. To be quite different from the sample mean, we choose $a_1 = 1$ and $a_2 = a_3 = \ldots = a_n = 0$. That is, we only consider the first draw X_1 and discard the remaining ones, $X_2, \ldots, X_n$. The estimator is then $\hat{\theta} = X_1$. Since we are only using one value out of the sample, the distribution of $\hat{\theta}$ will always look like the distribution of the random variable Y from which we are obtaining our sample because a larger sample size has obviously no effect on $\hat{\theta}$. An estimator of this type is definitely questionable when we maintain the assumption of independence between draws, as we are doing in this chapter. Its use, however, may become justifiable when there is dependence between a certain number of successive draws, which we can exclude by introducing the *lag* of size n between observations, such that we virtually observe only every nth value. Daily stock returns, for example, often reveal so-called *serial dependence*; that is, there is dependence between successive returns up to a certain lag. If that lag is, say 10, we can be certain that today's return still has some influence on the following nine returns while it fails to do so with respect to the return in 10 days.

QUALITY CRITERIA OF ESTIMATORS

The question related to each estimation problem should be what estimator would be best suited for the problem at hand. Estimators suitable for the very same parameters can vary quite remarkably when it comes to quality of their estimation. Here we will explain some of the most commonly employed *quality criteria*.

Bias

An important consideration in the selection of an estimator is the average behavior of that estimator over all possible scenarios. Depending on the sample outcome, the estimator may not equal the parameter value and, instead, be quite remote from it. This is a natural consequence of the variability of the underlying sample. However, the average value of the estimator is something we can control.

Let us begin by considering the *sampling error* that is the difference between the estimate and the population parameter. This distance is random due to the uncertainty associated with each individual draw. For the parameter θ and the estimator $\hat{\theta}$, we define the sample error as $(\hat{\theta} - \theta)$. Now, a most often preferred estimator should yield an expected sample error of zero. This expected value is defined as

$$E_\theta\left(\hat{\theta} - \theta\right) \tag{17.5}$$

and referred to as *bias*. (We assume in this chapter that the estimators and the elements of the sample have finite variance, and in particular the expression in equation (17.5) is well-defined.) If the expression in equation (17.5) is different from zero, the estimator is said to be a *biased estimator* while it is an *unbiased estimator* if the expected value in equation (17.5) is zero.

The subscript θ in equation (17.5) indicates that the expected value is computed based on the distribution with parameter θ whose value is unknown. Technically, however, the computation of the expected value is feasible for a general θ.

For example, the linear estimator $\hat{\theta} = X_1$ is an unbiased estimator for the mean $E(Y) = \theta$ since

$$E_\theta\left(\hat{\theta}\right) = E_\theta\left(X_1\right) = \theta$$

However, for many other reasons such as not utilizing the information in the remaining sample (i.e., $X_2, ..., X_n$), it may not be a very reasonable estimator.

On the other hand, the constant estimator $\hat{\theta} = c$ where c is any value no matter what is the outcome of the sample is not an unbiased estimator. Its bias is given by

$$E_\theta\left(\hat{\theta} - \theta\right) = E_\theta\left(c - \theta\right) = c - \theta$$

which can become arbitrarily large depending on the value of θ. It is zero, however, if the true value of θ happens to be c.

Bias of the Sample Mean

In this illustration, we analyze the sample mean. Whenever a population mean μ (i.e., the expected value) has to be estimated, a natural estimator of choice is the sample mean. Let us examine its bias. According to equation (17.5), the bias is given as

$$E(\bar{X} - \mu) = E\left(\frac{1}{n}\sum_{i=1}^{n} X_i - \mu\right)$$

$$= \frac{1}{n}\sum_{i=1}^{n} E(X_i) - \mu$$

$$= \frac{1}{n}\sum_{i=1}^{n} \mu - \mu = \frac{1}{n} \cdot n \cdot \mu - \mu$$

$$= 0$$

So, we see that the sample mean is an unbiased estimator, which provides a lot of support for its widespread use regardless of the probability distribution.[4]

Bias of the Sample Variance

In the next illustration, let the parameter to be estimated be the variance σ^2. For the estimation, we use the sample variance given by

$$s^2 = 1/n \sum_{i=1}^{n} (x_i - \bar{x})^2$$

introduced in Chapter 3 assuming that we do not know the population mean μ and therefore have to replace it with the sample mean, as done for s^2. Without presenting all the intermediate steps, its bias is computed as

$$E(s^2 - \sigma^2) = \frac{n-1}{n}\sigma^2 - \sigma^2$$

$$= -\frac{1}{n}\sigma^2$$

So, we see that the sample variance s^2 is slightly biased downward. The bias is $-1/n \cdot \sigma^2$, which is increasing in the population variance σ^2. However, note that it is almost negligible when the sample size n is large.

If instead of the sample variance, consider the following estimator that was also introduced in Chapter 3:

$$s^{*2} = 1/(n-1) \sum_{i=1}^{n} (X_i - \bar{X})^2$$

---

[4]The third line follows since the X_i are identically distributed with mean μ.

Point Estimators

We would obtain an unbiased estimator of the population variance since

$$E(s^{*2} - \sigma^2) = \frac{n-1}{n-1}\sigma^2 - \sigma^2$$
$$= \sigma^2 - \sigma^2$$
$$= 0$$

Because s^{*2} is unbiased, it is referred to as the *bias-corrected* (or simply *corrected*) *sample variance*.

We compare these two variance estimators with our daily return observations for GE between April 24, 1980 and March 30, 2009. First, the sample variance $s^2 = 0.0003100$. In contrast, the bias-corrected sample variance $s^{*2} = 0.0003101$. We see that the difference is negligible. Second, the theoretical bias of the sample variance is $= -1/7300 \cdot \sigma^2 = -0.0001 \cdot \sigma^2$. For this sample size, it does not make any practical difference whether we use the sample variance or the bias corrected sample variance.

As another example, consider the variance estimator

$$1/n \cdot \sum_{i=1}^{n}(x_i - \mu)^2$$

when the population mean μ is known. Its mean is

$$E\left(1/n \cdot \sum_{i=1}^{n}(x_i - \mu)^2\right) = \sigma^2$$

and, hence, it is an unbiased estimator for the population variance, σ^2, which follows from the computations

$$E\left(\frac{1}{n}\sum_{i=1}^{n}(X_i - \mu)^2\right) = \frac{1}{n}\sum_{i=1}^{n}E\left((X_i - \mu)^2\right)$$
$$= \frac{1}{n}\sum_{i=1}^{n}E(X_i^2 - 2\mu X_i + \mu^2)$$
$$= \frac{1}{n}\sum_{i=1}^{n}\left[E(X_i^2) - 2\mu E(X_i) + \mu^2\right]$$
$$= \frac{1}{n}\left[\sum_{i=1}^{n}E(X_i^2) - 2\mu\sum_{i=1}^{n}E(X_i) + n\mu^2\right]$$
$$= \frac{1}{n}\left[\sum_{i=1}^{n}\left(Var(X_i) + \mu^2\right) - 2n\mu^2 + n\mu^2\right]$$

$$= \frac{1}{n}\left[\sum_{i=1}^{n} Var(X_i) + n\mu^2 - 2n\mu^2 + n\mu^2\right]$$

$$= \frac{1}{n}\sum_{i=1}^{n} Var(X_i)$$

$$= \frac{1}{n} n Var(Y)$$

$$= Var(Y)$$

Note that in the fifth line, we used the equality $Var(X_i) = E(X_i^2) - \mu^2$.

Mean-Square Error

As just explained, bias as a quality criterion tells us about the expected deviation of the estimator from the parameter. However, the bias fails to inform us about the variability or spread of the estimator. For a reliable inference for the parameter value, we should prefer an estimator with rather small variability or, in other words, high precision.

Assume once more that we repeatedly, say m times, draw samples of given size n. Using estimator $\hat{\theta}$ for each of these samples, we compute the respective estimate $\hat{\theta}_t$ of parameter θ, where $t = 1, 2, \ldots, m$. From these m estimates, we then obtain an empirical distribution of the estimates including an empirical spread given by the sample distribution of the estimates

$$s_\theta^2 = 1/m \sum_{k=1}^{m} (\hat{\theta}_k - \bar{\hat{\theta}})^2$$

where $\bar{\hat{\theta}}$ is the sample mean of all estimates $\hat{\theta}_k$.

We know that with increasing sample length n, the empirical distribution will eventually look like the normal distribution for most estimators. However, regardless of any empirical distribution of estimates, an estimator has a theoretical sampling distribution for each sample size n. So, the random estimator is, as a random variable, distributed by the law of the sampling distribution. The empirical and the sampling distribution will look more alike the larger is n. The sampling distribution provides us with a theoretical measure of spread of the estimator, namely, its variance, $Var_{\theta,n}(\hat{\theta})$ where the subscript indicates that it is valid only for samples of size n. The square root of the variance,

$$\sqrt{Var_{\theta,n}(\hat{\theta})}$$

is called the *standard error* (S.E.). This is a measure that can often be found listed together with the observed estimate. In the remainder of this chapter, we will drop the subscript n wherever unambiguous since this dependence on the sample size will be obvious.

To completely eliminate the variance, one could simply take a constant $\hat{\theta} = c$ as the estimator for some parameter. However, this not reasonable since it is insensitive to sample information and thus remains unchanged for whatever the true parameter value θ may be. Hence, we stated the bias as an ultimately preferable quality criterion. Yet, a bias of zero may be too restrictive a criterion if an estimator $\hat{\theta}$ is only slightly biased but has a favorably small variance compared to all possible alternatives, biased or unbiased. So, we need some quality criterion accounting for both bias and variance.

That criterion can be satisfied by using the *mean-square error* (MSE). Taking squares rather than the loss itself incurred by the deviation, the *MSE* is defined as the expected square loss

$$MSE(\hat{\theta}) = E_\theta\left[(\hat{\theta} - \theta)^2\right] \qquad (17.6)$$

where the subscript θ indicates that the mean depends on the true but unknown parameter value. The *MSE* in equation (17.6) permits the following alternative representation

$$MSE(\hat{\theta}) = Var_\theta(\hat{\theta}) + (E_\theta(\hat{\theta}) - \theta)^2 \qquad (17.7)$$

The second term on the right-hand side of equation (17.7) is the squared bias. So, we see that the mean-square error is decomposed into the variance of the estimator and a transform (i.e., square) of the bias. If the transform is zero (i.e., the estimator is unbiased), the mean-square error equals the estimator variance.

It is interesting to note that *MSE*-minimal estimators are not available for all parameters. That is, we may have to face a trade-off between reducing either the bias or the variance over a set of possible estimators. As a consequence, we simply try to find a minimum-variance estimator of all unbiased estimators, which is called the *minimum-variance unbiased estimator*. We do this because in many applications, unbiasedness has priority over precision.

Mean-Square Error of the Sample Mean

As an illustration, we consider the population mean μ, which we endeavor to estimate with the sample mean $\bar{X}$. Since we know that it is unbiased, its *MSE* is simply the variance; that is, $MSE(\bar{X}) = Var_\mu(\bar{X})$. Now suppose that we analyze the observations of a normally distributed stock return Y with parameter (μ,σ^2). Given a sample of size n, the sampling distribution of $\bar{X}$ is the normal distribution $N(\mu,\sigma^2/n)$. More specifically, we consider a sample of size $n = 1{,}000$ observed between April 11, 2005 and March 30, 2009 of the returns of the stock of GE from our earlier example. The sample mean for this period is computed to be $\bar{x} = -0.0014$. Now, if the true distribution parameter was $\mu = 0$ (i.e., we had a mean return of zero and a variance of $\sigma^2 = 0.0003$), the sampling distribution of the sample mean is given by $N(0,\sigma^2/1000) = N(0,0.0000003)$.

In Figure 17.6 we illustrate the position of $\bar{x} = -0.0014$ relative to the population mean μ and the bounds given by one standard error of 0.000548 about μ. With the given return distribution, this observation of the sample mean is not all that common in the sense that a value of -0.0014 or less occurs only with probability 0.53% (i.e., $P(\bar{X} \leq -0.0014) = 0.0053$). Now, if we were not certain as to the true population parameters $\mu = 0$ and $\sigma^2 = 0.0003$, with this observation one might doubt these parameter values. This last statement will be discussed in Chapter 19 where we discuss hypothesis testing.

FIGURE 17.6 Sample Mean $\bar{x} = -0.0014$ of the GE Return Data. Population Mean $\mu = 0$ and Standard Error $S.E.(\theta) = 0.000548$ are Shown for Comparison

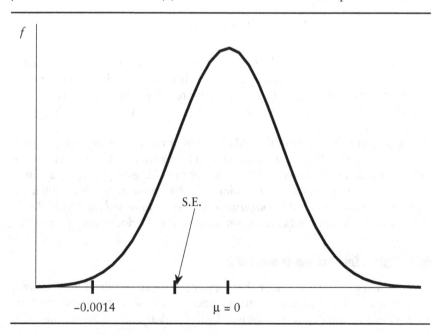

Mean-Square Error of the Variance Estimator

Let us now consider the estimation of the population variance σ^2. As we know, the sample variance is biased so that the second term in equation (17.7) will not be negative. To be exact, with variance[5]

$$Var(s^2) = \left(\frac{n-1}{n}\right)^2 \left(\frac{2\sigma^4}{n-1}\right)$$

and bias

$$-\frac{1}{n}\sigma^2$$

the MSE will be equal to

$$MSE(s^2) = \left(\frac{n-1}{n}\right)^2 \left(\frac{2\sigma^4}{n-1}\right) + \left(\frac{1}{n}\right)^2 \sigma^4$$

If, instead, we had used the bias corrected sample variance s^{*2}, the last term would vanish. However, the variance term

$$MSE(s^{*2}) = Var(s^{*2}) = \frac{2\sigma^4}{n-1}$$

is slightly larger than that of the sample variance.

We can see that there is a trade-off between variance (or standard error) and bias. With our GE return sample data observed between April 11, 2005 and March 30, 2009, n is equal to 1,000. Furthermore, the sample variance is $MSE(s^2) = 2.4656 \times 10^{-11}$, whereas $MSE(s^{*2}) = 2.4661 \times 10^{-11}$, such that the mean-square error of the bias-corrected sample variance is 0.02% larger than that of sample variance. So, despite the bias, we can use the sample variance in this situation.

LARGE SAMPLE CRITERIA

The treatment of the estimators thus far has not included their possible change in behavior as the sample size n varies. This is an important aspect of estimation, however. For example, it is possible that an estimator that is biased for any given finite n, gradually loses its bias as n increases. Here we will analyze the estimators as the sample size approaches infinity. In techni-

---
[5]The variance is given here without further details of its derivation.

cal terms, we focus on the so-called *large-sample* or *asymptotic properties of estimators*.

Consistency

Some estimators display stochastic behavior that changes as we increase the sample size. It may be that their exact distribution including parameters is unknown as long as the number of draws n is small or, to be precise, finite. This renders the evaluation of the quality of certain estimators difficult. For example, it may be impossible to give the exact bias of some estimator for finite n, in contrast to when n approaches infinity.

If we are concerned about some estimator's properties, we may reasonably have to remain undecided about the selection of the most suitable estimator for the estimation problem we are facing. In the fortunate cases, the uncertainty regarding an estimator's quality may vanish as n goes to infinity, so that we can base conclusions concerning its applicability for certain estimation tasks on its large-sample properties.

The Central Limit Theorem plays a crucial role in assessing the properties of estimators. This is because normalized sums turn into standard normal random variables, which provide us with tractable quantities. The asymptotic properties of normalized sums may facilitate deriving the large sample behavior of more complicated estimators.

Convergence in Probability

At this point, we need to think about a rather technical concept that involves controlling the behavior of estimators in the limit. Here we will analyze an estimator's *convergence* characteristics. That means we consider whether the distribution of an estimator approaches some particular probability distribution as the sample sizes increase. To proceed, we state the following definition

> *Convergence in probability*: We say that a random variable such as an estimator built on a sample of size n, $\hat{\theta}_n$, *converges in probability* to some constant c if
>
> $$\lim_{n \to \infty} P\left(\left|\hat{\theta}_n - c\right| > \varepsilon\right) = 0 \qquad (17.8)$$
>
> holds for any $\varepsilon > 0$.

The property (17.8) states that as the sample size becomes arbitrarily large, the probability that our estimator will assume a value that is more than ε away from c will become increasingly negligible, even as ε becomes smaller. Instead of the rather lengthy form of equation (17.8), we usually state that $\hat{\theta}_n$ converges in probability to c more briefly as

$$\text{plim}\hat{\theta}_n = c \qquad (17.9)$$

Here, we introduce the index n to the estimator $\hat{\theta}_n$ to indicate that it depends on the sample size n. Convergence in probability does not mean that an estimator will eventually be equal to c, and hence constant itself, but the chance of a deviation from it will become increasingly unlikely.

Suppose now that we draw several samples of size n. Let the number of these different samples be N. Consequently, we obtain N estimates $\hat{\theta}_n^{(1)}, \hat{\theta}_n^{(2)}, \ldots, \hat{\theta}_n^{(N)}$ where $\hat{\theta}_n^{(1)}$ is estimated on the first sample, $\hat{\theta}_n^{(2)}$ on the second, and so on. Utilizing the prior definition, we formulate the following law.

Law of large Numbers: Let $X^{(1)} = \left(X_1^{(1)}, X_2^{(1)}, \ldots, X_n^{(1)}\right), X^{(2)} = \left(X_1^{(2)}, X_2^{(2)}, \ldots, X_n^{(2)}\right)$, and $X^{(N)} = \left(X_1^{(N)}, X_2^N, \ldots, X_n^N\right)$ be a series of N independent samples of size n. For each of these samples, we apply the estimator $\hat{\theta}_n$ such that we obtain N independent and identically distributed as $\hat{\theta}_n$ random variables $\hat{\theta}_n^{(1)}, \hat{\theta}_n^{(2)}, \ldots, \hat{\theta}_n^{(N)}$. Further, let $E(\hat{\theta}_n)$ denote the expected value of $\hat{\theta}_n$ and $\hat{\theta}_n^{(1)}, \hat{\theta}_n^{(2)}, \ldots, \hat{\theta}_n^{(N)}$. Because they are identically distributed, then it holds that[6]

$$\text{plim}\frac{1}{N}\sum_{k=1}^{N}\hat{\theta}_n^{(k)} = E(\hat{\theta}_n) \qquad (17.10)$$

The law given by equation (17.10) states that the average over all estimates obtained from the different samples (i.e., their sample mean) will eventually approach their expected value or population mean. According to equation (17.8), large deviations from $E(\hat{\theta}_n)$ will become ever less likely the more samples we draw. So, we can say with a high probability that if N is large, the sample mean

$$1/N\sum_{k=1}^{N}\hat{\theta}_n^{(k)}$$

will be near its expected value. This is a valuable property since when we have drawn many samples, we can assert that it will be highly unlikely that the average of the observed estimates such as

$$1/N\sum_{k=1}^{N}\bar{x}_k$$

---

[6]Formally, equation (17.10) is referred to as the *weak law of large numbers*. Moreover, for the law to hold, we need to assure that the $\hat{\theta}_n^{(k)}$ have identical finite variance. Then by virtue of the Chebychev inequality discussed below, we can derive equation (17.10).

for example, will be a realization of some distribution with very remote parameter $E(\bar{X}) = \mu$.

An important aspect of the convergence in probability becomes obvious now. Even if the expected value of $\hat{\theta}_n$ is not equal to θ (i.e., $\hat{\theta}_n$ is biased for finite sample lengths n), it can still be that $\text{plim}\,\hat{\theta}_n = \theta$. That is, the expected value $E(\hat{\theta}_n)$ may gradually become closer to and eventually indistinguishable from θ, as the sample size n increases. To account for these and all unbiased estimators, we introduce the next definition.

Consistency: An estimator $\hat{\theta}_n$ is a *consistent estimator* for θ if it converges in probability to θ, as given by equation (17.9), that is,

$$\text{plim}\,\hat{\theta}_n = \theta \qquad (17.11)$$

The consistency of an estimator is an important property since we can rely on the consistent estimator to systematically give sound results with respect to the true parameter. This means that if we increase the sample size n, we will obtain estimates that will deviate from the parameter θ only in rare cases.

Consistency of the Sample Mean of Normally Distributed Data

As our first illustration, consider the sample mean $\bar{X}$ of samples of a normally distributed random variable Y, with $Y \sim N(\mu, \sigma^2)$, to infer upon the parameter component μ. As used throughout the chapter, the sample size is n. Then, the probability of the sample mean deviating from μ by any given amount ε is computed as

$$P(|\bar{X} - \mu| > \varepsilon) = 2\Phi\left(\frac{-\varepsilon}{\sigma/\sqrt{n}}\right)$$

We show this by the following computations.

$$P(|\bar{X} - \mu| > \varepsilon) = 1 - P(|\bar{X} - \mu| < \varepsilon)$$
$$= 1 - P(-\varepsilon < \bar{X} - \mu < \varepsilon)$$
$$= 1 - P\left(\frac{-\varepsilon}{\sigma/\sqrt{n}} < \frac{\bar{X} - \mu}{\sigma/\sqrt{n}} < \frac{\varepsilon}{\sigma/\sqrt{n}}\right)$$

$$= 1 - P\left(\frac{-\varepsilon}{\sigma/\sqrt{n}} < \frac{\bar{X}-\mu}{\sigma/\sqrt{n}} < \frac{\varepsilon}{\sigma/\sqrt{n}}\right)$$

$$= 1 - \left[P\left(\frac{\bar{X}-\mu}{\sigma/\sqrt{n}} < \frac{\varepsilon}{\sigma/\sqrt{n}}\right) - P\left(\frac{\bar{X}-\mu}{\sigma/\sqrt{n}} < \frac{-\varepsilon}{\sigma/\sqrt{n}}\right)\right]$$

$$= 1 - \left[1 - P\left(\frac{\bar{X}-\mu}{\sigma/\sqrt{n}} < \frac{-\varepsilon}{\sigma/\sqrt{n}}\right) - P\left(\frac{\bar{X}-\mu}{\sigma/\sqrt{n}} < \frac{-\varepsilon}{\sigma/\sqrt{n}}\right)\right]$$

$$= 2P\left(\frac{\bar{X}-\mu}{\sigma/\sqrt{n}} < \frac{-\varepsilon}{\sigma/\sqrt{n}}\right)$$

$$= 2\Phi\left(\frac{-\varepsilon}{\sigma/\sqrt{n}}\right)$$

Next, let's consider the daily return data from GE that we previously used, which was distributed $N(0, 0.0003)$. Proceeding with sample sizes of $n = 10, 100, 1{,}000$, and $7{,}300$, we have the following series of probabilities: $2\Phi(-\varepsilon \cdot 117.83)$, $2\Phi(-\varepsilon \cdot 372.60)$, $2\Phi(-\varepsilon \cdot 1178.30)$, and $2\Phi(-\varepsilon \cdot 3183.50)$ for $P(|\bar{X}-\mu| > \varepsilon)$, each depending on the value of ε. For $\varepsilon \to 0$, we display the corresponding probabilities in Figure 17.7. As can be seen from the solid graph, for $n = 10$ the probability $P(|\bar{X}-\mu| > \varepsilon)$ rises to 1 quite steadily. In contrast, for large sample sizes such as $N = 7{,}300$, the dotted graph is close to 0, even for very small ε, while it abruptly jumps to 1 as $\varepsilon \approx 0$. So, we see that, theoretically, the sample mean actually does converge to its expected value (i.e., the population mean $\mu = 0$). We compare this to the observed sample means of GE return data for sample sizes $n = 10, 100$, and $1{,}000$ in Figure 17.8. The period of observation is April 24, 1980 through March 30, 2009, providing 7,300 daily observations. In all three cases, 30 equidistant classes are used. The histogram with the largest bin size is related to the two histograms with the smaller bin sizes. It is wider than the visible range in this graphic (i.e., [−0.02, 0.02]). The narrower but higher bins belong to the histogram for $n = 100$, whereas the tightest and highest bins, virtually appearing as lines, are from the histogram corresponding to $n = 1{,}000$. We notice a convergence that is consistent with the theoretical convergence visualized in Figure 17.7.

FIGURE 17.7 Visualization of $\lim_{\varepsilon \to 0} P\left(\left|\hat{\theta}_n - \mu\right| > \varepsilon\right)$ as n Increases from 10 to 7,300. Here $\mu = 0$ and $\hat{\theta}_n = 1/n \sum_{i=1}^{n} x_i$

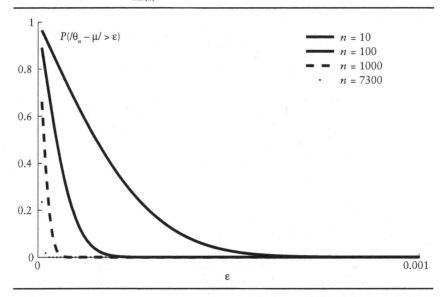

FIGURE 17.8 Comparison of Histograms of GE Sample Means as Sample Size Increases from $n = 10$ to $n = 100$ to $n = 1,000$

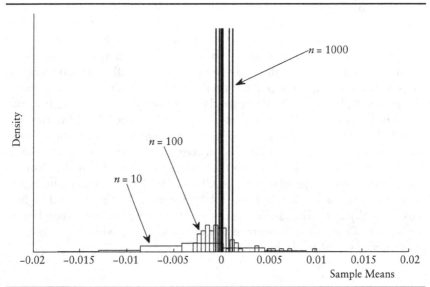

Consistency of the Variance Estimator

For our next illustration, we present an example of some estimator that is biased for finite samples, but is consistent for the population parameter. Suppose the parameter to be estimated is the population variance σ^2, which has to be inferred upon without knowledge of the population mean. Instead of the unbiased corrected sample variance, we simply use the sample variance s^2. First, we will show that as $n \to \infty$, the bias disappears. Let us look at the finite sample mean of s^2.

As we know from some previous example, its expected value is

$$E(s^2) = \frac{n-1}{n} \sigma^2$$

revealing that it is slightly biased downward. However, as $n \to \infty$, the factor $(n-1)/n$ approaches 1, so that for infinitely large samples, the mean of the sample variance is actually σ^2. Next, with the estimator's expected value for some finite n and its variance of

$$Var(s^2) = \left(\frac{n-1}{n}\right)^2 \left(\frac{2\sigma^4}{n-1}\right)$$

by virtue of the Chebyshev inequality,[7] we can state that for any $\varepsilon > 0$ that

$$P\left(\left|s^2 - \frac{n-1}{n}\sigma^2\right| > \varepsilon\right) \leq \frac{\left(\frac{n-1}{n}\right)^2 \left(\frac{2\sigma^4}{n-1}\right)}{\varepsilon^2} \xrightarrow{n \to \infty} 0$$

This proves that eventually any arbitrarily small deviation from $E(s^2) = [(n-1)/n] \cdot \sigma^2$ becomes unlikely. And since $[(n-1)/n] \cdot \sigma^2$ becomes arbitrarily close to σ^2 as n goes to infinity, the variance estimator s^2 will most likely be very close to the parameter σ^2. Thus, we have proven consistency of the sample variance s^2 for σ^2.

For comparison, we also check the behavior of the sample mean for our empirical daily GE return data observed between April 24, 1980 and March 30, 2009. For sample sizes of $n = 10$, 100, and 1,000, we display the histograms of the corresponding distributions of the sample variances in Figures 17.9, 17.10, and 17.11, respectively. By looking at the range in each figure, we can see that the variability of the sample variances about the population parameter $\sigma^2 = 0.0003$ decreases as n increases.

---

[7]See Chapter 11.

442 INDUCTIVE STATISTICS

FIGURE 17.9   Histogram of the Sample Variances s^2, for Sample Size $n = 10$, of the Observed GE Daily Return Data Between April 20, 1980 and March 30, 2009

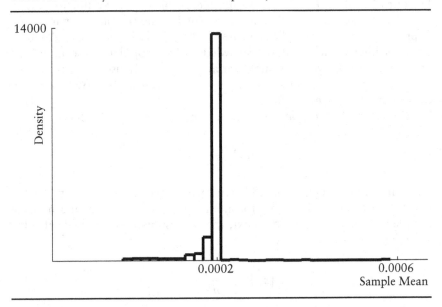

FIGURE 17.10   Histogram of the Sample Variances s^2, for Sample Size $n = 100$, of the Observed GE Daily Return Data Between April 20, 1980 and March 30, 2009

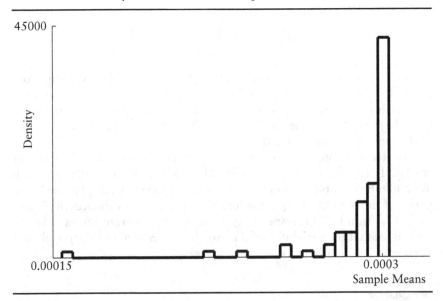

FIGURE 17.11 Histogram of the Sample Variances s^2, for Sample Size $n = 1{,}000$, of the Observed GE Daily Return Data Between April 20, 1980 and March 30, 2009

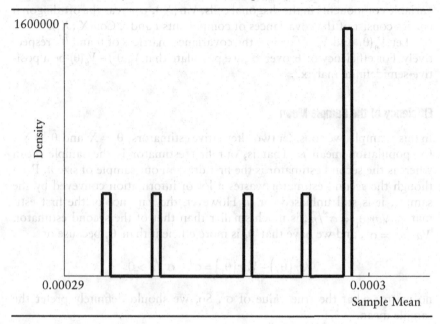

Unbiased Efficiency

In the previous discussions in this section, we tried to determine where the estimator tends to. This analysis, however, left unanswered the question of how fast does the estimator get there. For this purpose, we introduce the notion of *unbiased efficiency*.

Let us suppose that two estimators $\hat{\theta}$ and $\hat{\theta}^*$ are unbiased for some parameter θ. Then, we say that $\hat{\theta}$ is a more *efficient estimator* than $\hat{\theta}^*$ if it has a smaller variance; that is,

$$Var_\theta\left(\hat{\theta}\right) < Var_\theta\left(\hat{\theta}^*\right) \qquad (17.12)$$

for any value of the parameter θ. Consequently, no matter what the true parameter value is, the standard error of $\hat{\theta}$ is always smaller than that of $\hat{\theta}^*$. Since they are assumed to be both unbiased, the first should be preferred.

If the parameter consists of more than one component (i.e., $k > 1$ such that the parameter space $\Theta \subset \mathbb{R}^k$), then the definition of efficiency in equation (17.12) needs to be extended to an expression that uses the covariance matrix of the estimators rather than only the variances.

Recall from Chapter 14 that the covariance matrix of some random variable $X = (X_1, X_2, \ldots, X_n)$ has the variances of the respective random variable components in the diagonal cells, $Var(X_i)$, whereas the off-diagonal entries consist of the covariances of components i and j, $Cov(X_i, X_j)$.

Let $V_\theta(\hat{\theta})$ and $V_\theta(\hat{\theta}^*)$ denote the covariance matrices of $\hat{\theta}$ and $\hat{\theta}^*$, respectively. For efficiency of $\hat{\theta}$ over $\hat{\theta}^*$, we postulate that $V_\theta(\hat{\theta}^*) - V_\theta(\hat{\theta})$ be a positive-semidefinite matrix.[8]

Efficiency of the Sample Mean

In this example, we consider two alternative estimators, $\hat{\theta}_1 = \bar{X}$ and $\hat{\theta}_2 = X_1$, for population mean μ. That is, our first estimator is the sample mean whereas the second estimator is the first draw in our sample of size n. Even though the second estimator wastes a lot of information conveyed by the sample, it is still unbiased for μ. However, the variance of the first estimator, $Var(\hat{\theta}_1) = \sigma^2/n$, is much smaller than that of the second estimator, $Var(\hat{\theta}_2) = \sigma^2$, and we have that $\hat{\theta}_1$ is more efficient than $\hat{\theta}_2$ because of

$$Var(\hat{\theta}_2) - Var(\hat{\theta}_1) = \sigma^2 - \sigma^2/n > 0$$

no matter what the true value of σ^2. So, we should definitely prefer the sample mean.

Efficiency of the Bias Corrected Sample Variance

For our next example, let us estimate the parameter $\theta = (\mu, \sigma^2)$ of the probability distribution of a normal random variable using a sample of length n. Suppose that we have two alternative estimators, namely $\hat{\theta}_1 = (\bar{X}, s^2)$ where s^2 is the sample variance, and $\hat{\theta}_2 = (\bar{X}, s^{*2})$ where s^{*2} is the corrected sample variance. Because of the fact that the random variable is normally distributed, we state without proof that the components of $\hat{\theta}_1$ and $\hat{\theta}_2$ are independent. That is, $\bar{X}$ is independent of s^2 while $\bar{X}$ is independent of s^{*2}. So, we obtain as covariance matrices

$$V(\hat{\theta}_1) = \begin{pmatrix} \dfrac{\sigma^2}{n} & 0 \\ 0 & \left(\dfrac{n-1}{n}\right)^2 \dfrac{2\sigma^2}{n-1} \end{pmatrix}$$

---

[8] The definition of positive-semidefinite matrices is given in Appendix B.

and

$$V(\hat{\theta}_2) = \begin{pmatrix} \dfrac{\sigma^2}{n} & 0 \\ 0 & \dfrac{2\sigma^2}{n-1} \end{pmatrix}$$

Note that the off-diagonal cells are zero in each matrix because of the independence of the components of the respective estimators due to the normal distribution of the sample. The difference $V(\hat{\theta}_2) - V(\hat{\theta}_1)$ always yields the positive-semidefinite matrix

$$\begin{pmatrix} 0 & 0 \\ 0 & \left(\dfrac{2n-1}{n^2}\right)\dfrac{2\sigma^2}{n-1} \end{pmatrix}$$

Hence, this shows that $\hat{\theta}_1$ is more efficient than $\hat{\theta}_2$. So, based on the slightly smaller variability, we should always prefer estimator $\hat{\theta}_1$ even though the estimator s^2 is slightly biased for finite sample size n.

Linear Unbiased Estimators

A particular sort of estimators are *linear unbiased estimators*. We introduce them separately from the linear estimators here because they often display appealing statistical features.

In general, linear unbiased estimators are of the form

$$\hat{\theta} = \sum_{i=1}^{n} a_i X_i$$

To meet the condition of zero bias, the weights a_i have to add to one. Due to their lack of bias, the *MSE* in (17.7) will only consist of the variance part. With sample size n, their variances can be easily computed as

$$Var_\theta(\hat{\theta}) = \sum_{i=1}^{n} a_i^2 \sigma_X^2$$

where σ_X^2 denotes the common variance of each drawing. This variance can be minimized with respect to the coefficients a_i and we obtain the *best linear unbiased estimator* (BLUE) or *minimum variance linear unbiased estimator* (MVLUE). We have to be aware, however, that we are not always able to find such an estimator for each parameter.

An example of a BLUE is given by the sample mean $\bar{x}$. We know that all the $a_i = 1/n$. This not only guarantees that the sample mean is unbiased for the population mean μ, but it also provides for the smallest variance of all unbiased linear estimators. Therefore, the sample mean is efficient among all linear estimators. By comparison, the first draw is also unbiased. However, its variance is n times greater than that of the sample mean.

MAXIMUM LIKEHOOD ESTIMATOR

The method we discuss next provides one of the most essential tools for parameter estimation. Due to its structure, it is very intuitive.

Suppose the distribution (discrete or continuous) of some random variable Y is characterized by the parameter θ. As usual, we draw a random sample of length n, that is, $X = (X_1, X_2, ..., X_n)$, where each X_i is drawn independently and distributed identically as Y. That is, in brief notation, the individual drawings X_i are i.i.d. So, the joint probability distribution of the random sample X is given by equation (17.1) in the case of discrete random variables. If the random variable is continuous, the joint density function of the sample will be given by equation (17.2) where all f_{X_i} and f_Y are identical; that is,

$$f_X(x_1,...,x_n) = f_Y(x_1) \cdot ... \cdot f_Y(x_n) \qquad (17.13)$$

Here we say the random variable is continuous, meaning that it is not only a continuous distribution function, but also has a density.

To indicate that the distribution of the sample X is governed by the parameter θ, we rewrite equations (17.1) and (17.13), respectively, as the so-called *likelihood functions*, that is,

$$L_x(\theta) = \begin{cases} P(X = x) & \text{if Y is discrete} \\ f_X(x) & \text{if Y is continuous} \end{cases} \qquad (17.14)$$

Note that as presented in equation (17.14), the likelihood function is a function only of the parameter θ, while the observed sample value x is treated as a constant.

Usually instead of the likelihood function, we use the natural logarithm, denoted by ln, such that equation (17.14) turns into the *log-likelihood function*, abbreviated by $l_x(\theta)$; that is,

FIGURE 17.12 Natural Logarithm (ln) as a Strictly Monotonic Function Preserves the Location of the Maximum of the Density Function f

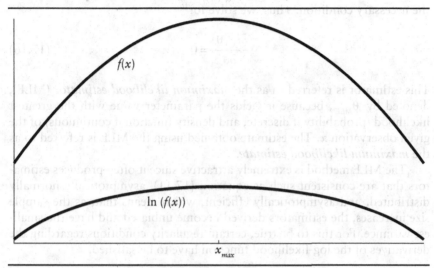

$$l_x(\theta) = \begin{cases} \ln P(X = x) & \text{if } Y \text{ is discrete} \\ \ln f_X(x) & \text{if } Y \text{ is continuous} \end{cases} \quad (17.15)$$

The log-likelihood function is, in many cases, easier to set up. We can do this transformation since the natural logarithm is a strictly monotonic increasing function and as such preserves the position of optima and maxima in particular.[9] We illustrate this in Figure 17.12.

Suppose we observe a particular value $x = (x_1, x_2, ..., x_n)$ in our sample. The question we now ask is: Which parameter values make the observation most plausible? Formally, that means we need to determine the very parameter value that maximizes the probability of the realized sample value x, or density function at x if the distribution is continuous.

Our task is now, regardless of whether the distribution is discrete or continuous, is to maximize the log-likelihood function $l_x(\theta)$ given in equation (17.15) with respect to all possible values of θ. We have to keep in mind that we do not know the true value of θ and that the value we then determine is only an estimate.

---

[9]For a discussion on monotonicity, see Appendix A.

We obtain the maximum value via the first derivatives of the log-likelihood function with respect to θ, $\partial l_x(\theta)/\partial\theta$, which we set equal to zero as the necessary condition. Thus, we solve for[10]

$$\frac{\partial l_x(\theta)}{\partial \theta} = 0 \qquad (17.16)$$

This estimator is referred to as the ***maximum likelihood estimator*** (MLE), denoted by $\hat{\theta}_{MLE}$, because it yields the parameter value with the greatest likelihood (probability if discrete, and density function if continuous) of the given observation x. The estimate obtained using the MLE is referred to as the ***maximum likelihood estimate***.

The MLE method is extremely attractive since it often produces estimators that are consistent such as equation (17.11), asymptotically normally distributed, and asymptotically efficient, which means that, as the sample size increases, the estimators derived become unbiased and have the smallest variance. For this to be true, certain regularity conditions regarding the derivatives of the log-likelihood function have to be satisfied.

MLE of the Parameter λ of the Poisson Distribution

We consider the parameter λ of the Poisson distribution for this example. From the sample of size n with observation $x = (x_1, x_2, ..., x_n)$, we obtain the likelihood function

$$L_x(\lambda) = P(X_1 = x_1, X_2 = x_2, ..., X_n = x_n)$$
$$= \prod_{i=1}^{n} \frac{\lambda^{x_i}}{x_i!} e^{-\lambda}$$

from which we compute the log-likelihood function

$$l_x(\lambda) = \ln\left[e^{-n\lambda} \prod_{i=1}^{n} \frac{\lambda^{x_i}}{x_i!} e^{-\lambda}\right]$$
$$= -n\lambda + \ln\left[\prod_{i=1}^{n} \frac{\lambda^{x_i}}{x_i!} e^{-\lambda}\right]$$

---

[10] Technically, after we have found the parameter value $\hat{\theta}$ that solves equation (17.16), we also have to compute the second derivative evaluated at $\hat{\theta}$ and make sure that it is negative for the value to be a maximum. We will assume that this is always the case in our discussion.

$$= -n\lambda + \ln \prod_{i=1}^{n} \lambda^{x_i} - \ln \prod_{i=1}^{n} x_i!$$

$$= -n\lambda + \sum_{i=1}^{n}(x_i \ln \lambda) - \sum_{i=1}^{n} \ln x_i!$$

Differentiating with respect to λ and setting equal to zero gives

$$\frac{\partial l_x(\lambda)}{\partial \lambda} = -n + \sum_{i=1}^{n} \frac{x_i}{\lambda} = 0 \Leftrightarrow \hat{\lambda}_{MLE} = \frac{1}{n}\sum_{i=1}^{n} x_i = \bar{x}$$

So, we see that the MLE of the Poisson parameter λ equals the sample mean.

Let's return to the example for the property and casualty insurance company we discussed in Chapter 9. We assumed that the number of claims received per year could be modeled as a Poisson random variable $Y \sim Poi(\lambda)$. As before, we are uncertain as to the true value of the parameter λ. For this reason, we have a look at the sample containing the number of claims of each of the previous 10 years, which is displayed below for convenience:

x_1	x_2	x_3	x_4	x_5
517,500	523,000	505,500	521,500	519,000
x_6	x_7	x_8	x_9	x_{10}
519,500	526,000	511,000	524,000	530,500

With these data, we obtain the estimate $\hat{\lambda}_{MLE} = 519{,}750$. Because the parameter λ equals the expected value of the Poisson random variable, we have to expect 519,750 claims per year.

As another illustration of the MLE of the Poisson parameter λ, we consider a portfolio consisting of risky bonds of, say 100 different companies. We are interested in the number of defaults expected to occur within the next year. For simplicity, we assume that each of these companies may default independently with identical probability p. So, technically, the default of a company is a Bernoulli random variable assuming value 1 in case of default and zero otherwise, such that the overall number of defaults is binomial, $B(100, p)$. Since the number of companies, 100, is large, we can approximate the binomial distribution with the Poisson distribution.[11] Now, the unknown parameter λ, which equals the population mean of the Poisson

---

[11] We know from Chapter 9 that the Poisson distribution approximates the hypergeometric distribution, which, in turn, is an approximation of the binomial distribution for large n.

random variable Y has to inferred from some sample.[12] Suppose that from the last 20 years, we observe the following annual defaults:

x_1	x_2	x_3	x_4	x_5	x_6	x_7	x_8	x_9	x_{10}
2	3	12	5	2	4	7	1	0	3
x_{11}	x_{12}	x_{13}	x_{14}	x_{15}	x_{16}	x_{17}	x_{18}	x_{19}	x_{20}
5	6	2	4	2	1	0	1	2	1

With these data, we compute as parameter estimate $\hat{\lambda}_{MLE} = 3.15$. So, we have to expect 3.15, or roughly 3 companies to default within the next year.

MLE of the Parameter λ of the Exponential Distribution

In our next illustration, we consider continuous random variables. Suppose that we are interested in the previous bond portfolio. The number of defaults per year was given as a Poisson random variable. Then, the time between two successive defaults follows an exponential law; that is, the interarrival time can be modeled as the random variable $Y \sim Exp(\lambda)$ with rate λ. We know that the expected time between two successive defaults is given by $E(Y) = 1/\lambda$, expressing the inverse relationship between the number of defaults and the time between them. Since the parameters λ from both the exponential and Poisson distributions coincide, we might just use the estimate obtained from the previous example. However, suppose instead that we decide to estimate the parameter via the MLE method under the exponential assumption explicitly.

From a sample observation $x = (x_1, x_2, ..., x_n)$ of an exponential random variable, we obtain the likelihood function

$$L_x(\lambda) = f_X(x_1, x_2, ..., x_n)$$

$$= \prod_{i=1}^{n} \lambda e^{-\lambda x_i}$$

$$= \lambda^n e^{-\lambda \sum_{i=1}^{n} x_i}$$

from which we compute the log-likelihood function

$$l_x(\lambda) = \ln\left[\lambda^n e^{-\lambda \sum_{i=1}^{n} x_i}\right]$$

$$= n \ln \lambda - \lambda \sum_{i=1}^{n} x_i$$

---

[12]Recall that the parameter $100 \cdot p$ of the binomial distribution also equals the population mean of Y and, hence, λ.

Computing the derivative of the above with respect to λ and setting the derivative equal to zero, we obtain

$$\frac{\partial l_x(\lambda)}{\partial \lambda} = \frac{n}{\lambda} - \sum_{i=1}^{n} x_i = 0 \Leftrightarrow \hat{\lambda}_{MLE} = \frac{n}{\sum_{i=1}^{n} x_i} = \frac{1}{\bar{x}}$$

Note that this MLE has a bias of $1/(n-1)$, which, as n goes to infinity, will vanish. Now, with our bond portfolio data from the previous example, we obtain the interarrival times:

x_1	x_2	x_3	x_4	x_5	x_6	x_7	x_8	x_9	x_{10}
0.500	0.333	0.083	0.200	0.500	0.250	0.143	1	0.318	0.333
x_{11}	x_{12}	x_{13}	x_{14}	x_{15}	x_{16}	x_{17}	x_{18}	x_{19}	x_{20}
0.200	0.167	0.500	0.250	0.500	1	0.318	1	0.500	1

A problem arose from x_9 and x_{17} being the interarrival times of years without default. We solved this by replacing the numbers of default in those two years by the sample average (i.e., 3.15 defaults per year). So, we compute the maximum likelihood estimate as $\hat{\lambda}_{MLE} = 2.1991$.[13] With this estimate, we have to expect to wait $1/\hat{\lambda}_{MLE} = 1/2.1991 = 0.4547$ years, or a little over five months, between two successive defaults. Note that $1/\hat{\lambda}_{MLE} = \bar{x}$ is an unbiased estimator of $E(Y) = 1/\lambda$, which itself, however, is not a parameter of the distribution.

MLE of the Parameter Components of the Normal Distribution

For our next example, we assume that the daily GE stock return can be modeled as a normally distributed random variable Y. Suppose that we do not know the true value of one of the components of the parameter (μ, σ^2). With the normal density function of Y, f_Y introduced in Chapter 11, we obtain as the likelihood function of the sample

$$L_X(\mu, \sigma^2) = \prod_{i=1}^{n} f_Y(x_i)$$
$$= f_Y(x_1) \cdot f_Y(x_2) \cdot \ldots \cdot f_Y(x_n)$$

---

[13]Note that because of the data replacement, the estimates here and in the previous example for the Poisson distribution diverge noticeably.

$$= \frac{1}{\sqrt{2\pi\sigma^2}} \cdot e^{-\frac{(x_1-\mu)^2}{2\sigma^2}} \cdot \frac{1}{\sqrt{2\pi\sigma^2}} \cdot e^{-\frac{(x_2-\mu)^2}{2\sigma^2}} \cdot \ldots \cdot \frac{1}{\sqrt{2\pi\sigma^2}} \cdot e^{-\frac{(x_n-\mu)^2}{2\sigma^2}}$$

$$= \left(\frac{1}{\sqrt{2\pi\sigma^2}}\right)^n \cdot e^{-\frac{\sum_{i=1}^{n}(x_i-\mu)^2}{2\sigma^2}}$$

Taking the natural logarithm, we obtain the log-likelihood function

$$l_X(\mu, \sigma^2) = \ln\left(\prod_{i=1}^{n} f_Y(x_i)\right)$$

$$= n \cdot \ln\left(\frac{1}{\sqrt{2\pi\sigma^2}}\right) - \frac{\sum_{i=1}^{n}(x_i-\mu)^2}{2\sigma^2}$$

If μ is the unknown component, we obtain as the first derivative of the log-likelihood function with respect to μ[14]

$$\frac{\partial l_X(\mu, \sigma^2)}{\partial \mu} = \frac{\sum_{i=1}^{n}(x_i-\mu)}{\sigma^2} = 0$$

such that

$$\hat{\mu}_{MLE} = \frac{1}{n}\sum_{i=1}^{n} x_i = \bar{x}$$

turns out to be the MLE for μ. Note that it exactly equals the sample mean. If, instead, we wanted to estimate σ^2 when μ is known, we obtain the derivative

$$\frac{\partial l_X(\mu, \sigma^2)}{\partial \sigma^2} = -\frac{n}{2\sigma^2} + \frac{\sum_{i=1}^{n}(x_i-\mu)^2}{2\sigma^4} = 0$$

and, from there, the maximum likelihood estimate is found to be

$$\hat{\sigma}^2_{MLE} = \frac{1}{n}\sum_{i=1}^{n}(x_i-\mu)^2$$

---

[14] Here we do not need to differentiate with respect to σ^2 since it is assumed to be known.

which, as we know, is unbiased for the population variance σ^2 if μ is known.

If no component of the parameter $\theta = (\mu, \sigma^2)$ is known, we obtain the same estimator for the mean μ (i.e., $\bar{x}$). However, the variance is estimated by the sample variance

$$1/n \sum_{i=1}^{n} (x_i - \bar{x})^2$$

which is biased for σ^2.

With the daily GE stock return data observed between April 24, 1980 and March 30, 2009, we obtain the maximum likelihood estimates $\bar{x} = 0.00029$ and $s^2 = 0.00031$, respectively, assuming both parameter components are unknown.

Cramér-Rao Lower Bound

In our discussion of the concept of consistency of an estimator, we learned that it is desirable to have an estimator that ultimately attains a value near the true parameter with high probability. The quantity of interest is *precision* with precision expressed in terms of the estimator's variance. If the estimator is efficient, we know that its variance shrinks and, equivalently, its precision increases most rapidly.

Here, we will concentrate on the fact that in certain cases there is a minimum bound, which we will introduce as the ***Cramér-Rao lower bound***, that no estimator variance will ever fall below. The distance of some estimator's variance from the respective lower bound can be understood as a quality criterion. The Cramér-Rao lower bound is based on the second derivative of log-likelihood function $l_X(\theta)$ with respect to the parameter θ, that is, $\partial^2 l_X(\theta)/\partial \theta^2$.

In our previous examples of the maximum likelihood estimation, we treated the log-likelihood function as a function of the parameter values θ for a given sample observation x. Now, if instead of the realization x, we take the random sample X, the log-likelihood function is a function of the parameter and the random sample. So, in general, the log-likelihood function is itself random. As a consequence, so is its second derivative $\partial^2 l_X(\theta)/\partial \theta^2$.

For any parameter value θ, we can compute the expected value of $\partial^2 l_X(\theta)/\partial \theta^2$, that is, $E(\partial^2 l_X(\theta)/\partial \theta^2)$, over all possible outcomes x of the sample X. The negative of this expected value of the second derivative we refer to as the *information number*, denoted as $I(\theta)$, which is given by

$$I(\theta) = \left[-E\left(\partial^2 l_X(\theta)/\partial \theta^2\right) \right] \qquad (17.17)$$

Let $\hat{\theta}$ be an unbiased estimator for θ. Without proof, we state the important result that its variance is at least as large as the inverse of the information number in equation (17.17); that is,

$$\frac{1}{I(\theta)} \qquad (17.18)$$

If the parameter θ consists of, say, $k > 1$ components, we can state the variance bounds in the multivariate situation. The second derivatives of the log-likelihood function $l_X(\theta)$ with respect to the parameter θ are now forming a matrix. This results from the fact that $l_X(\theta)$ first has to be differentiated with respect to every component $\theta_i, i = 1, 2, \ldots, k$ and these derivatives have to be differentiated with respect to every component again. Thus, we need to compute $k \times k$ partial second derivatives. This matrix of all second derivatives looks like

$$J = \left[\frac{\partial^2 l_X(\theta)}{\partial \theta \, \partial \theta^T} \right]$$

$$= \begin{bmatrix} \dfrac{\partial^2 l_X(\theta)}{\partial \theta_1 \, \partial \theta_1} & \dfrac{\partial^2 l_X(\theta)}{\partial \theta_1 \, \partial \theta_2} & \cdots & \dfrac{\partial^2 l_X(\theta)}{\partial \theta_1 \, \partial \theta_k} \\ \dfrac{\partial^2 l_X(\theta)}{\partial \theta_2 \, \partial \theta_1} & \dfrac{\partial^2 l_X(\theta)}{\partial \theta_2 \, \partial \theta_2} & \cdots & \vdots \\ \vdots & & \ddots & \vdots \\ \dfrac{\partial^2 l_X(\theta)}{\partial \theta_k \, \partial \theta_1} & \cdots & \cdots & \dfrac{\partial^2 l_X(\theta)}{\partial \theta_k \, \partial \theta_k} \end{bmatrix} \qquad (17.19)$$

where the $\partial^2 l_X(\theta)/\partial \theta_i \partial \theta_j$ in equation (17.19) denote the derivatives with respect to the i-th and j-th components of θ.

We restate that since X is random, so is the log-likelihood function and all its derivatives. Consequently, matrix J is also random. The negative expected value of J is the **information matrix** $I(\theta) = -E(J)$, which contains the expected values of all elements of J. With $\mathrm{Var}(\hat{\theta})$ being the covariance matrix of some k-dimensional unbiased estimator $\hat{\theta}$, the difference

$$\mathrm{Var}(\hat{\theta}) - \left[I(\theta)\right]^{-1} \qquad (17.20)$$

which is a matrix itself, is always positive-semidefinite. The expression $\left[I(\theta)\right]^{-1}$ denotes the matrix inverse of $I(\theta)$ as a generalization of equation (17.18).

If the difference matrix in (17.20) is zero in some components, the estimator $\hat{\theta}$ is efficient in these components with respect to any other unbiased estimator and, hence, always preferable. If the bound is not attained for some components and, hence, the difference (17.20) contains elements greater than zero, we do not know, however, if there might be an estimator more efficient than the one we are using.

Cramér-Rao Bound of the MLE of Parameter λ of the Exponential Distribution

As an illustration, consider once again the bond portfolio example. Suppose as before that the time between two successive defaults of firms, Y, is $Exp(\lambda)$. As we know, the log-likelihood function of the exponential distribution with parameter λ for a sample X is

$$l_X(\lambda) = n\ln\lambda - \lambda\sum_{i=1}^{n} X_i$$

such that the second derivative turns out to be

$$\frac{\partial^2 l_X(\lambda)}{\partial \lambda^2} = -\frac{n}{\lambda^2}$$

Then the expected value is

$$E(-n/\lambda^2) = -n/\lambda^2$$

since it is constant for any value of λ. Now, for some unbiased estimator λ, according to equation (17.18), the variance is at least λ^2/n. Obviously, the variance bound approaches zero, such that efficient unbiased estimators—if they should exist—can have arbitrarily small variance for ever larger sample sizes. With 20 observations, we have $n = 20$ and, consequently, the variance of our estimator is, at least, $\lambda^2/20$.

We know that the maximum likelihood estimator $\hat{\lambda}_{MLE} = 1/\bar{x}$ is approximately unbiased as the sample size goes to infinity. So, we can apply the lower bounds approximately. Therefore, with the given lower bound, the standard error will be at least $SE(\hat{\lambda}_{MLE}) = \lambda/\sqrt{20}$. With approximate normality of the estimator, roughly 68% of the estimates fall inside of an interval extending at least $\lambda/\sqrt{20}$ in each direction from λ. If the true parameter value λ is equal to 3, then this interval approximately contains the smaller interval [2.33, 3.67]. As the sample size increases, the standard error $3/\sqrt{n}$ of $\hat{\lambda}_{MLE}$ will vanish so that its precision increases and hardly any estimate will deviate from $\lambda = 3$ by much.

What this means is that if we have an unbiased estimator, we should check its variance first and see whether it attains the lower bounds exactly. In that case, we should use it. However, even though the MLE is not exactly unbiased, its bias becomes more negligible the larger the sample and its variance approaches the decreasing lower bound arbitrarily closely. As can be shown, the MLE is closest to the lower bound among most estimators that are only slightly biased such as the MLE.

Cramér-Rao Bounds of the MLE of the Parameters of the Normal Distribution

For our next illustration, we provide an illustration of a 2-dimensional parameter. Consider a normally distributed stock return whose parameters μ and σ^2 we wish to infer based on some sample of size n. With the respective second derivatives[15]

$$\frac{\partial l_X^2(\mu, \sigma^2)}{\partial \mu^2} = -\frac{n}{\sigma^2}$$

$$\frac{\partial l_X^2(\mu, \sigma^2)}{\partial (\sigma^2)^2} = \frac{n}{2\sigma^4} - \frac{1}{\sigma^6} \sum_{i=1}^{n}(x_i - \mu)^2$$

$$\frac{\partial l_X^2(\mu, \sigma^2)}{\partial \mu \, \partial \sigma^2} = \frac{\partial l_X^2(\mu, \sigma^2)}{\partial \sigma^2 \partial \mu} = -\frac{1}{\sigma^4} \sum_{i=1}^{n}(x_i - \mu)^2$$

we obtain

$$\left[I(\mu, \sigma^2)\right]^{-1} = \begin{bmatrix} \dfrac{\sigma^2}{n} & 0 \\ 0 & \dfrac{2\sigma^4}{n} \end{bmatrix} \qquad (17.21)$$

as the inverse of the information matrix.

Suppose further that we use the unbiased estimator with the respective components

$$\hat{\mu} = \bar{x} = \sum_{i=1}^{n} x_i \text{ and } s^{*2} = 1/(n-1) \sum_{i=1}^{n}(x_i - \bar{x})^2$$

Then, we obtain as the covariance matrix

---

[15]Note that from the last row of the table, the order of differentiating is not essential. Consequently, J is actually a symmetric matrix.

$$Var(\hat{\mu}, s^{*2}) = \Sigma = \begin{bmatrix} \dfrac{\sigma^2}{n} & 0 \\ 0 & \dfrac{2\sigma^4}{n-1} \end{bmatrix} \quad (17.22)$$

such that the difference between equations (17.21) and (17.22) becomes the positive-semidefinite matrix given by

$$\begin{bmatrix} 0 & 0 \\ 0 & \dfrac{2\sigma^4}{n(n-1)} \end{bmatrix} \quad (17.23)$$

Note that in equation (17.23) there is only one element different from zero, namely that for the variance estimator. This indicates that $\bar{x}$ is an efficient estimator since it meets its bound given by the top-left element in equation (17.21), whereas s^{*2} is not. So there might be a more efficient estimator than s^{*2}. However, as the sample size n increases, the variance becomes arbitrarily close to the lower bound, rendering the difference an unbiased estimator attaining the bound and s^{*2} meaningless.

EXPONENTIAL FAMILY AND SUFFICIENCY

Next we will derive a method to retrieve statistics that reveal positive features, which will become valuable particularly in the context of testing. Estimators based on these statistics often fulfill many of the quality criteria discussed before.

Exponential Family

Let Y be some random variable with its probability distribution depending on some parameter θ consisting of one or more components. If Y is continuous, the cumulative distribution function Y and density function $f(y)$ vary with θ.[16]

Now, suppose we have a sample $X = (X_1, X_2, ..., X_n)$ of n independent and identical draws from the population of Y. According to equation

---

[16] In literature, to express this dependence explicitly, one occasionally finds that the probability measure as well the cumulative distribution function and density function carry the subscript θ so that the notations $P_\theta(Y = y)$, $F_\theta(y)$, and $f_\theta(y)$ are used instead.

(17.15), the log-likelihood function $l_\theta(X)$ of this sample itself depends on the parameter θ. If we can present this log-likelihood function in the form

$$l_X(\theta) = a(X) + b(\theta) + \sum_{j=1}^{k} c_j(X) \cdot \tau_j(\theta) \qquad (17.24)$$

we say that it is an *exponential family of distributions*.

From equation (17.24), we see that the likelihood function has a term, namely $a(X)$, that is a function of the sample only, and one that is a function merely of the parameter, $b(\theta)$. The property of exponential families to allow for this separation often facilitates further use of the log-likelihood function since these terms can be discarded under certain conditions.

Exponential Family of the Poisson Distribution

We consider for our first illustration a Poisson random variable such as the number of defaults per year in a portfolio of risky bonds. The log-likelihood function can now be represented as

$$l_\lambda(X) = -n\lambda + \ln\lambda \cdot \sum_{i=1}^{n} X_i - \sum_{i=1}^{n} \ln(X_i!)$$

so that we obtain for the exponential family

$$a(X) = -\sum_{i=1}^{n} \ln(X_i!)$$
$$b(\lambda) = -n\lambda$$
$$\tau(\lambda) = \ln\lambda$$
$$c(X) = \sum_{i=1}^{n} X_i$$

Exponential Family of the Exponential Distribution

Consider for our next illustration an exponential random variable such as the interarrival time between two successive defaults in a bond portfolio. With the corresponding log-likelihood function

$$l_\lambda(X) = n\lambda + \lambda \cdot \sum_{i=1}^{n} X_i$$

we obtain the exponential family with

$$a(X) = 0$$
$$b(\lambda) = n \ln \lambda$$
$$\tau(\lambda) = \lambda$$
$$c(X) = \sum_{i=1}^{n} X_i$$

Exponential Family of the Normal Distribution

As our final illustration, we give the coefficients of the exponential family of the normal distribution. With the log-likelihood function

$$l_\lambda(X) = -n \ln\left(\sqrt{2\pi\sigma^2}\right) - \sum_{i=1}^{n} \frac{(X_i - \mu)^2}{2\sigma^2}$$

$$= -n \ln\left(\sqrt{2\pi\sigma^2}\right) - \frac{\sum_{i=1}^{n} X_i^2}{2\sigma^2} + \frac{\mu \sum_{i=1}^{n} X_i}{\sigma^2} - \frac{n\mu^2}{2\sigma^2}$$

we obtain

$$a(X) = 0$$
$$b(\mu, \sigma^2) = -n\left[\ln\left(\sqrt{2\pi\sigma^2}\right) + \frac{n\mu^2}{2\sigma^2}\right]$$
$$\tau_1(\mu, \sigma^2) = \frac{\mu}{\sigma^2}$$
$$c_1(X) = \sum_{i=1}^{n} X_i$$
$$\tau_2(\mu, \sigma^2) = -\frac{1}{2\sigma^2}$$
$$c_2(X) = \sum_{i=1}^{n} X_i^2$$

We see that we have a parameter with two components since we have two statistics of X, namely $c_1(X)$ and $c_2(X)$.

Sufficiency

Let's focus more on the statistics rather than the estimator, even though the two are closely related. In equation (17.3), we defined the statistic as a function that maps the samples to a k-dimensional real number. The dimension k usually coincides with the number of components of the parameter θ and the cases where this is not true will not be considered here.

Our focus is on how the statistic processes the information given in the sample. A statistic ought to maintain relevant information in a sample and dispose of any information not helpful in inference about the parameter. For example, two samples with observed values x and x^*, that are just permutations of each other such as $(x_1, x_2, x_3) = (0.5, 0, -1)$ and $\left(x_1^*, x_2^*, x_3^*\right) = (-1, 0.5, 0)$ should yield the same value of the statistic t. For example, we should obtain $t(0.5, 0, -1) = t(-1, 0.5, 0)$ and our decision with respect to the unknown parameter value is identical in both cases. That is, the statistic reduces the information contained in the sample to the part sufficient for the estimation.

To verify whether a statistic is sufficient, we resort to the following theorem by Neyman and Fisher

Factorization Theorem: Let Y be a random variable with probability distribution governed by parameter θ of one or more components. Moreover, we have a sample $X = (X_1, X_2, ..., X_n)$ of size n where each of the X_i is independently drawn from the population of Y and thus identically distributed as Y. The statistic t is said to be *sufficient* if we can factorize the probability of $X = x$ (Y discrete) or density function of X evaluated at the observation x (Y continuous) into two terms according to

$$\left.\begin{array}{r} P(X = x) \\ f_X(x) \end{array}\right\} = g(\theta, t(x)) \cdot h(x) \qquad (17.25)$$

That is, due to the sufficiency of t, we can find a function of the parameter and the statistic, g, and a function of the sample value, h.

Suppose that for some observation x, $t(x) = c$ where c is some real number. The factorization of equation (17.25) can then be interpreted as follows. If Y is a discrete random variable, the probability of the sample X assuming value x can be decomposed into the product of the probability $g(\theta, t(x))$ that the statistic assumes value c where θ is the true parameter value, and the conditional probability $h(x)$ that $X = x$ given the value c of the statistic and parameter θ. Now, when t is sufficient, the latter probability should not depend on the parameter θ anymore since t was able to retrieve all essential information concerning θ already, and the only additional information that is still contained in x compared to $t(x)$ is irrelevant for the estimation of θ. Consequently, the function h should only be a function of x.

In many cases, the only additional information conveyed by x concerns the order of appearance of the successively drawn values x_j. Then, the function h merely assigns probability to the observed order of appearance (i.e., $x_1, x_2, ..., x_n$) relative to any order of appearance of the x_j.

Estimators based on sufficient statistics often exhibit positive features such as being unbiased, yield minimal variance, have known probability distributions, and lead to optimal tests.

The importance of the representation of the exponential family of distributions may now be apparent. If we can give the exponential family of some distribution, then the $C_i(X)$, $i = 1, 2, ..., k$ in representation (17.24) are the sufficient statistics for the respective parameter θ.

Sufficient Statistic for the Parameter λ of the Poisson Distribution

To illustrate, consider a Poisson random sample. A sufficient statistic for the parameter λ is then given by

$$\sum_{i=1}^{n} X_i$$

as we see by its exponential family. An estimator for λ using this sufficient statistic is the well-known sample mean $\bar{X}$.

CONCEPTS EXPLAINED IN THIS CHAPTER
(IN ORDER OF PRESENTATION)

Parameter space
Sample
Statistic
Population parameter
Sampling distribution
Estimate
Linear estimator
Generalized Central Limit Theorem
Limiting distribution
Serial dependence
Quality criteria
Sampling error
Bias
Biased estimator
Unbiased estimator
Bias-corrected sample variance (Corrected sample variance)

Standard error
Mean-square error
Minimum-variance unbiased estimator
Large-sample properties of estimators
Asymptotic properties of estimators
Convergence
Convergence in probability
Law of Large Numbers
Consistency
Consistent estimator
Unbiased efficiency
Efficient estimator
Linear unbiased estimators
Best linear unbiased estimator
Minimum variance linear unbiased estimator
Likelihood functions
Log-likelihood function
Maximum likelihood estimator
Maximum likelihood estimate
Precision
Cramér-Rao lower bound
Information number
Information matrix
Exponential family of distributions
Factorization Theorem
Sufficient

CHAPTER 18
Confidence Intervals

Portfolio managers and risk managers must monitor the behavior of certain random variables such as the daily returns of the stocks contained in the portfolio or the default rate of bonds comprising the bond portfolio under management. In both cases, we need to know the population parameters characterizing the respective random variable's probability distribution. However, in most realistic situations, this information will not be available.

In the previous chapter, we dealt with this problem by estimating the unknown parameter with a point estimator to obtain a single number from the information provided by a sample. It will be highly unlikely, however, that this estimate—obtained from a finite sample—will be exactly equal to the population parameter value even if the estimator is consistent—a notion introduced in the previous chapter. The reason is that estimates most likely vary from sample to sample. However, for any realization, we do not know by how much the estimate will be off.

To overcome this uncertainty, one might think of computing an interval or, depending on the dimensionality of the parameter, an area that contains the true parameter with high probability. That is, we concentrate in this chapter on the construction of confidence intervals.

We begin with the presentation of the confidence level. This will be essential in order to understand the confidence interval that will be introduced subsequently. We then present the probability of error in the context of confidence intervals, which is related to the confidence level. In this chapter, we identify the most commonly used confidence intervals.

CONFIDENCE LEVEL AND CONFIDENCE INTERVAL

We begin with the one-dimensional parameter θ. As in the previous chapter, we will let Θ denote the parameter space; that is, it is the set containing all possible values for the parameter of interest. So, for the one-dimensional parameter, the parameter space is a subset of the real numbers (i.e., $\Theta \subseteq \mathbb{R}$). For example, the parameter θ of the exponential distribution, denoted $Exp(\theta)$, is a real number greater than zero.

In Chapter 17, we inferred the unknown parameter with a single estimate $\hat{\theta} \in \Theta$. The likelihood of the estimate exactly reproducing the true parameter value may be negligible. Instead, by estimating an interval, which we may denote by I_θ, we use a greater portion of the parameter space, that is, $I_\theta \subset \Theta$, and not just a single number. This may increase the likelihood that the true parameter is one of the many values included in the interval.

If, as one extreme case, we select as an interval the entire parameter space, the true parameter will definitely lie inside of it. Instead, if we choose our interval to consist of one value only, the probability of this interval containing the true value approaches zero and we end up with the same situation as with the point estimator. So, there is a trade-off between a high probability of the interval I_θ containing the true parameter value, achieved through increasing the interval's width, and the precision gained by a very narrow interval.

As in Chapter 17, we should use the information provided by the sample. Hence, it should be reasonable that the interval bounds depend on the sample in some way. Then technically each interval bound is a function that maps the sample space, denoted by $\mathbf{X}$, into the parameter space since the sample is some outcome in the sample space and the interval bound transforms the sample into a value in the parameter space representing the minimum or maximum parameter value suggested by the interval.

Formally, let $l: \mathbf{X} \to \Theta$ denote the lower bound and $u: \mathbf{X} \to \Theta$ the upper bound of the interval, respectively. Then, for some particular sample outcome $x = (x_1, x_2, \ldots, x_n)$, the interval is given by $[l(x), u(x)]$. Naturally, the bounds should be constructed such that the lower bound is never greater than the upper bound for any sample outcome x; that is, it is reasonable to require always that $l(x) \leq u(x)$.[1]

As mentioned, the interval depends on the sample $X = (X_1, X_2, \ldots, X_n)$, and since the sample is random, the interval $[l(X), u(X)]$ is also random. We can derive the probability of the interval lying beyond the true parameter (i.e., either completely below or above) from the sample distribution. These two possible errors occur exactly if either $u(x) < \theta$ or $\theta < l(x)$.

Our objective is then to construct an interval so as to minimize the probability of these errors occurring. That is,

$$P(\theta \notin [l(X), u(X)]) = P(\theta < l(X)) + P(u(X) < \theta)$$

Suppose we want this probability of error to be equal to α. For example, we may select $\alpha = 0.05$ such that in 5% of all outcomes, the true parameter will not be covered by the interval. Let

---

[1] It suffices to only require that this holds P-almost surely. For the introduction of P-almost sure events, see Chapter 8.

$$p_l = P(\theta < l(X)) \text{ and } p_u = P(u(X) < \theta)$$

Then, it must be that

$$P(\theta \notin [l(X), u(X)]) = p_l + p_u = \alpha$$

Now let's provide two important definitions: a confidence level and confidence interval.

Definition of a Confidence Level

For some parameter θ, let the probability of the interval not containing the true parameter value be given by the probability of error α. Then, with probability $1 - \alpha$, the true parameter is covered by the interval $[l(X), u(X)]$. The probability

$$P(\theta \in [l(X), u(X)]) = 1 - \alpha$$

is called *confidence level*.

It may not be possible to find bounds to obtain a confidence level exactly. We, then, simply postulate for the confidence level $1 - \alpha$ that

$$P(\theta \in [l(X), u(X)]) \geq 1 - \alpha$$

is satisfied, no matter what the value θ may be.

Definition and Interpretation of a Confidence Interval

Given the definition of the confidence level, we can refer to an interval $[l(X), u(X)]$ as $1 - \alpha$ *confidence interval* if

$$P(\theta \in [l(X), u(X)]) = 1 - \alpha$$

holds no matter what is the true but unknown parameter value θ. (Note that if equality cannot be exactly achieved, we take the smallest interval for which the probability is greater than $1 - \alpha$.)

The interpretation of the confidence interval is that if we draw an increasing number of samples of constant size n and compute an interval, from each sample, $1 - \alpha$ of all intervals will eventually contain the true parameter value θ.

As we will see in the examples, the bounds of the confidence interval are often determined by some standardized random variable composed of both the parameter and point estimator, and whose distribution is known. Furthermore, for a symmetric density function such as that of the normal distribution, it can be shown that with given α, the confidence interval is the tightest if we have $p_l = \alpha/2$ and $p_u = \alpha/2$ with p_l and p_u as defined before. That corresponds to bounds l and u with distributions that are symmetric to each other with respect to the the true parameter θ. This is an important property of a confidence interval since we seek to obtain the maximum precision possible for a particular confidence level.

Often in discussions of confidence intervals the statement is made that with probability $1 - \alpha$, the parameter falls inside of the confidence interval and is outside with probability α. This interpretation can be misleading in that one may assume that the parameter is a random variable. Recall that only the confidence interval bounds are random. The position of the confidence interval depends on the outcome x. By design, as we have just shown, the interval is such that in $(1 - \alpha) \times 100\%$ of all outcomes, the interval contains the true parameter and in $\alpha \times 100\%$, it does not cover the true parameter value.

The parameter is invariant, only the interval is random. We illustrate this in Figure 18.1. In the graphic, we display the confidence intervals $[l^{(1)}, u^{(1)}]$ through $[l^{(5)}, u^{(5)}]$ as results of the five sample realizations $x^{(1)} = \left(x_1^{(1)}, x_n^{(1)}, \ldots, x_n^{(1)}\right)$ through $x^{(1)} = \left(x_1^{(5)}, x_n^{(5)}, \ldots, x_n^{(5)}\right)$. As we can see, only the first two confidence intervals, that is, $[l^{(1)}, u^{(1)}]$ and $[l^{(2)}, u^{(2)}]$, contain the parameter θ. The remaining three are either above or below θ.

CONFIDENCE INTERVAL FOR THE MEAN OF A NORMAL RANDOM VARIABLE

Let us begin with the normal random variable Y with known variance σ^2 but whose mean μ is unknown. For the inference process, we draw a sample X of independent $X_1, X_2, \ldots, X_n$ observations that are all identically distributed as Y. Now, a sufficient and unbiased estimator for μ is given by the sample mean, which is distributed as

$$\bar{X} = \sum_{i=1}^{n} X_i \sim N\left(\mu, \frac{\sigma^2}{n}\right)$$

If we standardize the sample mean, we obtain the standard normally distributed random variable

$$Z = \sqrt{n}\frac{\bar{X} - \mu}{\sigma} \sim N(0,1)$$

Confidence Intervals

FIGURE 18.1 Five Realizations of Confidence Intervals of Equal Width for Parameter θ

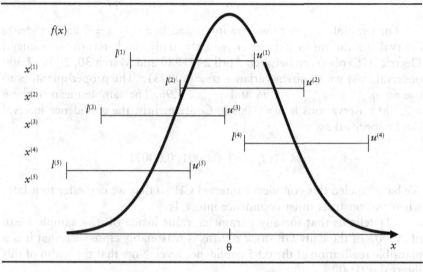

For this Z, it is true that

$$P\left(q_{\alpha/2} \leq Z \leq q_{1-\alpha/2}\right) = P\left(q_{\alpha/2} \leq \sqrt{n}\,\frac{\bar{X}-\mu}{\sigma} \leq q_{1-\alpha/2}\right)$$

$$= P\left(\frac{\sigma}{\sqrt{n}}\,q_{\alpha/2} \leq \bar{X}-\mu \leq \frac{\sigma}{\sqrt{n}}\,q_{1-\alpha/2}\right)$$

$$= P\left(\frac{\sigma}{\sqrt{n}}\,q_{\alpha/2} \leq \mu-\bar{X} \leq \frac{\sigma}{\sqrt{n}}\,q_{1-\alpha/2}\right)$$

$$= P\left(\bar{X}+\frac{\sigma}{\sqrt{n}}\,q_{\alpha/2} \leq \mu \leq \bar{X}+\frac{\sigma}{\sqrt{n}}\,q_{1-\alpha/2}\right)$$

$$= P\left(\bar{X}-\frac{\sigma}{\sqrt{n}}\,q_{1-\alpha/2} \leq \mu \leq \bar{X}+\frac{\sigma}{\sqrt{n}}\,q_{1-\alpha/2}\right)$$

$$= 1-\alpha$$

where $q_{\alpha/2}$ and $q_{1-\alpha/2}$ are the $\alpha/2$ and $1-\alpha/2$ quantiles of the standard normal distribution, respectively. Due to the symmetry of the standard normal distribution, the third equation follows and we have the identity $q_{\alpha/2} = q_{1-\alpha/2}$ from which we obtain the second-to-the-last equation above. The resulting $1-\alpha$ confidence interval for μ is then given as

$$I_{1-\alpha} = \left[\bar{X} - \frac{\sigma}{\sqrt{n}} q_{1-\alpha/2}, \bar{X} + \frac{\sigma}{\sqrt{n}} q_{1-\alpha/2} \right] \quad (18.1)$$

For example, suppose we are interested in a $1 - \alpha = 0.95$ confidence interval for the mean parameter μ of the daily stock returns of General Electric (GE) observed between April 24, 1980 and March 30, 2009 (7,300 observations) with known variance $\sigma^2 = 0.00031$.[2] The proper quantiles to use are $q_{0.025} = -q_{0.975} = -1.96$ and $q_{0.975} = 1.96$. The sample mean of the $n = 7{,}300$ observations is $\bar{x} = 0.0003$. Consequently, the confidence interval can be specified as

CI1: $I_{0.95} = [-0.0001, 0.0007]$

We have labeled this confidence interval CI1 so that we can refer to it later when we construct other confidence intervals.

CI1 tells us that for any parameter value inside of it, a sample mean of 0.0003 of the daily GE stock returns is reasonably close such that it is a plausible realization at the 0.95 confidence level. Note that the width of this interval is 0.0008.

Suppose we had only used 1,000 observations rather than 7,300 observations and everything else is the same. Then, the corresponding confidence interval would be

CI2: $I_{0.95} = [-0.0008, 0.0014]$

CI2 has a width of 0.0022, making it wider than CI1 by a factor of 2.75. We see that increasing the sample size reduces the interval width, leading to higher precision given that everything else remains unchanged including the confidence level. This is a result of the consistency of the sample mean.

Note that if the data are not truly normal, we can still apply equation (18.1) for μ if σ^2 is known. This is the result of the Central Limit Theorem discussed in Chapter 11. If we drop the assumption of normally distributed daily GE stock returns, we still use the same confidence interval since the sample size of $n = 7{,}300$ is sufficiently large to justify the use of the normal distribution.

Now let's construct an asymmetric confidence interval for the same data as in the previous illustration. Suppose we preferred to obtain the interval's lower bound by

$$l(X) = \bar{X} + q_{\alpha/4} \cdot \frac{\sigma}{\sqrt{n}}$$

---

[2] We assume here that the daily GE stock return Y follows a normal distribution.

because of the requirement $P(\mu < l(X)) = \alpha/4$. Consequently, we obtain $P(\mu < u(X)) = 1 - 3\alpha/4$ resulting in an upper bound of

$$u(X) = \bar{X} + q_{1-3\alpha/4} \cdot \frac{\sigma}{\sqrt{n}}$$

Again, $\alpha = 0.05$, so that we obtain $q_{0.0125} = -2.2414$ and $q_{0.9625} = 1.7805$. The confidence interval is now equal to [−0.00017, 0.00066], yielding a width of 0.00083, which is slightly larger than 0.0008. This is due to the fact that, for any given confidence level and random variable with symmetric density function and unimodal distribution, the confidence interval is the smallest when the probability of error is attributed symmetrically (i.e., when the probability of the interval being located below the parameter value is the same as that of the interval being positioned beyond the parameter value).

CONFIDENCE INTERVAL FOR THE MEAN OF A NORMAL RANDOM VARIABLE WITH UNKNOWN VARIANCE

Let's once again construct a confidence interval for a normal random variable Y but this time we assume that the variance (σ^2) is unknown as well. As before, we draw a sample of size n from which the sample mean $\bar{X}$ is computed. We obtain the new standardized random variable by (1) subtracting the mean parameter μ, even though unknown, (2) multiplying this difference by $\sqrt{n}$, and (3) dividing the result by the square root of the bias corrected sample variance from Chapter 17. That is, the new standardized random variable is

$$t = \sqrt{n}\,\frac{\bar{X} - \mu}{s^*} \qquad (18.2)$$

where

$$s^{*2} = 1/(n-1)\sum_{i=1}^{n}(X_i - \bar{X})^2$$

is the biased corrected sample variance.

The new standardized random variable given by equation (18.2) is Student's t-distributed with $n - 1$ degrees of freedom.[3] Therefore, we can state that

$$P\!\left(t_{\alpha/2}(n-1) \le t \le t_{1-\alpha/2}(n-1)\right) = 1 - \alpha \qquad (18.3)$$

---

[3] The Student's t-distribution is introduced in Chapter 11.

The quantities

$$t_{\alpha/2}(n-1) \text{ and } t_{1-\alpha/2}(n-1)$$

are the α/2- and 1 − α/2-quantiles of the Student's *t*-distribution with $n-1$ degrees of freedom, respectively. Using the definition (18.2) in equation (18.3), as follows

$$P\left(t_{\alpha/2}(n-1) \leq t \leq t_{1-\alpha/2}(n-1)\right)$$

$$= P\left(t_{\alpha/2}(n-1) \leq \sqrt{n}\frac{\bar{X}-\mu}{s^*} \leq t_{1-\alpha/2}(n-1)\right)$$

$$= P\left(\frac{s^*}{\sqrt{n}}t_{\alpha/2}(n-1) \leq \bar{X}-\mu \leq \frac{s^*}{\sqrt{n}}t_{1-\alpha/2}(n-1)\right)$$

$$= P\left(\frac{s^*}{\sqrt{n}}t_{\alpha/2}(n-1) \leq \mu-\bar{X} \leq \frac{s^*}{\sqrt{n}}t_{1-\alpha/2}(n-1)\right)$$

$$= P\left(\bar{X}+\frac{s^*}{\sqrt{n}}t_{\alpha/2}(n-1) \leq \mu \leq \bar{X}+\frac{s^*}{\sqrt{n}}t_{1-\alpha/2}(n-1)\right)$$

$$= P\left(\bar{X}-\frac{s^*}{\sqrt{n}}t_{1-\alpha/2}(n-1) \leq \mu \leq \bar{X}+\frac{s^*}{\sqrt{n}}t_{1-\alpha/2}(n-1)\right)$$

where we have taken advantage of the symmetry of the Student's *t*-distribution, in the third and fifth equations above, we obtain as the 1 − α confidence interval for the mean parameter μ

$$I_{1-\alpha} = \left[\bar{X} - \frac{s^*}{\sqrt{n}}t_{1-\alpha/2}(n-1), \bar{X} + \frac{s^*}{\sqrt{n}}t_{1-\alpha/2}(n-1)\right] \quad (18.4)$$

Let's return to our illustration involving the daily GE stock returns between April 24, 1980 and March 30, 2009. We assumed the daily stock returns to be normally distributed. Once again let's select a confidence level of 1 − α = 0.95. Therefore, the corresponding quantiles of the Student's *t*-distribution with 7,299 degrees of freedom are given as

$$t_{\alpha/2}(n-1) = -1.9603 \text{ and } t_{1-\alpha/2}(n-1) = 1.9603$$

Given the sample mean of 0.0003 computed earlier, the 0.95 confidence interval becomes

$$\text{CI3: } I_{0.95} = [-0.00017, 0.0007]$$

The width of this interval is equal to 0.00083 and is basically as wide as that obtained in CI1 in the previous illustration when the variance was known. For 7,299 degrees of freedom, the Student's t-distribution is virtually indistinguishable from the standard normal distribution and, hence, their confidence intervals coincide.

Recall that the degrees of freedom express the number of independent drawings of the sample of size n we have left for the estimation of the bias-corrected sample variance s^{*2} after the independence of one variable is lost in the computation of $\bar{X}$. So, technically, the confidence interval given by CI3 is slightly wider than CI1 when σ^2 is known. This is due to the additional uncertainty stemming from the estimation of s^{*2}. However, the difference as we can see from our two illustrations is small (given the precision of five digits).

CONFIDENCE INTERVAL FOR THE VARIANCE OF A NORMAL RANDOM VARIABLE

In our next illustration, we construct a confidence interval for the variance parameter σ^2 of a normal random variable Y given that the mean parameter μ is known to be 0.0003. Again, we consider drawing a sample $X = (X_1, X_2, ..., X_n)$ to base our inference on. Recall from Chapter 11 that if the random variables $X_1, X_2, ..., X_n$ are independently and identically distributed as the normal random variable Y, that is, $X_i \sim N(\mu, \sigma^2)$, then the sum

$$\sum_{i=1}^{n} \frac{(X_i - \mu)^2}{\sigma^2} \sim \chi^2(n) \qquad (18.5)$$

That is, the sum of the squared standard normal observations follows a chi-square distribution with n degrees of freedom. Note that for equation (18.5) we used the variance σ^2 even though it is unknown. In Figure 18.2, we display the probability density function as well as the $\alpha/2$- and $1 - \alpha/2$–quantiles,

$$\chi^2_{\alpha/2}(n) \text{ and } \chi^2_{1-\alpha/2}(n)$$

of the distribution of equation (18.5). The corresponding tail probabilities of $\alpha/2$ on either end of the density function are indicated by the gray shaded areas.

We can now see that the density function of the random variable in equation (18.5) is skewed to the right. Consequently, it is not a mandatory step to construct a confidence interval such that the probability of error α is attributed symmetrically, as was the case in our the previous illustrations, in order

FIGURE 18.2 Position of α/2 and 1 − α/2–Quantiles of the Chi-Square Distribution with n Degrees of Freedom of the Random Variable $\sum_{i=1}^{n}(X_i-\mu)^2/\sigma^2$

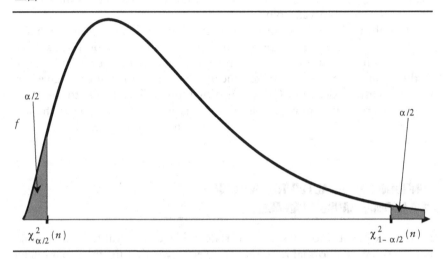

to obtain the smallest confidence intervals for a given confidence level of 1 − α. In general, the smallest confidence interval for σ^2 given a confidence level 1 − α will result in an asymmetric attribution of the probability of error. Thus, the probability that the interval will be located below the parameter will not equal that of the interval overestimating the parameter. However, it is standard to construct the confidence intervals analogous to the case where the standardized random variable has a symmetric unimodal distribution (i.e., symmetric distribution of the probability of error) because it is intuitively appealing. So, we begin with the situation of *equal tails* as displayed in Figure 18.2. That is, we consider the values that the random variable

$$\sum_{i=1}^{n}(X_i-\mu)^2/\sigma^2$$

either falls short of or exceeds with probability α/2 each. This yields

$$= P\left(\chi^2_{\alpha/2}(n) \leq \sum_{i=1}^{n}\frac{(X_i-\mu)^2}{\sigma^2} \leq \chi^2_{1-\alpha/2}(n)\right)$$

$$= P\left(\frac{1}{\chi^2_{\alpha/2}(n)} \geq \frac{\sigma^2}{\sum_{i=1}^{n}(X_i-\mu)^2} \geq \frac{1}{\chi^2_{1-\alpha/2}(n)}\right)$$

$$= P\left(\frac{\sum_{i=1}^{n}(X_i-\mu)^2}{\chi^2_{\alpha/2}(n)} \geq \sigma^2 \geq \frac{\sum_{i=1}^{n}(X_i-\mu)^2}{\chi^2_{1-\alpha/2}(n)}\right)$$
$$= 1-\alpha$$

where in the second equation we simply inverted the inequalities. Thus, we obtain as $1-\alpha$ confidence interval for the variance parameter σ^2 of a normal random variable

$$I_{1-\alpha} = \left[\frac{\sum_{i=1}^{n}(X_i-\mu)^2}{\chi^2_{1-\alpha/2}(n)}, \frac{\sum_{i=1}^{n}(X_i-\mu)^2}{\chi^2_{\alpha/2}(n)}\right] \qquad (18.6)$$

With our normally distributed daily GE stock returns from the previous examples, the random variable

$$\sum_{i=1}^{n}(X_i-\mu)^2/\sigma^2$$

is chi-square distributed with 7,300 degrees of freedom. Moreover,

$$\sum_{i=1}^{n}(x_i-\mu)^2 = 2.2634$$

Setting $1-\alpha = 0.95$, we have $\chi^2_{0.025}(7300) = 7065.1$ and $\chi^2_{0.975}(7300) = 7538.7$, and consequently from equation (18.6), we obtain the 0.95 confidence interval for σ^2 by

CI4: $I_{0.95} = [0.00030, 0.00032]$

So, given the mean parameter $\mu = 0.0003$, for any variance parameter between 0.00030 and 0.00032 for the daily GE stock returns, the observed

$$\sum_{i=1}^{n}(x_i-\mu)^2 = 2.2634$$

is a plausible result at the $1-\alpha$ confidence level. Note that the value of $\sigma^2 = 0.0003$—or, using five digits, $\sigma^2 = 0.00031$—we assumed known in previous illustrations, is included by this interval.[4]

---

[4]Note that the interval bounds are rounded. The exact interval is slightly larger.

CONFIDENCE INTERVAL FOR THE VARIANCE OF A NORMAL RANDOM VARIABLE WITH UNKNOWN MEAN

For our next illustration, let us again consider the construction of a confidence interval for the variance parameter σ^2 of a normal random variable, but this time we assume that the mean parameter μ is not known. We can use the same data as before. However, instead of equation (18.5), we use

$$\sum_{i=1}^{n} \frac{(X_i - \bar{X})^2}{\sigma^2} \sim \chi^2(n-1) \qquad (18.7)$$

The random variable given by equation (18.7) has one degree of freedom less due to the computation of the sample mean $\bar{X}$ as estimator for μ. Besides that, however, this standardized random variable resembles the one used for the case of known μ given by equation (18.5). Therefore, we use the same construction for the confidence interval for σ^2 as when μ is known. The $\alpha/2$- and $1 - \alpha/2$- quantiles, of course, have to be taken from the chi-square distribution with $n - 1$ degrees of freedom. Then, the $1 - \alpha/2$ confidence interval becomes

$$I_{1-\alpha} = \left[\frac{\sum_{i=1}^{n}(X_i - \bar{X})^2}{\chi^2_{1-\alpha/2}(n-1)}, \frac{\sum_{i=1}^{n}(X_i - \bar{X})^2}{\chi^2_{\alpha/2}(n-1)} \right] \qquad (18.8)$$

Suppose once again we are interested in a confidence interval for the variance parameter σ^2 of the normally distributed daily GE stock returns. Let us again select as the confidence level $1 - \alpha = 0.95$ such that $\alpha/2 = 0.025$. The corresponding quantiles are given by

$$\chi^2_{0.025}(7,299) = 7,064.1 \text{ and } \chi^2_{0.975}(7,299) = 7,537.7$$

respectively. With

$$\sum_{i=1}^{7,300}(x_i - \bar{x})^2 = 2.2634$$

we obtain from equation (18.8) the 0.95 confidence interval

CI5: $I_{0.95} = [0.00030, 0.00032]$

which equals that given by CI4, at least when rounded to five digits. This is to be expected since the $\chi^2(n - 1)$ and $\chi^2(n)$ distributions are very similar.[5]

---

[5]As a matter of fact, for these large degrees of freedom, the distributions are very close to a normal distribution.

TABLE 18.1 Confidence Intervals for the Parameters of a Normal Distribution

---

H2 $1 - \alpha$ confidence interval for μ when σ^2 is known

$$I_{1-\alpha} = \left[\bar{X} - \frac{\sigma}{\sqrt{n}} q_{1-\alpha/2}, \bar{X} + \frac{\sigma}{\sqrt{n}} q_{1-\alpha/2} \right]$$

H2 $1 - \alpha$ confidence interval for μ when is σ^2 unknown

$$I_{1-\alpha} = \left[\bar{X} - \frac{s^*}{\sqrt{n}} t_{1-\alpha/2}(n-1), \bar{X} + \frac{s^*}{\sqrt{n}} t_{1-\alpha/2}(n-1) \right]$$

with $s^{*2} = \frac{1}{n-1} \sum_{i=1}^{n} (X_i - \bar{X})^2$

H2 $1 - \alpha$ confidence interval for σ^2 when μ is known

$$I_{1-\alpha} = \left[\frac{\sum_{i=1}^{n}(X_i - \mu)^2}{\chi^2_{1-\alpha/2}(n)}, \frac{\sum_{i=1}^{n}(X_i - \mu)^2}{\chi^2_{\alpha/2}(n)} \right]$$

H2 $1 - \alpha$ confidence interval for σ^2 when μ is unknown

$$I_{1-\alpha} = \left[\frac{\sum_{i=1}^{n}(X_i - \bar{X})^2}{\chi^2_{1-\alpha/2}(n-1)}, \frac{\sum_{i=1}^{n}(X_i - \bar{X})^2}{\chi^2_{\alpha/2}(n-1)} \right]$$

Moreover, we know that $\bar{X}$ is a consistent estimator for μ and, for 7,300 observations, most likely very close to μ. Thus, the two confidence intervals are virtually constructed from the same data.

A summary of the confidence intervals for the parameters of a normal distribution is given by Table 18.1.

CONFIDENCE INTERVAL FOR THE PARAMETER *P* OF A BINOMIAL DISTRIBUTION

In our next illustration, we revisit the binomial stock price model we introduced in Chapters 8 and 9. Recall that, in period $t + 1$, the stock price S_t either goes up to S_{t+1}^u or down to S_{t+1}^d. The Bernoulli random variable $Y \sim B(p)$ expresses whether an up-movement, that is, $Y = 1$, or a down-movement, that is, $Y = 0$, has occurred.[6]

---

[6]The Bernoulli distribution is introduced in Chapter 9.

Suppose the parameter p that expresses the probability of an up-movement is unknown. Therefore, we draw a sample $(X_1, X_2, ..., X_n)$ of n independent Bernoulli random variables with parameter p. So if the stock price moves up in period i we have $X_i = 1$, but if the stock price moves down in the same period then, $X_i = 0$. As an estimator for p, we have the unbiased sample mean $\bar{X}$.

Since $X_i \sim B(p)$, the mean and variance are equal to p and $p \cdot (1-p)$, respectively. Furthermore, the sufficient statistic

$$\sum_{i=1}^{n} X_i$$

is $B(n,p)$ distributed such that it has mean $n \cdot p$ and variance $n \cdot p \cdot (1-p)$.[7] Then, the mean and variance of

$$\bar{X} = 1/n \cdot \sum_{i=1}^{n} X_i$$

are p and

$$\frac{p \cdot (1-p)}{n}$$

respectively.

For a sample size n sufficiently large, by the Central Limit Theorem, the sample mean $\bar{X}$ is approximately normally distributed; in particular,

$$\bar{X} \sim N\left(p, \frac{p \cdot (1-p)}{n}\right)$$

Since the parameter p is unknown, so is the variance and therefore we instead use the estimator

$$\hat{\sigma}^2 = \frac{\bar{X} \cdot (1-\bar{X})}{n}$$

where $\bar{X}$ is used as an estimator for p.[8] The random variable

$$\frac{p - \bar{X}}{\sqrt{\frac{\bar{X} \cdot (1-\bar{X})}{n}}}$$

---

[7]The binomial distribution is introduced in Chapter 9.
[8]This estimator is biased, however, for finite n, in contrast to the bias-corrected sample variance used previously.

that we will use is approximately standard normally distributed. Analogous to equation (18.1), the appropriate $(1-\alpha)$-confidence interval for the mean parameter p is given by

$$\left[\bar{X} - q_{1-\alpha/2}\sqrt{\frac{\bar{X}\cdot(1-\bar{X})}{n}}, \bar{X} + q_{1-\alpha/2}\sqrt{\frac{\bar{X}\cdot(1-\bar{X})}{n}}\right] \qquad (18.9)$$

Note that we do not use the confidence interval given by equation (18.4) for the mean parameter of normal random variables when the variance is unknown. The reason is that even for large sample sizes, $\bar{X}$ is only approximately normally distributed while, theoretically, equation (18.4) is designed for random variables following the normal law exactly. Ignoring this would lead to additional imprecision.

Now, suppose we had observed $x = (1, 1, 1, 0, 0, 1, 0, 0, 0, 1)$ such that $n = 10$. Then, $\bar{x} = 0.5$ and

$$\sqrt{\bar{x}\cdot(1-\bar{x})/n} = 0.1581$$

Consequently, from equation (18.9), we obtain the 0.95 confidence interval

$$\text{CI6: } I_{0.95} = [0.1901, 0.8099]$$

As a consequence, any value between 0.1901 and 0.8099 provides a plausible population parameter p to generate the sample x at the 0.95 confidence level.

CONFIDENCE INTERVAL FOR THE PARAMETER λ OF AN EXPONENTIAL DISTRIBUTION

Now consider an exponential random variable Y with parameter λ, that is, $Y \sim \text{Exp}(\lambda)$.[9] For the construction of a confidence interval for parameter λ, it will be helpful to look for a standardized random variable as we have done in the previous illustrations. Multiplying the random variable Y by the parameter λ yields a standard exponential random variable

$$\lambda \cdot Y \sim \text{Exp}(1)$$

This is true since

$$P(\lambda \cdot Y \leq t) = P\left(Y \leq \frac{t}{\lambda}\right) = 1 - e^{-\lambda\frac{t}{\lambda}} = 1 - e^{-t}$$

[9] This distribution is introduced in Chapter 11.

Next, we obtain a sample $X = (X_1, X_2, \ldots, X_n)$ of n independent and identical drawings from an $\text{Exp}(\lambda)$ distribution. Then, the $\lambda \cdot X_i$ are n independent $\text{Exp}(1)$ random variables and, thus,

$$\sum_{i=1}^{n} \lambda \cdot X_i \sim Erl(n, \lambda)$$

That is, by summation of the n independent standard exponential random variables, we have created an Erlang random variable with parameter $(1,n)$.[10] For the construction of the confidence interval, we have to bear in mind that the Erlang distribution (1) has positive density only on the nonnegative real numbers and (2) is skewed to the right. So, by convention, we design the confidence interval such that the probability of error is divided equally between the probability of the interval being below the parameter as well as the probability of the interval exceeding it. That is, we created the probability of error from equal tails.[11] Therefore, from

$$1 - \alpha = P\left(e_{\alpha/2}(n) \leq \lambda \cdot \sum_{i=1}^{n} X_i \leq e_{1-\alpha/2}(n)\right)$$

$$= P\left(\frac{e_{\alpha/2}(n)}{\sum_{i=1}^{n} X_i} \leq \lambda \leq \frac{e_{1-\alpha/2}(n)}{\sum_{i=1}^{n} X_i}\right)$$

we obtain as $1 - \alpha$ confidence interval

$$I_{1-\alpha} = \left[\frac{e_{\alpha/2}(n)}{\sum_{i=1}^{n} X_i}, \frac{e_{1-\alpha/2}(n)}{\sum_{i=1}^{n} X_i}\right] \tag{18.10}$$

where

$$e_{\alpha/2}(n) \text{ and } e_{1-\alpha/2}(n)$$

are the $\alpha/2$ and $1 - \alpha/2$–quantiles of the Erlang distribution, respectively.

---

[10] The Erlang distribution is introduced in Chapter 11.
[11] Note that this design is not a natural consequence of minimization of the width of the confidence interval given a specific confidence level.

Confidence Intervals

Suppose we are interested in the interarrival times between defaults in the bond portfolio we analyzed in Chapter 9. The (corrected) sample is reproduced below where we have displayed the average interarrival times (in years) between two successive defaults for the last 20 years:

x_1	x_2	x_3	x_4	x_5	x_6	x_7	x_8	x_9	x_{10}
0.500	0.333	0.083	0.200	0.500	0.250	0.143	1	0.318	0.333
x_{11}	x_{12}	x_{13}	x_{14}	x_{15}	x_{16}	x_{17}	x_{18}	x_{19}	x_{20}
0.200	0.167	0.500	0.250	0.500	1	0.318	1	0.500	1

If we let $1 - \alpha = 0.95$, then we have $e_{0.025}(20) = 0.5064$ and

$$e_{1-\alpha/2}(n) = 73.7776$$

Moreover, we compute

$$\sum_{i=1}^{20} X_i = 9.0950$$

Using equation (18.10), we obtain as the 0.95 confidence interval for λ

$$\text{CI7: } I_{0.95} = [0.0557, 8.1119]$$

So, for any parameter value λ between 0.0557 and 8.1119, the outcome 9.0950 of our statistic

$$\sum_{i=1}^{20} X_i$$

is not too extreme at the 0.95 confidence level. Notice that the maximum likelihood estimate $\hat{\lambda}_{MLE} = 2.1991$ from Chapter 17 lies inside the interval.

CONCEPTS EXPLAINED IN THIS CHAPTER (IN ORDER OF PRESENTATION)

Confidence level
Confidence interval
Equal tails

CHAPTER 19
Hypothesis Testing

Thus far in this book, inference on some unknown parameter meant that we had no knowledge of its value and therefore we had to obtain an estimate. This could either be a single point estimate or an entire confidence interval. However, sometimes, one already has some idea of the value a parameter might have or used to have. Thus, it might not be important for a portfolio manager, risk manager, or financial manager to obtain a particular single value or range of values for the parameter, but instead gain sufficient information to conclude that the parameter more likely either belongs to a particular part of the parameter space or not. So, instead we need to obtain information to verify whether some assumption concerning the parameter can be supported or has to be rejected.

This brings us to the field of *hypothesis testing*. Next to parameter estimation that we covered in Chapters 17 and 18, it constitutes the other important part of statistical inference; that is, the procedure for gaining information about some parameter. To see its importance, consider, for example, a portfolio manager who might have in mind a historical value such as the expected value of the daily return of the portfolio under management and seeks to verify whether the expected value can be supported. It might be that if the parameter belongs to a particular set of values, the portfolio manager incurs extensive losses.

In this chapter, we learn how to perform hypothesis testing. To do this, it is essential to express the competing statements about the value of a parameter as hypotheses. To test for these, we develop a *test statistic* for which we set up a *decision rule*. For a specific sample, this test statistic then either assumes a value in the acceptance region or the rejection region, regions that we describe in this chapter. Furthermore, we see the two error types one can incur when testing. We see that the hypothesis test structure allows one to control the probability of error through what we see to be the test size or significance level. We discover that each observation has a certain p-value expressing its significance. As a quality criterion of a test, we introduce the power from which the uniformly most powerful test can be defined. Furthermore, we learn what is meant by an unbiased test—unbiasedness provides another important quality criterion—as well as whether a test is consistent. We conclude with a set of examples intended to illustrate the issues discussed in the chapter.

HYPOTHESES

Before being able to test anything, we need to express clearly what we intend to achieve with the help of the test. For this task, it is essential that we unambiguously formulate the possible outcomes of the test. In the realm of hypothesis testing, we have two competing statements to decide upon. These statements are the *hypotheses* of the test.

Setting Up the Hypotheses

Since in statistical inference we intend to gain information about some unknown parameter θ, the possible results of the test should refer to the parameter space Θ containing all possible values that θ can assume. More precisely, to form the hypotheses, we divide the parameter space into two disjoint sets Θ_0 and Θ_1 such that $\Theta = \Theta_0 \cup \Theta_1$. We assume that the unknown parameter is either in Θ_0 or Θ_1 since it cannot simultaneously be in both. Usually, the two alternative parameter sets either divide the parameter space into two disjoint intervals or regions (depending on the dimensionality of the parameter), or they contrast a single value with any other value from the parameter space.

Now, with each of the two subsets Θ_0 and Θ_1, we associate a hypothesis. In the following two definitions, we present the most commonly applied denominations for the hypotheses.

> *Null hypothesis:* The **null hypothesis**, denoted by H_0, states that the parameter θ is in Θ_0.

The null hypothesis may be interpreted as the assumption to be maintained if we do not find ample evidence against it.

> *Alternative hypothesis:* The **alternative hypothesis**, denoted by H_1, is the statement that the parameter θ is in Θ_1.

We have to be aware that only one hypothesis can hold and, hence, the outcome of our test should only support one. We will return to this later in the chapter when we discuss the test in more detail.

FIGURE 19.1 Null Hypothesis H_0: $\lambda \leq 1$ versus Alternative Hypothesis H_1: $\lambda > 1$ for Parameter λ of the Exponential Distribution

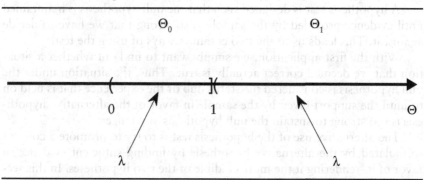

When we test for a parameter or a single parameter component, we usually encounter the following two constructions of hypothesis tests.

In the first construction, we split the parameter space Θ into a lower half up to some boundary value $\tilde{\theta}$ and an upper half extending beyond this boundary value. Then, we set the lower half either equal to Θ_0 or Θ_1. Consequently, the upper half becomes the counterpart Θ_1 or Θ_0, respectively. The boundary value $\tilde{\theta}$ is usually added to Θ_0; that is, it is the set valid under the null hypothesis. Such a test is referred to as a *one-tailed test*.

In the second construction, we test whether some parameter is equal to a particular value or not. Accordingly, the parameter space is once again divided into two sets Θ_0 and Θ_1. But this time, Θ_0 consists of only one value (i.e., $\Theta_0 = \tilde{\theta}$) while the set Θ_1, corresponding to the alternative hypothesis, is equal to the parameter space less the value belonging to the null hypothesis (i.e., $\Theta_1 = \Theta \setminus \tilde{\theta}$). This version of a hypothesis test is termed a *two-tailed test*.

At this point, let's consider as an example the exponential distribution with parameter λ. We know that the parameter space is the set of positive real numbers (i.e., $\Theta = (0, \infty)$). Suppose we were interested in whether λ is, at most, 1 or greater. So, our corresponding sets should be $\Theta_0 = (0, 1]$ and $\Theta_1 = (1, \infty)$ with associated hypotheses H_0: $\lambda \leq 1$ and H_1: $\lambda > 1$, respectively.[1] This is an example of a one-tailed test. We demonstrate this in Figure 19.1.

Decision Rule

The task of hypothesis testing is to make a decision about these hypotheses. So, we either cannot reject the null hypothesis and, consequently, have to

---

[1] Note that λ has to be greater than zero since we test for the parameter of the exponential distribution.

reject the alternative hypothesis, or we reject the null hypothesis and decide in favor of the alternative hypothesis.

A hypothesis test is designed such that the null hypothesis is maintained until evidence provided by the sample is so strong that we have to decide against it. This leads us to the two common ways of using the test.

With the first application, we simply want to find out whether a situation that we deemed correct actually is true. Thus, the situation under the null hypothesis is considered the status quo or the experience that is held on to until the support given by the sample in favor of the alternative hypothesis is too strong to sustain the null hypothesis any longer.

The alternative use of the hypothesis test is to try to promote a concept formulated by the alternative hypothesis by finding sufficient evidence in favor of it, rendering it the more credible of the two hypotheses. In this second approach, the aim of the tester is to reject the null hypothesis because the situation under the alternative hypothesis is more favorable.

In the realm of hypothesis testing, the decision is generally regarded as the process of following certain rules. We denote our decision rule by δ. The decision δ is designed to either assume value d_0 or value d_1. Depending on which way we are using the test, the meaning of these two values is as follows. In the first case, the value d_0 expresses that we hold on to H_0 while the contrarian value d_1 expresses that we reject H_0. In the second case, we interpret d_0 as being undecided with respect to H_0 and H_1 and that proof is not strong enough in favor of H_1. On the other hand, by d_1, we indicate that we reject H_0 in favor of H_1.

In general, d_1 can be interpreted as the result we obtain from the decision rule when the sample outcome is highly unreasonable under the null hypothesis.

So, what makes us come up with either d_0 or d_1? As in the previous chapters, we infer by first drawing a sample of some size n, $X = (X_1, X_2, ..., X_n)$. Our decision then should be based on this sample. That is, it would be wise to include in our decision rule the sample X such that the decision becomes a function of the sample, (i.e., $\delta(X)$). Then, $\delta(X)$ is a random variable due to the randomness of X. A reasonable step would be to link our test $\delta(X)$ to a statistic, denoted by $t(X)$, that itself is related or equal to an estimator $\hat{\theta}$ for the parameter of interest θ. Such estimators have been introduced in Chapter 17.

From now on, we will assume that our test rule δ is synonymous with checking whether the statistic $t(X)$ is assuming certain values or not from which we derive decision d_0 or d_1.

Acceptance and Rejection Region

As we know from Chapter 17, by drawing a sample X, we select a particular value x from the sample space $\mathbf{X}$.[2] Depending on this realization x, the statistic $t(x)$ either leads to rejection of the null hypothesis (i.e., $\delta(x) = d_0$), or not (i.e., $\delta(x) = d_1$).

To determine when we have to make a decision d_0 or, alternatively, d_1, we split the state space Δ of $t(X)$ into two disjoint sets that we denote by Δ_A and Δ_C.[3] The set Δ_A is referred to as the *acceptance region* while Δ_C is the *critical region* or *rejection region*.

When the outcome of the sample x is in Δ_A, we do not reject the null hypothesis (i.e., the result of the test is $\delta(x) = d_0$). If, on the other hand, x should be some value in Δ_C, the result of the test is now the contrary (i.e., $\delta(x) = d_1$), such that we decide in favor of the alternative hypothesis.

ERROR TYPES

We have to be aware that no matter how we design our test, we are at risk of committing an error by making the wrong decision. Given the two hypotheses, H_0 and H_1, and the two possible decisions, d_0 and d_1, we can commit two possible errors. These errors are discussed next.

Type I and Type II Error

The two possible errors we can incur are characterized by unintentionally deciding against the true hypothesis. Each error related to a particular hypothesis is referred to using the following standard terminology.

Type I error: The error resulting from rejection of the null hypothesis (H_0) (i.e., $\delta(X) = d_1$) given that it is actually true (i.e., $\theta \in \Theta_0$) is referred to as a *type I error*.

Type II error: The error resulting from not rejecting the null hypothesis (H_0) (i.e., $\delta(X) = d_0$) even though the alternative hypothesis (H_1) holds (i.e., $\theta \in \Theta_1$) is referred to as a *type II error*.

In the following table, we show all four possible outcomes from a hypothesis test depending on the respective hypothesis:

---

[2]This was introduced in Chapter 17.
[3]Recall from Chapter 8 that the state space is the particular set a random variable assumes values in.

		H_0: θ in Θ_0	H_1: θ in Θ_1
Decision	d_0	Correct	Type II error
	d_1	Type I error	Correct

So, we see that in two cases, we make the correct decision. The first case occurs if we do not reject the null hypothesis (i.e., $\delta(X) = d_0$) when it actually holds. The second case occurs if we correctly decide against the null hypothesis (i.e., $\delta(X) = d_1$), when it is not true and, consequently, the alternative hypothesis holds. Unfortunately, however, we do not know whether we commit an error or not when we are testing. We do have some control, though, as to the probability of error given a certain hypothesis as we explain next.

Test Size

We just learned that depending on which hypothesis is true, we can commit either a type I or a type II error. Now, we will concentrate on the corresponding probabilities of incurring these errors.

> *Test size*: The test size is the probability of committing a type I error. This probability is denoted by $P_I(\delta)$ for test δ.[4]

We illustrate this in Figure 19.2, where we display the density function $f_{(t(X),\Theta_0)}$ of the test statistic $t(X)$ under the null hypothesis. The horizontal axis along which $t(X)$ assumes values is subdivided into the acceptance Δ_A and the critical region Δ_C. The probability of this statistic having a value in the critical region is indicated by the shaded area.

Since, in general, the set Θ_0 belonging to the null hypothesis consists of several values (e.g., $\Theta_0 \subset \Theta$) the probability of committing a type I error, $P_I(\delta)$, may vary for each parameter value θ in Θ_0. By convention, we set the test size equal to the $P_I(\delta)$ computed at that value θ in Θ_0 for which this probability is maximal.[5] We illustrate this for some arbitrary test in Figure 19.3, where we depict the graph of the probability of rejection of the null hypothesis depending on the parameter value θ. Over the set Θ_0, as indicated by the solid line in the figure, this is equal to the probability of a type I error while, over Θ_1, this is the probability of a correct decision (i.e., d_1). The latter is given by the dashed line.

---

[4] The probability $P_I(\delta)$ could alternatively be written as $P_{\Theta_0}(d_1)$ to indicate that we erroneously reject the null hypothesis even though H_0 holds.
[5] Theoretically, this may not be possible for any test.

FIGURE 19.2 Determining the Size $P_I(\delta)$ of Some Test δ via the Density Function of the Test Statistic $t(X)$

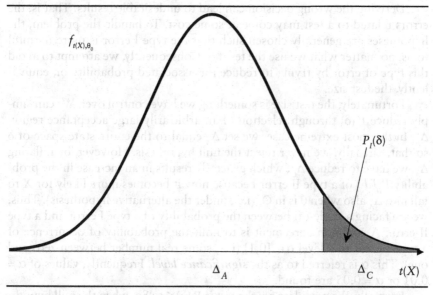

FIGURE 19.3 Determining the Test Size α by Maximizing the Probability of a Type I Error over the Set Θ_0 of Possible Parameter Values under the Null Hypothesis

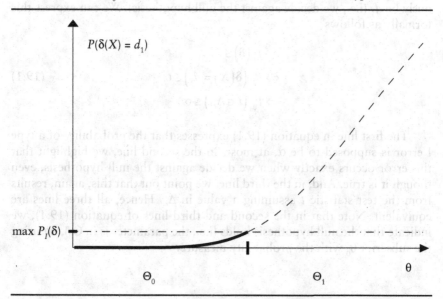

Analogously to the probability $P_I(\delta)$, we denote the probability of committing a type II error as $P_{II}(\delta)$.

Deriving the wrong decision can lead to undesirable results. That is, the errors related to a test may come at some cost. To handle the problem, the hypotheses are generally chosen such that the type I error is more harmful to us, no matter what we use the test for. Consequently, we attempt to avoid this type of error by trying to reduce the associated probability or, equivalently, the test size.

Fortunately, the test size is something we have control over. We can simply reduce $P_I(\delta)$ through selection of an arbitrarily large acceptance region Δ_A. In the most extreme case, we set Δ_A equal to the entire state space of δ so that, virtually, we never reject the null hypothesis. However, by inflating Δ_A, we have to reduce Δ_C, which generally results in an increase in the probability $P_{II}(d_0)$ of a type II error because now it becomes more likely for X to fall into Δ_A also when θ is in Θ_1 (i.e., under the alternative hypothesis). Thus, we are facing a trade-off between the probability of a type I error and a type II error. A common agreement is to limit the probability of occurrence of a type I error to a level $\alpha \in [0,1]$ (i.e., some real number between zero and one). This α is referred to as the *significance level*. Frequently, values of $\alpha = 0.01$ or $\alpha = 0.05$ are found.

Formally, the postulate for the test is $P_I(\delta) \leq \alpha$. So, when the null hypothesis is true, in at most α of all outcomes, we will obtain a sample value x in Δ_C. Consequently, in at most α of the test runs, the test result will erroneously be d_1 (i.e., we decide against the null hypothesis). We can express this formally as follows

$$P_I(\delta) \leq \alpha$$
$$\Leftrightarrow P_{\theta_0}(\delta(X) = d_1) \leq \alpha \quad (19.1)$$
$$\Leftrightarrow P_{\theta_0}(t \in \Delta_C) \leq \alpha$$

The first line in equation (19.1) expresses that the probability of a type I error is supposed to be α, at most. In the second line, we highlight that this error occurs exactly when we decide against the null hypothesis, even though it is true. And, in the third line, we point out that this, again, results from the test statistic t assuming a value in Δ_C. Hence, all three lines are equivalent. Note that in the second and third lines of equation (19.1), we indicate that the null hypothesis holds (i.e., the parameter is in Θ_0), by using the subscript θ_0 with the probability measure.

The p-Value

Suppose we had drawn some sample x and computed the value $t(x)$ of the statistic from it. It might be of interest to find out how significant this test result is or, in other words, at which significance level this value $t(x)$ would still lead to decision d_0 (i.e., no rejection of the null hypothesis), while any value greater than $t(x)$ would result in its rejection (i.e., d_1). This concept brings us to the next definition.

p-value: Suppose we have a sample realization given by $x = (x_1, x_2, ..., x_n)$. Furthermore, let $\delta(X)$ be any test with test statistic $t(X)$ such that the test statistic evaluated at x, $t(x)$, is the value of the acceptance region Δ_A closest to the rejection region Δ_C. The *p-value* determines the probability, under the null hypothesis, that in any trial X the test statistic $t(X)$ assumes a value in the rejection region Δ_C; that is,

$$p = P_{\theta_0}\left(t(X) \in \Delta_C\right) = P_{\theta_0}\left(\delta(X) = d_1\right)$$

We can interpret the *p*-value as follows. Suppose we obtained a sample outcome x such that the test statistics assumed the corresponding value $t(x)$. Now, we want to know what is the probability, given that the null hypothesis holds, that the test statistic might become even more extreme than $t(x)$. This probability is equal to the *p*-value.

If $t(x)$ is a value pretty close to the median of the distribution of $t(X)$, then the chance of obtaining a more extreme value, which refutes the null hypothesis more strongly might be fairly feasible. Then, the *p*-value will be large. However, if, instead, the value $t(x)$ is so extreme that the chances will be minimal under the null hypothesis that, in some other test run we obtain a value $t(X)$ even more in favor of the alternative hypothesis, this will correspond to a very low *p*-value. If p is less than some given significance level α, we reject the null hypothesis and we say that the test result is *significant*.

We demonstrate the meaning of the *p*-value in Figure 19.4. The horizontal axis provides the state space of possible values for the statistic $t(X)$. The figure displays the probability, given that the null hypothesis holds, of this $t(X)$ assuming a value greater than c, for each c of the state space, and in particular also at $t(x)$ (i.e., the statistic evaluated at the observation x). We can see that, by definition, the value $t(x)$ is the boundary between the acceptance region and the critical region, with $t(x)$ itself belonging to the acceptance region. In that particular instance, we happened to choose a test with $\Delta_A = (-\infty, t(x)]$ and $\Delta_C = (t(x), \infty)$.

FIGURE 19.4 Illustration of the p-Value for Some Test δ with Acceptance Region Δ_A = $(-\infty, t(x))$ and Critical Region $\Delta_C = (t(x), \infty)$

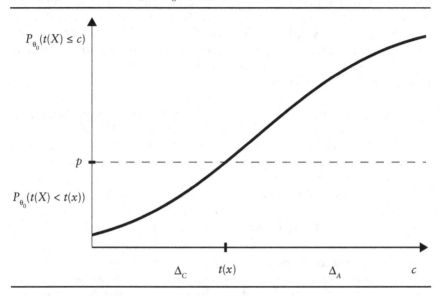

QUALITY CRITERIA OF A TEST

So far, we have learned how to construct a test for a given problem. In general, we formulate the two competing hypotheses and look for an appropriate test statistic to base our decision rule on and we are then done. However, in general, there is no unique test for any given pair of hypotheses. That is, we may find tests that are more suitable than others for our endeavor. How can we define what we mean by "suitable"? To answer this question, we will discuss the following quality criteria.

Power of a Test

Previously, we were introduced to the size of a test that may be equal to α. As we know, this value controls the probability of committing a type I error. So far, however, we may have several tests meeting a required test size α. The criterion selecting the most suitable ones among them involves the type II error. Recall that the type II error describes the failure of rejection of the null hypothesis when it actually is wrong. So, for parameter values $\theta \in \Theta_1$, our test should produce decision d_1 with as high a probability as possible in order to yield as small as possible a probability of a type II error, $P_{II}(d_0)$. In

Hypothesis Testing

FIGURE 19.5 The Solid Line of the Probability $P(\delta(X) = d_1)$, over the Set Θ_1, Indicates the Power of the Test δ

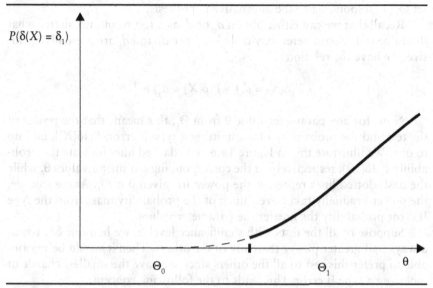

the following definition, we present a criterion that accounts for this ability of a test.

> *Power of a test* The **power of a test** is the probability of rejecting the null hypothesis when it is actually wrong (i.e., when the alternative hypothesis holds). Formally, this is written as $P_{\theta_1}(\delta(X) = d_1)$.[6]

For illustrational purposes, we focus on Figure 19.5 where we depict the parameter-dependent probability $P(\delta(X) = d_1)$ of some arbitrary test δ, over the parameter space Θ. The solid part of the figure, computed over the set Θ_1, represents the power of the test. As we can see, here, the power increases for parameter values further away from Θ_0 (i.e., increasing θ). If the power were rising more steeply, the test would become more powerful. This brings us to the next concept.

Uniformly Most Powerful Test

In the following, let us only consider tests of size α. That is, none of these tests incurs a type I error with greater probability than α. For each of these

---

[6]The index θ_1 of the probability measure P indicates that the alternative hypothesis holds (i.e., the true parameter is a *value* in Θ_1).

tests, we determine the respective power function (i.e., the probability of rejecting the null hypothesis, $P(\delta(X) = d_1)$, computed for all values θ in the set Θ_1 corresponding to the alternative hypothesis.

Recall that we can either obtain d_0 or d_1 as a test result, no matter what the value of the parameter may truly be. Since d_0 and d_1 are mutually exclusive, we have the relation

$$P(\delta(X) = d_0) + P(\delta(X) = d_1) = 1$$

Now, for any parameter value θ from Θ_1, this means that the power of the test and the probability of committing a type II error, $P_{II}(\delta(X))$, add up to one. We illustrate this in Figure 19.6. The dashed lines indicate the probability $P_{II}(\delta(X))$, respectively, at the corresponding parameter values θ, while the dash-dotted lines represent the power for given $\theta \in \Theta_1$. As we can see, the power gradually takes over much of the probability mass from the type II error probability the greater the parameter values.

Suppose of all the tests with significance level α, we had one δ^*, which always had greater power than any of the others. Then it would be reasonable to prefer this test to all the others since we have the smallest chance of incurring a type II error. This leads to the following concept.

FIGURE 19.6 Decomposition $1 = P_{II}(\delta) + P_{\theta_1}(\delta(X) = d_1)$, over Θ_1

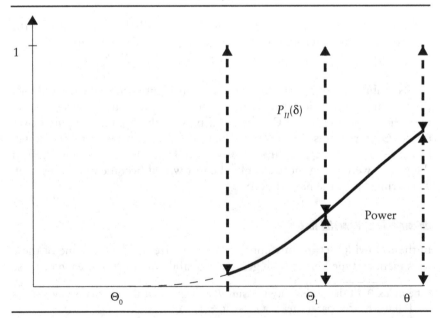

Hypothesis Testing

Uniformly most powerful (UMP) *test of size* α: A test δ^* of size α is **uniformly most powerful**, if of all the tests δ of size α, this test δ^* has greatest power for any $\theta \in \Theta_1$.

Formally, this is

$$P_{II}(\delta^*(X)) \leq P_{II}(\delta(X)), \quad \theta \in \Theta_1$$

of all tests with

$$P_I(\delta) \leq \alpha$$

We illustrate this definition in Figure 19.7 for some arbitrary tests of size α. As we can see, the test δ^* has power always greater than the alternatives δ_1 and δ_2. Consequently, of all the choices, δ^* is the most powerful test in this example.

FIGURE 19.7 Uniformly Most Powerful Test δ^* in Comparison to the Tests of Size α, δ_1 and δ_2

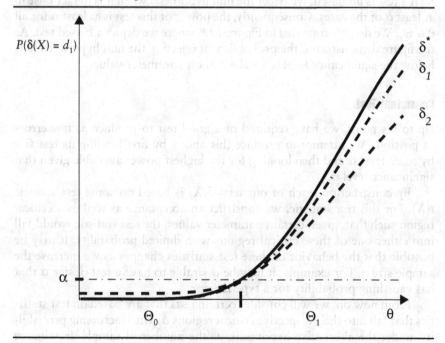

Unbiased Test

We know that when a test is of size α, the probability of it causing a type I error is never greater than α. And when the design of the test is reasonable, the power of the test should increase quickly once we are considering parameter values in Θ_1. In both cases (i.e., when we compute the probability of a type I error for $\theta \in \Theta_0$, as well as when we compute the power, for $\theta \in \Theta_1$), we are dealing with the probability to reject the null hypothesis (i.e., $P(\delta(X) = d_1)$). In case $P(\delta(X) = d_1)$ should be smaller than α, then for certain parameter values $\theta \in \Theta_1$ it is more likely to accept the null hypothesis when it is wrong than when it holds. This certainly does not appear useful and we should try to avoid it when designing our test. This concept is treated in the following definition.

Unbiased test: A test of size α is unbiased if the probability of a type II error is never greater than $1 - \alpha$; formally,

$$P_{II}(\delta(X)) \leq 1 - \alpha \text{ for any } \theta \in \Theta_1$$

So if a test is unbiased, we reject the null hypothesis when it is in fact false in at least α of the cases. Consequently, the power of this test is at least α for all $\theta \in \Theta_1$. We demonstrate this in Figure 19.8 where we depict a biased test. As the figure demonstrates, the probability of rejecting the null hypothesis falls below the significance level α for the highest parameter values.

Consistent Test

Up to his point, we have required of a good test to produce as few errors as possible. We attempt to produce this ability by first limiting its test size by some level α and then looking for the highest power available given that significance level α.

By construction, each of our tests $\delta(X)$ is based on some test statistic $t(X)$. For this test statistic, we construct an acceptance as well as a critical region such that, given certain parameter values, the test statistic would fall into either one of these critical regions with limited probability. It may be possible that the behavior of these test statistics changes as we increase the sample size n. For example, it may be desirable to have a test of size α that has vanishing probability for a type II error.

From now on, we will consider certain tests that are based on test statistics that fall into their respective critical regions Δ_C with increasing probability, under the alternative hypothesis, as the number of sample drawings n tends to infinity. That is, these tests reject the null hypothesis more and more

FIGURE 19.8 Graph of the Probability to Reject the Null Hypothesis for Some Biased Test δ

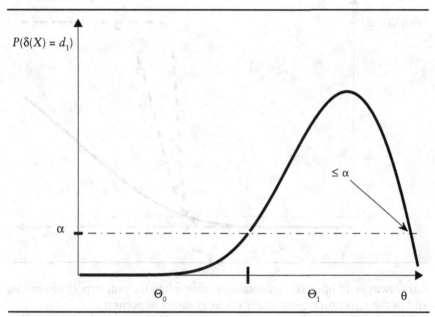

reliably when they actually should (i.e., $\theta \in \Theta_1$) for ever larger samples. In the optimal situation, these tests reject the null hypothesis (i.e., $\delta(X) = d_1$) with 100% certainty when the alternative hypothesis holds. This brings us to the next definition.

Consistent test: A test of size α is consistent if its power grows to one for increasing sample size; that is,

$$P(\delta(X) = d_1) \xrightarrow{n \to \infty} 1, \text{ for all } \theta \in \Theta_1$$

Recall that in Chapter 17 we introduced the consistent estimator that had the positive feature that it varied about its expected value with vanishing probability. So, with increasing probability, it assumed values arbitrarily close to this expected value such that eventually it would become virtually indistinguishable from it. The use of such a statistic for the test leads to the following desirable characteristic: The test statistic will cease to assume values that are extreme under the respective hypothesis such that it will basi-

FIGURE 19.9 $P(\delta(X) = d_1)$ for a Consistent Test, for $n = 10$ (solid), $n = 100$ (dash-dotted), and $n = 1,000$ (dashed)

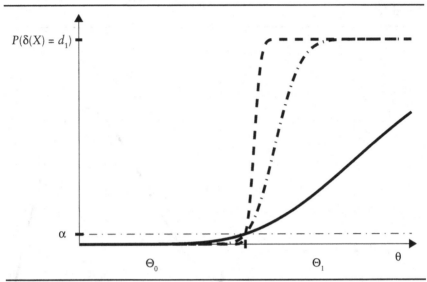

cally always end up in the acceptance region when the null hypothesis holds, and in the rejection region under the alternative hypothesis.

We demonstrate the consistency criterion in Figure 19.9 where we display the probability $P(\delta(X) = d_1)$ that equals the power over Θ_1 for three different sample sizes, $n = 10$, $n = 100$, and $n = 1,000$. For $n = 10$, the probability is depicted by the solid line, while for $n = 100$ and $n = 1,000$, the probabilities are displayed by the dash-dotted and the dashed line, respectively. The power can be seen to become greater and gradually close to 1 for increasing sample size.

EXAMPLES

In this concluding section of the chapter, we provide examples that apply the concepts introduced in the chapter.

Simple Test for Parameter λ of the Poisson Distribution

As our first example, let us consider a portfolio consisting of risky bonds where the number of defaulting bonds within one year is modeled as a Poisson random variable. In the past, the parameter λ has been assumed to be λ_0

= 1, which accounts for the null hypothesis H_0. Because of perceived higher risk, the bond portfolio manager may have reason to believe that the true parameter might instead be $\lambda_1 = 2$, representing the alternative hypothesis H_1.

The corresponding test is now designed as follows. Let us draw a sample of size n, $X = (X_1, X_2, ..., X_n)$. Next, we compute the probability of the observation $x = (x_1, x_2, ..., x_n)$, first using λ_0 and then λ_1. We indicate the respective probabilities as $P_{\lambda_0}(X = x)$ and $P_{\lambda_1}(X = x)$. Now, if λ_1 is the true parameter, then, in most drawings, we should obtain an outcome x that is more likely to occur under λ_1 than under λ_0. For these outcomes, it follows that the probability $P_{\lambda_1}(X = x)$ is greater than $P_{\lambda_0}(X = x)$. Consequently, the ratio

$$\rho(x, \lambda_0, \lambda_1) = \frac{P_{\lambda_1}(X = x)}{P_{\lambda_0}(X = x)} \quad (19.2)$$

should be large for these outcomes and only in a few outcomes, the ratio will be smaller. On the other hand, if λ_0 is the true parameter, then for most outcomes the ratio (19.2) will assume small values since these outcomes will be more likely under the null hypothesis than under the alternative hypothesis, resulting in a greater probability $P_{\lambda_0}(X = x)$ relative to $P_{\lambda_1}(X = x)$.

Depending on our desired significance level α, we will accept values for the ratio (19.2) less than or equal to some number k before we reject the null hypothesis. That is, we let the probability of the sample, evaluated at λ_1, be up to k times as large as when evaluated at λ_0. The benchmark k is chosen so that, under the null hypothesis (i.e., when λ_0 is true), $\rho(X, \lambda_0, \lambda_1)$ exceeds k with probability α at most. We see that the ratio itself is random.

Now let us insert the probability of the Poisson law into equation (19.2) considering independence of the individual drawings. Then for any sample X, the ratio turns into

$$\rho(X, \lambda_0, \lambda_1) = \frac{\lambda_1^{X_1} \cdot e^{-\lambda_1} \cdot \lambda_1^{X_2} \cdot e^{-\lambda_1} \cdot ... \cdot \lambda_1^{X_n} \cdot e^{-\lambda_1}}{\lambda_0^{X_1} \cdot e^{-\lambda_0} \cdot \lambda_0^{X_2} \cdot e^{-\lambda_0} \cdot ... \cdot \lambda_0^{X_n} \cdot e^{-\lambda_0}}$$

$$= \frac{e^{-n\lambda_1}}{e^{-n\lambda_0}} \left(\frac{\lambda_1}{\lambda_0}\right)^{\sum_{i=1}^{n} X_i} \quad (19.3)$$

where in the second line of equation (19.3) we have simply applied the rules for exponents when multiplying identical values. This ratio is required to be greater than k for the test to reject the null hypothesis. Instead of using the rather complicated form of equation (19.3) for the test, we can transform the second line using the natural logarithm to obtain the following equivalent test rule that may be simpler to compute

$$\sum_{i=1}^{n} X_i \leq \frac{\ln k + n(\lambda_1 - \lambda_0)}{\ln \lambda_1 - \ln \lambda_0}, \quad \lambda_1 > \lambda_0$$

$$\sum_{i=1}^{n} X_i \geq \frac{\ln k + n(\lambda_1 - \lambda_0)}{\ln \lambda_1 - \ln \lambda_0}, \quad \lambda_1 < \lambda_0$$

Note that here we have to pay attention to whether $\lambda_1 > \lambda_0$ or $\lambda_1 < \lambda_0$ since these two alternatives lead to opposite directions of the inequality.

Suppose we wish to have a test of size $\alpha = 0.05$. Then we have to find k such that the sum

$$\sum_{i=1}^{n} X_i$$

either exceeds the boundary (if $\lambda_1 > \lambda_0$) or falls below it (if $\lambda_1 < \lambda_0$) with probability of at most 0.05 when the null hypothesis holds. So, for $\lambda_1 > \lambda_0$, the boundary has to be equal to the 95% quantile of the distribution of

$$t(X) = \sum_{i=1}^{n} X_i$$

while for $\lambda_1 < \lambda_0$, we are looking for the 5% quantile. Since under the null hypothesis the number of defaults, Y, is Poisson distributed with parameter λ_0, the sum is

$$t(X) = \sum_{i=1}^{n} X_i \sim Poi(n \cdot \lambda_0)$$

As a sample, we refer to the observations of the bond portfolio used in Chapter 17 that we produce below

x_1	x_2	x_3	x_4	x_5	x_6	x_7	x_8	x_9	x_{10}
2	3	12	5	2	4	7	1	0	3
x_{11}	x_{12}	x_{13}	x_{14}	x_{15}	x_{16}	x_{17}	x_{18}	x_{19}	x_{20}
5	6	2	4	2	1	0	1	2	1

Since $\lambda_1 = 2 > \lambda_0 = 1$ and $n = 20$, the boundary becomes $\ln k + 20/0.6931$. In particular, for $\lambda_0 = 1$, the boundary is $\ln k + l20/0.6931 = 28$ because of

$$t(X) = \sum_{i=1}^{n} X_i \sim Poi(20)$$

which corresponds to $k = 0.4249$.[7]

---
[7] Here we have used the 0.95-quantile of the Poisson distribution with parameter 20.

Together with our hypotheses, we finally formulate the test as

$$H_0: \lambda_0 = 1 \text{ versus } H_1: \lambda_1 = 2$$

$$\delta(X) = \begin{cases} t(X) = \sum_{i=1}^{n} X_i \leq 28 & \Rightarrow d_0 \\ t(X) = \sum_{i=1}^{n} X_i > 28 & \Rightarrow d_1 \end{cases}$$

since we have $\lambda_1 > \lambda_0$. Consequently, our acceptance region and critical region are Δ_A and Δ_C, respectively.

Summing up $x_1, x_2, ..., x_{20}$ for the 20 observations, we obtain

$$t(x) = \sum_{i=1}^{n} x_i = 63$$

for our test statistic and, hence, reject the null hypothesis of $H_0: \lambda_0 = 1$ in favor of $H_1: \lambda_1 = 2$.

The design of the test with the probability ratio $\rho(X, \lambda_0, \lambda_1)$ for any probability distribution is known as the **Neyman-Pearson test**. This test is *UMP* as well as consistent. Therefore, it is a very powerful test for the simple testing situation where there are two single competing parameter values, λ_0 and λ_1.

One-Tailed Test for Parameter λ of Exponential Distribution

Let us again consider the previous bond portfolio. As we know, the number of defaults per year can be modeled by a Poisson random variable. Then, the time between two successive defaults is given by an exponential random variable that we denote by Y. The portfolio manager would like to believe that the default rate given by parameter λ is low such that the average time between defaults is large. From experience, suppose that the portfolio manager knows that values for λ of less than 1 are favorable for the bond portfolio while larger values would impose additional risk. So, we can formulate the two hypotheses as $H_0: \lambda \geq 1$ and $H_1: \lambda < 1$.[8] Here, the portfolio manager would like to test if H_0 is wrong.

For the test, which we will design to have size $\alpha = 0.05$, we draw a sample of size n, $X = (X_1, X_2, ..., X_n)$. Since Y is exponentially distributed with parameter λ, the sum of the X_i is Erlang distributed with parameter (λ, n), which, as mentioned in Chapter 11, is a particular form of the gamma distribution. Here we will use it as the test statistic

---
[8] We have to keep in mind, though, that λ has to be some positive real number.

FIGURE 19.10 Probability of Rejecting the Null Hypothesis Computed over the Parameter Space of λ of the Exponential Distribution

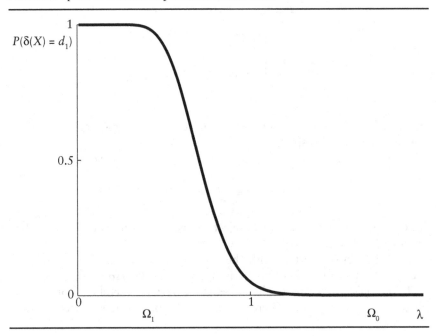

$$t(X) = \sum_{i=1}^{n} X_i \sim Ga(\lambda, n)$$

which represents the combined waiting time between successive defaults of all n years. Now, if the null hypothesis holds (i.e., $\lambda \in [1,\infty)$), then, only in rare instances should we realize large values for $t(X)$, such that an observed value $t(x)$ beyond some threshold should lead to rejection of the null hypothesis. As the threshold, we take the 95% quantile of the $Ga(1,n)$ distribution. Why did we take $\lambda = 1$? Let us have a look at Figure 19.10. We see that for this particular situation of an Erlang distributed statistic, the probability of falsely rejecting the null hypothesis (i.e., committing a type I error) is the greatest when $\lambda = 1$ for all $\lambda \in \Theta_0$. Therefore, by computing the critical region Δ_C for $\lambda = 1$ it is guaranteed that the statistic $t(X)$ will assume a value in Δ_C with probability α or less.

As a sample, we refer to the 20 observations of inter-arrival times of the bond portfolio given in Chapter 17 and reproduced below:

x_1	x_2	x_3	x_4	x_5	x_6	x_7	x_8	x_9	x_{10}
0.500	0.333	0.083	0.200	0.500	0.250	0.143	1	0.318	0.333

x_{11}	x_{12}	x_{13}	x_{14}	x_{15}	x_{16}	x_{17}	x_{18}	x_{19}	x_{20}
0.200	0.167	0.500	0.250	0.500	1	0.318	1	0.500	1

For $\lambda = 1$ and $n = 20$, the threshold is $c = 27.88$, which is the 0.95-quantile of the $Ga(1,20)$ distribution. So, we have $\Delta_A = [0, 27.88]$ as well as $\Delta_C = (27.88, \infty)$. Then, we can formulate our test as

$$H_0: \lambda \geq 1 \text{ versus } H_1: \lambda < 1$$

$$\delta(X) = \begin{cases} t(X) = \sum_{i=1}^{n} X_i \leq 27.88 & \Rightarrow d_0 \\ t(X) = \sum_{i=1}^{n} X_i > 27.88 & \Rightarrow d_1 \end{cases}$$

With the given sample, the statistic is $t(x) = 9.095$. Thus, we cannot reject the null hypothesis, which corresponds to the higher risk of default in our bond portfolio.

One-Tailed Test for μ of the Normal Distribution When σ^2 Is Known

Suppose we are interested in whether there is strong evidence that the mean of the daily returns Y of the Standard and Poor's 500 (S&P 500) index has increased from a $\mu = 0.0003$ to some higher value. As holder of assets that are positively correlated with the S&P 500 index, higher returns would be more favorable to us than smaller ones. So, an increase in the mean return would be positive in this respect. Hence, our hypotheses are $H_0: \mu \leq 0.0003$ and $H_1: \mu > 0.0003$.

To infer, we draw some sample $X = (X_1, X_2, ..., X_n)$. Since the daily return Y is $N(\mu, \sigma^2)$ distributed with some known variance σ^2, the sum

$$\sum_{i=1}^{n} X_i$$

is $N(n\mu, n\sigma^2)$ distributed, assuming independence of the individual drawings. Consequently, the standardized statistic

$$t(X) = \frac{\bar{X} - \mu}{\sigma / \sqrt{n}} \tag{19.4}$$

is a standard normal random variable. Let our test size be $\alpha = 0.05$. We should look for a benchmark that $t(X)$ from equation (19.4) exceeds with probability 0.05 under the null hypothesis. This is given by the standard normal 95% quantile, $q_{0.95} = 1.6449$. So, we test whether

$$t(X) = \frac{\bar{X} - \mu}{\sigma/\sqrt{n}} \leq 1.6449 \qquad (19.5)$$

is fulfilled or not. We can rewrite the condition (19.5) as

$$\bar{X} \leq \mu + \sigma/\sqrt{n} \cdot 1.6449 \qquad (19.6)$$

Thus, instead of using the statistic (19.5), we can use the sample mean of our observations to test for condition (19.6). The statistic of interest for us is now $\tau(\bar{X})$.

To evaluate equation (19.6), under the null hypothesis, which value $\mu \in (-\infty, 0.0003]$ should we actually use? Look at Figure 19.11 where we display the probability $P(\bar{X} \leq c)$ for varying values of μ and some arbitrary real number c. As we can see, the probability of exceeding c increases for larger values of μ as we might have expected. So, if we compute the test boundary at $\mu = 0.0003$, we guarantee that the boundary is exceeded with maximum probability of 0.05 for any $\mu \in (-\infty, 0.0003]$.

FIGURE 19.11 Probability of $P(\bar{X} > c)$ Computed over the Entire Parameter Space

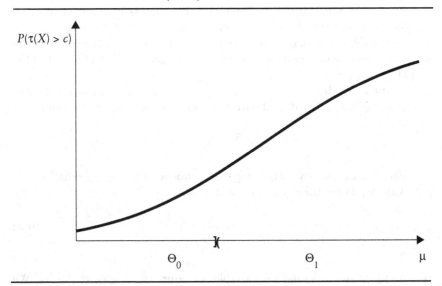

As a sample, we take the daily returns of the S&P 500 index between April 24, 1980 and March 30, 2009. This sample accounts for $n = 7,300$ observations. Suppose we knew that the variance was $\sigma^2 = 0.00013$. Then, our test boundary computed for $\mu = 0.0003$ becomes 0.00049. So, we have $\Delta_A = (-\infty, 0.00049]$ and $\Delta_C = (0.00049, \infty)$. Consequently, our test is given as

$$H_0: \mu \leq 0.0003 \text{ versus } H_1: \mu > 0.0003$$

$$\delta(X) = \begin{cases} \tau(X) = \bar{X} \leq 0.00049 & \Rightarrow d_0 \\ \tau(X) = \bar{X} > 0.00049 & \Rightarrow d_1 \end{cases}$$

For this sample, we obtain 0.00027, which yields d_0. Hence, we cannot reject the null hypothesis. Note that this test is *UMP*.

One-Tailed Test for σ^2 of the Normal Distribution When μ Is Known

Suppose that a portfolio manager is interested in whether the variance of the daily returns of the S&P 500 index is 0.00015 or actually larger. Since a smaller variance is more favorable for the portfolio manager than a larger one and the harm from not realizing the variance actually is larger, we select as hypotheses $H_0: \sigma^2 \in [0.00015, \infty)$ and $H_1: \sigma^2 \in (0, 0.00015)$. In the following, we assume that we can model the daily returns by the normal random variable Y for which we assume $\mu = 0$ is known.

Drawing a sample $X = (X_1, X_2, \ldots, X_n)$, we know from Chapter 17 that the statistic for the variance estimator is given by

$$t(X) = \sum_{i=1}^{n}(X_i - \mu)^2 \overset{\mu=0}{=} \sum_{i=1}^{n} X_i^2 \qquad (19.7)$$

If the true variance σ^2 is large, then the random variable in equation (19.7) should tend to be large as well. If, instead, σ^2 is small, then the very same random variable should predominantly assume small values.

What is large and what is small? Given an arbitrary test size α and the relationship between the hypotheses, then under the null hypothesis of large σ^2, the random variable

$$\sum_{i=1}^{n} X_i^2$$

should fall below some boundary k with probability α and assume values equal to or above k with probability $1 - \alpha$. Since we do not know the distribution of equation (19.7), we have some difficulty finding the corresponding boundary. However dividing equation (19.7) by the variance, though unknown, we obtain the random variable

$$\frac{\sum_{i=1}^{n} X_i^2}{\sigma^2} \qquad (19.8)$$

which, as we know from Chapter 11, is chi-square distributed with n degrees of freedom. So, we simply have to find the value c that, under the null hypothesis, (19.8) falls below with probability α (i.e., the 5% quantile of the chi-square distribution) with n degrees of freedom.

As our sample, let's use the data from the previous example with $n = 7{,}300$ observations. Furthermore, we set the test size equal to $\alpha = 0.05$. So, as a benchmark we are looking for the 95% quantile of the chi-square distribution given a test size 0.05. The resulting boundary is $\chi^2_{0.95}(7{,}300) = 7{,}499$.

Returning to our test statistic $t(X)$, we can derive the benchmark, or critical level k, by

$$\sum_{i=1}^{n} X_i^2 < \sigma^2 \cdot \chi^2_{0.95}(7300) = k$$

which $t(X)$ needs to fall below in order to yield d_1 (i.e., reject the null hypothesis). To compute this $k = \sigma^2 \cdot \chi^2_{0.95}(7{,}300)$, given the null hypothesis holds, which of the $\sigma^2 \in [0.00015, \infty)$ should we use? We see that k becomes smallest for $\sigma^2 = 0.00015$. So, for any σ^2 greater than 0.00015, the probability for equation (19.7) to fall below this value should be smaller than $\alpha = 0.05$. We illustrate this in Figure 19.12. Conversely, k is smallest for $\sigma^2 = 0.00015$. Thus, our threshold is $\sigma^2 \cdot \chi^2_{0.95}(7300) = 1.1250$. Hence, the acceptance region is $\Delta_A = [1.1250, \infty)$ with critical region equal to $\Delta_C = (-\infty, 1.1250)$.

Now let us formulate the test as

$$H_0: \sigma^2 \geq 0.00015 \text{ versus } H_1: \sigma^2 < 0.00015$$

$$\delta(X) = \begin{cases} t(X) \geq 1.1250 & \Rightarrow d_0 \\ t(X) < 1.1250 & \Rightarrow d_1 \end{cases}$$

With our given observations, we obtain $t(X) = 0.9559$. Thus, we can reject the null hypothesis of a variance greater than 0.00015.

This test just presented is *UMP* for the given hypotheses.

FIGURE 19.12 Probability of $\sum_{i=1}^{n}(X_i-\mu)^2/\sigma^2$ Exceeding Some Benchmark c for Various Parameter Values σ^2

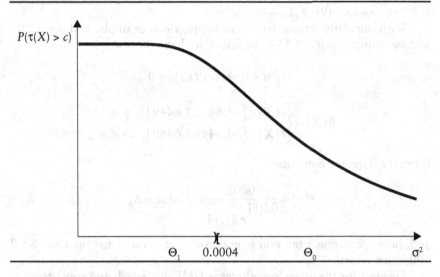

Two-Tailed Test for the Parameter μ of the Normal Distribution When σ^2 Is Known

In this example, we demonstrate how we can test the hypothesis that the parameter μ assumes some particular value μ_0 against the alternative hypothesis that $\mu_0 \neq \mu_0 (H_1)$.

Suppose, again, that a portfolio manager is interested in whether the daily returns of the S&P 500 index have zero mean or not. As before, we model the daily return as the normal random variable Y with known variance σ^2. To be able to infer upon the unknown μ, we draw a sample $X = (X_1, X_2, ..., X_n)$. As test statistic, we take the random variable

$$t(X) = \frac{\bar{X}-\mu_0}{\sigma/\sqrt{n}} \qquad (19.9)$$

which is standard normal under the null hypothesis. So, if the true parameter is μ_0, the random variable (19.9) should not deviate much from zero.

How do we determine how much it must deviate to reject the null hypothesis? First, let us set the test size equal to $\alpha = 0.05$. Since $t(X) \sim N(0,1)$, then $t(X)$ assumes values outside of the interval $[-1.6449, 1.6449]$ with 5% probability since -1.6449 and 1.6449 are the 2.5% and 97.5%

quantiles of the standard normal distribution, respectively. So, the acceptance region is $\Delta_A = [-1.6449, 1.6449]$ and, consequently, the critical region is $\Delta_C = (-\infty, -1.6449) \cup (1.6449, \infty)$.

With our daily return data from the previous example, and $n = 7{,}300$ and assuming $\sigma^2 = 0.00015$ to be know, we have the test

$$H_0: \mu = 0 \text{ versus } H_1: \mu \neq 0$$

$$\delta(X) = \begin{cases} t(X) \in [-1.6449, 1.6449] & \Rightarrow d_0 \\ t(X) \notin [-1.6449, 1.6449] & \Rightarrow d_1 \end{cases}$$

From the data, we compute

$$t(x) = \frac{0.00027}{0.0114/85.44} = 2.0236 \in \Delta_A$$

and, hence, can reject the null hypothesis of zero mean for the daily S&P 500 index return.

This test for the given hypotheses is *UMP*, unbiased, and consistent.

Equal Tails Test for the Parameter σ^2 of the Normal Distribution When μ Is Known

In this illustration, we seek to infer the variance parameter σ^2 of a normal random variable with mean parameter μ assumed known. The hypotheses are as follows. The null hypothesis states that the variance has a certain value while the alternative hypothesis represents the opinion that the variance is anything but this value.

Once more, let us focus on the daily S&P 500 index return modeled by the normal random variable Y. Moreover, suppose, again, that the mean parameter $\mu = 0$. Furthermore, let us assume that the portfolio manager believed that $\sigma^2 = 0.00015$. However, the risk manager responsible for the portfolio has found reason to believe that the real value is different from 0.00015. The portfolio management team is indifferent to any value and simply wishes to find out what is true. So, the hypotheses are $H_0: \sigma^2 = 0.00015$ and $H_1: \sigma^2 \neq 0.00015$. Suppose, $\alpha = 0.05$ is chosen as in the previous examples.

Having drawn the sample $X = (X_1, X_2, \ldots, X_n)$, we can again use the test statistic

$$\sum_{i=1}^{n}(X_i - \mu)^2$$

Since $\mu = 0$, this simplifies to

$$t(X) = \sum_{i=1}^{n} X_i^2 \qquad (19.10)$$

which will be our test statistic henceforth.

What result should now lead to rejection of the null hypothesis (i.e., d_1)? If $t(X)$ from equation (19.10) either assumes very small values or becomes very large, the result should refute the null hypothesis. Suppose we designed the test so it would have size α. Thus, we should determine a lower bound c_l and an upper bound c_u that, when the null hypothesis holds, $t(X)$ lies between these bounds with probability $1 - \alpha$. To determine these bounds, recall that dividing $t(X)$ by its variance σ^2, though unknown, results in a chi-square distributed random variable

$$\frac{\sum_{i=1}^{n} X_i^2}{\sigma^2}$$

with n degrees of freedom. Since the chi-square distribution is not symmetric, we artificially set the bounds so that under the null hypothesis, $t(X)$ may fall short of the lower bound with probability $\alpha/2$ and with probability $\alpha/2$, $t(X)$ might exceed the upper bound. In Figure 19.13, we display this for a probability of a type I error of 5%. So, we have

$$P\left(\chi^2_{0.025}(n) \leq \frac{\sum_{i=1}^{n} X_i^2}{\sigma^2} \leq \chi^2_{0.975}(n)\right) = 0.05 \qquad (19.11)$$

This probability in equation (19.11) translates into the equivalent relationship

$$P\left(\chi^2_{0.025}(n) \cdot \sigma^2 \leq \sum_{i=1}^{n} X_i^2 \leq \chi^2_{0.975}(n) \cdot \sigma^2\right) = 0.05 \qquad (19.12)$$

The bounds used in equation (19.12) have to be computed using the value for σ^2 under the null hypothesis (i.e., 0.00015). Since our sample is of length $n = 7{,}300$, the corresponding quantiles are $\chi^2_{0.025}(7{,}300) = 7{,}065$ and $\chi^2_{0.975}(7{,}300) = 7{,}539$. Finally, from (19.12), we obtain

$$P\left(7065 \cdot 0.00015 \leq \sum_{i=1}^{n} X_i^2 \leq 7539 \cdot 0.00015\right) = 0.05$$

FIGURE 19.13 Partitioning of Test Size α = 0.05 of Equal-Tails Test

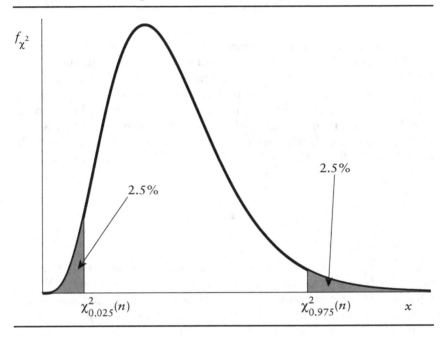

$$P\left(1.0598 \le \sum_{i=1}^{n} X_i^2 \le 1.1308\right) = 0.05$$

Hence, the corresponding acceptance region is $\Delta_A = [1.0598, 1.1308]$ while the critical region is the composed set $\Delta_C = (-\infty, 1.0598) \cup (1.1308, \infty)$. Our test $\delta(X)$ is then

$$H_0: \sigma^2 = 0.00015 \text{ versus } H_1: \sigma^2 \ne 0.00015$$

$$\delta(X) = \begin{cases} t(X) \in [1.0598, 1.1308] & \Rightarrow d_0 \\ t(X) \notin [1.0598, 1.1308] & \Rightarrow d_1 \end{cases}$$

With

$$t(X) = \sum_{i=1}^{n} X_i^2 = 0.9571$$

computed from our sample data, we make the decision d_1 and, hence, reject the null hypothesis of $\sigma^2 = 0.00015$. Since the sum falls below the lower bound of the acceptance region, the true variance is probably smaller.

Test for Equality of Means

Suppose a portfolio manager is interested in whether the daily stock returns of General Electric (GE) and International Business Machines (IBM) had identical means. By Y_{GE} and Y_{IBM}, let us denote the random variables representing the daily stock returns of GE and IBM, respectively. Both are assumed to be normally distributed with known variances σ^2_{GE} and σ^2_{IBM}, such that $N(\mu_{GE}, \sigma^2_{GE})$ and $N(\mu_{IBM}, \sigma^2_{IBM})$. Furthermore, we conjecture that the expected values of both are equal (i.e., $\mu_{GE} = \mu_{IBM}$).

With this, we formulate our null hypothesis. Consequently, the alternative hypothesis states that the two means are not equal. So, we formally state the testing hypotheses as

$$H_0: \mu_{GE} = \mu_{IBM} \text{ versus } H_1: \mu_{GE} \neq \mu_{IBM}$$

Next we approach the construction of a suitable test statistic. For this, we draw two independent samples, one from each stock return. Let

$$X^{GE} = \left(X_1^{GE}, X_2^{GE}, \ldots, X_{n_{GE}}^{GE}\right)$$

denote the GE sample of independent drawings and

$$X^{IBM} = \left(X_1^{IBM}, X_2^{IBM}, \ldots, X_{n_{IBM}}^{IBM}\right)$$

that of the IBM stock return. We do not require that the two samples have to be of equal length. This means n_{GE} and n_{IBM} can be different.

We now compute the difference $\bar{X}_{GE} - \bar{X}_{IBM}$ of the two sample means, $\bar{X}_{GE}$ and $\bar{X}_{IBM}$. This difference is a normal random variable with mean $\mu_{GE} - \mu_{IBM}$. Now, if we subtract from $\bar{X}_{GE} - \bar{X}_{IBM}$, the theoretical mean $\mu_{GE} - \mu_{IBM}$, the resulting random variable

$$\bar{X}_{GE} - \bar{X}_{IBM} - (\mu_{GE} - \mu_{IBM}) \tag{19.13}$$

is normal with zero mean. Because of the independence of the two samples, X^{GE} and X^{IBM}, the variance of the random variable equation (19.13) is

$$Var\left(\bar{X}_{GE} - \bar{X}_{IBM} - (\mu_{GE} - \mu_{IBM})\right) = \frac{\sigma^2_{GE}}{n_{GE}} + \frac{\sigma^2_{IBM}}{n_{IBM}} \tag{19.14}$$

Dividing equation (19.13) by the standard deviation equal to the square root of equation (19.14), we obtain the test statistic

$$t\left(X^{GE}, X^{IBM}\right) = \frac{\bar{X}_{GE} - \bar{X}_{IBM} - \left(\mu_{GE} - \mu_{IBM}\right)}{\sqrt{\dfrac{\sigma_{GE}^2}{n_{GE}} + \dfrac{\sigma_{IBM}^2}{n_{IBM}}}} \qquad (19.15)$$

which is a standard normal random variable. Under the null hypothesis of equal means, we have $\mu_{GE} - \mu_{IBM} = 0$ such that equation (19.15) becomes

$$t_0\left(X^{GE}, X^{IBM}\right) = \frac{\bar{X}_{GE} - \bar{X}_{IBM}}{\sqrt{\dfrac{\sigma_{GE}^2}{n_{GE}} + \dfrac{\sigma_{IBM}^2}{n_{IBM}}}}$$

which follows the same probability law as equation (19.15).

Consequently, we can formulate the test as follows. For a given significance level α, we reject the null hypothesis if the test statistic t_0 either falls below the $\alpha/2$-quantile of the standard normal distribution or exceeds the corresponding $1 - \alpha/2$-quantile. In any other case, we cannot reject the null hypothesis. Formally, this test is given by

$$\delta\left(X^{GE}, X^{IBM}\right) = \begin{cases} d_0, & t_0\left(X^{GE}, X^{IBM}\right) \in \left[q_{\alpha/2}, q_{1-\alpha/2}\right] \\ d_1, & t_0\left(X^{GE}, X^{IBM}\right) \notin \left[q_{\alpha/2}, q_{1-\alpha/2}\right] \end{cases} \qquad (19.16)$$

where $q_{\alpha/2}$ and $q_{1-\alpha/2}$ denote the $\alpha/2$- and $1 - \alpha/2$-quantile of the standard normal distribution, respectively. With a significance level $\alpha = 0.05$, the test given by (19.16) becomes

$$\delta\left(X^{GE}, X^{IBM}\right) = \begin{cases} d_0, & t_0\left(X^{GE}, X^{IBM}\right) \in \left[-1.9600, 1.9600\right] \\ d_1, & t_0\left(X^{GE}, X^{IBM}\right) \notin \left[-1.9600, 1.9600\right] \end{cases}$$

So, the acceptance region of our test is $\Delta_A = [-1.9600, 1.9600]$ while the corresponding rejection region is $\Delta_C = \mathbb{R}\setminus[-1.9600, 1.9600]$ (i.e., any real number that does not belong to the acceptance region).

For our test, we examine the daily GE and IBM returns observed between April 24, 1980 and March 30, 2009 such that $n_{GE} = n_{IBM} = 7,300$. The two sample means are $X^{GE} = 0.00029$ and $X^{IBM} = 0.00027$. Suppose the corresponding known variances are $\sigma_{GE}^2 = 0.00027$ and $\sigma_{IBM}^2 = 0.00032$. Then, from equation (19.15), the test statistic is equal to $t_0(X^{GE}, X^{IBM}) = 0.0816$, which leads to decision d_0; that is, we cannot reject the null hypothesis of equal means.

Note that even if the daily returns were not normally distributed, by the central limit theorem, we could still apply equation (19.15) for these hypotheses.

In the following, we maintain the situation

$$H_0: \mu_{GE} = \mu_{IBM} \text{ versus } H_1: \mu_{GE} \neq \mu_{IBM}$$

except that the variances of the daily stock returns of GE and IBM are assumed to be equal, albeit unknown. So, if their means are not equal, the one with the lesser mean is dominated by the other return with the higher mean and identical variance.[9]

To derive a suitable test statistic, we refer to equation (19.15) with the means cancelling each other where we have to replace the individual variances, σ_{GE}^2 and σ_{IBM}^2, by a single variance σ^2 to express equality of the two individual stocks, that is,

$$\frac{\bar{X}_{GE} - \bar{X}_{IBM}}{\sigma \cdot \sqrt{\dfrac{1}{n_{GE}} + \dfrac{1}{n_{IBM}}}} \qquad (19.17)$$

As we know, this statistic is a standard normal random variable under the null hypothesis of equal means. We could use it if we knew σ. However, the latter is assumed unknown, and we have to again infer from our samples X^{GE} and X^{IBM}. Using the respective sample means, we can compute the two intermediate statistics

$$\sum_{i=1}^{n_{GE}} \frac{\left(X_i^{GE} - \bar{X}_{GE}\right)^2}{\sigma^2} \qquad (19.18)$$

and

$$\sum_{i=1}^{n_{IBM}} \frac{\left(X_i^{IBM} - \bar{X}_{IBM}\right)^2}{\sigma^2} \qquad (19.19)$$

which are chi-square distributed with $n_{GE} - 1$ and $n_{IBM} - 1$ degrees of freedom, respectively.[10] Adding equations (19.18) and (19.19), we obtain a chi-square distributed random variable with $n_{GE} + n_{IBM} - 2$ degrees of freedom,

---
[9] Mean-variance portfolio optimization is explained in Chapter 14.
[10] Note that in theory we can compute these random variables even though we do not know the variance σ^2.

$$\sum_{i=1}^{n_{GE}} \frac{\left(X_i^{GE} - \bar{X}_{GE}\right)^2}{\sigma^2} + \sum_{i=1}^{n_{IBM}} \frac{\left(X_i^{IBM} - \bar{X}_{IBM}\right)^2}{\sigma^2} \sim \chi^2\left(n_{GE} + n_{IBM} - 2\right) \quad (19.20)$$

Dividing the statistic in equation (19.17) by the square root of the ratio of statistic (19.20) and its degrees of freedom, we obtain the new test statistic

$$\tau_0\left(X^{GE}, X^{IBM}\right) = \frac{\dfrac{\bar{X}_{GE} - \bar{X}_{IBM}}{\sqrt{\dfrac{1}{n_{GE}} + \dfrac{1}{n_{IBM}}}}}{\sqrt{\dfrac{\sum_{i=1}^{n_{GE}}\left(X_i^{GE} - \bar{X}_{GE}\right)^2 + \sum_{i=1}^{n_{IBM}}\left(X_i^{IBM} - \bar{X}_{IBM}\right)^2}{n_{GE} + n_{IBM} - 2}}} \quad (19.21)$$

In equation (19.21), we took advantage of the fact that the variance terms in the numerator and denominator cancel each other out. The statistic τ_0 is Student's t-distributed with $n_{GE} + n_{IBM} - 2$ degrees of freedom under the null hypothesis of equal means.

Now the test is formulated as follows. Given significance level α, we reject the null hypothesis if the test statistic τ_0 either falls below the $\alpha/2$-quantile of the Student's t-distribution or exceeds the corresponding $1 - \alpha/2$-quantile. Any other case does not lead to rejection of the null hypothesis. Formally, this test is

$$\delta\left(X^{GE}, X^{IBM}\right)$$

$$= \begin{cases} d_0, & \tau_0\left(X^{GE}, X^{IBM}\right) \in \left[t_{\alpha/2}\left(n_{GE} + n_{IBM} - 2\right), t_{1-\alpha/2}\left(n_{GE} + n_{IBM} - 2\right)\right] \\ d_1, & \tau_0\left(X^{GE}, X^{IBM}\right) \notin \left[t_{\alpha/2}\left(n_{GE} + n_{IBM} - 2\right), t_{1-\alpha/2}\left(n_{GE} + n_{IBM} - 2\right)\right] \end{cases} \quad (19.22)$$

where

$$t_{\alpha/2}\left(n_{GE} + n_{IBM} - 2\right) \text{ and } t_{1-\alpha/2}\left(n_{GE} + n_{IBM} - 2\right)$$

denote the $\alpha/2$- and $1 - \alpha/2$-quantile of the Student's t-distribution, respectively, with $n_{GE} + n_{IBM} - 2$ degrees of freedom.

With a significance level $\alpha = 0.05$, the test statistic given by equation (19.22) becomes

$$\delta\left(X^{GE}, X^{IBM}\right) = \begin{cases} d_0, & \tau_0\left(X^{GE}, X^{IBM}\right) \in [-1.9601 \quad 1.9601] \\ d_1, & \tau_0\left(X^{GE}, X^{IBM}\right) \notin [-1.9601 \quad 1.9601] \end{cases} \quad (19.23)$$

The test has as acceptance region $\Delta_A = [-1.9601, 1.9601]$ with rejection region equal to $\Delta_C = \mathbb{R}\setminus[-1.9601, 1.9601]$ (i.e., all real numbers except for the interval [-1.9601,1.9601]).

With $n_{GE} = n_{IBM} = 7{,}300$ observations each from the period between April 14, 1980 and March 30, 2009, we obtain the corresponding sample means $\overline{X}^{GE} = 0.00029$ and $\overline{X}^{IBM} = 0.00027$. With these data, we are ready to compute the test statistic (19.21) as

$$\tau_0\left(X^{GE}, X^{IBM}\right) = 0.0816$$

such that, according to (19.23), the decision is, again, d_0 such that we cannot reject the null hypothesis of equal means.

Note that for a large number of observations used in the test, the two test statistics (19.15) and (19.21) are almost the same and with almost identical acceptance regions. So, the tests are virtually equivalent.

Two-Tailed Kolmogorov-Smirnov Test for Equality of Distribution

Suppose we were interested in whether the assumption of a certain distribution for the random variable of interest is justifiable or not. For this particular testing situation, we formulate the hypotheses as "random variable Y has distribution function F_0" (H_0) and "random variable Y does not have distribution function F_0" (H_1) or vice versa depending on our objectives.

Suppose that we draw our sample $X = (X_1, X_2, ..., X_n)$ as usual. From this, we compute the empirical cumulative relative frequency distribution function $F_{emp}^f(x)$ as explained in Chapter 2. To simplify the terminology, we will refer to $F_{emp}^f(x)$ as the empirical distribution function. This function will be compared to the distribution assumed under the null hypothesis, which we will denote by $F_0(x)$. By comparing, we mean that we will compute the absolute value of the difference $F_{emp}^f(x) - F_0(x)$ (i.e., $\left|F_{emp}^f(x) - F_0(x)\right|$) at each real number x between the minimum observation $X_{(1)}$ and the maximum observation $X_{(n)}$.[11] Of all such obtained quantities $\left|F_{emp}^f(x) - F_0(x)\right|$, we select the ***supremum***

---

[11] An index in parentheses indicates order statistics (i.e., the random variable $X_{(i)}$ indicates the i-th smallest value in the sample).

$$t = \sup_x \left| F_{emp}^f(x) - F_0(x) \right| \qquad (19.24)$$

What does expression (19.24) mean? Look at Figure 19.14 where we depict an arbitrary theoretical distribution F_0 (dotted curve) plotted against the empirical counterpart, F_{emp}^f, for the order statistics $x_{(1)}$ through $x_{(5)}$. At each ordered value $x_{(i)}$, we compute the distance between the empirical distribution function and the hypothesized distribution function. These distances are indicated by the solid vertical lines. In the figure they are D_2, D_5, and D_7. Note that at each ordered value $x_{(i)}$, the empirical distribution function jumps and, hence, assumes the next higher value that we indicate in the figure by the solid dot at each level. The empirical distribution function is right-continuous but not left-continuous. As a result, between ordered values $x_{(i)}$ and the next one, $x_{(i+1)}$, F_{emp}^f remains constant and keeps this level as a limit as x approaches $x_{(i+1)}$ from the left. This limit, however, is not the same as the function value $F_{emp}^f(x_{(i+1)})$ since, as explained, it is already assuming the next higher level at $x_{(i+1)}$. Nonetheless, at each $x_{(i+1)}$, we still compute the distance between the hypothesized value $F_0(x_{(i+1)})$ and the left side limit $F_{emp}^f(x_{(i)})$. In the figure, we indicate this by the dash-dotted lines accounting for the distances D_1, D_3, D_4, and D_6. For this illustration, the maximum dis-

FIGURE 19.14 Computing the Two-Tailed Kolmogorov-Smirnov Distance

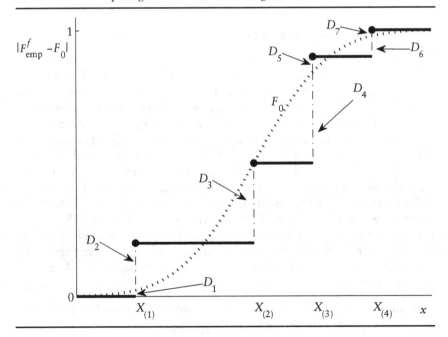

Hypothesis Testing

tance is D_2 (solid vertical lines). If we include the distances computed using limits as well (i.e., the dash-dotted lines), we obtain the supremum. The supremum may be greater than the maximum as is the case here because D_4 is greater than D_2, the maximum.

This may have just been a little too much mathematics for one's liking. But do not get frustrated since we only have to remember to compute distances between the empirical distribution function levels on either end, and the hypothesized distribution function values at these positions. From these distances, we determine the largest one—our supremum—which was D_4, in the previous illustration, and compare it to quantiles of the distribution of the so-called **Kolmogorov-Smirnov** (KS) **statistic**.[12] To look up whether the result is significant enough to reject the null hypothesis, we need the significance level α, the number of observations n, and the specification two-tailed or two-sided since the KS distribution can also be determined for one-tailed tests.

Let us use the S&P 500 daily return data with $n = 7,300$ observations from the previous examples and set the test size equal to $\alpha = 0.05$. We know that $\mu = 0.00015$ and $\sigma^2 = 0.00013$. We now compute the empirical distribution function of the standardized observations, that is,

$$\frac{x_i - \mu}{\sigma}$$

Suppose that the portfolio manager used to believe that these observations were generated from a standard normal population but now wants to verify if this is in fact the case by means of a hypothesis test. Then the test is given by[13]

$$H_0 : F^f_{emp} = N(0,1) \text{ versus } H_1 : F^f_{emp} \neq N(0,1)$$

$$\delta(X) = \begin{cases} \sup_x \left| F^f_{emp}(x) - F_0(x) \right| \leq 0.0159 & \Rightarrow d_0 \\ \sup_x \left| F^f_{emp}(x) - F_0(x) \right| > 0.0159 & \Rightarrow d_1 \end{cases}$$

With our data, we obtain as the supremum a value of

$$\sup_x \left| F^f_{emp}(x) - F_0(x) \right| = 0.4906$$

---

[12] Tables with values for the KS distribution can be found in textbooks on non-parametric estimation. Moreover, most statistical standard software packages include the KS test.

[13] For samples as large as the one used, 0.0159 is merely an approximation and the exact test boundary may be slightly different.

such that we have to reject the null hypothesis (d_1) of a standard normally distributed daily return of the S&P 500 index. The computed value 0.4906 is highly significant since the *p*-value is virtually 0.

Likelihood Ratio Test

Now let's consider a particular testing situation by assuming some model where certain parameter components are set equal to zero. We refer to this model as being a *restricted model*. Alternatively, these particular parameter components of interest may be different from zero. The first case (i.e., the restricted model) represents the null hypothesis while the more general model represents the alternative hypothesis. Since the model given by the null hypothesis is a special case of the more general model of the alternative hypothesis, we say the competing models are *nested models*.

The testing procedure is then as follows. Of course, we need to base our inference on some sample $X = (X_1, X_2, \ldots, X_n)$. The null hypothesis is rejected if the likelihood function with the parametrization under the null hypothesis is significantly lower than the likelihood function as given by the alternative hypothesis, both evaluated for the same sample X.[14]

How do we determine if the likelihood under the null hypothesis is significantly lower? This leads to the introduction of a new test statistic, the so-called *likelihood ratio*. It is defined as

$$\rho(X) = \frac{L_X(\theta_0)}{L_X(\theta_1)} \qquad (19.25)$$

where $L_X(\theta_0)$ denotes the likelihood function of the restricted model as given by the null hypothesis and $L_X(\theta_1)$ the likelihood function of the general model.

If we have a realization $x = (x_1, x_2, \ldots, x_n)$ such that the ratio in equation (19.25) is smaller than 1, this could indicate that the restriction is inaccurate and we should prefer the alternative, the more general model. It then appears that the parameter components held constant at zero under the null hypothesis may, in truth, have different values.

Since the likelihood ratio of equation (19.25) depends on the random sample X, it is random itself. In particular, if the sample size n becomes ever larger such that we are in the realm of large sample behavior, then the test statistic $-2 \cdot \ln(\rho(X))$ obtained from the likelihood ratio $\rho(X)$ in equation (19.25) is chi-square distributed with degrees of freedom equal to the number of restricted parameter components under the null hypothesis. That is, formally,

---

[14] The likelihood function was introduced in Chapter 17.

Hypothesis Testing

$$-2\ln(\rho(X)) \sim \chi^2(k) \qquad (19.26)$$

where the degrees of freedom k indicate the number of restricted parameters.

A mistake often encountered is that different models are tested against each other that are based on completely different parameters and, hence, not nested. In that case, the test statistic $\rho(X)$ is no longer chi-square distributed and, generally, becomes virtually meaningless.

Now, suppose we wanted to test whether the daily S&P 500 return Y is normally distributed with mean $\mu = 0$ and variance $\sigma^2 = 1$ (H_0) or rather with a general value for σ^2. In the following, we therefore have to use the likelihood function for the normal distribution introduced in Chapter 17. So under the null hypothesis, we have to compute the likelihood function inserting $\mu = 0$ and $\sigma^2 = 1$ such that we obtain

$$L_X^0\left(\sigma^2\right) = \frac{1}{(2\pi)^{n/2}} \cdot e^{-\frac{\sum_{i=1}^n X_i^2}{2}}$$

On the other hand, under the alternative hypothesis, we have to compute the likelihood function with μ set equal to 0 and the maximum likelihood estimator

$$\hat{\sigma}_{MLE} = \sum_{i=1}^n X_i^2 / n$$

replacing the unknown σ^2. This yields

$$L_X^1\left(\sigma^2\right) = \frac{1}{\left(2\pi \frac{\sum_{i=1}^n X_i^2}{n}\right)^{n/2}} \cdot e^{-\frac{n}{2}}$$

and, consequently, we obtain as the test statistic from equation (19.26)

$$\rho(X) = \left[n \cdot \ln(n) - n \cdot \ln\left(\sum_{i=1}^n X_i^2\right) + n - \sum_{i=1}^n X_i^2\right] \qquad (19.27)$$

This statistic is chi-square distributed with one degree of freedom since the difference in the number of unrestricted parameters between the two hypotheses is one. Computing equation (19.27) for the 7,300 daily S&P 500 index return observations we have been using in previous examples, we obtain such a large value for the statistic, it is beyond any significance level. Hence, we have to reject the null hypothesis of $\sigma^2 = 1$ given $\mu = 0$.

CONCEPTS EXPLAINED IN THIS CHAPTER (IN ORDER OF PRESENTATION)

Hypothesis testing
Test statistic
Decision rule
Hypotheses
Null hypothesis
Alternative hypothesis
One-tailed test
Two-tailed test
Acceptance region
Critical rejection region (rejection region)
Type I error
Type II error
Test size
Statistically significant
p-value
Power of a test
Uniformly most powerful
Unbiased test
Consistent test
Neyman-Pearson test
Supremum
Kolmogorov-Smirnov (KS) statistic
Restricted model
Nested model
Likelihood ratio

PART Four

Multivariate Linear Regression Analysis

CHAPTER 20

Estimates and Diagnostics for Multivariate Linear Regression Analysis

It is often the case in finance that we find it necessary to analyze more than one variable simultaneously. In Chapter 5, for example, we considered the association between two variables through correlation in order to express linear dependence and in Chapter 6 explained how to estimate a linear dependence between two variables using the linear regression method. When there is only one independent variable, the regression model is said to be a *simple linear regression* or a *univariate regression*.

Univariate modeling in many cases is not sufficient to deal with real problems in finance. The behavior of a certain variable of interest sometimes needs to be explained by two or more variables. For example, suppose that we want to determine the financial or macroeconomic variables that affect the monthly return on the Standard & Poor's 500 (S&P 500) index. Let's suppose that economic and financial theory suggest that there are 10 such explanatory variables. Thus we have a multivariate setting of 11 dimensions—the return on the S&P 500 and the 10 explanatory variables.

In this chapter and in the next, we explain the multivariate linear regression model (also called the multiple linear regression model) to explain the linear relationship between several independent variables and some dependent variable we observe. As in the univariate case (i.e., simple linear regression) discussed in Chapter 6, the relationship between the variables of interest may not be linear. However, that can be handled by a suitable transformation of the variables.

Because the regression model is evaluated by goodness-of-fit measures, we will have to draw on topics from probability theory, parameter estimation, and hypothesis testing that we have explained in Parts Two and Three. In particular, we will use the normal distribution, t-statistic, and F-statistic. Moreover, at certain points in this chapter, it will be necessary to understand some matrix algebra. All the needed essentials for this topic are presented in the Appendix B.

THE MULTIVARIATE LINEAR REGRESSION MODEL

The *multivariate linear regression model* for the population is of the form

$$y = \beta_0 + \beta_1 x_1 + \beta_2 x_2 + \ldots + \beta_k x_k + \varepsilon \qquad (20.1)$$

where we have

β_0 = Constant intercept.
$\beta_1, \ldots, \beta_k$ = Regression coefficients of k independent variables.
ε = Model error.

In vector notation, given samples of dependent and independent variables, we can represent equation (20.1) as

$$y = X\beta + \varepsilon \qquad (20.2)$$

where y is an $n \times 1$ column vector consisting of the n observations of the dependent variable, that is,

$$y = \begin{pmatrix} y_1 \\ \vdots \\ y_n \end{pmatrix} \qquad (20.3)$$

X is a $n \times (k + 1)$ matrix consisting of n observations of each of the k independent variables and a column of ones to account for the vertical intercepts β_0 such that

$$X = \begin{pmatrix} 1 & x_{11} & \cdots & x_{1k} \\ 1 & x_{21} & & x_{2k} \\ \vdots & \vdots & & \vdots \\ 1 & x_{n1} & \cdots & x_{nk} \end{pmatrix} \qquad (20.4)$$

The $(k + 1)$ regression coefficients including intercept are given by the $k + 1$ column vector

$$\beta = \begin{pmatrix} \beta_0 \\ \vdots \\ \beta_k \end{pmatrix} \qquad (20.5)$$

Each observation's residual is represented in the column vector ε

$$\varepsilon = \begin{pmatrix} \varepsilon_1 \\ \vdots \\ \varepsilon_n \end{pmatrix} \qquad (20.6)$$

The regression coefficient of each independent variable given in equation (20.5) represents the average change in the dependent variable per unit change in the independent variable with the other independent variables held constant.

ASSUMPTIONS OF THE MULTIVARIATE LINEAR REGRESSION MODEL

For the multivariate linear regression model, we make the following three assumptions about the error terms:

Assumption 1. The regression errors are normally distributed with zero mean.

Assumption 2: The variance of the regression errors (σ_ε^2) is constant.

Assumption 3. The error terms from different points in time are independent such that $\varepsilon_t \neq \varepsilon_{t+d}$ for any $d \neq 0$ are independent for all t.

Formally, we can restate the above assumptions in a concise way as

$$\varepsilon_i \overset{i.i.d.}{\sim} N(0, \sigma^2)$$

Furthermore, the residuals are assumed to be uncorrelated with the independent variables. In Chapter 22, we briefly describe how to deal with situations when these assumptions are violated.

ESTIMATION OF THE MODEL PARAMETERS

Since the model is not generally known for the population, we need to estimate it from some sample. Thus, the estimated regression is

$$\hat{y} = b_0 + b_1 x_1 + b_2 x_2 + \cdots + b_k x_k \qquad (20.7)$$

The matrix notation analogue of equation (20.7) is

$$y = \hat{y} + e = Xb + e \quad (20.8)$$

which is similar to equation (20.2) except the model's parameters and error terms are replaced by their corresponding estimates, b and e.

The independent variables $x_1, ..., x_k$ are thought to form a space of dimension k. Then, with the y-values, we have an additional dimension such that our total dimensionality is $k + 1$. The estimated model generates values on a k-multidimensional *hyperplane*, which expresses the functional linear relationship between the dependent and independent variables. The estimated hyperplane is called a *regression hyperplane*. In the univariate case, this is simply the regression line of the $\hat{y}$-estimates stemming from the one single independent variable x.[1]

Each of the k coefficients determines the slope in the direction of the corresponding independent variable. In the direction of the $k+1$st dimension of the y values, we extend the estimated errors, $e = y - \hat{y}$. At each y value, these errors denote the distance between the hyperplane and the observation of the corresponding y value.

To demonstrate this, we consider some variable y. Suppose, we also have a two-dimensional variable x with independent components x_1 and x_2. Hence, we have a three-dimensional space as shown in Figure 20.1. For y, we have three observations, y_1, y_2, and y_3. The hyperplane for equation (20.7) formed by the regression is indicated by the gray plane. The intercept b_0 is indicated by the dashed arrow while the slopes in the directions of x_1 and x_2 are indicated by the arrows b_1 and b_2, respectively.[2] Now, we extend vertical lines between the hyperplane and the observations, e_1, e_2, and e_3, to show by how much we have missed approximating the observations with the hyperplane.

Generally, with the ordinary least squares regression method described in Chapter 6, the estimates are, again, such that $\Sigma(y - \hat{y})^2$ is minimized with respect to the regression coefficients. For the computation of the regression estimates, we need to indulge somewhat in matrix computation. If we write the minimization problem in matrix notation, finding the vector β that minimizes the squared errors looks like[3]

$$\sum_{i=1}^{n} (y_i - \hat{y})^2 = (y - X\beta)^T (y - X\beta) \quad (20.8)$$

---

[1] In general, the hyperplane formed by the linear combination of the x values is always one dimension less than the overall dimensionality.
[2] The arrow b_0 is dashed to indicate that it extends from our point of view vertically from the point (0,0,0) behind the hyperplane.
[3] Make sure you are familiar with the concept of transpose and matrix inverse from Appendix B. When we use the matrix inverse, we implicitly assume the matrix of interest to be full rank, a requirement for the inversion of a matrix.

FIGURE 20.1 Vector Hyperplane and Residuals

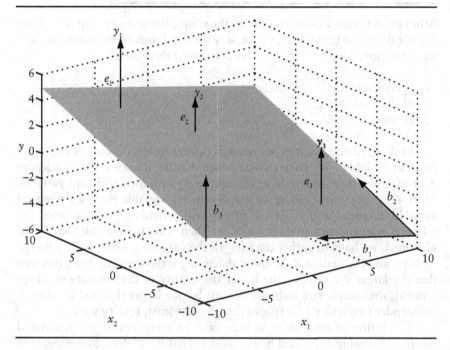

Differential calculus and matrix algebra lead to the optimal regression coefficient estimates and estimated residuals given by

$$b = \left(X^T X\right)^{-1} X^T y \tag{20.9}$$

and

$$e = y - X^T b = y - \hat{y} \tag{20.10}$$

where b in equation (20.9) and e in equation (20.10) are $(k+1) \times 1$ and $n \times 1$ column vectors, respectively. One should not worry however if this appears rather complicated and very theoretical. Most statistical software have these computations implemented and one has to just insert the data for the variables and select some least squares regression routine to produce the desired estimates according to equation (20.9).

DESIGNING THE MODEL

Athough in Chapter 6 we introduced the simple linear regression model, we did not detail the general steps necessary for the design of the regression and its evaluation. This building process consists of three steps:

1. Specification
2. Fitting/estimating
3. Diagnosis

In the specification step, we need to determine the dependent and independent variables. We have to make sure that we do not include independent variables that seem to have nothing to do with the dependent variable. More than likely, in dealing with a dependent variable that is a financial variable, financial and economic theory will provide a guide to what the relevant independent variables might be. Then, after the variables have been identified, we have to gather data for all the variables. Thus, we obtain the vector y and the matrix X. Without defending it theoretically here, it is true that the larger the sample, the better the quality of the estimation. Theoretically, the sample size n should at least be one larger than the number of independent variables k. A rule of thumb is, at least, four times k.

The fitting or estimation step consists of constructing the functional linear relationship expressed by the model. That is, we need to compute the correlation coefficients for the regression coefficients. We perform this even for the independent variables to test for possible interaction between them as explained later in the next chapter. The estimation, then, yields so-called point estimates of the dependent variable for given values of the independent variables.[4]

Once we have obtained the estimates for equation (20.9), we can move on to evaluating the quality of the regression with respect to the given data. This is the diagnosis step.

DIAGNOSTIC CHECK AND MODEL SIGNIFICANCE

As just explained, diagnosing the quality of some model is essential in the building process. Thus we need to set forth criteria for determining model quality. If, according to some criteria the fit is determined to be insufficient, we might have to redesign the model by including different independent variables.

---

[4]This is in contrast to a range or interval of values as given by a confidence interval. For an explicit treatment of point estimators and confidence intervals, see Chapters 17 and 18.

We know from Chapter 6 the goodness-of-fit measure is the coefficient of determination (denoted by R^2). We will use that measure here as well. As with the univariate regression, the coefficient of determination measures the percentage of variation in the dependent variable explained by all of the independent variables employed in the regression. The R^2 of the multivariate linear regression is referred to as the ***multiple coefficient of determination***. We reproduce its definition from Chapter 6 below:

$$R^2 = \frac{SSR}{SST} \quad (20.11)$$

where

SSR = sum of squares explained by the regression model
SST = total sum of squares

Following the initial assessment, one needs to verify the model by determining its statistical significance.[5] To do so, we compute the overall model's significance and also the significance of the individual regression coefficients. The estimated regression errors play an important role as well. If the standard deviation of the regression errors is found to be too large, the fit could be improved by an alternative. The reason is that too much of the variance of the dependent y is put into the residual variance s^2. Some of this residual variance may, in fact, be the result of variation in some independent variable not considered in the model so far. And a final aspect is testing for the interaction of the independent variables that we discuss in the next chapter.

Testing for the Significance of the Model

To test whether the entire model is significant, we consider two alternative hypotheses. The first, our null hypothesis H_0, states that all regression coefficients are equal to zero, which means that none of the independent variables play any role. The alternative hypothesis H_1, states that, at least, one coefficient is different from zero. More formally,

$$H_0 : \beta_0 = \beta_1 = \ldots = \beta_k = 0$$
$$H_1 : \beta_j \neq 0 \text{ for at least one } j \in \{1,2,\ldots,k\}$$

In case of a true null hypothesis, the linear model with the independent variables we have chosen does not describe the behavior of the depen-

---

[5]Statistical significance of some test statistic value is explained in Chapter 19.

dent variable. To perform the test, we carry out an *analysis of variance* (ANOVA) *test*. In this context, we compute the F-statistic defined by

$$F = \frac{\frac{SSR}{k}}{\frac{SSE}{n-k-1}} = \frac{MSR}{MSE} \qquad (20.12)$$

where SSR was defined above and

SSE = unexplained sum of squares

SSE was defined in Chapter 6 but in the multivariate case $\hat{y}$ is given by (20.7) and the error terms by (20.10).

The degrees of freedom of the SSR equal the number of independent variables, $d_n = k$, while the degrees of freedom of the SSE are $d_d = n - k - 1$.[6] The MSR and MSE are the *mean squares of regression* and *mean squared of errors*, respectively, obtained by dividing the sum of squared deviations by their respective degrees of freedom. All results necessary for the ANOVA are shown in Table 20.1.

If the statistic is found significant at some predetermined level (i.e., $p_F < \alpha$), the model does explain some variation of the dependent variable y.[7]

We ought to be careful not to overdo it; that is, we should not create a model more complicated than necessary. A good guideline is to use the simplest model suitable. Complicated and refined models tend to be inflexible and fail to work with different samples. In most cases, they are poor models for forecasting purposes. So, the best R^2 is not necessarily an indicator of

TABLE 20.1 ANOVA Component Pattern

	df	SS	MS	F	p-Value of F
Regression	k	SSR	$MSR = \frac{SSR}{k}$	$\frac{MSR}{MSE}$	
Residual	n − k − 1	SSE	$MSE = \frac{SSE}{n-k-1}$		
Total	n − 1	SST			

---

[6]In total, the SST is chi-square distributed with $n - 1$ degrees of freedom. The chi-square distribution is covered in Chapter 11.

[7]Alternatively, one can check whether the test statistic is greater than the critical value, that is, $F > F_\alpha$.

the most useful model. The reason is that one can artificially increase R^2 by including additional independent variables into the regression. But the resulting seemingly better fit may be misleading. One will not know the true quality of the model if one evaluates it by applying it to the same data used for the fit. However, often if one uses the fitted model for a different set of data, the weakness of the over-fitted model becomes obvious.

It is for this reason a redefined version of the coefficient of determination is needed and is called the *adjusted R-squared* (or *adjusted R^2*) given by

$$R_{adj}^2 = 1 - \left(1 - R^2\right)\left(\frac{n-1}{n-k-1}\right) \qquad (20.13)$$

This adjusted goodness-of-fit measure incorporates the number of observations, n, as well as the number of independent variables, k, plus the constant term in the denominator $(n - k - 1)$. For as long as the number of observations is very large compared to k, R^2 and R_{adj}^2 are approximately the same.[8] However, if the number k of independent variables included increases, the R_{adj}^2 drops noticeably compared to the original R^2. One can interpret this new measure of fit as penalizing excessive use of independent variables. Instead, one should set up the model as parsimonious as possible. To take most advantage of the set of possible independent variables, one should consider those that contribute a maximum of explanatory variation to the regression. That is, one has to balance the cost of additional independent variables and reduction in the adjusted R^2.

Testing for the Significance of the Independent Variables

Suppose we have found that the model is significant. Now, we turn to the test of significance for individual independent variables. Formally, for each of the k independent variables, we test

$$H_0 : \beta_j = 0 \qquad H_1 : \beta_j \neq 0$$

conditional on the other independent variables already included in the regression model.

The appropriate test would be the *t*-test given by

---

[8]For instance, inserting $k = 1$ into (20.13) we obtain

$$R_{adj}^2 = \frac{(n-2)R^2 + R^2 - 1}{n-2} = R^2 - \frac{1-R^2}{n-2}$$

which, for large n, is only slightly less than R^2.

$$t_j = \frac{b_j - 0}{s_{b_j}} \qquad (20.14)$$

with $n - k - 1$ degrees of freedom. The value b_j is the sample estimate of the j-th regression coefficient and s_{b_j} is the **standard error of the coefficient estimate**.

The standard error of each coefficient is determined by the following estimate of the variance matrix of the entire vector β given by

$$s^2 \left(X^T X \right)^{-1} \qquad (20.15)$$

which is a matrix multiplied by the univariate **standard error of the regression**, s^2. The latter is given by

$$s^2 = \frac{e^T e}{n - k - 1} = \frac{SSE}{n - k - 1} \qquad (20.16)$$

SSE was previously defined and the degrees of freedom are determined by the number of observations, n, minus the number of independent parameters, k, and minus one degree of freedom lost on the constant term. Hence, we obtain $n - k - 1$ degrees of freedom. The j-th diagonal element of equation (20.15), then, is the standard error of the j-th regression coefficient used in equation (20.14).[9] This test statistic in equation (20.14) needs to be compared to the critical values of the tabulated t-distribution with $n - k - 1$ degrees of freedom at some particular significance level α, say 0.05. So, if the test statistic should exceed the critical value then the independent variable is said to be statistically significant. Equivalently, the p-value of equation (20.14) would then be less than α.

The F-Test for Inclusion of Additional Variables

Suppose we have $k - 1$ independent variables in the regression. The goodness-of-fit is given by R_1^2. If we want to check whether it is appropriate to add another independent variable to the regression model, we need a test statistic measuring the improvement in the goodness-of-fit due to the additional variable. Let R^2 denote the goodness-of-fit of the regression after the additional independent variable has been included into the regression. Then the improvement in the explanatory power is given by $R^2 - R_1^2$, which is chi-square distributed with one degree of freedom. Because $1 - R^2$ is chi-square distributed with $n - k - 1$ degrees of freedom, the statistic

---

[9] Typically one does not have to worry about all these rather mathematical steps because statistical software performs these calculations. The interpretation of that output must be understood.

$$F_1 = \frac{R^2 - R_1^2}{\frac{1-R^2}{n-k-1}} \qquad (20.17)$$

is F-distributed with 1 and $n - k - 1$ degrees of freedom under the null hypothesis that the true model consists of $k - 1$ independent variables only.[10]

APPLICATIONS TO FINANCE

We conclude this chapter with two real-world finance problems:

- Estimation of empirical duration
- Predicting the 10-year Treasury yield

Estimation of Empirical Duration

A commonly used measure of the interest-rate sensitivity of a financial asset's value is its *duration*. For example, if a financial asset has a duration of 5, this means that the financial asset's value or price will change by roughly 5% for a 100 basis point change in interest rates. The direction of the change is determined by the sign of the duration. Specifically, if the duration is positive, the price will decline when the relevant interest rate increases but will increase if the relevant interest rate declines. If the duration is negative, the price will increase if the relevant interest rate increases and fall if the relevant interest rate decreases.

So suppose that a common stock selling at a price of $80 has a duration of +5 and that the relevant interest rate that affects the value of the common stock is currently 6%. This means that if that relevant interest rate increases by 100 basis points (from 6% to 7%), the price of the financial asset will decrease by 5%. Since the current price is $80, the price will decline by about $4. On the other hand, if the relevant interest rate decreases from 6% to 5% (a decline of 100 basis points), the price will increase by roughly 5% to $84.

Duration can be estimated by using a valuation model or empirically by estimating from historical returns the sensitivity of the asset's value to changes in interest rates. When duration is measured in the latter way, it is referred to as *empirical duration*. Since it is estimated using regression analysis, it is sometimes referred to as *regression-based duration*.

The dependent variable in the regression model is the percentage change in the value of the asset. We will not use in our illustration individual assets. Rather we will use sectors of the financial market and refer to them as assets.

---

[10] The chi-square and the F-distribution are covered in Chapter 11.

Effectively, these sectors can be viewed as portfolios that are comprised of the components of the index representing the sector. The assets we will estimate the duration for are the (1) electric utility sector of the S&P 500 index, (2) commercial bank sector of the S&P 500 index, and (3) Lehman U.S. Aggregate Bond Index.[11] For each of these indexes the dependent variable is the monthly return in the value of the index. The time period covered is from October 1989 to November 2003 (170 observations) and the monthly return observations are given in the last three columns of Table 20.2.[12]

Let's begin with just one independent variable, an interest rate index. We will use the monthly change in the U.S. Treasury yield index as measured by the Lehman Treasury Index as the relevant interest rate variable. The monthly values are given in the second column of Table 20.2. Notice that the data are reported as the percentage difference between two months. So, if in one month the value of the Treasury yield index is 7.20% and in the next month it is 7.70%, the value for the observation is 0.50%. In finance, a basis point is equal to 0.0001 or 0.01% so that 0.50% is equal to 50 basis points. A 100 basis point change in interest rates is 1% or 1.00. We'll need to understand this in order to interpret the regression results.

The simple linear regression model (i.e., the univariate case) is

$$y = b_0 + b_1 x_1 + e$$

where

y = the monthly return of an index.
x_1 = the monthly change in the Treasury yield.

The estimated regression coefficient b_1 is the empirical duration. To understand why, if we substitute for 100 basis points in the above equation for the monthly change in the Treasury yield, the regression coefficient b_1 tells us that the estimated change in the monthly return of an index will be b_1. This is precisely the definition of empirical duration: the approximate change in the value of an asset for a 100 basis point change in interest rates.

The estimated regression coefficient and other diagnostic values are reported in Table 20.3. Notice that negative values for the estimated empirical duration are reported. In practice, however, the duration is quoted as a positive value. Let's look at the results for all three assets.

---

[11] The Lehman U.S. Aggregate Bond Index is now the Barclays Capital U.S. Aggregate Bond Index.
[12] The data for this illustration were supplied by David Wright of Northern Illinois University.

TABLE 20.2 Data for Empirical Duration Illustration

Month	Change in Lehman Bros Treasury Yield	S&P 500 Return	Monthly Returns for		
			Electric Utility Sector	Commercial Bank Sector	Lehman U.S. Aggregate Bond Index
Oct-89	−0.46	−2.33	2.350	−11.043	2.4600
Nov-89	−0.10	2.08	2.236	−3.187	0.9500
Dec-89	0.12	2.36	3.794	−1.887	0.2700
Jan-90	0.43	−6.71	−4.641	−10.795	−1.1900
Feb-90	0.09	1.29	0.193	4.782	0.3200
Mar-90	0.20	2.63	−1.406	−4.419	0.0700
Apr-90	0.34	−2.47	−5.175	−4.265	−0.9200
May-90	−0.46	9.75	5.455	12.209	2.9600
Jun-90	−0.20	−0.70	0.966	−5.399	1.6100
Jul-90	−0.21	−0.32	1.351	−8.328	1.3800
Aug-90	0.37	−9.03	−7.644	−10.943	−1.3400
Sep-90	−0.06	−4.92	0.435	−15.039	0.8300
Oct-90	−0.23	−0.37	10.704	−10.666	1.2700
Nov-90	−0.28	6.44	2.006	18.892	2.1500
Dec-90	−0.23	2.74	1.643	6.620	1.5600
Jan-91	−0.13	4.42	−1.401	8.018	1.2400
Feb-91	0.01	7.16	4.468	12.568	0.8500
Mar-91	0.03	2.38	2.445	5.004	0.6900
Apr-91	−0.15	0.28	−0.140	7.226	1.0800
May-91	0.06	4.28	−0.609	7.501	0.5800
Jun-91	0.15	−4.57	−0.615	−7.865	−0.0500
Jul-91	−0.13	4.68	4.743	7.983	1.3900
Aug-91	−0.37	2.35	3.226	9.058	2.1600
Sep-91	−0.33	−1.64	4.736	−2.033	2.0300
Oct-91	−0.17	1.34	1.455	0.638	1.1100
Nov-91	−0.15	−4.04	2.960	−9.814	0.9200
Dec-91	−0.59	11.43	5.821	14.773	2.9700
Jan-92	0.42	−1.86	−5.515	2.843	−1.3600
Feb-92	0.10	1.28	−1.684	8.834	0.6506
Mar-92	0.27	−1.96	−0.296	−3.244	−0.5634
Apr-92	−0.10	2.91	3.058	4.273	0.7215
May-92	−0.23	0.54	2.405	2.483	1.8871
Jun-92	−0.26	−1.45	0.492	1.221	1.3760
Jul-92	−0.41	4.03	6.394	−0.540	2.0411
Aug-92	−0.13	−2.02	−1.746	−5.407	1.0122
Sep-92	−0.26	1.15	0.718	1.960	1.1864
Oct-92	0.49	0.36	−0.778	2.631	−1.3266
Nov-92	0.26	3.37	−0.025	7.539	0.0228
Dec-92	−0.24	1.31	3.247	5.010	1.5903
Jan-93	−0.36	0.73	3.096	4.203	1.9177
Feb-93	−0.29	1.35	6.000	3.406	1.7492
Mar-93	0.02	2.15	0.622	3.586	0.4183
Apr-93	−0.10	−2.45	−0.026	−5.441	0.6955

TABLE 20.3 Estimation of Regression Parameters for Empirical Duration—Simple Linear Regression

	Electric Utility Sector	Commercial Bank Sector	Lehman U.S. Aggregate Bond Index
Intercept			
b_0	0.6376	1.1925	0.5308
t-statistic	1.8251	2.3347	21.1592
p-value	0.0698	0.0207	0.0000
Change in the Treasury yield			
b_1	−4.5329	−2.5269	−4.1062
t-statistic	−3.4310	−1.3083	−43.2873
p-value	0.0008	0.1926	0.0000
Goodness of Fit			
R^2	0.0655	0.0101	0.9177
F-value	11.7717	1.7116	1873.8000
p-value	0.0007	0.1926	0.0000

For the Electric Utility sector, the estimated regression coefficient for b_1 is −4.5329 suggesting that for a 100 basis point change in Treasury yields, the percentage change in the value of the stocks comprising this sector will be roughly 4.53%. Moreover, as expected the change will be in the opposite direction to the change in interest rates—when interest rates increase (decrease) the value of this sector decreases (increases). The regression coefficient is statistically significant at the 1% level as can be seen from the t-statistic and p-value. The R^2 for this regression is 6.5%. Thus although statistically significant, this regression only explains 6.5% of the variation is the movement of the Electric Utility sector, suggesting that there are other variables that have not been considered.

Moving on to the Commercial Bank sector, the estimated regression coefficient is not statistically significant at any reasonable level of significance. The regression explains only 1% of the variation in the movement of the stocks in this sector.

Finally, the Lehman U.S. Aggregate Bond Index is, not unexpectedly, highly statistically significant explaining almost 92% of the movement in this index. The reason is obvious. This is a bond index that includes all bonds including Treasury securities.

Now let's move on to add another independent variable that moves us from the univariate case to the multivariate linear regression case. The new independent variable we shall add is the return on the Standard & Poor's

500 Stock Index (S&P 500 hereafter). The observations are given in Table 20.2. So, in this illustration we have $k = 2$. The multivariate linear regression to be estimated is

$$y = b_0 + b_1 x_1 + b_2 x_2 + e$$

where

y = the monthly return of an index
x_1 = the monthly change in the Treasury yield
x_2 = the monthly return on the S&P 500

In a simple linear regression involving only x_2 and y, the estimated regression coefficient b_2 would be the beta of the asset. In the multivariate linear regression model above, b_2 is the asset beta taking into account changes in the Treasury yield.

The regression results including the diagnostic statistics are shown in Table 20.4. Looking first at the independent variable x_1, we reach the same conclusion as to its significance for all three assets as in the univariate case. Note also that the estimated value of the regression coefficients are not much different than in the univariate case. As for our new independent variable, x_2, we see that it is statistically significant at the 1% level of significance for all three asset indexes. While we can perform statistical tests discussed earlier for the contribution of adding the new independent variable, for the two stock sectors, the contribution to explaining the movement in the return in the sector indexes are clearly significant. The R^2 for the Electric Utility sector increased from around 7% in the univariate case to 13% in the multivariate linear regression case. The increase was obviously more dramatic for the Commercial Bank sector, the R^2 increasing from 1% to 49%.

Next we analyze the regression of the Lehman U.S. Aggregate Bond Index. Using only one independent variable, we have $R_1^2 = 91.77\%$. If we include the additional independent variable, we obtain the improved $R^2 = 93.12\%$. For the augmented regression, we compute with $n = 170$ and $k = 2$ the adjusted R^2 as

$$R_{adj}^2 = 1 - \left(1 - R^2\right)\left(\frac{n-1}{n-k-1}\right)$$

$$= 1 - \left(1 - 0.9312\right)\left(\frac{170-1}{170-2-1}\right)$$

$$= 0.9304$$

TABLE 20.3 Estimation of Regression Parameters for Empirical Duration—Multivariate Linear Regression

	Electric Utility Sector	Commercial Bank Sector	Lehman U.S. Aggregate Bond Index
Intercept			
b_0	0.3937	0.2199	0.5029
t-statistic	1.1365	0.5835	21.3885
p-value	0.2574	0.5604	0.0000
Change in the Treasury yield			
b_1	−4.3780	−1.9096	−4.0885
t-statistic	−3.4143	−1.3686	−46.9711
p-value	0.0008	0.1730	0.0000
Return on the S&P 500			
b_2	0.2664	1.0620	0.0304
t-statistic	3.4020	12.4631	5.7252
p-value	0.0008	0.0000	0.0000
Goodness of Fit			
R_2	0.1260	0.4871	0.9312
F-value	12.0430	79.3060	1130.5000
p-value	0.00001	0.00000	0.00000

Let's apply the F-test to Lehman U.S. Aggregate Bond Index to see if the addition of the new independent variable increasing the R^2 from 91.77% to 93.12% is statistically significant.
From (20.17), we have

$$F_1 = \frac{R^2 - R_1^2}{\dfrac{1-R^2}{n-k-1}} = \frac{0.9312 - 0.9177}{\dfrac{1-0.9312}{170-2-1}} = 32.7689$$

This value is highly significant with a p-value of virtually zero. Hence, the inclusion of the additional variable is statistically reasonable.

Predicting the 10-Year Treasury Yield[13]

The U.S. Treasury securities market is the world's most liquid bond market. The U.S. Department of the Treasury issues two types of securities: zero-

---
[13] We are grateful to Robert Scott of the Bank for International Settlement for suggesting this illustration and for providing the data.

coupon securities and coupon securities. Securities issued with one year or less to maturity are called Treasury bills; they are issued as zero-coupon instruments. Treasury securities with more than one year to maturity are issued as coupon-bearing securities. Treasury securities from more than one year up to 10 years of maturity are called Treasury notes; Treasury securities with a maturity in excess of 10 years are called Treasury bonds. The U.S. Treasury auctions securities of specified maturities on a regular calendar basis. The Treasury currently issues 30-year Treasury bonds but had stopped issuance of them from October 2001 to January 2006.

An important Treasury note is the 10-year Treasury note. In this illustration, we try to forecast this rate based on two independent variables suggested by economic theory. A well-known theory of interest rates, known as the Fisher's Law, is that the interest rate in any economy consists of two components. The first is the expected rate of inflation. The second is the real rate of interest. We use regression analysis to produce a model to forecast the yield on the 10-year Treasury note (simply, the 10-year Treasury yield) —the dependent variable—and the expected rate of inflation (simply, expected inflation) and the real rate of interest (simply, real rate).

The 10-year Treasury yield is observable, but we need a proxy for the two independent variables (i.e., the expected rate of inflation and the real rate of interest) because they are not observable at the time of the forecast. Keep in mind that since we are forecasting, we do not use as our independent variable information that is unavailable at the time of the forecast. Consequently, we need a proxy available at the time of the forecast.

The inflation rate is available from the U.S. Department of Commerce. However, we need a proxy for expected inflation. We can use some type of average of past inflation as a proxy. In our model, we use a five-year moving average. There are more sophisticated methodologies for calculating expected inflation, but the five-year moving average is sufficient for our illustration.[14] For the real rate, we use the rate on three-month certificates of deposit (CDs). Again, we use a five-year moving average.

The monthly data for the three variables from November 1965 to December 2005 (482 observations) are provided in Table 20.5. The regression results are reported in Table 20.6. As can be seen, the regression coefficients of both independent variables are positive (as would be predicted by economic theory) and highly significant. The R^2 and adjusted R^2 are 0.90 and 0.83, respectively. The ANOVA table is also shown in the table. The results suggest a good fit for forecasting the 10-year rate.

---

[14]For example, one can use an exponential smoothing of actual inflation, a methodology used by the Organisation for Economic Co-operation and Development (OECD).

TABLE 20.5 Monthly Data for 10-Year Treasury Yield, Expected Inflation, and Real Rate: November 1965–December 2005

Date	10-Yr. Trea. Yield	Exp. Infl.	Real Rate	Date	10-Yr. Trea. Yield	Exp. Infl.	Real Rate	Date	10-Yr. Trea. Yield	Exp. Infl.	Real Rate
1965											
Nov	4.45	1.326	2.739								
Dec	4.62	1.330	2.757								
1966				1969				1972			
Jan	4.61	1.334	2.780	Jan	6.04	2.745	2.811	Jan	5.95	4.959	2.401
Feb	4.83	1.348	2.794	Feb	6.19	2.802	2.826	Feb	6.08	4.959	2.389
Mar	4.87	1.358	2.820	Mar	6.3	2.869	2.830	Mar	6.07	4.953	2.397
Apr	4.75	1.372	2.842	Apr	6.17	2.945	2.827	Apr	6.19	4.953	2.403
May	4.78	1.391	2.861	May	6.32	3.016	2.862	May	6.13	4.949	2.398
June	4.81	1.416	2.883	June	6.57	3.086	2.895	June	6.11	4.941	2.405
July	5.02	1.440	2.910	July	6.72	3.156	2.929	July	6.11	4.933	2.422
Aug	5.22	1.464	2.945	Aug	6.69	3.236	2.967	Aug	6.21	4.924	2.439
Sept	5.18	1.487	2.982	Sept	7.16	3.315	3.001	Sept	6.55	4.916	2.450
Oct	5.01	1.532	2.997	Oct	7.1	3.393	3.014	Oct	6.48	4.912	2.458
Nov	5.16	1.566	3.022	Nov	7.14	3.461	3.045	Nov	6.28	4.899	2.461
Dec	4.84	1.594	3.050	Dec	7.65	3.539	3.059	Dec	6.36	4.886	2.468
1967				1970				1973			
Jan	4.58	1.633	3.047	Jan	7.80	3.621	3.061	Jan	6.46	4.865	2.509
Feb	4.63	1.667	3.050	Feb	7.24	3.698	3.064	Feb	6.64	4.838	2.583
Mar	4.54	1.706	3.039	Mar	7.07	3.779	3.046	Mar	6.71	4.818	2.641
Apr	4.59	1.739	3.027	Apr	7.39	3.854	3.035	Apr	6.67	4.795	2.690
May	4.85	1.767	3.021	May	7.91	3.933	3.021	May	6.85	4.776	2.734
June	5.02	1.801	3.015	June	7.84	4.021	3.001	June	6.90	4.752	2.795
July	5.16	1.834	3.004	July	7.46	4.104	2.981	July	7.13	4.723	2.909
Aug	5.28	1.871	2.987	Aug	7.53	4.187	2.956	Aug	7.40	4.699	3.023
Sept	5.3	1.909	2.980	Sept	7.39	4.264	2.938	Sept	7.09	4.682	3.110
Oct	5.48	1.942	2.975	Oct	7.33	4.345	2.901	Oct	6.79	4.668	3.185
Nov	5.75	1.985	2.974	Nov	6.84	4.436	2.843	Nov	6.73	4.657	3.254
Dec	5.7	2.027	2.972	Dec	6.39	4.520	2.780	Dec	6.74	4.651	3.312
1968				1971				1974			
Jan	5.53	2.074	2.959	Jan	6.24	4.605	2.703	Jan	6.99	4.652	3.330
Feb	5.56	2.126	2.943	Feb	6.11	4.680	2.627	Feb	6.96	4.653	3.332
Mar	5.74	2.177	2.937	Mar	5.70	4.741	2.565	Mar	7.21	4.656	3.353
Apr	5.64	2.229	2.935	Apr	5.83	4.793	2.522	Apr	7.51	4.657	3.404
May	5.87	2.285	2.934	May	6.39	4.844	2.501	May	7.58	4.678	3.405
June	5.72	2.341	2.928	June	6.52	4.885	2.467	June	7.54	4.713	3.419
July	5.5	2.402	2.906	July	6.73	4.921	2.436	July	7.81	4.763	3.421
Aug	5.42	2.457	2.887	Aug	6.58	4.947	2.450	Aug	8.04	4.827	3.401
Sept	5.46	2.517	2.862	Sept	6.14	4.964	2.442	Sept	8.04	4.898	3.346
Oct	5.58	2.576	2.827	Oct	5.93	4.968	2.422	Oct	7.9	4.975	3.271
Nov	5.7	2.639	2.808	Nov	5.81	4.968	2.411	Nov	7.68	5.063	3.176
Dec	6.03	2.697	2.798	Dec	5.93	4.964	2.404	Dec	7.43	5.154	3.086

TABLE 20.5 (Continued)

Date	10-Yr. Trea. Yield	Exp. Infl.	Real Rate	Date	10-Yr. Trea. Yield	Exp. Infl.	Real Rate	Date	10-Yr. Trea. Yield	Exp. Infl.	Real Rate
1975				1978				1981			
Jan	7.5	5.243	2.962	Jan	7.96	6.832	1.068	Jan	12.57	8.520	1.132
Feb	7.39	5.343	2.827	Feb	8.03	6.890	0.995	Feb	13.19	8.594	1.242
Mar	7.73	5.431	2.710	Mar	8.04	6.942	0.923	Mar	13.12	8.649	1.336
Apr	8.23	5.518	2.595	Apr	8.15	7.003	0.854	Apr	13.68	8.700	1.477
May	8.06	5.585	2.477	May	8.35	7.063	0.784	May	14.1	8.751	1.619
June	7.86	5.639	2.384	June	8.46	7.124	0.716	June	13.47	8.802	1.755
July	8.06	5.687	2.311	July	8.64	7.191	0.598	July	14.28	8.877	1.897
Aug	8.4	5.716	2.271	Aug	8.41	7.263	0.482	Aug	14.94	8.956	2.037
Sept	8.43	5.738	2.241	Sept	8.42	7.331	0.397	Sept	15.32	9.039	2.155
Oct	8.15	5.753	2.210	Oct	8.64	7.400	0.365	Oct	15.15	9.110	2.256
Nov	8.05	5.759	2.200	Nov	8.81	7.463	0.322	Nov	13.39	9.175	2.305
Dec	8	5.761	2.186	Dec	9.01	7.525	0.284	Dec	13.72	9.232	2.392
1976				1979				1982			
Jan	7.74	5.771	2.166	Jan	9.1	7.582	0.254	Jan	14.59	9.285	2.497
Feb	7.79	5.777	2.164	Feb	9.1	7.645	0.224	Feb	14.43	9.334	2.612
Mar	7.73	5.800	2.138	Mar	9.12	7.706	0.174	Mar	13.86	9.375	2.741
Apr	7.56	5.824	2.101	Apr	9.18	7.758	0.108	Apr	13.87	9.417	2.860
May	7.9	5.847	2.060	May	9.25	7.797	0.047	May	13.62	9.456	2.958
June	7.86	5.870	2.034	June	8.91	7.821	-0.025	June	14.3	9.487	3.095
July	7.83	5.900	1.988	July	8.95	7.834	-0.075	July	13.95	9.510	3.183
Aug	7.77	5.937	1.889	Aug	9.03	7.837	-0.101	Aug	13.06	9.524	3.259
Sept	7.59	5.981	1.813	Sept	9.33	7.831	-0.085	Sept	12.34	9.519	3.321
Oct	7.41	6.029	1.753	Oct	10.3	7.823	0.011	Oct	10.91	9.517	3.363
Nov	7.29	6.079	1.681	Nov	10.65	7.818	0.079	Nov	10.55	9.502	3.427
Dec	6.87	6.130	1.615	Dec	10.39	7.818	0.154	Dec	10.54	9.469	3.492
1977				1980				1983			
Jan	7.21	6.176	1.573	Jan	10.8	7.825	0.261	Jan	10.46	9.439	3.553
Feb	7.39	6.224	1.527	Feb	12.41	7.828	0.418	Feb	10.72	9.411	3.604
Mar	7.46	6.272	1.474	Mar	12.75	7.849	0.615	Mar	10.51	9.381	3.670
Apr	7.37	6.323	1.427	Apr	11.47	7.879	0.701	Apr	10.4	9.340	3.730
May	7.46	6.377	1.397	May	10.18	7.926	0.716	May	10.38	9.288	3.806
June	7.28	6.441	1.340	June	9.78	7.989	0.702	June	10.85	9.227	3.883
July	7.33	6.499	1.293	July	10.25	8.044	0.695	July	11.38	9.161	3.981
Aug	7.4	6.552	1.252	Aug	11.1	8.109	0.716	Aug	11.85	9.087	4.076
Sept	7.34	6.605	1.217	Sept	11.51	8.184	0.740	Sept	11.65	9.012	4.152
Oct	7.52	6.654	1.193	Oct	11.75	8.269	0.795	Oct	11.54	8.932	4.204
Nov	7.58	6.710	1.154	Nov	12.68	8.356	0.895	Nov	11.69	8.862	4.243
Dec	7.69	6.768	1.119	Dec	12.84	8.446	1.004	Dec	11.83	8.800	4.276

TABLE 20.5 (Continued)

Date	10-Yr. Trea. Yield	Exp. Infl.	Real Rate	Date	10-Yr. Trea. Yield	Exp. Infl.	Real Rate	Date	10-Yr. Trea. Yield	Exp. Infl.	Real Rate
1984				1987				1990			
Jan	11.67	8.741	4.324	Jan	7.08	4.887	4.607	Jan	8.418	4.257	3.610
Feb	11.84	8.670	4.386	Feb	7.25	4.793	4.558	Feb	8.515	4.254	3.595
Mar	12.32	8.598	4.459	Mar	7.25	4.710	4.493	Mar	8.628	4.254	3.585
Apr	12.63	8.529	4.530	Apr	8.02	4.627	4.445	Apr	9.022	4.260	3.580
May	13.41	8.460	4.620	May	8.61	4.551	4.404	May	8.599	4.264	3.586
June	13.56	8.393	4.713	June	8.4	4.476	4.335	June	8.412	4.272	3.589
July	13.36	8.319	4.793	July	8.45	4.413	4.296	July	8.341	4.287	3.568
Aug	12.72	8.241	4.862	Aug	8.76	4.361	4.273	Aug	8.846	4.309	3.546
Sept	12.52	8.164	4.915	Sept	9.42	4.330	4.269	Sept	8.795	4.335	3.523
Oct	12.16	8.081	4.908	Oct	9.52	4.302	4.259	Oct	8.617	4.357	3.503
Nov	11.57	7.984	4.919	Nov	8.86	4.285	4.243	Nov	8.252	4.371	3.493
Dec	12.5	7.877	4.928	Dec	8.99	4.279	4.218	Dec	8.067	4.388	3.471
1985				1988				1991			
Jan	11.38	7.753	4.955	Jan	8.67	4.274	4.180	Jan	8.007	4.407	3.436
Feb	11.51	7.632	4.950	Feb	8.21	4.271	4.149	Feb	8.033	4.431	3.396
Mar	11.86	7.501	4.900	Mar	8.37	4.268	4.104	Mar	8.061	4.451	3.360
Apr	11.43	7.359	4.954	Apr	8.72	4.270	4.075	Apr	8.013	4.467	3.331
May	10.85	7.215	5.063	May	9.09	4.280	4.036	May	8.059	4.487	3.294
June	10.16	7.062	5.183	June	8.92	4.301	3.985	June	8.227	4.504	3.267
July	10.31	6.925	5.293	July	9.06	4.322	3.931	July	8.147	4.517	3.247
Aug	10.33	6.798	5.346	Aug	9.26	4.345	3.879	Aug	7.816	4.527	3.237
Sept	10.37	6.664	5.383	Sept	8.98	4.365	3.844	Sept	7.445	4.534	3.223
Oct	10.24	6.528	5.399	Oct	8.8	4.381	3.810	Oct	7.46	4.540	3.207
Nov	9.78	6.399	5.360	Nov	8.96	4.385	3.797	Nov	7.376	4.552	3.177
Dec	9.26	6.269	5.326	Dec	9.11	4.384	3.787	Dec	6.699	4.562	3.133
1986				1989				1992			
Jan	9.19	6.154	5.284	Jan	9.09	4.377	3.786	Jan	7.274	4.569	3.092
Feb	8.7	6.043	5.249	Feb	9.17	4.374	3.792	Feb	7.25	4.572	3.054
Mar	7.78	5.946	5.225	Mar	9.36	4.367	3.791	Mar	7.528	4.575	3.014
Apr	7.3	5.858	5.143	Apr	9.18	4.356	3.784	Apr	7.583	4.574	2.965
May	7.71	5.763	5.055	May	8.86	4.344	3.758	May	7.318	4.571	2.913
June	7.8	5.673	4.965	June	8.28	4.331	3.723	June	7.121	4.567	2.864
July	7.3	5.554	4.878	July	8.02	4.320	3.679	July	6.709	4.563	2.810
Aug	7.17	5.428	4.789	Aug	8.11	4.306	3.644	Aug	6.604	4.556	2.757
Sept	7.45	5.301	4.719	Sept	8.19	4.287	3.623	Sept	6.354	4.544	2.682
Oct	7.43	5.186	4.671	Oct	8.01	4.273	3.614	Oct	6.789	4.533	2.624
Nov	7.25	5.078	4.680	Nov	7.87	4.266	3.609	Nov	6.937	4.522	2.571
Dec	7.11	4.982	4.655	Dec	7.84	4.258	3.611	Dec	6.686	4.509	2.518

TABLE 20.5 (Continued)

Date	10-Yr. Trea. Yield	Exp. Infl.	Real Rate	Date	10-Yr. Trea. Yield	Exp. Infl.	Real Rate	Date	10-Yr. Trea. Yield	Exp. Infl.
1993				1996				1999		
Jan	6.359	4.495	2.474	Jan	5.58	3.505	1.250	Jan	4.651	2.631
Feb	6.02	4.482	2.427	Feb	6.098	3.458	1.270	Feb	5.287	2.621
Mar	6.024	4.466	2.385	Mar	6.327	3.418	1.295	Mar	5.242	2.605
Apr	6.009	4.453	2.330	Apr	6.67	3.376	1.328	Apr	5.348	2.596
May	6.149	4.439	2.272	May	6.852	3.335	1.359	May	5.622	2.586
June	5.776	4.420	2.214	June	6.711	3.297	1.387	June	5.78	2.572
July	5.807	4.399	2.152	July	6.794	3.261	1.417	July	5.903	2.558
Aug	5.448	4.380	2.084	Aug	6.943	3.228	1.449	Aug	5.97	2.543
Sept	5.382	4.357	2.020	Sept	6.703	3.195	1.481	Sept	5.877	2.527
Oct	5.427	4.333	1.958	Oct	6.339	3.163	1.516	Oct	6.024	2.515
Nov	5.819	4.309	1.885	Nov	6.044	3.131	1.558	Nov	6.191	2.502
Dec	5.794	4.284	1.812	Dec	6.418	3.102	1.608	Dec	6.442	2.490
1994				1997				2000		
Jan	5.642	4.256	1.739	Jan	6.494	3.077	1.656	Jan	6.665	2.477
Feb	6.129	4.224	1.663	Feb	6.552	3.057	1.698	Feb	6.409	2.464
Mar	6.738	4.195	1.586	Mar	6.903	3.033	1.746	Mar	6.004	2.455
Apr	7.042	4.166	1.523	Apr	6.718	3.013	1.795	Apr	6.212	2.440
May	7.147	4.135	1.473	May	6.659	2.990	1.847	May	6.272	2.429
June	7.32	4.106	1.427	June	6.5	2.968	1.899	June	6.031	2.421
July	7.111	4.079	1.394	July	6.011	2.947	1.959	July	6.031	2.412
Aug	7.173	4.052	1.356	Aug	6.339	2.926	2.016	Aug	5.725	2.406
Sept	7.603	4.032	1.315	Sept	6.103	2.909	2.078	Sept	5.802	2.398
Oct	7.807	4.008	1.289	Oct	5.831	2.888	2.136	Oct	5.751	2.389
Nov	7.906	3.982	1.278	Nov	5.874	2.866	2.189	Nov	5.468	2.382
Dec	7.822	3.951	1.278	Dec	5.742	2.847	2.247	Dec	5.112	2.374
1995				1998				2001		
Jan	7.581	3.926	1.269	Jan	5.505	2.828		Jan	5.114	2.368
Feb	7.201	3.899	1.261	Feb	5.622	2.806		Feb	4.896	2.366
Mar	7.196	3.869	1.253	Mar	5.654	2.787		Mar	4.917	2.364
Apr	7.055	3.840	1.240	Apr	5.671	2.765		Apr	5.338	2.364
May	6.284	3.812	1.230	May	5.552	2.744		May	5.381	2.362
June	6.203	3.781	1.222	June	5.446	2.725		June	5.412	2.363
July	6.426	3.746	1.223	July	5.494	2.709		July	5.054	2.363
Aug	6.284	3.704	1.228	Aug	4.976	2.695		Aug	4.832	2.365
Sept	6.182	3.662	1.232	Sept	4.42	2.680		Sept	4.588	2.365
Oct	6.02	3.624	1.234	Oct	4.605	2.666		Oct	4.232	2.366
Nov	5.741	3.587	1.229	Nov	4.714	2.653		Nov	4.752	2.368
Dec	5.572	3.549	1.234	Dec	4.648	2.641		Dec	5.051	2.370

TABLE 20.5 (Continued)

Date	10-Yr. Trea. Yield	Exp. Infl.	Real Rate	Date	10-Yr. Trea. Yield	Exp. Infl.	Real Rate
2002				2004			
Jan	5.033	2.372	2.950	Jan	4.134	2.172	1.492
Feb	4.877	2.372	2.888	Feb	3.973	2.157	1.442
Mar	5.396	2.371	2.827	Mar	3.837	2.149	1.385
Apr	5.087	2.369	2.764	Apr	4.507	2.142	1.329
May	5.045	2.369	2.699	May	4.649	2.136	1.273
June	4.799	2.367	2.636	June	4.583	2.134	1.212
July	4.461	2.363	2.575	July	4.477	2.129	1.156
Aug	4.143	2.364	2.509	Aug	4.119	2.126	1.097
Sept	3.596	2.365	2.441	Sept	4.121	2.124	1.031
Oct	3.894	2.365	2.374	Oct	4.025	2.122	0.966
Nov	4.207	2.362	2.302	Nov	4.351	2.124	0.903
Dec	3.816	2.357	2.234	Dec	4.22	2.129	0.840
2003				2005			
Jan	3.964	2.351	2.168	Jan	4.13	2.131	0.783
Feb	3.692	2.343	2.104	Feb	4.379	2.133	0.727
Mar	3.798	2.334	2.038	Mar	4.483	2.132	0.676
Apr	3.838	2.323	1.976	Apr	4.2	2.131	0.622
May	3.372	2.312	1.913	May	3.983	2.127	0.567
June	3.515	2.300	1.850	June	3.915	2.120	0.520
July	4.408	2.288	1.786	July	4.278	2.114	0.476
Aug	4.466	2.267	1.731	Aug	4.016	2.107	0.436
Sept	3.939	2.248	1.681	Sept	4.326	2.098	0.399
Oct	4.295	2.233	1.629	Oct	4.553	2.089	0.366
Nov	4.334	2.213	1.581	Nov	4.486	2.081	0.336
Dec	4.248	2.191	1.537	Dec	4.393	2.075	0.311

Note:
Expected Infl. (%) = Expected rate of inflation as proxied by the five-year moving average of the actual inflation rate.
Real Rate (%) = Real rate of interest as proxied by the five-year moving average of the interest rate on three-month certificates of deposit.

TABLE 20.6 Results of Regression for Forecasting 10-Year Treasury Yield

Regression Statistics	
Multiple R^2	0.9083
R^2	0.8250
Adjusted R^2	0.8243
Standard Error	1.033764
Observations	482

Analysis of Variance

	df	SS	MS	F	Significance F
Regression	2	2413.914	1206.957	1129.404	4.8E-182
Residual	479	511.8918	1.068668		
Total	481	2925.806			

	Coefficients	Standard Error	t	Statistics p-value
Intercept	1.89674	0.147593	12.85	1.1E-32
Expected Inflation	0.996937	0.021558	46.24	9.1E-179
Real Rate	0.352416	0.039058	9.02	4.45E-18

CONCEPTS EXPLAINED IN THIS CHAPTER (IN ORDER OF PRESENTATION)

Simple linear regression
Univariate regression
Multivariate linear regression model
Hyperplane
Regression hyperplane
Multiple coefficient of determination
Analysis of variance (ANOVA) test
Mean squares of regression
Mean squared of errors
Adjusted R-squared (or adjusted R^2)
Standard error of the coefficient estimate
Standard error of the regression
Duration
Empirical duration
Regression-based duration

CHAPTER 21

Designing and Building a Multivariate Linear Regression Model

In this chapter we continue with our coverage of multivariate linear regression analysis. The three topics covered in this chapter are the problem of multicollinearity, incorporating dummy variables into a regression model, and model building techniques using stepwise regression analysis.

THE PROBLEM OF MULTICOLLINEARITY

When discussing the suitability of a model, an important issue is the structure or interaction of the independent variables. This is referred to as *multicollinearity*. Tests for the presence of multicollinearity must be performed after the model's significance has been determined and all significant independent variables to be used in the final regression have been determined. Investigation for the presence of multicollinearity involves the correlation between the independent variables and the dependent variable.

A good deal of intuition is helpful in assessing if the regression coefficients make any sense. For example, one by one, select each independent variable and let all other independent variables be equal to zero. Now, estimate a regression merely with this particular independent variable and see if the regression coefficient of this variable seems unreasonable because if its sign is counterintuitive or its value appears too small or large, one may want to consider removing that independent variable from the regression. The reason may very well be attributable to multicollinearity. Technically, multicollinearity is caused by independent variables in the regression model that contain common information. The independent variables are highly intercorrelated; that is, they have too much linear dependence. Hence the presence of multicollinear independent variables prevents us from obtaining insight into the true contribution to the regression from each independent variable.

Formally, multicollinearity coincides with the following mathematical statement

$$\text{rank of } (X^T X) < k \tag{21.1}$$

Equation (21.1) can be interpreted as X not consisting of vectors X_i, $i = 1$, ..., k that jointly can reproduce any vector in the k-dimensional real numbers, R^k.

In a very extreme case, two or more variables may be perfectly correlated (i.e., their pairwise correlations are equal to one), which would imply that some vectors of observations of these variables are merely linear combinations of others. The result of this would be that some variables are fully explained by others and, thus, provide no additional information. This is a very extreme case, however. In most problems in finance, the independent data vectors are not perfectly correlated but may be correlated to a high degree. In any case, the result is that, roughly speaking, the regression estimation procedure is confused by this ambiguity of data information such that it cannot produce distinct regression coefficients for the variables involved. The β_i, $i = 1, ..., k$ cannot be identified; hence, an infinite number of possible values for the regression coefficients can serve as a solution. This can be very frustrating in building a reliable regression model.

We can demonstrate the problem with an example. Consider a regression model with three independent variables—X_1, X_2, and X_3. Also assume the following regarding these three independent variables

$$X_1 = 2X_2 = 4X_3$$

such that there is, effectively, just one independent variable, either X_1, X_2, or X_3. Now, suppose all three independent variables are erroneously used to model the following regression

$$y = \beta_1 X_1 + \beta_2 X_2 + \beta_3 X_3$$
$$= 4\beta_1 X_3 + 2\beta_2 X_3 + \beta_3 X_3$$

Just to pick one possibility of ambiguity, the same effect is achieved by either increasing β_1 by, for example, 0.25 or by increasing β_3 by 1, and so forth. In this example, the rank would just be 1. This is also intuitive since, generally, the rank of $(X^T X)^{-1}$ indicates the number of truly independent sources.[1]

---

[1] One speaks of "near multicollinearity" if the rank is less than k.

Procedures for Mitigating Multicollinearity

While it is quite impossible to provide a general rule to eliminate the problem of multicollinearity, there are some techniques that can be employed to mitigate the problem.

Multicollinearity might be present if there appears to be a mismatch between the sign of the correlation coefficient and the regression coefficient of that particular independent variable. So, the first place to always check is the correlation coefficient for each independent variable and the dependent variable.

Other indicators of multicollinearity are:

1. The sensitivity of regression coefficients to the inclusion of additional independent variables.
2. Changes from significance to insignificance of already included independent variables after new ones have been added.
3. An increase in the model's standard error of the regression.

A consequence of the above is that the regression coefficient estimates vary dramatically as a result of only minor changes in the data X.

A remedy most commonly suggested is to try to single out independent variables that are likely to cause the problems. This can be done by excluding those independent variable so identified from the regression model. It may be possible to include other independent variables, instead, that provide additional information.

In general, due to multicollinearity, the standard error of the regression increases, rendering the t-ratios of many independent variables too small to indicate significance despite the fact that the regression model, itself is highly significant.

To find out whether the variance error of the regression is too large, we present a commonly employed tool. We measure multicollinearity by computing the impact of the correlation between some independent variables and the j-th independent variable. Therefore, we need to regress the j-th variable on the remaining $k-1$ variables. The resulting regression would look like

$$x_j = c + b_1^{(j)} x_1 + \ldots + b_{j-1}^{(j)} x_{j-1} + b_{j+1}^{(j)} x_{j+1} + \ldots + b_k^{(j)} x_k, \quad j = 1, 2, \ldots, k.$$

Then we obtain the coefficient of determination of this regression, R_j^2. This, again, is used to divide the original variance of the j-th regression coefficient estimate by a correction term. Formally, this correction term, called the *variance inflation factor* (VIF), is given by

$$VIF = \frac{1}{(1-R_j^2)} \qquad (21.2)$$

So, if there is no correlation present between independent variable j and the other independent variables, the variance of b_j will remain the same and the t-test results will be unchanged. On the contrary, in the case of more intense correlation, the variance will increase and most likely reject variable x_j as significant for the overall regression.

Consequently, prediction for the j-th regression coefficient becomes less precise since its confidence interval increases due to equation (21.2).[2] We know from the theory of estimation in Chapter 18 that the confidence interval for the regression coefficient at the level α is given by

$$\left[b_j - t_{\alpha/2} \cdot s_{b_j}, b_j + t_{\alpha/2} \cdot s_{b_j} \right] \qquad (21.3)$$

where $t_{\alpha/2}$ is the critical value at level α of the t-distribution with $n-k$ degrees of freedom. This means that with probability $1-\alpha$, the true coefficient is inside of this interval.[3] Naturally, the result of some $VIF > 1$ leads to a widening of the confidence interval given by equation (21.3).

As a rule of thumb, a benchmark for the VIF is often given as 10. A VIF that exceeds 10 indicates a severe impact due to multicollinearity and the independent variable is best removed from the regression.

INCORPORATING DUMMY VARIABLES AS INDEPENDENT VARIABLES

So far, we have considered quantitative data only. There might be a reason to perform some regression that includes some independent variables that are qualitative. For example, one may wish to determine whether the industry sector has some influence on the performance of stocks in general. The independent variable of interest entering the regression model would be one that assumes integer (code) values, one for each sector taken into consideration. The solution is provided by the inclusion of so-called *dummy variables*.

---

[2]The concept of the confidence interval is explained in Chapter 18. The confidence level is often chosen $1-\alpha = 0.99$ or $1-\alpha = 0.95$ such that the parameter is inside of the interval with 0.95 or 0.99 probability, respectively.
[3]This is based on the assumptions stated in the context of estimation.

These dummy variables are designed such that they either assume the value 1 or 0; for that reason they are referred to as *binary variables*.[4] The value 1 is used if some case is true (e.g., the company analyzed belongs to a certain sector) and 0 else. Hence, the dummy variable shifts the regression line by some constant such that we actually work with two parallel lines.

Should the independent variable of interest assume several values, for example, seven industry sectors, we would then need dummy variables accounting for all the different possibilities. To be precise, it will be necessary, in this example, to include six different dummy variables. The resulting regression—if we only consider industry sectors in the regression model—will look like

$$y = \beta_0 + \beta_1 x_1 + \ldots + \beta_6 x_6 + \varepsilon \qquad (21.4)$$

with

y = return on the stock of some company
$x_1, \ldots, x_6$ = industry sectors 1 through 6

Note that there are only six industry sectors accounted for. The seventh is left out as an explicit variable. However, the unique case of $x_1 = \ldots = x_6 = 0$ is equivalent to the company belonging to the seventh sector. So, the sector values are mutually exclusive (i.e., a company cannot belong to more than one industry sector).

We can, alternatively, decide to model the regression without a constant intercept. In our example with industry sectors, this means we have to use seven dummy variables. However, we cannot have a constant and the number of dummy variables equal to the number of different values the variables of interest can assume. In our example, we could not have a constant intercept and seven dummy variables. The reason for this is what is referred to as the *dummy variable trap*. Our data matrix X would contain redundant (i.e., linearly dependent data vectors), which means perfect multicollinearity. In that case, we could not compute the regression coefficients.

Another case where dummy variables are commonly employed is when we want to discern if some data originated from some particular period or not. We set the variable equal to one if the date of the observation is within that period, and zero elsewhere. Consider again the simple regression model

$$y = \beta_0 + \beta_1 x_1 + \varepsilon \qquad (21.5)$$

---

[4]Since they only assume two different values, they are also sometimes referred to as *dichotomous variables*.

Suppose we can divide our data into two groups. For each group, its own functional relationship between the independent and the dependent variables appears reasonable. So since the regression equations for both are potentially different from one another, we use a dummy variable as an indicator of the group. The effect is that we can switch between one and the other model.

How do we obtain the two models? Consider that equation (21.5) is only true for one group. For the second group, equation (21.5) changes to another simple regression model with different regression coefficients. This new regression line is assembled from what is already there, namely equation (21.5), plus something that is unique to the second group, which is

$$\beta_2 + \beta_3 x_1 \ . \tag{21.6}$$

Combining equations (21.5) and (21.6) and incorporating the "switch" dummy variable to discern between the two groups, the resulting regression line is of the form

$$y = \beta_0 + \beta_1 x_1 + \beta_2 x_2 + \beta_3 x_1 x_2 + \varepsilon \ . \tag{21.7}$$

Now, it can be easily seen how equation (21.7) functions. If we have group one, the dummy variable is set to $x_2 = 0$. On the other hand, if we happen to observe the dependent variable y of the second group, we switch to the alternative model by setting $x_2 = 1$, and, hence, the additional terms kick in.

We will refer to the additional terms as **common terms** compared to equation (21.5). The entire model (21.6) is a **composite model**. The model is estimated in the usual fashion by additionally indicating which group the data from X came from.

The more general version of the regression model (21.7) would permit any type of variable for x_2. We restrict ourselves to dummy variables for the composite model. Here, we can see how helpful the composite model can be in differentiating time series with respect to certain periods.

The t-statistic applied to the regression coefficients of dummy variables offer a set of important tests to judge which independent variables are significant. There is also an F-test that can be used to test the significance of all the dummy variables in the regression. This test, known as the **Chow test**,[5] involves running a regression with and without that variable and an F-test to gauge if all the dummy variables are collectively irrelevant. We will not present the Chow test here.

---

[5] See Chow (1960).

Application to Testing the Mutual Fund Characteristic Lines in Different Market Environments

In Chapter 6, we calculated the characteristic line of two large-cap mutual funds. Let's now perform a simple application of the use of dummy variables by determining if the slope (beta) of the two mutual funds is different in a rising stock market ("up market") and a declining stock market ("down market"). To test this, we can write the following multiple regression model:

$$y_t = b_0 + b_1 x_t + b_2(D_t x_t) + e_t$$

where D_t is the dummy variable that can take on a value of 1 or 0. We will let

$D_t = 1$ if period t is classified as an up market
$D_t = 0$ if period t is classified as a down market

The regression coefficient for the dummy variable is b_2. If that regression coefficient is statistically significant, then for the mutual fund:

- In an up market beta is equal to $b_1 + b_2$.
- In a down market beta is equal to b_1.

If b_2 is not statistically significant, then there is no difference in beta for up and down markets, both being equal to b_1.

In our illustration, we have to define what we mean by an up and a down market. We will define an up market precisely as one where the average excess return (market return over the risk-free rate or $(r_M - r_{ft})$) for the prior three months is greater than zero. Then

$D_t = 1$ if the average $(r_{Mt} - r_{ft})$ for the prior three months > 0
$D_t = 0$ otherwise

The independent variable will then be

$D_t x_t = x_t$ if $(r_M - r_{ft})$ for the prior three months > 0
$D_t x_t = 0$ otherwise

We use the S&P 500 Stock Index as a proxy for the market returns and the 90-day Treasury rate as a proxy for the risk-free rate. The data are presented in Table 21.1, which shows each observation for the variable $D_t x_t$.

The regression results for the two mutual funds are shown in Table 21.2. The adjusted R^2 is 0.93 and 0.83 for mutual funds A and B, respectively. For both funds, b_2 is statistically significantly different from zero. Hence, for

TABLE 21.1 Data for Estimating Mutual Fund Characteristic Line with a Dummy Variable

Month Ended	r_M	r_{ft}	Dummy D_t	$r_M - r_{ft}$ x_t	$D_t x_t$	Mutual Fund A r_t	B r_t	A y_t	B y_t
01/31/1995[1]	2.60	0.42	0	2.18	0	0.65	1.28	0.23	0.86
02/28/1995[1]	3.88	0.40	0	3.48	0	3.44	3.16	3.04	2.76
03/31/1995[1]	2.96	0.46	1	2.50	2.5	2.89	2.58	2.43	2.12
04/30/1995	2.91	0.44	1	2.47	2.47	1.65	1.81	1.21	1.37
05/31/1995	3.95	0.54	1	3.41	3.41	2.66	2.96	2.12	2.42
06/30/1995	2.35	0.47	1	1.88	1.88	2.12	2.18	1.65	1.71
07/31/1995	3.33	0.45	1	2.88	2.88	3.64	3.28	3.19	2.83
08/31/1995	0.27	0.47	1	−0.20	−0.2	−0.40	0.98	−0.87	0.51
09/30/1995	4.19	0.43	1	3.76	3.76	3.06	3.47	2.63	3.04
10/31/1995	−0.35	0.47	1	−0.82	−0.82	−1.77	−0.63	−2.24	−1.10
11/30/1995	4.40	0.42	1	3.98	3.98	4.01	3.92	3.59	3.50
12/31/1995	1.85	0.49	1	1.36	1.36	1.29	1.73	0.80	1.24
01/31/1996	3.44	0.43	1	3.01	3.01	3.36	2.14	2.93	1.71
02/29/1996	0.96	0.39	1	0.57	0.57	1.53	1.88	1.14	1.49
03/31/1996	0.96	0.39	1	0.57	0.57	0.59	1.65	0.20	1.26
04/30/1996	1.47	0.46	1	1.01	1.01	1.46	1.83	1.00	1.37
05/31/1996	2.58	0.42	1	2.16	2.16	2.17	2.20	1.75	1.78
06/30/1996	0.41	0.40	1	0.01	0.01	−0.63	0.00	−1.03	−0.40
07/31/1996	−4.45	0.45	1	−4.90	−4.9	−4.30	−3.73	−4.75	−4.18
08/31/1996	2.12	0.41	0	1.71	0	2.73	2.24	2.32	1.83
09/30/1996	5.62	0.44	0	5.18	0	5.31	4.49	4.87	4.05
10/31/1996	2.74	0.42	1	2.32	2.32	1.42	1.34	1.00	0.92
11/30/1996	7.59	0.41	1	7.18	7.18	6.09	5.30	5.68	4.89
12/31/1996	−1.96	0.46	1	−2.42	−2.42	−1.38	−0.90	−1.84	−1.36
01/31/1997	6.21	0.45	1	5.76	5.76	4.15	5.73	3.70	5.28
02/28/1997	0.81	0.39	1	0.42	0.42	1.65	−1.36	1.26	−1.75
03/31/1997	−4.16	0.43	1	−4.59	−4.59	−4.56	−3.75	−4.99	−4.18
04/30/1997	5.97	0.43	1	5.54	5.54	4.63	3.38	4.20	2.95
05/31/1997	6.14	0.49	1	5.65	5.65	5.25	6.05	4.76	5.56
06/30/1997	4.46	0.37	1	4.09	4.09	2.98	2.90	2.61	2.53
07/31/1997	7.94	0.43	1	7.51	7.51	6.00	7.92	5.57	7.49
08/31/1997	−5.56	0.41	1	−5.97	−5.97	−4.40	−3.29	−4.81	−3.70
09/30/1997	5.48	0.44	1	5.04	5.04	5.70	4.97	5.26	4.53
10/31/1997	−3.34	0.42	1	−3.76	−3.76	−2.76	−2.58	−3.18	−3.00
11/30/1997	4.63	0.39	0	4.24	0	3.20	2.91	2.81	2.52
12/31/1997	1.72	0.48	1	1.24	1.24	1.71	2.41	1.23	1.93
01/31/1998	1.11	0.43	1	0.68	0.68	−0.01	−0.27	−0.44	−0.70
02/28/1998	7.21	0.39	1	6.82	6.82	5.50	6.84	5.11	6.45
03/31/1998	5.12	0.39	1	4.73	4.73	5.45	3.84	5.06	3.45
04/30/1998	1.01	0.43	1	0.58	0.58	−0.52	1.07	−0.95	0.64
05/31/1998	−1.72	0.40	1	−2.12	−2.12	−1.25	−1.30	−1.65	−1.70
06/30/1998	4.06	0.41	1	3.65	3.65	3.37	4.06	2.96	3.65
07/31/1998	−1.06	0.40	1	−1.46	−1.46	0.10	−1.75	−0.30	−2.15
08/31/1998	−14.46	0.43	1	−14.89	−14.89	−15.79	−13.44	−16.22	−13.87

TABLE 21.1 (Continued)

Month Ended	r_M	r_{ft}	Dummy D_t	$r_M - r_{ft}$ x_t	$D_t x_t$	Mutual Fund A r_t	B r_t	A y_t	B y_t
09/30/1998	6.41	0.46	0	5.95	0	5.00	4.86	4.54	4.40
10/31/1998	8.13	0.32	0	7.81	0	5.41	4.56	5.09	4.24
11/30/1998	6.06	0.31	0	5.75	0	5.19	5.56	4.88	5.25
12/31/1998	5.76	0.38	1	5.38	5.38	7.59	7.18	7.21	6.80
01/31/1999	4.18	0.35	1	3.83	3.83	2.60	3.11	2.25	2.76
02/28/1999	-3.11	0.35	1	-3.46	-3.46	-4.13	-3.01	-4.48	-3.36
03/31/1999	4.00	0.43	1	3.57	3.57	3.09	3.27	2.66	2.84
04/30/1999	3.87	0.37	1	3.50	3.5	2.26	2.22	1.89	1.85
05/31/1999	-2.36	0.34	1	-2.70	-2.7	-2.12	-1.32	-2.46	-1.66
06/30/1999	5.55	0.40	1	5.15	5.15	4.43	5.36	4.03	4.96
07/31/1999	-3.12	0.38	1	-3.50	-3.5	-3.15	-1.72	-3.53	-2.10
08/31/1999	-0.50	0.39	0	-0.89	0	-1.05	-2.06	-1.44	-2.45
09/30/1999	-2.74	0.39	1	-3.13	-3.13	-2.86	-1.33	-3.25	-1.72
10/31/1999	6.33	0.39	0	5.94	0	5.55	2.29	5.16	1.90
11/30/1999	2.03	0.36	1	1.67	1.67	3.23	3.63	2.87	3.27
12/31/1999	5.89	0.44	1	5.45	5.45	8.48	7.09	8.04	6.65
01/31/2000	-5.02	0.41	1	-5.43	-5.43	-4.09	-0.83	-4.50	-1.24
02/29/2000	-1.89	0.43	1	-2.32	-2.32	1.43	2.97	1.00	2.54
03/31/2000	9.78	0.47	0	9.31	0	6.84	5.86	6.37	5.39
04/30/2000	-3.01	0.46	1	-3.47	-3.47	-4.04	-4.55	-4.50	-5.01
05/31/2000	-2.05	0.50	1	-2.55	-2.55	-2.87	-4.47	-3.37	-4.97
06/30/2000	2.46	0.40	1	2.06	2.06	0.54	6.06	0.14	5.66
07/31/2000	-1.56	0.48	0	-2.04	0	-0.93	1.89	-1.41	1.41
08/31/2000	6.21	0.50	0	5.71	0	7.30	6.01	6.80	5.51
09/30/2000	-5.28	0.51	1	-5.79	-5.79	-4.73	-4.81	-5.24	-5.32
10/31/2000	-0.42	0.56	0	-0.98	0	-1.92	-4.84	-2.48	-5.40
11/30/2000	-7.88	0.51	0	-8.39	0	-6.73	-11.00	-7.24	-11.51
12/31/2000	0.49	0.50	0	-0.01	0	2.61	3.69	2.11	3.19
01/31/2001	3.55	0.54	0	3.01	0	0.36	5.01	-0.18	4.47
02/28/2001	-9.12	0.38	0	-9.50	0	-5.41	-8.16	-5.79	-8.54
03/31/2001	-6.33	0.42	0	-6.75	0	-5.14	-5.81	-5.56	-6.23
04/30/2001	7.77	0.39	0	7.38	0	5.25	4.67	4.86	4.28
05/31/2001	0.67	0.32	0	0.35	0	0.47	0.45	0.15	0.13
06/30/2001	-2.43	0.28	1	-2.71	-2.71	-3.48	-1.33	-3.76	-1.61
07/31/2001	-0.98	0.30	1	-1.28	-1.28	-2.24	-1.80	-2.54	-2.10
08/31/2001	-6.26	0.31	0	-6.57	0	-4.78	-5.41	-5.09	-5.72
09/30/2001	-8.08	0.28	0	-8.36	0	-6.46	-7.27	-6.74	-7.55
10/31/2001	1.91	0.22	0	1.69	0	1.01	2.30	0.79	2.08
11/30/2001	7.67	0.17	0	7.50	0	4.49	5.62	4.32	5.45
12/31/2001	0.88	0.15	1	0.73	0.73	1.93	2.14	1.78	1.99
01/31/2002	-1.46	0.14	1	-1.60	-1.6	-0.99	-3.27	-1.13	-3.41
02/28/2002	-1.93	0.13	1	-2.06	-2.06	-0.84	-2.68	-0.97	-2.81
03/31/2002	3.76	0.13	0	3.63	0	3.38	4.70	3.25	4.57
04/30/2002	-6.06	0.15	0	-6.21	0	-4.38	-3.32	-4.53	-3.47
05/31/2002	-0.74	0.14	0	-0.88	0	-1.78	-0.81	-1.92	-0.95
06/30/2002	-7.12	0.13	0	-7.25	0	-5.92	-5.29	-6.05	-5.42

554 MULTIVARIATE LINEAR REGRESSION ANALYSIS

TABLE 21.1 (Continued)

Month Ended	r_M	r_{ft}	Dummy D_t	$r_M - r_{ft}$ x_t	$D_t x_t$	Mutual Fund A r_t	Mutual Fund B r_t	A y_t	B y_t
07/31/2002	−7.80	0.15	0	−7.95	0	−6.37	−7.52	−6.52	−7.67
08/31/2002	0.66	0.14	0	0.52	0	−0.06	1.86	−0.20	1.72
09/30/2002	−10.87	0.14	0	−11.01	0	−9.38	−6.04	−9.52	−6.18
10/31/2002	8.80	0.14	0	8.66	0	3.46	5.10	3.32	4.96
11/30/2002	5.89	0.12	0	5.77	0	3.81	1.73	3.69	1.61
12/31/2002	−5.88	0.11	1	−5.99	−5.99	−4.77	−2.96	−4.88	−3.07
01/31/2003	−2.62	0.10	1	−2.72	−2.72	−1.63	−2.34	−1.73	−2.44
02/28/2003	−1.50	0.09	0	−1.59	0	−0.48	−2.28	−0.57	−2.37
03/31/2003	0.97	0.10	0	0.87	0	1.11	1.60	1.01	1.50
04/30/2003	8.24	0.10	0	8.14	0	6.67	5.44	6.57	5.34
05/31/2003	5.27	0.09	1	5.18	5.18	4.96	6.65	4.87	6.56
06/30/2003	1.28	0.10	1	1.18	1.18	0.69	1.18	0.59	1.08
07/31/2003	1.76	0.07	1	1.69	1.69	1.71	3.61	1.64	3.54
08/31/2003	1.95	0.07	1	1.88	1.88	1.32	1.13	1.25	1.06
09/30/2003	−1.06	0.08	1	−1.14	−1.14	−1.34	−1.12	−1.42	−1.20
10/31/2003	5.66	0.07	1	5.59	5.59	5.30	4.21	5.23	4.14
11/30/2003	0.88	0.07	1	0.81	0.81	0.74	1.18	0.67	1.11
12/31/2003	5.24	0.08	1	5.16	5.16	4.87	4.77	4.79	4.69
01/31/2004	1.84	0.07	1	1.77	1.77	0.87	2.51	0.80	2.44
02/29/2004	1.39	0.06	1	1.33	1.33	0.97	1.18	0.91	1.12
03/31/2004	−1.51	0.09	1	−1.60	−1.6	−0.89	−1.79	−0.98	−1.88
04/30/2004	−1.57	0.08	1	−1.65	−1.65	−2.59	−1.73	−2.67	−1.81
05/31/2004	1.37	0.06	0	1.31	0	0.66	0.83	0.60	0.77
06/30/2004	1.94	0.08	0	1.86	0	1.66	1.56	1.58	1.48
07/31/2004	−3.31	0.10	1	−3.41	−3.41	−2.82	−4.26	−2.92	−4.36
08/31/2004	0.40	0.11	0	0.29	0	−0.33	0.00	−0.44	−0.11
09/30/2004	1.08	0.11	0	0.97	0	1.20	1.99	1.09	1.88
10/31/2004	1.53	0.11	0	1.42	0	0.33	1.21	0.22	1.10
11/30/2004	4.05	0.15	1	3.90	3.9	4.87	5.68	4.72	5.53
12/31/2004	3.40	0.16	1	3.24	3.24	2.62	3.43	2.46	3.27

Notes:
1. The following information is used for determining the value of the dummy variable for the first three months:

	r_m	r_f	$r_m - r_f$
Sep–94	−2.41	0.37	−2.78
Oct–94	2.29	0.38	1.91
Nov–94	−3.67	0.37	−4.04
Dec–94	1.46	0.44	1.02

2. The dummy variable is defined as follows:

   $D_t x_t = x_t$ if the average $(r_M - r_{ft})$ for the prior three months > 0
   $D_t x_t = 0$ otherwise

TABLE 21.2 Regression Results for the Mutual Fund Characteristic Line with Dummy Variable

Regression Coefficient	Coefficient Estimate	Standard Error	t-statistic	p-value
Fund A				
b_0	−0.23	0.10	−2.36	0.0198
b_1	0.75	0.03	25.83	4E-50
b_2 (dummy)	0.18	0.04	4.29	4E-05
Fund B				
b_0	0.00	0.14	−0.03	0.9762
b_1	0.75	0.04	18.02	2E-35
b_2 (dummy)	0.13	0.06	2.14	0.0344

these two mutual funds, there is a difference in the beta for up and down markets.[6] From the results reported in Table 21.2, we would find that:

	Mutual Fund A	Mutual Fund B
Beta down market b_1	0.75	0.75
Beta p market (= $b_1 + b_2$)	0.93 (= 0.75 + 0.18)	0.88 (= 0.75 + 0.13)

Application to Predicting High-Yield Corporate Bond Spreads

As a second application of the use of dummy variables, we estimate a model to predict corporate bond spreads. A bond spread is the difference between the yield on two bonds. Typically, in the corporate bond market participants are interested in the difference between the yield spread between a corporate bond and an otherwise comparable (i.e., maturity or duration) U.S. Treasury bond. Managers of institutional bond portfolios formulate their investment strategy based on expected changes in corporate bond spreads.

The model presented in this illustration was developed by FridsonVision.[7] The unit of observation is a corporate bond issuer at a given point in time. The bonds in the study are all high-yield corporate bonds. A high-yield bond, also called a noninvestment grade bond or junk bond, is one that has

---

[6] We actually selected funds that had this characteristic so one should not infer that all mutual funds exhibit this characteristic.
[7] The model is described in "Focus Issues Methodology," *Leverage World* (May 30, 2003). The data for this illustration were provided by Greg Braylovskiy of Fridson-Vision. The firm uses about 650 companies in its analysis. Only 100 observations were used in this illustration.

a credit rating below Ba (referred to as being minimum investment grade) as assigned by the rating agencies. Within the high-yield bond sector of the corporate bond market there are different degrees of credit risk. Specifically, there are bonds classified as low grade, very speculative grade, substantial risk, very poor quality, and default (or imminent default).

The multivariate regression equation to be estimated is

$$y = b_0 + b_1 x_1 + b_2 x_2 + b_3 x_3 + e$$

where

y = yield spread (in basis points) for the bond issue of a company[8]
x_1 = coupon rate for the bond of a company, expressed without considering percentage sign (i.e., 7.5% = 7.5)
x_2 = coverage ratio that is found by dividing the earnings before interest, taxes, depreciation and amortization (EBITDA) by interest expense for the company issuing the bond
x_3 = logarithm of earnings (earnings before interest and taxes, EBIT, in millions of dollars) for company issuing the bond

Financial theory would suggest the following sign for each of the estimated regression coefficients:

- The higher the coupon rate (x_1), the greater the issuer's default risk and hence the larger the spread. Therefore, a positive coefficient for the coupon rate is expected.
- A coverage ratio (x_2) is a measure of a company's ability to satisfy fixed obligations, such as interest, principal repayment, and lease payments. There are various coverage ratios. The one used in this illustration is the ratio of the (EBITDA) divided by interest expense. Since the higher the coverage ratio the lower the default risk, an inverse relationship is expected between the spread and the coverage ratio; that is, the estimated coefficient for the coverage ratio is expected to be negative.
- There are various measures of earnings reported in financial statements. Earnings (x_3) in this illustration is defined as the trailing 12-months earnings before interest and taxes (EBIT). Holding other factors constant, it is expected that the larger the EBIT, the lower the default risk and therefore an inverse relationship (negative coefficient) is expected.

---

[8]This dependent variable is not measured by the typical nominal spread but by the option-adjusted spread. This spread measure adjusts for any embedded options in a bond such as a bond being callable.

We use 100 observations at two different dates, June 6, 2005 and November 28, 2005. For both dates there are the same 100 bond issues in the sample. Our purpose is to investigate two important empirical questions that a portfolio manager may have. First, we test whether there is a difference between bonds classified as low grade and speculative grade, on the one hand, and bonds classified as substantial risk, very poor quality, and default (or imminent default) on the other. Corporate bonds that fall into the latter category have a rating of CCC+ or below. Second, we test whether there is a structural shift in the corporate high-yield corporate bond market between June 6, 2005 and November 28, 2005.

We organize the data in matrix form as usual. Data are shown in Table 21.3. The table is divided into two panels, one for each date. The first column of the table indicates the corporate bond issue identified by number.

TABLE 21.3 Regression Data for the Bond Spread Application: 11/28/2005 and 06/06/2005

Issue #	Spread, 11/28/05	CCC+ and below	Coupon	Coverage Ratio	Logged EBIT	Spread, 6/6/05	CCC+ and below	Coupon	Coverage Ratio	Logged EBIT
1	509	0	7.400	2.085	2.121	473	0	7.400	2.087	2.111
2	584	0	8.500	2.085	2.121	529	0	8.500	2.087	2.111
3	247	0	8.375	9.603	2.507	377	0	8.375	5.424	2.234
4	73	0	6.650	11.507	3.326	130	0	6.650	9.804	3.263
5	156	0	7.125	11.507	3.326	181	0	7.125	9.804	3.263
6	240	0	7.250	2.819	2.149	312	0	7.250	2.757	2.227
7	866	1	9.000	1.530	2.297	852	1	9.000	1.409	1.716
8	275	0	5.950	8.761	2.250	227	0	5.950	11.031	2.166
9	515	0	8.000	2.694	2.210	480	0	8.000	2.651	2.163
10	251	0	7.875	8.289	1.698	339	0	7.875	8.231	1.951
11	507	0	9.375	2.131	2.113	452	0	9.375	2.039	2.042
12	223	0	7.750	4.040	2.618	237	0	7.750	3.715	2.557
13	71	0	7.250	7.064	2.348	90	0	7.250	7.083	2.296
14	507	0	8.000	2.656	1.753	556	0	8.000	2.681	1.797
15	566	1	9.875	1.030	1.685	634	1	9.875	1.316	1.677
16	213	0	7.500	11.219	3.116	216	0	7.500	10.298	2.996
17	226	0	6.875	11.219	3.116	204	0	6.875	10.298	2.996
18	192	0	7.750	11.219	3.116	201	0	7.750	10.298	2.996
19	266	0	6.250	3.276	2.744	298	0	6.250	3.107	2.653
20	308	0	9.250	3.276	2.744	299	0	9.250	3.107	2.653
21	263	0	7.750	2.096	1.756	266	0	7.750	2.006	3.038
22	215	0	7.190	7.096	3.469	259	0	7.190	6.552	3.453
23	291	0	7.690	7.096	3.469	315	0	7.690	6.552	3.453
24	324	0	8.360	7.096	3.469	331	0	8.360	6.552	3.453
25	272	0	6.875	8.612	1.865	318	0	6.875	9.093	2.074

TABLE 21.3 (Continued)

Issue #	Spread, 11/28/05	CCC+ and below	Coupon	Coverage Ratio	Logged EBIT	Spread, 6/6/05	CCC+ and below	Coupon	Coverage Ratio	Logged EBIT
26	189	0	8.000	4.444	2.790	209	0	8.000	5.002	2.756
27	383	0	7.375	2.366	2.733	417	0	7.375	2.375	2.727
28	207	0	7.000	2.366	2.733	200	0	7.000	2.375	2.727
29	212	0	6.900	4.751	2.847	235	0	6.900	4.528	2.822
30	246	0	7.500	19.454	2.332	307	0	7.500	16.656	2.181
31	327	0	6.625	3.266	2.475	365	0	6.625	2.595	2.510
32	160	0	7.150	3.266	2.475	237	0	7.150	2.595	2.510
33	148	0	6.300	3.266	2.475	253	0	6.300	2.595	2.510
34	231	0	6.625	3.266	2.475	281	0	6.625	2.595	2.510
35	213	0	6.690	3.266	2.475	185	0	6.690	2.595	2.510
36	350	0	7.130	3.266	2.475	379	0	7.130	2.595	2.510
37	334	0	6.875	4.310	2.203	254	0	6.875	5.036	2.155
38	817	1	8.625	1.780	1.965	635	0	8.625	1.851	1.935
39	359	0	7.550	2.951	3.078	410	0	7.550	2.035	3.008
40	189	0	6.500	8.518	2.582	213	0	6.500	13.077	2.479
41	138	0	6.950	25.313	2.520	161	0	6.950	24.388	2.488
42	351	0	9.500	3.242	1.935	424	0	9.500	2.787	1.876
43	439	0	8.250	2.502	1.670	483	0	8.250	2.494	1.697
44	347	0	7.700	4.327	3.165	214	0	7.700	4.276	3.226
45	390	0	7.750	4.327	3.165	260	0	7.750	4.276	3.226
46	149	0	8.000	4.327	3.165	189	0	8.000	4.276	3.226
47	194	0	6.625	4.430	3.077	257	0	6.625	4.285	2.972
48	244	0	8.500	4.430	3.077	263	0	8.500	4.285	2.972
49	566	1	10.375	2.036	1.081	839	1	10.375	2.032	1.014
50	185	0	6.300	7.096	3.469	236	0	6.300	6.552	3.453
51	196	0	6.375	7.096	3.469	221	0	6.375	6.552	3.453
52	317	0	6.625	3.075	2.587	389	0	6.625	2.785	2.551
53	330	0	8.250	3.075	2.587	331	0	8.250	2.785	2.551
54	159	0	6.875	8.286	3.146	216	0	6.875	7.210	3.098
55	191	0	7.125	8.286	3.146	257	0	7.125	7.210	3.098
56	148	0	7.375	8.286	3.146	117	0	7.375	7.210	3.098
57	112	0	7.600	8.286	3.146	151	0	7.600	7.210	3.098
58	171	0	7.650	8.286	3.146	221	0	7.650	7.210	3.098
59	319	0	7.375	3.847	1.869	273	0	7.375	4.299	1.860
60	250	0	7.375	12.656	2.286	289	0	7.375	8.713	2.364
61	146	0	5.500	5.365	3.175	226	0	5.500	5.147	3.190
62	332	0	6.450	5.365	3.175	345	0	6.450	5.147	3.190
63	354	0	6.500	5.365	3.175	348	0	6.500	5.147	3.190
64	206	0	6.625	7.140	2.266	261	0	6.625	5.596	2.091
65	558	0	7.875	2.050	2.290	455	0	7.875	2.120	2.333
66	190	0	6.000	2.925	3.085	204	0	6.000	3.380	2.986

TABLE 21.3 (Continued)

Issue #	Spread, 11/28/05	CCC+ and below	Coupon	Coverage Ratio	Logged EBIT	Spread, 6/6/05	CCC+ and below	Coupon	Coverage Ratio	Logged EBIT
67	232	0	6.750	2.925	3.085	244	0	6.750	3.380	2.986
68	913	1	11.250	2.174	1.256	733	0	11.250	2.262	1.313
69	380	0	9.750	4.216	1.465	340	0	9.750	4.388	1.554
70	174	0	6.500	4.281	2.566	208	0	6.500	4.122	2.563
71	190	0	7.450	10.547	2.725	173	0	7.450	8.607	2.775
72	208	0	7.125	2.835	3.109	259	0	7.125	2.813	3.122
73	272	0	6.500	5.885	2.695	282	0	6.500	5.927	2.644
74	249	0	6.125	5.133	2.682	235	0	6.125	6.619	2.645
75	278	0	8.750	6.562	2.802	274	0	8.750	7.433	2.785
76	252	0	7.750	2.822	2.905	197	0	7.750	2.691	2.908
77	321	0	7.500	2.822	2.905	226	0	7.500	2.691	2.908
78	379	0	7.750	4.093	2.068	362	0	7.750	4.296	2.030
79	185	0	6.875	6.074	2.657	181	0	6.875	5.294	2.469
80	307	0	7.250	5.996	2.247	272	0	7.250	3.610	2.119
81	533	0	10.625	1.487	1.950	419	0	10.625	1.717	2.081
82	627	0	8.875	1.487	1.950	446	0	8.875	1.717	2.081
83	239	0	8.875	2.994	2.186	241	0	8.875	3.858	2.161
84	240	0	7.375	8.160	2.225	274	0	7.375	8.187	2.075
85	634	0	8.500	2.663	2.337	371	0	8.500	2.674	2.253
86	631	1	7.700	2.389	2.577	654	1	7.700	2.364	2.632
87	679	1	9.250	2.389	2.577	630	1	9.250	2.364	2.632
88	556	1	9.750	1.339	1.850	883	1	9.750	1.422	1.945
89	564	1	9.750	1.861	2.176	775	1	9.750	1.630	1.979
90	209	0	6.750	8.048	2.220	223	0	6.750	7.505	2.092
91	190	0	6.500	4.932	2.524	232	0	6.500	4.626	2.468
92	390	0	6.875	6.366	1.413	403	0	6.875	5.033	1.790
93	377	0	10.250	2.157	2.292	386	0	10.250	2.057	2.262
94	143	0	5.750	11.306	2.580	110	0	5.750	9.777	2.473
95	207	0	7.250	2.835	3.109	250	0	7.250	2.813	3.122
96	253	0	6.500	4.918	2.142	317	0	6.500	2.884	1.733
97	530	1	8.500	0.527	2.807	654	1	8.500	1.327	2.904
98	481	0	6.750	2.677	1.858	439	0	6.750	3.106	1.991
99	270	0	7.625	2.835	3.109	242	0	7.625	2.813	3.122
100	190	0	7.125	9.244	3.021	178	0	7.125	7.583	3.138

Notes:
Spread = option-adjusted spread (in basis points)
Coupon = coupon rate, expressed without considering percentage sign (i.e., 7.5% = 7.5)
Coverage Ratio = EBITDA divided by interest expense for company
Logged EBIT = logarithm of earnings (EBIT in millions of dollars)

The second column and seventh column show the yield spread for November 28, 2005 and June 6, 2005, respectively. Look now at the third and eighth columns. This is where we begin to use the dummy variable. Since we are interested in testing if there is a differential impact on the yield spread between the two categories of high-yield bonds, there is a zero or one in these two columns depending on if the bond issue has a credit rating of CCC+ and below, in which case a one is assigned, and zero otherwise.

Let's first estimate the regression equation for the fully pooled data, that is, all data without any distinction by credit rating and date. That is, there are 200 data points consisting of 100 observation for each of the two dates. The estimated regression coefficients for the model and their corresponding t-statistics are shown in Table 21.4. The R^2 for the regression is 0.57. The coefficients for the three independent variables are statistically significant and each has the expected sign. However, the intercept term is not statistically significant.

Let's now test if there is a difference in the yield spread based on the two categories of high-yield bonds. It should be emphasized that this is only an exercise to show the application of regression analysis using dummy variables. The conclusions we reach are not meaningful from a statistical point of view given the small size of the database. The regression model we want to estimate is

$$y = b_0 + b_1 D1 + b_2 x_1 + b_3 (D1\ x_1) + b_4 x_2 + b_5 (D1\ x_2) + b_6 x_3 + b_7 (D1\ x_3) + e$$

where $D1$ is the dummy variable that takes on the value of one if the bond issue has a credit rating of CCC+ and below, and zero otherwise.

There are now seven variables and eight parameters to estimate. The regression coefficients for the dummy variables are b_1, b_3, b_5, and b_7. The estimated regression coefficients for the model and the t-statistics are shown in Table 21.5. The R^2 for the regression is 0.73. From the table we see that none of the regression coefficients for the dummy variables are statistically significant at the 10% level.

TABLE 21.4 Regression Results for High-Yield Corporate Bond Spreads

Regression Coefficient	Estimated Coefficient	Standard Error	t-Statistic	p-Value
b_0	157.01	89.56	1.753	0.081
b_1	61.27	8.03	7.630	9.98E-13
b_2	–13.20	2.27	–5.800	2.61E-08
b_3	–90.88	16.32	–5.568	8.41E-08

TABLE 21.5 Regression Results for High-Yield Corporate Bond Spread Using Dummy Variables for Credit Rating Classification

Regression Coefficient	Estimated Coefficient	Standard Error	t-Statistic	p-Value
b_0	284.52	73.63	3.86	0.00
b_1 (dummy)	597.88	478.74	1.25	0.21
b_2	37.12	7.07	5.25	3.96E-07
b_3 (dummy)	−45.54	38.77	−1.17	0.24
b_4	−10.33	1.84	−5.60	7.24E-08
b_5 (dummy)	50.13	40.42	1.24	0.22
b_6	−83.76	13.63	−6.15	4.52E-09
b_7 (dummy)	−0.24	62.50	−0.00	1.00

Now let's use dummy variables to test if there is a regime shift between the two dates. This is a common use for dummy variables in practice. To this end we create a new dummy variable that has the value zero for the first date 11/28/05 and one for the second date 6/6/05. The regression model to be estimated is then

$$y = b_0 + b_1 D2 + b_2 x_1 + b_3 (D2\ x_1) + b_4 x_2 + b_5 (D2\ x_2) + b_6 x_3 + b_7 (D2\ x_3) + e$$

where $D2$ is a dummy variable that takes on the value of zero if the observation is on November 28, 2005 and one if the observation is on June 6, 2005. Again, there are four dummy variables: b_1, b_3, b_5, and b_7.

The estimated model regression coefficients and t-statistics are reported in Table 21.6. The R^2 for the regression is 0.71. Notice that only one of the dummy variable regression coefficients is statistically significant at the 10% level, b_3. This suggests that market participants reassessed the impact of the coupon rate (x_1) on the yield spread. The estimated regression coefficient for x_1 on November 28, 2005 was 33.25 (b_2) whereas on June 6, 2005 it was 61.29 $(b_2 + b_3 = 33.25 + 28.14)$.

MODEL BUILDING TECHNIQUES

We now turn our attention to the model building process in the sense that we attempt to find the independent variables that best explain the variation in the dependent variable y. At the outset, we do not know how many and which independent variables to use. Increasing the number of independent

TABLE 21.6 Regression Results for High-Yield Corporate Bond Spread Using Dummy Variables for Time

Regression Coefficient	Estimated Coefficient	Standard Error	t-Statistic	p-Value
b_0	257.26	79.71	3.28	0.00
b_1 (dummy)	82.17	61.63	1.33	0.18
b_2	33.25	7.11	4.67	5.53E-06
b_3 (dummy)	28.14	2.78	10.12	1.45E-19
b_4	−10.79	2.50	−4.32	2.49E-05
b_5 (dummy)	0.00	3.58	0.00	1.00
b_6	−63.20	18.04	−3.50	0.00
b_7 (dummy)	−27.48	24.34	−1.13	0.26

variables does not always improve regressions. The econometric theorem known as *Pyrrho's lemma* relates to the number of independent variables.[9] Pyrrho's lemma states that by adding one special independent variable to a linear regression, it is possible to arbitrarily change the size and sign of regression coefficients as well as to obtain an arbitrary goodness of fit. This tells us that if we add independent variables without a proper design and testing methodology, we risk obtaining spurious results.

The implications are especially important for those financial models that seek to forecast prices, returns, or rates based on regressions over economic or fundamental variables. With modern computers, by trial and error, one might find a complex structure of regressions that give very good results in-sample but have no real forecasting power.

There are three methods that are used for the purpose of determining the suitable independent variables to be included in a final regression model. They are [10]

1. Stepwise inclusion method
2. Stepwise exclusion method
3. Standard stepwise regression method

We explain each next.

Stepwise Inclusion Method

In the *stepwise inclusion method* we begin by selecting a single independent variable. It should be the one most highly correlated (positive or negative)

---
[9]Dijkstra (1995).
[10]See, for example, Rachev et al. (2007).

with the dependent variable.[11] After inclusion of this independent variable, we perform an F-test to determine whether this independent variable is significant for the regression. If not, then there will be no independent variable from the set of possible choices that will significantly explain the variation in the dependent variable y. Thus, we will have to look for a different set of variables.

If, on the other hand, this independent variable, say x_1, is significant, we retain x_1 and consider the next independent variable that best explains the remaining variation in y. We require that this additional independent variable, say x_2, be the one with the highest *coefficient of partial determination*. This is a measure of the goodness-of-fit given that the first x_1 is already in the regression. It is defined to be the ratio of the remaining variation explained by the second independent variable to the total of unexplained variation before x_2 was included. Formally, we have

$$\frac{SSE_1 - SSE_2}{SSE_1} \qquad (21.9)$$

where

SSE_1 = the variation left unexplained by x_1
SSE_2 = the variation left unexplained after both, x_1 and x_2 have been included

This is equivalent to requiring that the additional variable is to be the one that provides the largest coefficient of determination once included in the regression. After the inclusion, an F-test with

$$F = \frac{SSE_1 - SSE_2}{\frac{SSE_1}{n-2}} \qquad (21.10)$$

is conducted to determine the significance of the additional variable.

The addition of independent variables included in some set of candidate independent variables is continued until either all independent variables are in the regression or the additional contribution to explain the remaining variation in y is not significant any more. Hence, as a generalization to equation (21.10), we compute

$$F = (SSE_i - SSE_{i+1}) / (SSE_i) \times (n - i - 1)$$

---

[11]The absolute value of the correlation coefficient should be used since we are only interested in the extent of linear dependence, not the direction.

after the inclusion of the $i + 1st$ variable and keep it included only if F is found to be significant. Accordingly, SSE_i denotes the sum of square residuals with i variables included while SSE_{i+1} is the corresponding quantity for $i + 1$ included variables.

Stepwise Exclusion Method

The ***stepwise exclusion method*** mechanically is basically the opposite of the stepwise inclusion method. That is, one includes all independent variables at the beginning. One after another of the insignificant variables are eliminated until all insignificant independent variables have been removed. The result constitutes the final regression model. In other words, we include all k independent variables into the regression at first. Then we consider all variables for exclusion on a stepwise removal basis.

For each independent variable, we compute

$$F = \left(SSE_{k-1} - SSE_k\right) / \left(SSE_{k-1}\right) \times \left(n - k\right) \quad (21.11)$$

to find the ones where F is insignificant. The one that yields the least significant value F is discarded. We proceed stepwise by alternatively considering all remaining variables for exclusion and, likewise, compute the F-test statistic given by equation (21.11) for the new change in the coefficient of partial determination.

In general, at each step i, we compute

$$F = \left(SSE_{k-i} - SSE_{k-i+1}\right) / \left(SSE_{k-i}\right) \times \left(n - k + i - 1\right) \quad (21.12)$$

to evaluate the coefficient of partial determination lost due to discarding the i-th independent variable.[12] If no variable with an insignificant F-test statistic can be found, we terminate the elimination process.

Standard Stepwise Regression Method

The ***standard stepwise regression method*** involves introducing independent variables based on significance and explanatory power and possibly eliminating some that have been included at previous steps. The reason for elimination of any such independent variables is that they have now become insignificant after the new independent variables have entered the model. Therefore, we check the significance of all coefficient statistics ac-

---

[12]The SSE_{k-i+1} is the sum of square residuals before independent variable i is discarded. After the i-th independent variable has been removed, the sum of square residuals of the regression with the remaining $k - i$ variables is given by SSE_{k-i}.

cording to equation (20.14) in Chapter 20. This methodology provides a good means for eliminating the influence from possible multicollinearity discussed earlier.

CONCEPTS EXPLAINED IN THIS CHAPTER
(IN ORDER OF PRESENTATION)

Multicollinearity
Variance inflation factor
Dummy variables
Binary variables
Dummy variable trap
Common terms
Composite model
Chow test
Pyrrho's lemma
Stepwise inclusion method
Coefficient of partial determination
Stepwise exclusion method
Standard stepwise regression method

CHAPTER 22

Testing the Assumptions of the Multivariate Linear Regression Model

As explained in Chapter 20, after we have come up with some regression model, we have to perform a diagnosis check. The question that must be asked is: How well does the model fit the data? This is addressed using diagnosis checks that include the coefficient of determination, R^2 as well as R^2_{adj}, and the standard error or square root of the mean-square error (MSE) of the regression. In particular, the diagnosis checks analyze whether the linear relationship between the dependent and independent variables is justifiable from a statistical perspective.

As we also explained in Chapter 20, there are several assumptions that are made when using the general multivariate linear regression model. The first assumption is the independence of the independent variables used in the regression model. This is the problem of multicollinearity that we discussed in Chapter 21 where we briefly described how to test and correct for this problem. The second assumption is that the model is in fact linear. The third assumption has to do with assumptions about the statistical properties of the error term for the general multivariate linear regression model. Furthermore, we assumed that the residuals are uncorrelated with the independent variables. In this chapter, we look at the assumptions regarding the linearity of the model and the assumptions about the error term. We discuss the implications of the violation of these assumptions, some tests used to detect violations, and provide a brief explanation of how to deal with any violations.

It is important to understand that in the area of economics, these topics fall into the realm of the field of econometrics and textbooks are devoted to the treatment of regression analysis in the presence of the violations of these assumptions. For this reason, we only provide the basics.

TESTS FOR LINEARITY

To test for linearity, a common approach is to plot the regression residuals on the vertical axis and values of the independent variable on the horizontal axis. This graphical analysis is performed for each independent variable. What we are looking for is a random scattering of the residuals around zero. If this should be the case, the model assumption with respect to the residuals is correct. If not, however, then there seems to be some systematic behavior in the residuals that depends on the values of the independent variables. The explanation is that the relationship between the independent and dependent variables is not linear.

The problem of a nonlinear functional form can be dealt with by transforming the independent variables or making some other adjustment to the variables. For example, suppose that we are trying to estimate the relationship between a stock's return as a function of the return on a broad-based stock market index such as the S&P 500 Index. Letting y denote the return on the stock and x the return on the S&P 500 Index, then we might assume the following bivariate regression model:

$$y = b_0 + b_1 x + e \qquad (22.1)$$

where e is the error term.

We have made the assumption that the functional form of the relationship is linear. Suppose that we find that a better fit appears to be that the return on the stock is related to the return on the broad-based market index as

$$y = b_0 + b_1 x + b_2 x^2 + e \qquad (22.2)$$

If we let $x = x_1$ and $x^2 = x_2$ and we adjust our table of observations accordingly, then we can rewrite equation (22.2) as

$$y = b_0 + b_1 x_1 + b_2 x_2 + e \qquad (22.3)$$

The model given by equation (22.3) is now a linear regression model despite the fact that the functional form of the relationship between y and x is nonlinear. That is, we are able to modify the functional form to create a linear regression model.

Let's see how a simple transformation can work as explained in Chapter 6. Suppose that the true relationship of interest is exponential, that is,

$$y = \beta e^{\alpha x} \qquad (22.4)$$

Taking the natural logarithms of both sides of equation (22.4) will result in

$$\ln y = \ln \beta + \alpha x \qquad (22.5)$$

which is again linear.

Now consider that the fit in equation (22.5) is not exact; that is, there is some random deviation by some residual. Then we obtain

$$\ln y = \ln \beta + \alpha x + \varepsilon \qquad (22.6)$$

If we let $z = \ln y$ and adjust the table of observations for y accordingly and let $\lambda = \ln \beta$, we can then rewrite equation (22.6) as

$$z = \lambda + \alpha x + \varepsilon \qquad (22.7)$$

This regression model is now linear with the parameters to be estimated being λ and α.

Now we transform equation (22.7) back into the shape of equation (22.4) ending up with

$$y = \beta e^{\alpha x} \cdot e^{\varepsilon} = \beta e^{\alpha x} \cdot \xi \qquad (22.8)$$

where in equation (22.8) the deviation is multiplicative rather than additive as would be the case in a linear model. This would be a possible explanation of the nonlinear function relationship observed for the residuals.

However, not every functional form that one might be interested in estimating can be transformed or modified so as to create a linear regression. For example, consider the following relationship:

$$y = (b_1 x)/(b_2 + x) + e \qquad (22.9)$$

Admittedly, this is an odd looking functional form. What is important here is that the regression parameters to be estimated (b_1 and b_2) cannot be transformed to create a linear regression model. A regression such as equation (22.9) is referred to a ***nonlinear regression*** and the estimation of nonlinear regressions is far more complex than that of a linear regression because they have no closed-form formulas for the parameters to be estimated. Instead, nonlinear regression estimation techniques require the use of optimization techniques to identify the parameters that best fit the model. Researchers in disciplines such as biology and physics often have to deal with nonlinear regressions.

ASSUMED STATISTICAL PROPERTIES ABOUT THE ERROR TERM

The third assumption about the general linear model concerns the three assumptions about the error term that we listed in Chapter 20 and repeat below:

Assumption 1. The regression errors are normally distributed with zero mean.

Assumption 2: The variance of the regression errors (σ_ε^2) is constant.

Assumption 3. The error terms from different points in time are independent such that ε_t are independent variables for all t.

Assumption 1 states that the probability distribution for the error term is that it is normally distributed. Assumption 2 says that the variance of the probability distribution for the error term does not depend on the level of any of the independent variables. That is, the variance of the error term is constant regardless of the level of the independent variable. If this assumption holds, the error terms are said to be *homoskedastic* (also spelled *homoscedastic*). If this assumption is violated, the variance of the error term is said to be *heteroskedastic* (also spelled *heteroscedastic*). Assumption 3 says that there should not be any statistically significant correlation between adjacent residuals. The correlation between error terms is referred to as *autocorrelation*. Recall that we also assume that the residuals are uncorrelated with the independent variables.

TESTS FOR THE RESIDUALS BEING NORMALLY DISTRIBUTED

An assumption of the general linear regression model is that the residuals are normally distributed. The implications of the violation of this assumption are:

1. The regression model is misspecified.
2. The estimates of the regression coefficients are also not normally distributed.
3. The estimates of the regression coefficients although still best linear unbiased estimators,[1] they are no longer efficient estimators.

From the second implication above we can see that violation of the normality assumption makes hypothesis testing suspect. More specifically, if the assumption is violated, the t-tests explained in Chapter 20 will not be applicable.

---

[1] See Chapter 17 for what is meant by best linear unbiased estimators.

Typically the following three methodologies are used to test for normality of the error terms (1) chi-square statistic, (2) Jarque-Bera test statistic, and (3) analysis of standardized residuals.

Chi-Square Statistic

The *chi-square statistic* is defined as

$$\chi^2 = \sum_{i=1}^{k} \frac{(n_i - n \cdot p_i)^2}{n \cdot p_i} \qquad (22.10)$$

where some interval along the real numbers is divided into k segments of possibly equal size.

The p_i indicate which percentage of all n values of the sample should fall into the i-th segment if the data were truly normally distributed.[2] Hence, the theoretical number of values that should be inside of segment i is $n \cdot p_i$. The n_i are the values of the sample that actually fall into that segment i. The test statistic given by equation (22.10) is approximately chi-square distributed with $k - 1$ degrees of freedom. As such, it can be compared to the critical values of the chi-square distribution at arbitrary α-levels.[3] If the critical values are surpassed or, equivalently, the p-value[4] of the statistic is less than α, then the normal distribution has to be rejected for the residuals.

Jarque-Bera Test Statistic

The *Jarque-Bera test statistic* is not quite simple to compute manually, but most computer software packages have it installed. Formally, it is

$$JB = \frac{n}{6}\left(S^2 + \frac{(K-3)^2}{4} \right) \qquad (22.11)$$

with

$$S = \frac{\frac{1}{n}\sum_{i=1}^{n}(x-\bar{x})^3}{\left(\frac{1}{n}\sum_{i=1}^{n}(x-\bar{x})^2\right)^{3/2}} \qquad (22.12)$$

---

[2] Standardized residuals would be compared with the standard normal distribution, $N(0,1)$.
[3] The chi-square distribution is covered in Chapter 11.
[4] The p-value is explained in Chapter 19.

$$K = \frac{\frac{1}{n}\sum_{i=1}^{n}(x-\bar{x})^4}{\left(\frac{1}{n}\sum_{i=1}^{n}(x-\bar{x})^2\right)^2} \qquad (22.13)$$

for a sample of size n.

The expression in equation (22.12) is the skewness statistic of some distribution, and equation (22.13) is the kurtosis. As explained in Chapter 13, kurtosis measures the peakedness of the probability density function of some distribution around the median compared to the normal distribution. Also, kurtosis estimates, relative to the normal distribution, the behavior in the extreme parts of the distribution (i.e., the tails of the distribution). For a normal distribution, $K = 3$. A value for K that is less than 3 indicates a so-called *light-tailed distribution* in that it assigns less weight to the tails. The opposite is a value for K that exceeds 3 and is referred to as a *heavy-tailed distribution*. The test statistics given by equation (22.11) is approximately distributed chi-square with two degrees of freedom.

Analysis of Standardized Residuals

Another means of determining the appropriateness of the normal distribution are the standardized residuals. Once computed, they can be graphically analyzed in histograms. Formally, each standardized residual at the i-th observation is computed according to

$$\tilde{e}_i = n \cdot \frac{e_i}{s_e\sqrt{(n+1)+\frac{(x_i-\bar{x})^2}{s_x^2}}} \qquad (22.14)$$

where s_e is the estimated standard error (as defined in Chapter 20) and n is the sample size. This is done in most statistical software.

If the histogram appears skewed or simply not similar to a normal distribution, the linearity assumption is very likely to be incorrect. Additionally, one might compare these standardized residuals with the normal distribution by plotting them against their theoretical normal counterparts in a normal probability plot. There is a standard routine in most statistical software that performs this analysis. We introduced the probability plot in Chapter 4. If the pairs lie along the line running through the sample quartiles, the regression residuals seem to follow a normal distribution and, thus, the assumptions of the regression model stated in Chapter 20 are met.

TESTS FOR CONSTANT VARIANCE OF THE ERROR TERM (HOMOSKEDASTICITY)

The second test regarding the residuals in a linear regression analysis is that the variance of all squared error terms is the same. As we noted earlier, this assumption of constant variance is referred to as homoskedasticity. However, many time series data exhibit heteroskedasticity, where the error terms may be expected to be larger for some observations or periods of the data than for others.

There are several tests that have been used to detect the presence of heteroskedasticity. These include the

- White's generalized heteroskedasticity test
- Park test
- Glejser Test
- Goldfeld-Quandt test
- Breusch-Pagan-Godfrey Test (Lagrangian multiplier test)

These tests are described in books on econometrics and will not be described here.

Modeling to Account for Heteroskedasticity

Once heteroskedasticity is detected, the issue is then how to construct models that accommodate this feature of the residual variance so that valid regression coefficient estimates and models are obtained for the variance of the error term. There are two methodologies used for dealing with heteroskedasticity: weighted least squares estimation technique and autoregressive conditional heteroskedastic models.

Weighted Least Squares Estimation Technique

A potential solution for correcting the problem of heteroskedasticity is to give less weight to the observations coming from the population with larger variances and more weight to the observations coming from observations with higher variance. This is the basic notion of the *weighted least squares* (WLS) *technique*.

To see how the WLS technique can be used, let's consider the case of the bivariate regression given by

$$y_t = \beta_0 + \beta_1 x_t + \varepsilon_t \qquad (22.15)$$

Let's now make the somewhat bold assumption that the variance of the error term for each time period is known. Denoting this variance by σ_t^2 (dropping the subscript ε for the error term), then we can deflate the terms in the bivariate linear regression given by equation (22.15) by the assumed known standard deviation of the error term as follows:

$$\left(\frac{y_t}{\sigma_t}\right) = \beta_0 \left(\frac{1}{\sigma_t}\right) + \beta_1 \left(\frac{x_t}{\sigma_t}\right) + \left(\frac{\varepsilon_t}{\sigma_t}\right) \quad (22.16)$$

We have transformed all the variables in the bivariate regression, including the original error term. It can be demonstrated that the regression with the transformed variables as shown in equation (22.16) no longer has heteroskedasticity. That is, the variance of the error term in equation (22.16), ε/σ_t, is homoskedastic.

Equation (22.16) can be estimated using ordinary least squares by simply adjusting the table of observations so that the variables are deflated by the known σ_t. When this is done, the estimates are referred to as **weighted least squares estimators**.

We simplified the illustration by assuming that the variance of the error term is known. Obviously, this is an extremely aggressive assumption. In practice, the true value for the variance of the error term is unknown. Other less aggressive assumptions are made but nonetheless are still assumptions. For example, the variance of the error term can be assumed to be proportional to one of the values of the independent variables. In any case, the WLS estimator requires that we make some assumption about the variance of the error term and then transform the value of the variables accordingly in order to apply the WLS technique.

Advanced Modeling to Account for Heteroskedasticity

Autoregressive conditional heteroskedasticity (ARCH) *model* and its generalization, the *generalized autoregressive conditional heteroskedasticity* (GARCH) *model*,[5] have proven to be very useful in finance to model the residual variance when the dependent variable is the return on an asset or an exchange rate. Understanding the behavior of the variance of the return process is important for forecasting as well as pricing option-type derivative instruments since the variance is a proxy for risk.

A widely observed phenomenon regarding asset returns in financial markets suggests that they exhibit *volatility clustering*. This refers to the tendency of large changes in asset returns (either positive or negative) to be

---

[5] The term "conditional" in the title of the two models means that the variance depends on or is conditional on the value of the random variable.

followed by large changes, and small changes in asset returns to be followed by small changes. Hence, there is temporal dependence in asset returns.[6] ARCH and GARCH models can accommodate volatility clustering.

While ARCH and GARCH models do not answer the question of what causes heteroskedasticity in the data, they do model the underlying time-varying behavior so that forecasting models can be developed. As it turns out, ARCH and GARCH models allow for not only volatility clustering but also heavy tails. The ARCH model is one of the pivotal developments in the field of econometrics and seems to be purposely built for applications in finance.[7] This model was introduced by Engle (1982) and subsequently extended by Bollerslev (1987). Since then other researchers have developed variants of the initial ARCH and GARCH models. All of these models have a common goal of modeling time-varying volatility, but at the same time they allow extensions to capture more detailed features of financial time series.[8]

Below we give a brief description of this model for time series data. Let's start with the simplest ARCH model,

$$\sigma_t^2 = a + b\,(x_{t-1} - x_M)^2 \qquad (22.17)$$

where

σ_t^2 = variance of the error term at time t[9]
x_M = mean of x
$(x_{t-1} - x_M)$ = deviation from the mean at time $t-1$

and a and b are parameters to be estimated using regression analysis.[10]

Equation (22.17) states that the estimate of the variance at time t depends on how much the observation at time $t-1$ deviates from the mean of the independent variable. Thus, the variance on time t is "conditional" on the deviation of the observation at time $t-1$. The reason for squaring the deviation is that it is the magnitude, not the direction, of the deviation that is important. By using the deviation at time $t-1$, recent information (as measured by the deviation) is being considered in the model.

---

[6]This pattern in the volatility of asset returns was first reported by Mandelbrot (1963).
[7]See Bollerslev (2001).
[8]In addition to ARCH/GARCH models, there are other models that deal with time-varying volatility, such as stochastic-volatility models, which are beyond the scope of this introductory chapter.
[9]For convenience, we have dropped the error term subscript.
[10]This simplest ARCH model is commonly referred to as ARCH(1) since it incorporates exactly one lag-term (i.e., the squared deviation from the mean at time $t-1$).

The ARCH model can be generalized in two ways. First, information for periods prior to $t-1$ can be included into the model by using the squared deviations for several time periods. For example, suppose three prior periods are used. Then equation (22.17) can be generalized to

$$\sigma_t^2 = a + b_1 (x_{t-1} - x_M)^2 + b_2 (x_{t-2} - x_M)^2 + b_3 (x_{t-3} - x_M)^2 \quad (22.18)$$

where a, b_1, b_2, and b_3 are parameters to be estimated statistically.[11]

A second way to generalize the ARCH model is to include not only squared deviations from prior time periods as a random variable that the variance is conditional on but also the estimated variance for prior time periods. This is the GARCH model described earlier. For example, the following equation generalizes equation (22.17) for the case where the variance at time t is conditional on the deviation squared at time $t-1$ and the variance at time $t-1$

$$\sigma_t^2 = a + b (x_{t-1} - x_M)^2 + c \sigma_{t-1}^2 \quad (22.19)$$

where a, b, and c are parameters to be estimated.

Suppose that the variance at time t is assumed to be conditional on three prior periods of squared deviations and two prior time period variances, then the GARCH model is

$$\sigma_t^2 = a + b_1(x_{t-1} - x_M)^2 + b_2(x_{t-2} - x_M)^2 + b_3(x_{t-3} - x_M)^2 \\ + c_1 \sigma_{t-1}^2 + c_2 \sigma_{t-2}^2 \quad (22.20)$$

where the parameters to be estimated are a, the b_i's ($i = 1, 2, 3$), and c_j's ($j = 1, 2$). In accordance with literature, we can write this as GARCH(2,3) where 2 indicates the number of prior time period variances and 3 refelects the number of prior periods of squard deviations.

As noted earlier, there have been further extensions of ARCH models. However, these extensions are beyond the scope of this chapter.

ABSENCE OF AUTOCORRELATION OF THE RESIDUALS

Assumption 3 is that there is no correlation between the residual terms. Simply put, this means that there should not be any statistically significant correlation between adjacent residuals. In time series analysis, this means no significant correlation between two consecutive time periods.

---

[11]Since here we use the squared deviations from three immediately prior periods (i.e., lag terms), that is commonly denoted as ARCH(3).

The correlation of the residuals is critical from the point of view of estimation. Autocorrelation of residuals is quite common in financial data where there are quantities that are time series. A time series is said to be autocorrelated if each term is correlated with its predecessor so that the variance of each term is partially explained by regressing each term on its predecessor.

Autocorrelation, which is also referred to as *serial correlation* and *lagged correlation* in time series analysis, like any correlation, can range from –1 to +1. Its computation is straightforward since it is simply a correlation using the residual pairs e_t and e_{t-1} as the observations. The formula is

$$\rho_{auto} = \frac{\sum_{t=2}^{n} e_t e_{t-1}}{\sum_{t=1}^{n} e_t^2} \qquad (22.21)$$

where ρ_{auto} means the estimated autocorrelation and e_t is the computed residual or error term for the t-th observation.

A *positive autocorrelation* means that if a residual t is positive (negative), then the residual that follows, $t + 1$, tends to be positive (negative). Positive autocorrelation is said to exhibit persistence. A *negative autocorrelation* means that a positive (negative) residual t tends to be followed by a negative (positive) residual $t + 1$.

The presence of significant autocorrelation in a time series means that, in a probabilistic sense, the series is predictable because future values are correlated with current and past values. From an estimation perspective, the existence of autocorrelation complicates hypothesis testing of the regression coefficients. This is because although the regression coefficient estimates are unbiased, they are not best linear unbiased estimates. Hence, the variances may be significantly underestimated and the resulting hypothesis test questionable.

Detecting Autocorrelation

How do we detect the autocorrelation of residuals? Suppose that we believe that there is a reasonable linear relationship between two variables, for instance stock returns and some fundamental variable. We then perform a linear regression between the two variables and estimate regression parameters using the OLS method. After estimating the regression parameters, we can compute the sequence of residuals. At this point, we can apply statistical tests. There are several tests for autocorrelation of residuals that can be used. Two such tests are the Durbin-Watson test and the Dickey-Fuller test. We discuss only the first below.

The most popular test is the Durbin-Watson test, or more specifically, the ***Durbin-Watson d-statistic*** computed as

$$d = \frac{\sum_{t=2}^{n}(e_t - e_{t-1})^2}{\sum_{t=1}^{n} e_t^2} \qquad (22.22)$$

The denominator of the test is simply the sum of the squares of the error terms; the numerator is the squared difference of the successive residuals.

It can be shown that if the sample size is large, then the Durbin-Watson d test statistic given by formula (22.22) is approximately related to the autocorrelation given by formula (22.21) as

$$d \approx 2\,(1 - \rho_{auto}) \qquad (22.23)$$

Since ρ_{auto} can vary between -1 and 1, this means that d can vary from 0 to 4 as shown:

Value of ρ_{auto}	Interpretation of ρ_{auto}	Approximate value of d
-1	Perfect negative autocorrelation	4
0	No autocorrelation	2
1	Perfect positive autocorrelation	0

From the above table we see that if d is close to 2 there is no autocorrelation. A d value less than 2 means there is potentially positive autocorrelation; the closer the value to 0 the greater the likelihood of positive autocorrelation. There is potentially negative autocorrelation if the computed d exceeds 2 and the closer the value is to 4, the greater the likelihood of negative autocorrelation.

In previous hypothesis tests discussed in this book, we stated that there is a critical value that a test statistic had to exceed in order to reject the null hypothesis. In the case of the Durbin-Watson d statistic, there is not a single critical value but two critical values, which are denoted by d_L and d_U. Moreover, there are ranges for the value of d where no decision can be made about the presence of autocorrelation. The general decision rule given the null hypothesis and the computed value for d is summarized in the following table:

Null hypothesis	Range for computed d	Decision rule
No positive autocorrelation	$0 < d < d_L$	Reject the null hypothesis
No positive autocorrelation	$d_L \leq d \leq d_U$	No decision
No negative autocorrelation	$4 - d_L < d < 4$	Reject the null hypothesis
No negative autocorrelation	$4 - d_U \leq d \leq 4 - d_L$	No decision
No autocorrelation	$d_U < d < 4 - d_U$	Accept the null hypothesis

Where does one obtain the critical values d_L and d_U? There are tables that report those values for the 5% and 1% levels of significance. The critical values also depend on the sample size and the number of independent variables in the multivariate regression.[12]

For example, suppose that there are 12 independent variables in a regression, there are 200 observations, and that the significance level selected is 5%. Then according to the Durbin-Watson critical value table, the critical values are

$$d_L = 1.643 \text{ and } d_U = 1.896$$

Then the tests in the previous table can be written as:

Null hypothesis	Range for computed d	Decision rule
No positive autocorrelation	$0 < d < 1.643$	Reject the null hypothesis
No positive autocorrelation	$1.643 \leq d \leq 1.896$	No decision
No negative autocorrelation	$2.357 < d < 4$	Reject the null hypothesis
No negative autocorrelation	$2.104 \leq d \leq 2.357$	No decision
No autocorrelation	$1.896 < d < 2.104$	Accept the null hypothesis

Modeling in the Presence of Autocorrelation

If residuals are autocorrelated, the regression coefficient can still be estimated without bias using the formula given by equation (20.9) in Chapter 20. However, this estimate will not be optimal in the sense that there are other estimators with lower variance of the sampling distribution. Fortunately, there is a way to deal with this problem. There is an optimal linear unbiased estimator called the *Aitken's generalized least squares* (GLS) *estimator* that can be used. The discussion about this estimator is beyond the scope of this chapter.

---

[12]See Savin and White (1977).

The principle underlying the use of such estimators is that in the presence of correlation of residuals, it is common practice to replace the standard regression models with models that explicitly capture autocorrelations and produce uncorrelated residuals. The key idea here is that autocorrelated residuals signal that the modeling exercise has not been completed. That is, if residuals are autocorrelated, this signifies that the residuals at a generic time t can be predicted from residuals at an earlier time.

Autoregressive Moving Average Models

There are models for dealing with the problem of autocorrelation in time series data. These models are called *autoregressive moving average* (ARMA) *models*. Although financial time series typically exhibit structures that are more complex than those provided by ARMA models, these models are a starting point and often serve as a benchmark to compare more complex approaches.

There are two components to an ARMA model: (1) autoregressive process and (2) moving average process.

An *autoregressive process* (AR) *of order p,* or briefly an AR(p) process, is a process where realization of the dependent variable in a time series, y_t, is a weighted sum of past p realizations of the dependent variable (i.e., y_{t-1}, y_{t-2}, ..., y_{t-p}) plus a disturbance term that is denoted by ε_t. The process can be represented as

$$y_t = a_0 + a_1 y_{t-1} + a_2 y_{t-2} + \ldots + a_p y_{t-p} + \varepsilon_t \qquad (22.24)$$

Equation (22.24) is referred to as a p-th order difference equation. In Chapter 7 we introduced the case of an autoregressive process of order 1, AR(1).

A *moving average* (MA) *process of order q,* in short, an MA(q) process, is the weighted sum of the preceding q lagged disturbances plus a contemporaneous disturbance term; that is,

$$y_t = b_0 \varepsilon_t + b_1 \varepsilon_{t-1} + \ldots + b_q \varepsilon_{t-q} \qquad (22.25)$$

Usually, and without loss of generality, we assume that $b_0 = 1$.

The AR and MA processes just discussed can be regarded as special cases of a mixed autoregressive moving average (ARMA) process, in short, an ARMA(p,q) process, given by combining the AR(p) given by equation (22.24) and the MA(q) given by equation (22.25) assuming b_0 is equal to 1. That is,

$$y_t = a_0 + a_1 y_{t-1} + a_2 y_{t-2} + \ldots + a_p y_{t-p} + \varepsilon_t + b_1 \varepsilon_{t-1} + \ldots + b_q \varepsilon_{t-q} \qquad (22.26)$$

The advantage of the ARMA process relative to AR and MA processes is that it gives rise to a more parsimonious model with relatively few unknown parameters. Instead of capturing the complex structure of time series with a relatively high-order AR or MA model, the ARMA model which combines the AR and MA presentation forms can be used.

CONCEPTS EXPLAINED IN THIS CHAPTER
(IN ORDER OF PRESENTATION)

Nonlinear regression
Homoskedastic (homoscedastic)
Heteroskedastic (heteroscedastic)
Autocorrelation
Chi-square statistic
Jarque-Bera test statistic
Light-tailed distribution
Heavy-tailed distribution
Weighted least squares technique
Weighted least squares estimators
Autoregressive conditional heteroskedasticity model
Generalized autoregressive conditional heteroskedasticity model
Volatility clustering
Serial correlation
Lagged correlation
Positive autocorrelation
Negative autocorrelation
Durbin-Watson d statistic
Aitken's generalized least squares estimator
Autoregressive moving average models
Autoregressive process of order p
Moving average process of order q

APPENDIX A

Important Functions and Their Features

In this appendix, we review the functions that are used in some of the chapters in this book. In particular, we introduce the concept of continuous functions, the indicator function, the derivative of a function, monotonic functions, and the integral. Moreover, as special functions, we get to know the factorial, the gamma, beta, and Bessel functions as well as the characteristic function of random variables.

CONTINUOUS FUNCTION

In this section, we introduce general continuous functions.

General Idea

Let $f(x)$ be a continuous function for some real-valued variable x. The general idea behind continuity is that the graph of $f(x)$ does not exhibit gaps. In other words, $f(x)$ can be thought of as being seamless. We illustrate this in Figure A.1. For increasing x, from $x = 0$ to $x = 2$, we can move along the graph of $f(x)$ without ever having to jump. In the figure, the graph is generated by the two functions $f(x) = x^2$ for $x \in [0,1)$, and $f(x) = \ln(x) + 1$ for $x \in [1,2)$.[1]

A function $f(x)$ is *discontinuous* if we have to jump when we move along the graph of the function. For example, consider the graph in Figure A.2. Approaching $x = 1$ from the left, we have to jump from $f(x) = 1$ to $f(1) = 0$. Thus, the function f is discontinuous at $x = 1$. Here, f is given by $f(x) = x^2$ for $x \in [0,1)$, and $f(x) = \ln(x)$ for $x \in [1,2)$.

---

[1] The function $f(x) = \ln(x)$ is the *natural logarithm*. It is the inverse function to the exponential function $g(x) = e^x$ where $e = 2.7183$ is the Euler constant. The inverse has the effect that $f(g(x)) = \ln(e^x) = x$, that is, ln and e cancel each other out.

FIGURE A.1 Continuous Function $f(x)$

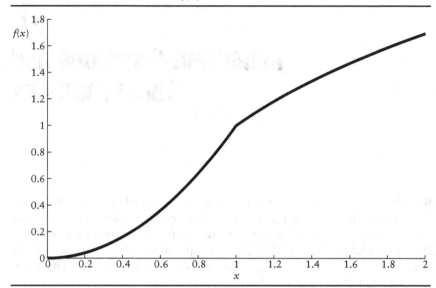

Note: For $x \in [0,1)$, $f(x) = x^2$ and for $x \in [1,2)$, $f(x) = 1 + \ln(x)$.

FIGURE A.2 Discontinuous Function $f(x)$

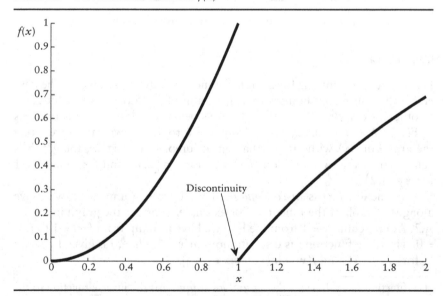

Note: For $x \in [0,1)$, $f(x) = x^2$ and for $x \in [1,2)$, $f(x) = \ln(x)$.

Formal Derivation

For a formal treatment of continuity, we first concentrate on the behavior of f at a particular value x^*.

We say that that a function $f(x)$ is *continuous at x^** if, for any positive distance δ, we obtain a related distance $\varepsilon(\delta)$ such that

$$f(x^* - \delta) \leq f(x) \leq f(x^* + \delta), \text{ for all } x \in \left(x^* - \varepsilon(\delta), x^* + \varepsilon(\delta)\right)$$

What does that mean? We use Figure A.3 to illustrate.[2] At x^*, we have the value $f(x^*)$. Now, we select a neighborhood around $f(x^*)$ of some arbitrary distance δ as indicated by the dashed horizontal lines through $f(x^* - \delta)$ and $f(x^* + \delta)$, respectively. From the intersections of these horizontal lines and the function graph (solid line), we extend two vertical dash-dotted lines down to the x-axis so that we obtain the two values x^L and x^U, respectively. Now, we measure the distance between x^L and x^* and also the distance between x^U and x^*. The smaller of the two yields the distance $\varepsilon(\delta)$. With this distance $\varepsilon(\delta)$ on the x-axis, we obtain the environment $(x^* - \varepsilon(\delta), x^* +$

FIGURE A.3   Continuity Criterion

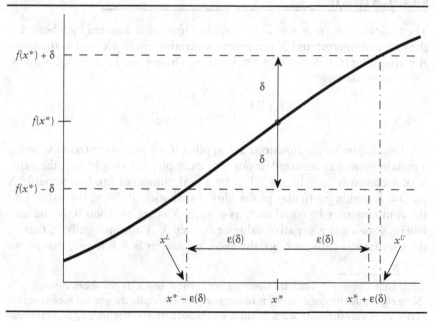

Note: Function $f = \sin(x)$, for $-1 \leq x \leq 1$.

---

[2] The function is $f(x) = \sin(x)$ with $x^* = 0.2$.

$\varepsilon(\delta))$ about x^*.[3] The environment is indicated by the dashed lines extending vertically above $x^* - \varepsilon(\delta)$ and $x^* + \varepsilon(\delta)$, respectively. We require that all x that lie in $\left(x^* - \varepsilon(\delta), x^* + \varepsilon(\delta)\right)$ yield values $f(x)$ inside of the environment $\left[f\left(x^* - \delta\right), f\left(x^* + \delta\right)\right]$. We can see by Figure A.3 that this is satisfied.

Let us repeat this procedure for a smaller distance δ. We obtain new environments $[f(x^* - \delta), f(x^* + \delta)]$ and $(x^* - \varepsilon(\delta), x^* + \varepsilon(\delta))$. If, for all x in $(x^* - \varepsilon(\delta), x^* + \varepsilon(\delta))$, the $f(x)$ are inside of $[f(x^* - \delta), f(x^* + \delta)]$, again, then we can take an even smaller δ. We continue this for successively smaller values of δ just short of becoming 0 or until the condition on the $f(x)$ is no longer satisfied. As we can easily see in Figure A.3, we could go on forever and the condition on the $f(x)$ would always be satisfied. Hence, the graph of f is seamless or continuous at x.

Finally, we say that the function f is *continuous* if it is continuous at all x for which f is defined, that is, in the *domain* of f.[4] In Figure A.3, the interval $-1 \leq x \leq 1$ is depicted.[5] Without formal proof, we see that the function f is continuous, at all single x between -1 and 1, and, thus, f is continuous on this interval.

INDICATOR FUNCTION

The *indicator function* acts like a switch. Often, it is denoted by where A is the event of interest and X is a random variable. So, $1_A(X)$ is 1 if the event A is true, that is, if X assumes a value in A. Otherwise, $1_A(X)$ is 0. Formally, this is expressed as

$$1_A(X) = \begin{cases} 1 & X \in A \\ 0 & \text{otherwise} \end{cases}$$

Usually, indicator functions are applied if we are interested in whether a certain event has occurred or not. For example, in a simple way, the value V of a company may be described by a real numbered random variable X on $\Omega = R$ with a particular probability distribution P. Now, the value V of the company may be equal to X as long as X is greater than 0. In the case where X assumes a negative value or 0, then V is automatically 0, that is, the company is bankrupt. So, the event of interest is $A = [0, \infty)$, that is, we

---

[3] Note that $x^L = x^* - \varepsilon_\delta$ since the distance between x^L and x^* is the shorter one.

[4] Note that only the domain of f is of interest. For example, the square root function $f(x) = \sqrt{x}$ is only defined for $x \leq 0$. Thus, we do not care about whether f is continuous for any x other than $x \leq 0$.

[5] For a better overview, we omitted the individual x-ticks between -1 and 1, on the horizontal axis.

FIGURE A.4 The Company Value V as a Function of the Random Variable X Using the Indicator Function $1_{[0,\infty)}(X) \cdot X$

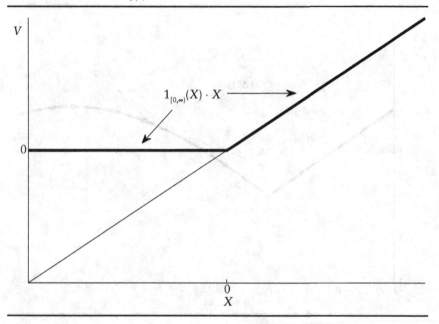

want to know whether X is still positive. Using the indicator function this can be expressed as

$$1_{[0,\infty)}(X) = \begin{cases} 1 & X \in [0,\infty) \\ 0 & \text{otherwise} \end{cases}$$

Finally, the company value can be given as

$$V = 1_{[0,\infty)}(X) \cdot X = \begin{cases} X & X \in [0,\infty) \\ 0 & \text{otherwise} \end{cases}$$

The company value V as a function is depicted in Figure A.4. We can clearly detect the kink at $x = 0$ where the indicator function becomes 1 and, hence, $V = X$.

DERIVATIVES

Suppose we have some continuous function f with the graph given by the solid line in Figure A.5. We now might be interested in the growth rate of

FIGURE A.5 Function f (solid) with Derivatives $f'(x)$ at x, for $0 < x < 0.5$ (dashed), $x = 1$ (dash-dotted), and $x = 1.571$ (dotted)

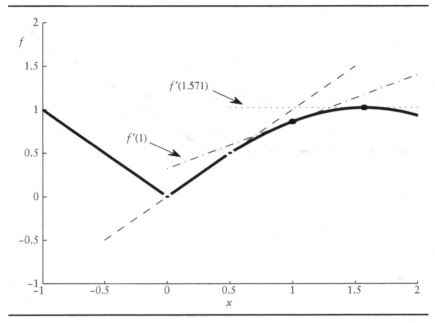

f at some position x. That is, we might want to know by how much f increases or decreases when we move from some x by a step of a given size, say Δx, to the right. This difference in f, we denote by Δf.[6]

Let us next have a look at the graphs given by the solid lines in Figure A.6. These represent the graphs of f and g. The important difference between f and g is that, while g is linear, f is not, as can be seen by f's curvature.

We begin the analysis of the graphs' slopes with function g on the top right of the figure. Let us focus on the point $(x^+, g(x^+))$ given by the solid circle at the lower end of graph g. Now, when we move to the right by Δx_4 along the horizontal dashed line, the corresponding increase in g is given by Δy_4, as indicated by the vertical dashed line. If, on the other hand, we moved to the right by the longer distance, Δx_5, the according increment of g would be given by Δy_5.[7] Since g is linear, it has constant slope everywhere and, hence, also at the point $(x^+, f(x^+))$. We denote that slope by s_4. This implies that the ratios representing the relative increments (i.e., the slopes) have to be equal. That is

---

[6]This Δ symbol is called *delta*.
[7]This vertical increment Δy_5 is also indicated by a vertical dashed line.

FIGURE A.6 Functions f and g with Slopes Measured at the Points $(x^*, f(x^*))$ and $(x^+, g(x^+))$ Indicated by the • Symbol

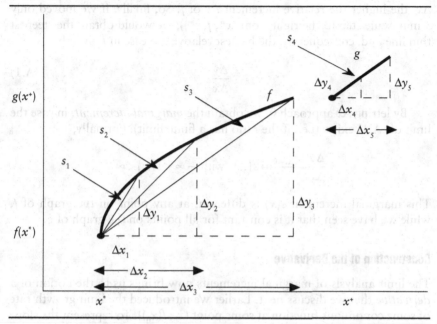

$$s_4 = \frac{\Delta y_4}{\Delta x_4} = \frac{\Delta y_5}{\Delta x_5}$$

Next, we focus on the graph of f on the lower left of Figure A.6. Suppose we measured the slope of f at the point $(x^*, f(x^*))$. If we extended a step along the dashed line to the right by Δx_1, the corresponding increment in f would be Δy_1, as indicated by the leftmost vertical dashed line. If we moved, instead, by the longer Δx_2 to the right, the corresponding increment in f would be Δy_2. And, a horizontal increment of Δx_3 would result in an increase of f by Δy_3.

In contrast to the graph of g, the graph of f does not exhibit the property of a constant increment Δy in f per unit step Δx to the right. That is, there is no constant slope of f, which results in the fact that the three ratios of the relative increase of f are different. To be precise, we have

$$\frac{\Delta y_1}{\Delta x_1} > \frac{\Delta y_2}{\Delta x_2} > \frac{\Delta y_3}{\Delta x_3}$$

as can be seen in Figure A.6. So, the shorter our step Δx to the right, the steeper the slopes of the thin solid lines through $(x^*, f(x^*))$ and the corresponding points on the curve, $(x^*+\Delta x_1, f(x^*+\Delta x_1))$, $(x^*+\Delta x_2, f(x^*+\Delta x_2))$, and $(x^*+\Delta x_2, f(x^*+\Delta x_2))$, respectively. That means that, the smaller the increment Δx, the higher the relative increment Δy of f. So, finally, if we moved only a minuscule step to the right from $(x^*, f(x^*))$, we would obtain the steepest thin line and, consequently, the highest relative increase in f given by

$$\frac{\Delta y}{\Delta x} \tag{A.1}$$

By letting Δx approach 0, we obtain the *marginal increment*, in case the limit of (A.1) exists (i.e., if the ratio has a finite limit). Formally,

$$\frac{\Delta y}{\Delta x} \xrightarrow{\Delta x \to 0} s(x) \text{ with } -\infty < s(x) < \infty$$

This marginal increment $s(x)$ is different, at any point on the graph of f, while we have seen that it is constant for all points on the graph of g.

Construction of the Derivative

The limit analysis of marginal increments now brings us to the notion of a *derivative* that we discuss next. Earlier we introduced the limit growth rate of some continuous function at some point $(x_0, f(x_0))$. To represent the slope of the line through $(x_0, f(x_0))$ and $(x_0+\Delta x, f(x_0+\Delta x))$, we define the *difference quotient*

$$\frac{f(x_0 + \Delta x) - f(x_0)}{\Delta x} \tag{A.2}$$

If we let $\Delta x \to 0$, we obtain the limit of the difference quotient (A.2). If this limit is not finite, then we say that it does not exist. Suppose, we were not only interested in the behavior of f when moving Δx to the right but also wanted to analyze the reaction by f to a step Δx to the left. We would then obtain two limits of (A.2). The first with $\Delta x^+ > 0$ (i.e., a step to the right) would be the *upper limit* L^U

$$\frac{f(x_0 + \Delta x^+) - f(x_0)}{\Delta x^+} \xrightarrow{\Delta x^+ \to 0} L^U$$

and the second with $\Delta x^- < 0$ (i.e., a step to the left), would be the *lower limit* L^L

$$\frac{f(x_0 + \Delta x^-) - f(x_0)}{|\Delta x^-|} \xrightarrow{\Delta x^- \to 0} L^U$$

If L^U and L^L are equal, $L^U = L^L = L$, then f is said to be *differentiable at* x_0. The limit L is the *derivative of f*. We commonly write the derivative in the fashion

$$f'(x_0) = \frac{df(x)}{dx}\bigg|_{x=x_0} = \frac{dy}{dx}\bigg|_{x=x_0} \tag{A.3}$$

On the right side of (A.3), we have replaced $f(x)$ by the variable y as we will often do, for convenience. If the derivative (A.3) exists for all x, then f is said to be *differentiable*.

Let us now return to Figure A.5. Recall that the graph of the continuous function f is given by the solid line. We start at $x = -1$. Since f is not continuous at $x = -1$, we omit this end point (1,1) from our analysis. For $-1 < x < 0$, we have that f is constant with slope $s = -1$. Consequently, the derivative $f'(x) = -1$, for these x.

At $x = 0$, we observe that f is linear to the left, with $f'(x) = -1$ and that it is also linear to the right, however, with $f'(x) = 1$, for $0 < x < 0.5$. So, at $x = 0$, $L^U = 1$ while $L^L = -1$. Since here $L^U \neq L^L$, the derivative of f does not exist at $x = 0$.

For $0 < x < 0.5$, we have the constant derivative $f'(x) = 1$. The corresponding slope of 1 through (0,0) and (0.5,0.5) is indicated by the dashed line. At $x = 0.5$, the left side limit $L^L = 1$ while the right side limit $L^U = 0.8776$.[8] Hence, the two limits are not equal and, consequently, f is not differentiable at $x = 0.5$.

Without formal proof, we state that f is differentiable for all $0.5 < x < 2$. For example, at $x = 1$, $L^L = L^U = 0.5403$ and, thus, the derivative $f'(0.5) = 0.5403$. The dash-dotted line indicating this derivative is called the *tangent of f* at $x = 0.5$.[9] As another example, we select $x = 1.571$ where f assumes its maximum value. Here, the derivative $f'(0) = 0$ and, hence, the tangent at $x = 1.571$ is flat as indicated by the horizontal dotted line.[10]

MONOTONIC FUNCTION

Suppose we have some function $f(x)$ for real-valued x. For example, the graph of f may look like that in Figure A.5. We see that on the interval [0,1], the graph is increasing from $f(0) = 0$ to $f(1) = 1$. For $1 \leq x \leq 2$, the graph

---

[8] This value of $\cos(0.5) = 0.8776$ is a result from calculus. We will not go into detail here.
[9] In Figure A.5, the arrow indexed $f'(1)$ points at this tangent.
[10] In Figure A.5, the arrow indexed $f'(1.571)$ points at this tangent.

remains at the level $f(1) = 1$ like a platform. And, finally, between $x = 2$ and $x = 3$, the graph is increasing, again, from $f(2) = 1$ to $f(3) = 2$.

In contrast, we may have another function, $g(x)$. Its graph is given by Figure A.6. It looks somewhat similar to the graph in Figure A.5, however, without the platform. The graph of g never remains at a level, but increases constantly. Even for the smallest increments from one value of x, say x_1, to the next higher, say x_2, there is always an upward slope in the graph.

Both functions, f and g, never decrease. The distinction is that f is *monotonically increasing* since the graph can remain at some level, while g is *strictly monotonic increasing* since its graph never remains at any level. If we can differentiate f and g, we can express this in terms of the derivatives of f and g. Let f' be the derivative of f and g' the derivative of g. Then, we have the following definitions of continuity for continuous functions with existing derivatives:

Monotonically increasing functions: A continuous function f with derivative f' is monotonically increasing if its derivative $f' \geq 0$.

Strictly monotonic increasing functions: A continuous function g with derivative g' is strictly monotonic increasing if its derivative $g' > 0$.

Analogously, a function $f(x)$ is monotonically decreasing if it behaves in the opposite manner. That is, f never increases when moving from some x to any higher value $x_1 > x$. When f is continuous with derivative f', then we say that f is *monotonically decreasing* if $f'(x) \leq 0$ and that it is ***strictly monotonic increasing*** if $f'(x) < 0$ for all x. For these two cases, illustrations are given by mirroring Figures A.7 and A.8 against their vertical axes, respectively.

INTEGRAL

Here we derive the concept of integration necessary to understand the probability density and continuous distribution function. The integral of some function over some set of values represents the area between the function values and the horizontal axis. To sketch the idea, we start with an intuitive graphical illustration.

We begin by analyzing the area A between the graph (solid line) of the function $f(t)$ and the horizontal axis between $t = 0$ and $t = T$ in Figure A.9. Looking at the graph, it appears quite complicate to compute this area A in comparison to, for example, the area of a rectangle where we would only need to know its width and length. However, we can approximate this area by rectangles as will be done next.

Important Functions and Their Features

FIGURE A.7 Monotonically Increasing Function f

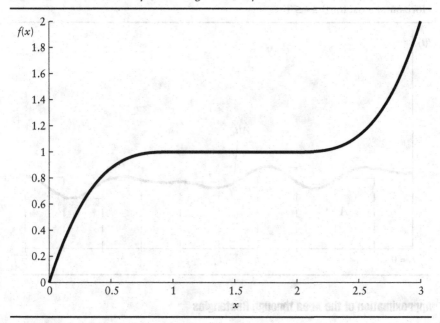

FIGURE A.8 Strictly Monotonic Increasing Function g

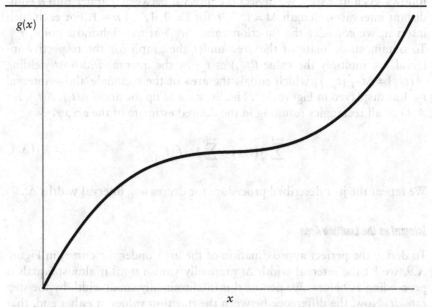

FIGURE A.9 Approximation of the Area A between Graph of $f(t)$ and the Horizontal axis, for $0 \leq t \leq T$

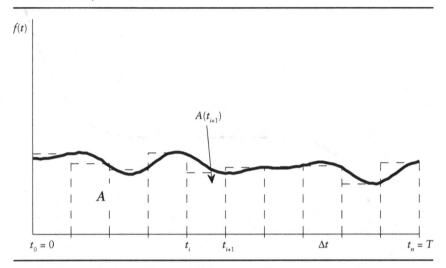

Approximation of the Area through Rectangles

Let's approximate the area A under the function graph in Figure A.9 as follows. As a first step, we dissect the interval between 0 and T into n equidistant intervals of length $\Delta t = t_{i+1} - t_i$ for $i = 0, 1, \ldots, n - 1$. For each such interval, we consider the function value $f(t_{i+1})$ at the rightmost point, t_{i+1}. To obtain an estimate of the area under the graph for the respective interval, we multiply the value $f(t_{i+1})$ at t_{i+1} by the interval width Δt yielding $A(t_{i+1}) = \Delta t \cdot f(t_{i+1})$, which equals the area of the rectangle above interval $i + 1$ as displayed in Figure A.9. Finally, we add up the areas $A(t_1), A(t_2), \ldots, A(T)$ of all rectangles resulting in the desired estimate of the area A

$$\sum_{i=0}^{n-1} A(t_{i+1}) = \sum_{i=0}^{n-1} \Delta t \cdot f(t_{i+1}) \tag{A.4}$$

We repeat the just described procedure for decreasing interval widths Δt.

Integral as the Limiting Area

To derive the perfect approximation of the area under the curve, in Figure A.9, we let the interval width Δt gradually vanish until it almost equals 0 proceeding as before. We denote this infinitesimally small width by the step rate dt. Now, the difference between the function values at either end, that

Important Functions and Their Features

is, $f(t_i)$ and $f(t_{i+1})$, of the interval $i + 1$ will be nearly indistinguishable since t_i and t_{i+1} almost coincide. Hence, the corresponding rectangle with area $A(t_{i+1})$ will turn into a dash with infinitesimally small base dt.

Summation as in equation (A.4) of the areas of the dashes becomes infeasible. For this purpose, the *integral* has been introduced as the limit of (A.4) as $\Delta t \to 0$.[11] It is denoted by

$$\int_0^T f(t)\,dt \qquad (A.5)$$

where the limits 0 and T indicate which interval the integration is performed on. In our case, the *integration variable* is t while the function $f(t)$ is called the *integrand*. In words, equation (A.5) is the *integral of the function $f(t)$ over t from 0 to T*. It is immaterial how we denote the integration variable. The same result as in equation (A.5) would result if we wrote

$$\int_0^T f(y)\,dy$$

instead. The important factors are the integrand and the integral limits.

Note that instead of using the function values of the right boundaries of the intervals $f(t_{i+1})$ in equation (A.4), referred to as the *right-point rule*, we might as well have taken the function values of the left boundaries $f(t_i)$, referred to as the *left-point rule*, which would have led to the same integral. Moreover, we might have taken the function $f\left(0.5\cdot(t_{i+1}+t_i)\right)$ values evaluated at the mid-points of the intervals and still obtained the same interval. This latter procedure is the called the *mid-point rule*.

If we keep 0 as the lower limit of the integral in equation (A.5) and vary T, then equation (A.5) becomes a function of the variable T. We may denote this function by

$$F(T) = \int_0^T f(t)\,dt \qquad (A.6)$$

Relationship Between Integral and Derivative

In equation (A.6) the relationship between $f(t)$ and $F(T)$ is as follows. Suppose we compute the derivative of $F(T)$ with respect to T.[12] The result is

$$F'(T) = \frac{dF(T)}{dT} = f(T) \qquad (A.7)$$

---

[11] Conditions under which these limits exist are omitted, here.
[12] We assume that $F(T)$ is differentiable, for $T > 0$.

Hence, from equation (A.7) we see that the marginal increment of the integral at any point (i.e., its derivative) is exactly equal to the integrand evaluated at the according value.[13]

The implication of this discussion for probability theory is as follows. Let P be a continuous probability measure with probability distribution function F and (probability) density function f. There is the unique link between f and P given through

$$P(X \leq x) = F(x) = \int_{-\infty}^{\infty} f(x)dx \qquad (A.8)$$

Formally, the integration of f over x is always from $-\infty$ to ∞, even if the support is not on the entire real line. This is no problem, however, since the density is zero outside the support and, hence, integration over those parts yields 0 contribution to the integral. For example, suppose that some density function were

$$f(x) = \begin{cases} h(x), & x \geq 0 \\ 0 & x < 0 \end{cases} \qquad (A.9)$$

where $h(x)$ is just some function such that f satisfies the requirements for a density function. That is, the support is only on the positive part of the real line. Substituting the function from equation (A.9) into equation (A.8) yields the equality

$$\int_{-\infty}^{\infty} f(x)dx = \int_{0}^{\infty} f(x)dx = \int_{0}^{\infty} h(x)dx \qquad (A.10)$$

SOME FUNCTIONS

Here we introduce some functions needed in probability theory to describe probability distributions of random variables: factorials, gamma function, beta function, Bessel function of the third kind, and characteristic function While the first four are functions of very special shape, the characteristic function is of a more general structure. It is the function characterizing the probability distribution of some random variable and, hence, is of unique form for each random variable.

---

[13]This need not generally be true. But in most cases and particularly those we will be dealing with, this statement is valid.

Factorial

Let $k \in N$ (i.e., $k = 1, 2, \ldots$). Then the *factorial* of this natural number k, denoted by the symbol !, is given by

$$k! = k \cdot (k-1) \cdot (k-2) \cdot \ldots \cdot 1 \qquad (A.11)$$

A factorial is the product of this number and all natural numbers smaller than k including 1. By definition, the factorial of zero is one (i.e., $0! \equiv 1$). For example, the factorial of 3 is $3! = 3 \cdot 2 \cdot 1 = 6$.

Gamma Function

The *gamma function* for nonnegative values x is defined by

$$\Gamma(x) = \int_0^\infty e^{-t} t^{x-1} dt, \quad x \geq 0 \qquad (A.12)$$

The gamma function has the following properties. If the x correspond with a natural number $n \in N$ (i.e., $n = 1, 2, \ldots$), then we have that equation (A.12) equals the factorial given by equation (A.11) of $n - 1$. Formally, this is

$$\Gamma(n) = (n-1)! = (n-1) \cdot (n-2) \cdot \ldots \cdot 1$$

Furthermore, for any $x \geq 0$, it holds that $\Gamma(x+1) = x\Gamma(x)$.

In Figure A.10, we have displayed part of the gamma function for x values between 0.1 and 5. Note that, for either $x \to 0$ or $x \to \infty$, $\Gamma(x)$ goes to infinity.

Beta Function

The *beta function* with parameters c and d is defined as

$$\begin{aligned} B(c,d) &= \int_0^1 u^{c-1}(1-u)^{d-1} du \\ &= \frac{\Gamma(c)\Gamma(d)}{\Gamma(c+d)} \end{aligned}$$

where Γ is the gamma function from equation (A.12).

FIGURE A.10 Gamma Function $\Gamma(x)$

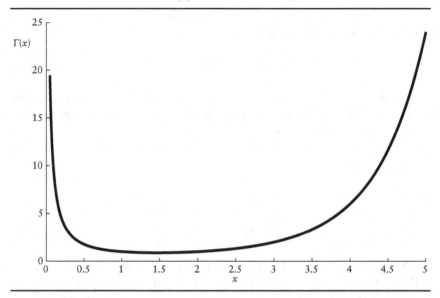

Bessel Function of the Third Kind

The *Bessel function of the third kind* is defined as

$$K_1(x) = \frac{1}{2}\int_0^\infty \exp\left\{-\frac{x}{2}\left(y+\frac{1}{y}\right)\right\} dy$$

This function is often a component of other, more complex functions such as the density function of the *NIG* distribution.

Characteristic Function

Before advancing to introduce the *characteristic function*, we briefly explain *complex numbers*.

Suppose we were to take the square root of the number –1, that is, $\sqrt{-1}$. So far, our calculus has no solution for this since the square root of negative numbers has not yet been introduced. However, by introducing the *imaginary* number *i*, which is defined as

$$i = \sqrt{-1}$$

Important Functions and Their Features

we can solve square roots of any real number. Now, we can represent any number as the combination of a real (*Re*) part *a* plus some units *b* of *i*, which we refer to as the *imaginary* (*Im*) *part*. Then, any number *z* will look like

$$z = a + i \cdot b \tag{A.13}$$

The number given by equation (A.13) is a complex number. The set of complex numbers is symbolized by **C**. This set contains the real numbers that are those complex numbers with $b = 0$. Graphically, we can represent the complex numbers on a two-dimensional space as given in Figure A.11.

Now, we can introduce the characteristic function as some function φ mapping real numbers into the complex numbers. Formally, we write this as φ: **R** → **C**. Suppose we have some random variable *X* with density function *f*. The characteristic function is then defined as

$$\phi(t) = \int_{-\infty}^{\infty} e^{itx} f(x) dx \tag{A.14}$$

FIGURE A.11 Graphical Representation of the Complex Number $z = 0.8 + 0.9i$

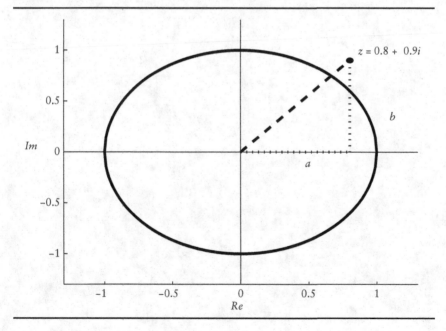

which transforms the density f into some complex number at any real position t. Equation (A.14) is commonly referred to as the *Fourier transformation of the density*.

The relationship between the characteristic function φ and the density function f of some random variable is unique. So, when we state either one, the probability distribution of the corresponding random variable is unmistakably determined.

APPENDIX B

Fundamentals of Matrix Operations and Concepts

For several topics in this book, we used principles, concepts, and results from the field of matrix algebra. In this appendix, we provide a review of matrix algebra.

THE NOTION OF VECTOR AND MATRIX

Consider a data series $x = \{x_1, x_2, ..., x_n\}$ of size n. Instead of the complicated notation using commas between the individual observations, we could write the n observations in terms of vector notation as follows

$$x = \begin{pmatrix} x_1 \\ x_2 \\ \vdots \\ x_n \end{pmatrix} \text{ or } x^T = \begin{pmatrix} x_1 & x_2 & \ldots & x_n \end{pmatrix} \quad (B.1)$$

where the left side in equation (B.1) denotes x as a *column vector* while the right side denotes it as a *row vector*. Both vectors are of length n (coordinates) since all observations have to be accounted for. Note that the right (row) vector is the *transpose* of the left (column) vector and vice versa. The transpose is obtained by simply turning the vector by 90 degrees, which is indicated by the upper index T. We could have defined x as a row vector instead and the transpose would then be the column vector. By doing so, we shift the *dimensionality* of the vectors. The column vector is often indicated by the size $n \times 1$ meaning n rows and 1 column. A row vector is indicated by size $1 \times n$ meaning just one row and n columns.

Suppose there is a second series of observations of the same length n as x, say $y = (y_1, y_2, ..., y_n)$. This second series of observation could be written as a vector. However, x and y could both be written as vectors and combined in a construct called *matrix A* of the form

$$A = \begin{pmatrix} x_1 & y_1 \\ x_2 & y_2 \\ \vdots & \vdots \\ x_n & y_n \end{pmatrix} \quad (B.2)$$

Since both vectors are entities of their own of dimension one, the combined result in matrix A is of dimension two. To be precise, A is an $n \times 2$ matrix with n coordinates in each of the two columns. The matrix is thus of size $n \times 2$. Its transpose would naturally be of size $2 \times n$.

MATRIX MULTIPLICATION

Now consider a third column vector of length n, say z. Moreover, it is hypothetically assumed that each component z_i is equal to a linear combination of the corresponding components of the vectors x and y. That is, for each i, we obtain

$$z_i = \beta_1 x_i + \beta_2 y_i \quad (B.3)$$

which presents vector z as the result of the weighted sum of the vectors x and y. The weight for x is β_1 and the weight for y is β_2. In other words, each component of the vectors x and y is weighted by the corresponding factor β_1 and β_2, respectively, before summation. Since we perform equation (B.3) for each $i = 1, ..., n$, we obtain the *equation system*

$$\begin{matrix} z_1 = \beta_1 x_1 + \beta_2 y_1 \\ \vdots \quad \quad \vdots \\ z_n = \beta_1 x_n + \beta_2 y_n \end{matrix} \quad (B.4)$$

The above can be written in a more compact way using vector and matrix notation by

$$z = (x \quad y)\beta \quad (B.5)$$

where z is the column vector consisting of all the components on the left side of the equations in equation (B.4). We have combined the two column vectors x and y in the matrix $(x \quad y)$. This is multiplied by the column vector

Fundamentals of Matrix Operations and Concepts

$\beta = (\beta_1 \ \beta_2)^T$. Note that the $(x \ y)$ is of size $n \times 2$ and β is of size 2×1. Since the number of columns of $(x \ y)$ (i.e., two) matches the number of rows of β, this works. Equation (B.5) is the result of *matrix multiplication*, which is only feasible if the number of columns of the matrix on the left matches the number of rows of the matrix on the right.[1]

Matrix multiplication is defined as follows. Given the $m \times n$ rectangular matrix $X = \{x_{ij}\}$ and the $n \times p$ matrix $Y = \{y_{rq}\}$, the product $Z = \{z_{iq}\}$ is the $m \times p$ matrix:

$$Z = \{z_{iq}\}$$

$$z_{iq} = \sum_{j=1}^{n} x_{ij} y_{jq}$$

For example, suppose we want to multiply $(x \ y)$ by the 2×2 matrix $(a \ b)$ where a and b both are 2×1 vectors. The result is the $n \times 2$ matrix

$$\begin{pmatrix} x_1 & y_1 \\ \vdots & \vdots \\ x_n & y_n \end{pmatrix} \cdot \begin{pmatrix} a_1 & b_1 \\ a_2 & b_2 \end{pmatrix} = \begin{pmatrix} x_1 a_1 + y_1 a_2 & x_1 b_1 + y_1 b_2 \\ \vdots & \vdots \\ x_n a_1 + y_n a_2 & x_n b_1 + y_n b_2 \end{pmatrix} \quad (B.6)$$

Note that each of the two column vectors in the matrix on the right side of equation (B.6) is of the form equation (B.4). We could extend the matrix containing the vectors a and b by inserting a third 2×1 vector, say c, such that the resulting matrix $(a \ b \ c)$ is 2×3. The matrix product of multiplying $(x \ y)$ and $(a \ b \ c)$ yields a $n \times 3$ matrix with one additional column vector compared to the matrix on the right side of equation (B.6). Analogously, we could create any $n \times k$ matrix for some positive integer k. Remember, we can multiply any two matrices A and B to yield AB provided A has the same number of columns as B has rows.

PARTICULAR MATRICES

When we have a matrix of size $n \times n$ (i.e., both the number of columns and rows are the same), the matrix is a square and referred to as a *square matrix*. For example, consider the 3×3 matrix A given by

$$A = \begin{pmatrix} a_{11} & a_{12} & a_{13} \\ a_{21} & a_{22} & a_{23} \\ a_{31} & a_{32} & a_{33} \end{pmatrix} \quad (B.7)$$

---

[1] Here we have implicitly used the fact that a vector is a particular matrix.

The diagonal elements a_{ii}, $i = 1, \ldots, n$ have the particular property that they remain at their positions after transposing the matrix.[2]

We consider a particular case of equation (B.7) next. That is, we have three particular column vectors (or, analogously, row vectors) of a special form. For example, let the first column vector of A $a_1^T = (a_{11}\ a_{21}\ a_{31})^T = (1\ 0\ 0)^T$, the second column vector of $a_2^T = (a_{12}\ a_{22}\ a_{32})^T = (0\ 1\ 0)^T$, and the third $a_3^T = (a_{13}\ a_{23}\ a_{33})^T = (0\ 0\ 1)^T$. Each vector takes one step in its particular direction and none in the direction of any of the other two. That is, they move *orthogonally* to one another along the coordinate axis. This is shown in Figure B.1.

As can be seen in the figure, together a_1, a_2, and a_3 reach into all three dimensions spanned by the coordinate axes. They are basically equivalent. By linearly combining these three, any vector in this three-dimensional space can be generated. Say, we want to "walk" in the direction of $b = \begin{pmatrix} 1 & 2 & 3 \end{pmatrix}$, that is, we step one unit in the direction of a_1, two units in the direction of a_2, and three units in the direction of a_3. In Figure B.2 it is demonstrated what vector b looks like. It is indicated by the solid dash-dotted line while the individual steps in each direction are given by the thin-dashed lines. More formally, we could recover b by finding a column vector $\beta = \begin{pmatrix} \beta_1 & \beta_2 & \beta_3 \end{pmatrix}^T$ to multiply from the left by A such that

FIGURE B.1 Direction of 3 × 3 Matrix A with Column Vectors a_1, a_2, and a_3

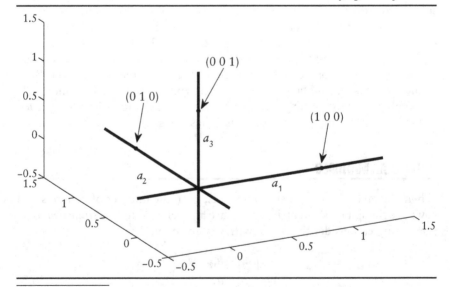

[2]In equation (B.7), we presented the indexes of the matrix components as follows. The first index number indicates the row while the second index value indicates the column. Together they yield the position in the matrix.

Fundamentals of Matrix Operations and Concepts

FIGURE B.2 Vector b as a Linear Composition of the Vectors a_1, a_2, and a_3

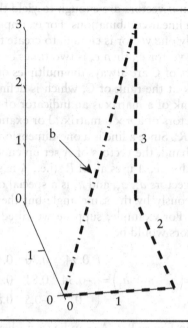

$$b = A\beta \qquad (B.8)$$

is solved.

Now, for each $i = 1, 2, 3$, the components of b are given by $b_i = a_{i1}\beta_1 + a_{i2}\beta_2 + a_{i3}\beta_3$. So we can retrieve as the solution $\beta = (1 \ 2 \ 3)^T$. The analogy to equation (B.4) is obvious. It is well-known that there exists a unique β as solution to equation (B.8) since the vectors of A span the whole three-dimensional space of real numbers (i.e., R^3). There is no redundant information in the vectors. In that case, A is said to have *full rank*.[3] That is, the rank of the square $n \times n$ matrix is equivalent to either the number of rows or the number of columns, which are the same. If matrix A were composed of vectors that contained redundant components, its rank would be less than three. For example, look at matrix C given by

$$C = (c_1 \ c_2 \ c_3) = \begin{pmatrix} 1 & 2 & 3 \\ 2 & 4 & 6 \\ 3 & 6 & 9 \end{pmatrix} \qquad (B.9)$$

---

[3] This holds for any $n \times n$ matrix with the same features as A, however in R^n.

Redundant components are the result of the fact that, in the case of a $n \times n$ matrix, less than n vectors are enough to yield the remaining vectors of the matrix through linear combinations. For example, with matrix C, we have the case that only one vector is enough to create the other two. To see this, suppose we take vector c_1, then c_2 is two times c_1, and c_3 is three times c_1. So, any two vectors of C are always the multiples of a third vector of C. When we have a look at the rank of C, which is definitely less than three, we consider that a rank of a matrix is an indicator of what shape is set up by the component vectors of a $n \times n$ matrix.[4] For example, the vectors of C are all along a line in R^3. Since a line is a one-dimensional structure, its rank is one. On the other hand, the vectors of A set up cubes in R^3. Cubes have dimension three and, hence, A uses all of R^3 (i.e., A has rank three).

The form of the vectors a_1, a_2, and a_3 is a special one.[5] One could turn all of them simultaneously by the same angle but they would still remain pairwise orthogonal. For example, suppose we tilted them such that the resulting column vectors would be

$$A^* = \begin{pmatrix} a_1^* & a_2^* & a_3^* \end{pmatrix} = \begin{pmatrix} 0.94 & 0.30 & 0.17 \\ -0.34 & 0.81 & 0.47 \\ 0. & -0.5 & 0.87 \end{pmatrix} \quad (B.10)$$

They are shown in Figure B.3. As can be seen, they still span the entire space of the three-dimensional real numbers. That is, they are pairwise orthogonal. A way of testing pairwise orthogonality of two vectors is to compute their *inner product*. The inner product requires that the two vectors, a and b, are both of the same length, say n.[6] The inner product then is defined as

$$\langle a, b \rangle = a^T b = a_1 b_1 + \ldots + a_n b_n \quad (B.11)$$

where both a and b are column vectors such that a^T is in row vector form. If not, then they have to be appropriately transposed such that in equation (B.11) after the first equality sign, the left vector is a row vector, and the vector on the right is column vector.

For orthogonal vectors, the inner product is zero. Now, with our example, all inner products of the component vectors of A^* are zero. So are, consequently, the inner products of the matrix A.

Now consider again equation (B.8) with matrix A such that it has full rank, which in our example is equal to three. The unique vector β can be

---

[4] We consider only square matrices; that is, $n \times n$ matrices for rank determination.
[5] That is true for all vectors of a full rank matrix.
[6] That is, they both have to have the same number of components.

Fundamentals of Matrix Operations and Concepts

FIGURE B.3 Turned Vectors a_1^*, a_2^*, and a_3^* Spanning the R^3

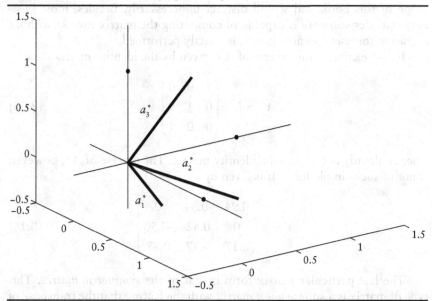

recovered by multiplying b by the *inverse* of A (i.e., A^{-1}) from the left. This is formally represented as

$$A^{-1}b = \beta \qquad (B.12)$$

The inverse of some full rank $n \times n$ matrix A is defined by

$$A^{-1}A = I_{n,n} \qquad (B.13)$$

where $I_{n,n}$ is the $n \times n$ *identity matrix*. It is defined by all ones on the diagonal and zeros else. Thus, the identity matrix is[7]

$$I_{n,n} = \begin{pmatrix} 1 & 0 & \cdots & 0 \\ 0 & 1 & & \vdots \\ \vdots & & \ddots & 0 \\ 0 & \cdots & 0 & 1 \end{pmatrix} \qquad (B.14)$$

The identity matrix has the *neutrality property*. That is, any properly sized matrix multiplied by the identity matrix from either left or right remains unchanged.

---
[7] When the dimension n is obvious, we simply write I.

We avoid the algebraic derivation of the inverse since it is beyond the scope of this book and would distract unnecessarily. Besides, most business statistics software is capable of computing the matrix inverse without requiring the user to know how it is exactly performed.

In our example, the inverse of A is given by the identity matrix

$$A^{-1} = I_3 = \begin{pmatrix} 1 & 0 & 0 \\ 0 & 1 & 0 \\ 0 & 0 & 1 \end{pmatrix} \quad (B.15)$$

since A already is equal to the identity matrix. The inverse of A^*, however, is not of such simple form. It is given by

$$(A^*)^{-1} = \begin{pmatrix} 0.94 & -0.34 & 0 \\ 0 & 0.82 & -0.50 \\ 0.17 & 0.47 & 0.87 \end{pmatrix} \quad (B.16)$$

The last particular matrix form is that of the *symmetric matrix*. This type of matrix is a square $n \times n$ matrix with the feature that the transpose of it is itself. Formally, let S be some symmetric $n \times n$ matrix, then

$$S^T = S \quad (B.17)$$

For example, we have S given by

$$S = \begin{pmatrix} 1 & 2 & 3 \\ 2 & 1 & 2 \\ 3 & 2 & 1 \end{pmatrix} \quad (B.18)$$

S is symmetric since its transpose is

$$S^T = \begin{pmatrix} 1 & 2 & 3 \\ 2 & 1 & 2 \\ 3 & 2 & 1 \end{pmatrix} = S \quad (B.19)$$

Next, we shall discuss the determinant, the eigenvalue, and the eigenvectors of a square matrix, say A. These give insight into important features of a matrix without having to know all its components.

Determinant of a Matrix

Geometrically, the *determinant* of the matrix A computes—or determines—the area or volume enclosed by the vectors of A. The determinant is usually denoted by either $\det(A)$ or $|A|$.

Suppose A is a 3×3 matrix, then we have three vectors setting up something cubic, a *parallelepiped*,[8] that is, the case when A is full rank (i.e., three). If it is short rank, however, the construct designed by the three vectors is not three dimensional since the vectors of A do not span the entire R^3. They are then linearly dependent. At least one dimension is lost of the parallelepiped. So, it is two dimensional, at most, depending on the vectors of A. The entire volume, hence, is zero since we have zero extension in at least one dimension. A more rigorous understanding of the theory behind it, however, is not necessary.

Since it may not hurt to have at least seen the algorithm according to which the determinant of a matrix is computed, we will present it here in as informal a way as possible. It is helpful to consider that the algorithm is such that, in the end, only determinants of 2×2 matrices are computed and summed up in a particular fashion to result in the determinant of the matrix of interest, again say A. Now, for any 2×2 matrix, say M, the determinant is computed according to

$$\det(M) = |M| = \begin{vmatrix} m_{11} & m_{12} \\ m_{21} & m_{22} \end{vmatrix} = m_{11} \cdot m_{22} - m_{12} \cdot m_{12} \qquad \text{(B.20)}$$

The algorithm for the computation of the determinant of an $n \times n$ matrix with n greater than two is a sequential summation of the determinants of a series of matrices contained in the original matrix A. Before starting the summation, we assign imaginary signs to each matrix component a_{ij} in the following way. If $i+j$ is even, then in the summation, a_{ij} is multiplied by the unaltered sign (i.e., 1). If, on the other hand, $i+j$ should be uneven, then a_{ij} is multiplied by -1. Formally, this is achieved by assigning signs according to $(-1)^{i+j}$. We compute these imaginary signs for all components of A.

This done as follows. In step 1, we select the first row (i.e., $i = 1$). Within row one, we select the first column (i.e., $j = 1$) such that we have singled out the component a_{11}. Now we create a new matrix, say $A_{1,1}^1$ out of A in that we discard row one and column one. The first component of the overall sum is then

$$(-1)^{1+1} a_{ij} \cdot \left| A_{1,1}^{(1)} \right| = a_{ij} \cdot \left| A_{1,1}^{(1)} \right| \qquad \text{(B.21)}$$

---

[8] A parallelepiped is in three-dimensional space such as a prism and the sides of the prism form parallelograms.

where $|A_{11}^{(1)}|$ is the determinant of the matrix obtained by discarding row one and column one from matrix A.[9] Still in step 1, we move one further along row one to select column two (i.e., $j = 2$). At this intermediate step, we single out a_{12} and generate the determinant of the matrix A_{12}^1 obtained by discarding row one and column two from matrix A. We obtain the second term of the overall sum by

$$(-1)^{1+2} a_{12} \cdot |A_{12}^1| = -a_{12} \cdot |A_{12}^1| \tag{B.22}$$

In step 1, we continue in this fashion through all columns while remaining in row one. After we have arrived at column n, we move one row further. That is in step 2, we remain in row two while successively selecting one row after another from one through n. We proceed as in step 1 column-by-column to obtain the terms that are contributing to the overall sum. Finally, we perform these equivalent steps until step n where we concentrate on row n. In general, in step i and for each column j, we contribute the term

$$(-1)^{i+j} a_{ij} \cdot |A_{ij}^i| \tag{B.23}$$

to the overall sum yielding the desired determinant $|A|$, at least formally. If the A_{ij}^i just generated are 2×2 matrices, we can compute the $|A_{ij}^i|$ according to (B.20), sum up, and we are done.[10] If not, however, we will have to carry through the procedure for each individual matrix A_{ij}^i just as for matrix A. We keep doing this recursively until by discarding rows and columns, we finally obtain 2×2 matrices.

We will demonstrate this by some fairly simple example employing the following matrix

$$A = \begin{pmatrix} 16 & 2 & 3 & 13 \\ 5 & 11 & 10 & 8 \\ 9 & 7 & 6 & 12 \\ 4 & 14 & 15 & 1 \end{pmatrix} \tag{B.24}$$

We begin by setting up the matrix of imaginary signs that we will call S for reference. This yields

---

[9] The indexes of $|A_{ij}^{(i)}|$ are to be understood as follows: We are in step i generating the determinant of the matrix $A_{ij}^{(i)}$ obtained from A by discarding row i and column j.

[10] The expressions $(-1)^{i+j} |A_{ij}^i|$ are commonly referred to as the *minors of A*.

Fundamentals of Matrix Operations and Concepts

$$S = \begin{pmatrix} \ddots & & \\ & (-1)^{i+j} & \\ & & \ddots \end{pmatrix} = \begin{pmatrix} 1 & -1 & 1 & -1 \\ -1 & 1 & -1 & 1 \\ 1 & -1 & 1 & -1 \\ -1 & 1 & -1 & 1 \end{pmatrix} \quad (B.25)$$

which we will only keep for a look-up reference. Now, in step 1, we focus on row one. We, then begin with column one to obtain the first term of the determinant of equation (B.24). According to the algorithm we just explained, this results in

$$(-1)^{1+1} a_{11} |A_{11}^1| = 16 \cdot \begin{vmatrix} 11 & 10 & 8 \\ 7 & 6 & 12 \\ 14 & 15 & 1 \end{vmatrix} \quad (B.26)$$

The second through fourth terms are, respectively,

$$(-1)^{1+2} a_{12} |A_{12}^1| = -2 \cdot \begin{vmatrix} 5 & 10 & 8 \\ 9 & 6 & 12 \\ 4 & 15 & 1 \end{vmatrix}$$

$$(-1)^{1+3} a_{13} |A_{13}^1| = 3 \cdot \begin{vmatrix} 5 & 11 & 8 \\ 9 & 7 & 12 \\ 4 & 14 & 1 \end{vmatrix}$$

and

$$(-1)^{1+4} a_{14} |A_{14}^1| = -13 \cdot \begin{vmatrix} 5 & 11 & 10 \\ 9 & 7 & 6 \\ 4 & 14 & 15 \end{vmatrix} \quad (B.27)$$

Now we see that none of the reduced matrices A_{1j}^1 are 2 × 2. Hence, we have to repeat the procedure for each of the determinants $|A_{1j}^1|$. For $|A_{11}^1|$ through $|A_{14}^1|$, we keep in mind as a reference the new imaginary sign matrix[11]

$$S_1 = \begin{pmatrix} 1 & -1 & 1 \\ -1 & 1 & -1 \\ 1 & -1 & 1 \end{pmatrix} \quad (B.28)$$

From equation (B.26), we obtain

---

[11] The same matrix S_1 is used for A_{11}^1 through A_{14}^1 since they are all 3 × 3 matrices.

$$|A^1_{11}| = 11 \cdot \begin{vmatrix} 6 & 12 \\ 15 & 1 \end{vmatrix} - 10 \cdot \begin{vmatrix} 7 & 12 \\ 14 & 1 \end{vmatrix} + 8 \cdot \begin{vmatrix} 7 & 6 \\ 14 & 15 \end{vmatrix} = -136 \qquad (B.29)$$

where we have taken advantage of the fact that the new reduced matrices are 2 × 2, which thereby enables us to compute them according to equation (B.20). Without explicit display of the computations, we list the results for the remaining reduced matrices

$$|A^1_{12}| = 408$$
$$|A^1_{13}| = 408 \qquad (B.30)$$
$$|A^1_{14}| = -136$$

Our next task is to perform the recursion by entering equations (B.29) and (B.30) into equations (B.26) and (B.27), respectively. This finally yields the determinant of the original matrix A

$$|A| = 16 \cdot (-136) - 2 \cdot (408) + 3 \cdot (408) - 13 \cdot (-136) = 0 \qquad (B.31)$$

Since the determinant is equal to zero, we know that A is not full rank. Hence, its component vectors do not span the entire four-dimensional reals (i.e., R^4). As a consequence, A does not have an inverse A^{-1}.

Eigenvalues and Eigenvectors

We now turn our attention to the *eigenvalues* and *eigenvectors* of square matrices.[12] We limit our focus to those matrices that are also symmetric as in equation (B.17). Before providing a formal definition, we give a hint as to what is the intuition behind all this.

An eigenvector has the feature that one obtains the same result from either multiplying the matrix from the right by the eigenvector or multiplying the corresponding eigenvalue by that very same eigenvector.[13] Formally, for a square $n \times n$ matrix A, we derive the eigenvectors, $\beta_1, \ldots, \beta_n$, from the equation

$$A \cdot \beta = \lambda \cdot \beta \qquad (B.32)$$

---

[12]The word *eigen* is a German word that literally means "own" in English, but as a prefix in this sense it is better translated as characteristic or unique. Eigenvalues and eigenvectors are also referred to as *characteristic values* and *vectors*.

[13]Without explicit mentioning, we limit ourselves to *right* eigenvectors since they will suffice in the context of this book.

where the β are obviously column vectors. We have n eigenvectors and, technically, the same number of eigenvalues λ since each eigenvector has a corresponding eigenvalue. While the eigenvectors are distinct and orthogonal, the n eigenvalues need not be. If the value of some eigenvalue occurs several times, then this eigenvalue is of certain *multiplicity*. For example, if three eigenvalues corresponding to three distinct eigenvectors have the value, say 2, then 2 is eigenvalues of multiplicity three.

For the derivation of the eigenvectors and eigenvalues solving equation (B.32), we rewrite the equality to obtain

$$A\beta - \lambda\beta = 0$$
$$\Leftrightarrow A\beta - \lambda I_{n,n}\beta = 0 \qquad (B.33)$$
$$\Leftrightarrow (A - \lambda I_{n,n})\beta = 0$$

where in the second row we have used the fact that multiplication by the identity matrix of appropriate dimension yields the original matrix or vector.[14] From equation (B.33), we see that when β is not a zero vector, $(A - \lambda I_{n,n})$ needs to contain linearly dependent component vectors.[15] Otherwise, multiplying $(A - \lambda I_{n,n})$ from the right by β would never yield zero. Thus, if $(A - \lambda I_{n,n})\beta = 0$ for some nontrivial β, $(A - \lambda I_{n,n})$ is not full rank and its determinant has to be zero. We show now how this insight will facilitate our retrieving of eigenvectors and eigenvalues by setting up the equality for the determinant

$$|A - \lambda I_{n,n}| = 0$$
$$\Leftrightarrow \begin{vmatrix} a_{11} - \lambda & a_{12} & & a_{1n} \\ a_{21} & a_{22} - \lambda & & \\ & & \ddots & \\ & & & \ddots \\ a_{n1} & & & a_{nn} - \lambda \end{vmatrix} = 0 \qquad (B.34)$$

Since we know how to compute the determinant according to the algorithm described earlier, we will obtain a polynomial in λ of degree n from equation (B.34) that will produce all n eigenvalues.[16] Once the eigenvalues are known including their multiplicity, then the corresponding eigenvectors can be obtained by solving the equations system in equation (B.33) for each

---

[14] The ⇔ sign indicates equivalence of expressions.
[15] The solution β = (0, ..., 0) is called a *trivial solution*.
[16] A polynomial in x of degree n is of the form $a_n x^n + a_{n-1} x^{n-1} + ... + a_1 x + c$.

value λ. For symmetric matrices, the eigenvectors β produced are orthogonal, meaning that for i and j with $i \neq j$, we have

$$\beta_i^T \beta_j = 0 \tag{B.35}$$

Moreover, we transform them such that they are *standardized*. Hence we have

$$\beta_i^T \beta_j = \begin{cases} 1, & i = j \\ 0, & i \neq j \end{cases} \tag{B.36}$$

which is equivalent to equation (B.13). This is a very important result since we are interested predominantly in symmetric matrices in this book.

POSITIVE SEMIDEFINITE MATRICES

In Chapter 17 where we explain point estimates, we use the concept of a positive definite matrix. We discuss that concept in this concluding section of the appendix.

As we already know, an $n \times n$ matrix A of the form

$$A = \begin{pmatrix} a_{11} & a_{12} & \cdots & a_{1n} \\ a_{21} & a_{22} & & \vdots \\ \vdots & & \ddots & \\ a_{n1} & \cdots & & a_{nn} \end{pmatrix}$$

—that is, a matrix with as many columns as rows—is a square matrix. This matrix A is *positive-semidefinite* if for any n-dimensional vector x, which is not all zeroes, the following is true

$$\omega^T A \omega \geq 0 \tag{B.37}$$

where ω^T denotes the vector transpose of vector ω. In other words, in equation (B.37), we postulate that for a positive-semidefinite matrix A, all so-called *quadratic forms*

$$a_1 \omega_1^2 + a_{12} \omega_1 \omega_2 + \ldots + a_{1n} \omega_1 \omega_n + a_{21} \omega_2 \omega_1 + a_{22} \omega_2^2 + \ldots + a_{2n} \omega_2 \omega_n + \ldots$$
$$\ldots + a_{n1} \omega_n \omega_1 + a_{n2} \omega_n \omega_2 + \ldots + a_{nn} \omega_n^2 = \omega^T A \omega$$

result in nonnegative numbers.

APPENDIX C

Binomial and Multinomial Coefficients

In this appendix, we explain the concept of the binomial and multinomial coefficients used in discrete probability distributions described in Chapter 9.

BINOMIAL COEFFICIENT

The *binomial coefficient* is defined as

$$\binom{n}{k} = \frac{n!}{k!(n-k)!} \qquad (C.1)$$

for some nonnegative integers k and n with $0 \leq k \leq n$. For the binomial coefficient, we use the *factorial* operator denoted by the "!" symbol. As explained in Appendix A, a factorial is defined in the set of natural numbers N that is $k = 1, 2, 3, \ldots$ as

$$k! = k \cdot (k-1) \cdot (k-2) \cdot \ldots \cdot 1$$

For $k = 0$, we define $0! \equiv 1$.

Derivation of the Binomial Coefficient

In the context of the binomial distribution described in Chapter 9, we form the sum X of n independent and identically distributed Bernoulli random variables Y_i with parameter p or, formally, $Y_i \stackrel{iid}{\sim} B(p)$, $i = 1, 2, \ldots, n$.[1] The

---
[1] By "$x_i \stackrel{iid}{\sim}$", we indicate that the random variables X_i, $i = 1, 2, \ldots, n$, follow the same distribution law, pairwise independently. A formal definition of independence in the context of probability theory is explained in Chapter 9. Here we refer to independence as having the meaning "uninfluenced by."

random variable is then distributed binomial with parameters n and p – $X \sim B(n,p)$. Since the random variables Y_i have either value 0 or 1, the resulting binomial random variable (i.e., the sum X) assumes some integer value between 0 and n. Let $X = k$ for $0 \leq k \leq n$. Depending on the exact value k, there may be several alternatives to obtain k since, for the sum X, it is irrelevant in which order the individual values of the Y_i appear.

Special Case n = 3

We illustrate the special case where $n = 3$ using a $B(3,0.4)$ random variable X; that is, X is the sum of three independent $B(0.4)$ distributed random variables Y_1, Y_2, and Y_3. All possible values for X are contained in the state space $\Omega' = \{0, 1, 2, 3\}$. As we will see, some of these $k \in \Omega'$ can be obtained in different ways.

We start with $k = 0$. This value can only be obtained when all Y_i are 0, for $i = 1, 2, 3$. So, there is only one possibility.

Next we consider $k = 1$. A sum of $X = 1$ can be the result of one $Y_i = 1$ while the remaining two Y_i are 0. We have three possibilities for $Y_i = 1$ since it could be either the first, the second, or the third of the Bernoulli random variables. Then we place the first 0. For this, we have two possibilities since we have two Y_i left that are not equal to 1. Next, we place the second 0, which we have to assign to the remaining Y_i. As an intermediate result, we have $3 \cdot 2 \cdot 1 = 6$ possibilities. However, we do not need to differentiate between the two 0 values because it does not matter which of the zeros is assigned first and which second. So, we divide the total number of options by the number of possibilities to place the 0 values (i.e., 2). The resulting number of possible ways to end up with $X = 1$ is

$$\frac{3 \cdot 2 \cdot 1}{2} = \frac{3!}{2! \cdot 1!} = 3 \tag{C.2}$$

For reasons we will make clear later, we introduced the middle term in equation (C.2).

Let us illustrate this graphically. In Figure C.1, a black ball represents a value $Y_i = 1$ at the i-th drawing while the white numbered circles represent a value of $Y_i = 0$ at the respective i-th drawing with i matching the number in the circle.

Now let $k = 2$. To yield the sum $X = 2$, we need two $Y_i = 1$ and one $Y_i = 0$. So, we have three different positions to place the 0, while the remaining two Y_i have to be equal to 1 automatically. Analogous to the prior case, $X = 1$, we do not need to differentiate between the two 1 values, once the 0 is positioned.

Binomial and Multinomial Coefficients

FIGURE C.1 Three Different Ways to Obtain a Total of $X = \sum_{i=1}^{3} Y_i = 1$

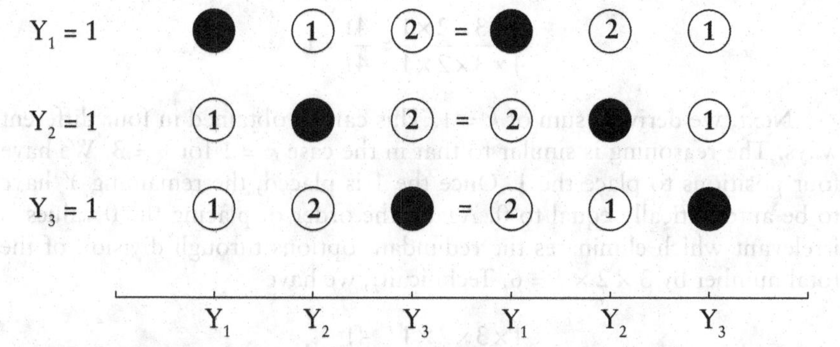

[Figure showing three arrangements with filled and open circles labeled 1 and 2, with rows $Y_1 = 1$, $Y_2 = 1$, $Y_3 = 1$, and columns labeled $Y_1, Y_2, Y_3, Y_1, Y_2, Y_3$]

Note: The alternatives matched by the = symbol lead to the same outcome, respectively.

TABLE C.1 Different Choices to Obtain $X = k$ when $n = 3$

$k = 0$	$k = 1$	$k = 2$	$k = 3$
$1 = \dfrac{3!}{0! \times 3!} = \binom{3}{0}$	$3 = \dfrac{3!}{1! \times 2!} = \binom{3}{1}$	$3 = \dfrac{3!}{2! \times 1!} = \binom{3}{2}$	$1 = \dfrac{3!}{3! \times 0!} = \binom{3}{3}$

Finally, let $k = 3$. This is accomplished by all three $Y_i = 1$. So, there is only one possibility to obtain $X = 3$.

We summarize these results in Table C.1.

Special Case n = 4

We extend the prior case to the case where the random variable X is the sum of four Bernoulli distributed random variables—that is, $Y_i \stackrel{iid}{\sim} B(p)$, $i = 1, 2, 3, 4$ —assuming either value 0 or 1 for each. The resulting sum X is then binomial distributed $B(4, p)$ assuming values k in the state space $\Omega' = \{0,1,2,3,4\}$. Again, we will analyze how the individual values of the sum X can be obtained.

To begin, let us consider the case $k = 0$. As in the prior case $n = 3$, we have only one possibility (i.e., all four Y_i equal to 0, that is, $Y_1 = Y_2 = Y_3 = Y_4 = 0$). This can be seen from the following. Technically, we have four positions to place the first 0. Then, we have three choices to place the second 0. For the third 0, we have two positions available, and one for the last 0. In total, we have

$$4 \times 3 \times 2 \times 1 = 24$$

But due to the fact that we do not care about the order of the 0 values, we divide by the total number of options (i.e., 24) and then obtain

$$\frac{4\times 3\times 2\times 1}{4\times 3\times 2\times 1} = \frac{4!}{4!} = 1$$

Next, we derive a sum of $k = 1$. This can be obtained in four different ways. The reasoning is similar to that in the case $k = 1$ for $n = 3$. We have four positions to place the 1. Once the 1 is placed, the remaining Y_i have to be automatically equal to 0. Again, the order of placing the 0 values is irrelevant which eliminates the redundant options through division of the total number by $3 \times 2 \times 1 = 6$. Technically, we have

$$\frac{4\times 3\times 2\times 1}{3\times 2\times 1} = \frac{4!}{3!} = 4$$

For a sum X equal to $k = 2$, we have four different positions to place the first 1. Then, we have three positions left to place the second 1. This yields $4 \times 3 = 12$ different options. However, we do not care which one of the 1 values is placed first since, again, their order is irrelevant. So, we divide the total number by 2 to indicate that the order of the two 1 values is unimportant. Next, we place the first 0, which offers us two possible positions for the remaining Y_i that are not equal to 1 already. For this, we have two options. In total, we then have

$$\frac{4\times 3\times 2\times 1}{2\times 1} = \frac{4!}{2!} = 6$$

possibilities. Then, the second 0 is placed on the remaining Y_i. So, there is only one choice for this 0. Because we do not care about the order of placement of the 2 values, we divide by 2. The resulting number of different ways to yield a sum X of $k = 2$ is

$$\frac{4\times 3\times 2\times 1}{2\times 1\times 2\times 1} = \frac{4!}{2!\times 2!} = 12$$

which is illustrated in Figures C.2 through C.7.

A sum of X equal to $k = 3$ is achieved by three 1 values and one 0 value. So, since the order of the 1 values is irrelevant due to the previous reasoning, we only care about where to place the 0 value. We have four possibilities, that is,

$$\frac{4\times 3\times 2\times 1}{3\times 2\times 1} = \frac{4!}{3!} = 4$$

Binomial and Multinomial Coefficients

FIGURE C.2 Four Different Ways to Obtain $Y_1 = Y_2 = 1$

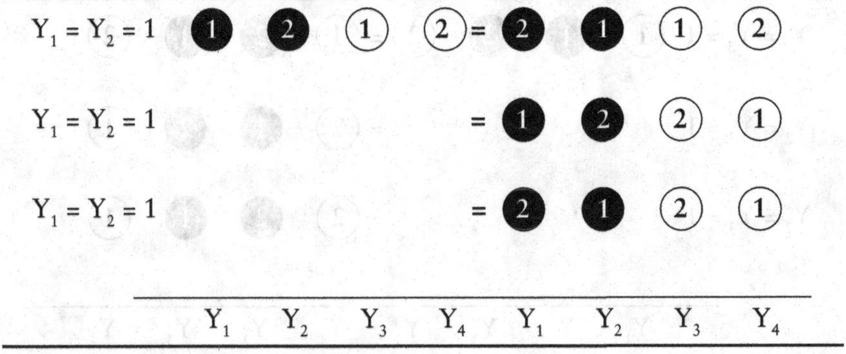

$Y_1 = Y_2 = 1$ ● ● ○ ○ = ● ● ○ ○

$Y_1 = Y_2 = 1$ = ● ● ○ ○

$Y_1 = Y_2 = 1$ = ● ● ○ ○

$\quad\quad\quad\quad Y_1 \; Y_2 \; Y_3 \; Y_4 \quad Y_1 \; Y_2 \; Y_3 \; Y_4$

FIGURE C.3 Four Different Ways to Obtain $Y_1 = Y_3 = 1$

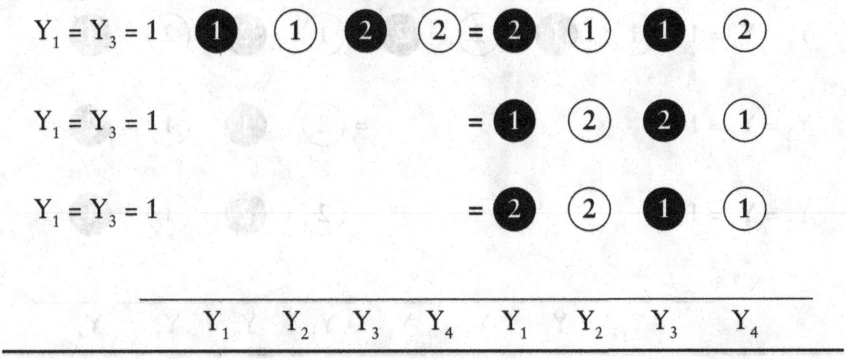

$Y_1 = Y_3 = 1$ ● ○ ● ○ = ● ○ ● ○

$Y_1 = Y_3 = 1$ = ● ○ ● ○

$Y_1 = Y_3 = 1$ = ● ○ ● ○

$\quad\quad\quad\quad Y_1 \; Y_2 \; Y_3 \; Y_4 \quad Y_1 \; Y_2 \; Y_3 \; Y_4$

FIGURE C.4 Four Different Ways to Obtain $Y_1 = Y_4 = 1$

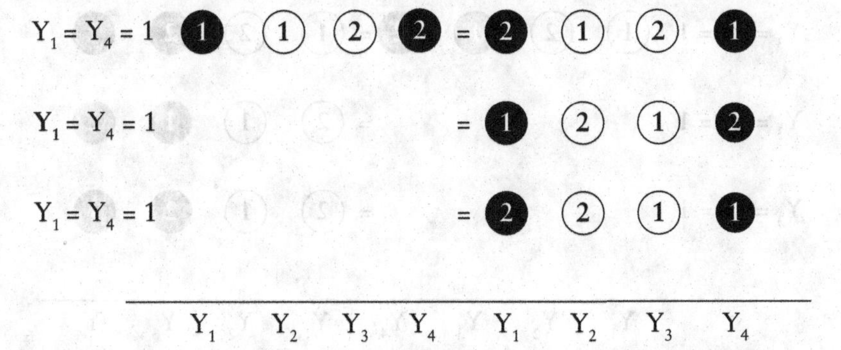

$Y_1 = Y_4 = 1$ ● ○ ○ ● = ● ○ ○ ●

$Y_1 = Y_4 = 1$ = ● ○ ○ ●

$Y_1 = Y_4 = 1$ = ● ○ ○ ●

$\quad\quad\quad\quad Y_1 \; Y_2 \; Y_3 \; Y_4 \quad Y_1 \; Y_2 \; Y_3 \; Y_4$

FIGURE C.5 Four Different Ways to Obtain $Y_2 = Y_3 = 1$

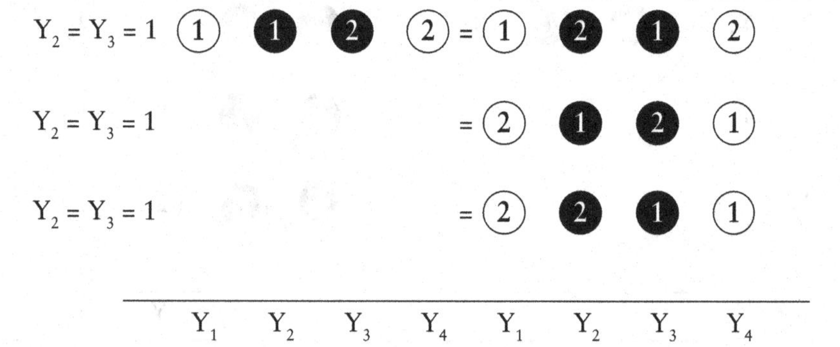

$Y_2 = Y_3 = 1$	①	**①**	**②**	②	=	①	**②**	**①**	②
$Y_2 = Y_3 = 1$					=	②	**①**	**②**	①
$Y_2 = Y_3 = 1$					=	②	**②**	**①**	①
	Y_1	Y_2	Y_3	Y_4		Y_1	Y_2	Y_3	Y_4

FIGURE C.6 Four Different Ways to Obtain $Y_2 = Y_4 = 1$

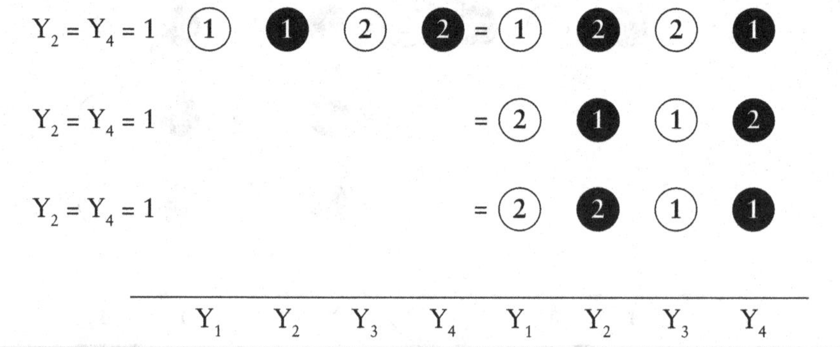

$Y_2 = Y_4 = 1$	①	**①**	②	**②**	=	①	**②**	②	**①**
$Y_2 = Y_4 = 1$					=	②	**①**	①	**②**
$Y_2 = Y_4 = 1$					=	②	**②**	①	**①**
	Y_1	Y_2	Y_3	Y_4		Y_1	Y_2	Y_3	Y_4

FIGURE C.7 Four Different Ways to Obtain $Y_3 = Y_4 = 1$

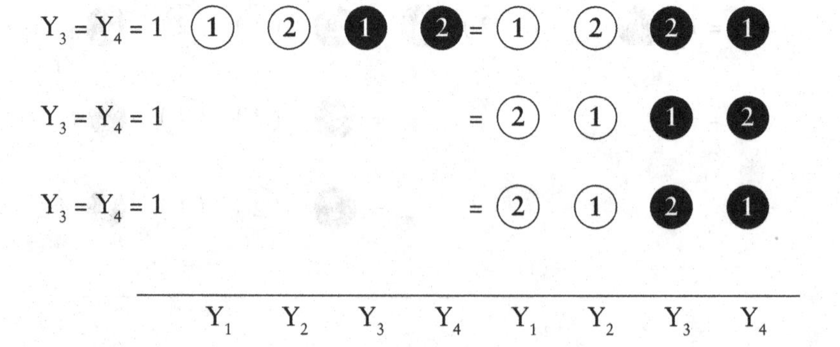

$Y_3 = Y_4 = 1$	①	②	**①**	**②**	=	①	②	**②**	**①**
$Y_3 = Y_4 = 1$					=	②	①	**①**	**②**
$Y_3 = Y_4 = 1$					=	②	①	**②**	**①**
	Y_1	Y_2	Y_3	Y_4		Y_1	Y_2	Y_3	Y_4

Binomial and Multinomial Coefficients

Finally, to obtain $k = 4$, we only have one possible way to do so, as in the case where $k = 0$. Mathematically, this is

$$\frac{4 \times 3 \times 2 \times 1}{4 \times 3 \times 2 \times 1} = \frac{4!}{4!} = 1$$

We summarize the results in Table C.2.

General Case

Now we generalize for any $n \in N$ (i.e., some nonnegative integer number). The binomial random variable X is hence the $B(n,p)$ distributed sum of n independent and identically distributed random variables Y_i.

From the two special cases (i.e., $n = 3$ and $n = 4$), it seems like to obtain the number of choices for some $0 \leq k \leq n$, we have $n!$ in the numerator to account for all the possibilities to assign the individual n values to the Y_i, no matter how many 1 values and 0 values we have. In the denominator we correct for the fact that the order of the 1 values and 0 values, is irrelevant. That is, we divide by the number of different orders to place the 1 values on the Y_i that are equal to 1, and also by the number of different orders to assign the 0 values to the Y_i being equal to 0. Therefore, we have $n!$ in the numerator and $k! \times (n-k)!$ in the denominator. The result is illustrated in Table C.3.

TABLE C.2 Different Choices to Obtain $X = k$ when $n = 4$

$k = 0$	$k = 1$	$k = 2$	$k = 3$	$k = 4$
$1 = \dfrac{4!}{0! \times 4!} = \binom{4}{0}$	$4 = \dfrac{4!}{1! \times 3!} = \binom{4}{1}$	$6 = \dfrac{4!}{2! \times 2!} = \binom{4}{2}$	$4 = \dfrac{4!}{3! \times 1!} = \binom{4}{3}$	$1 = \dfrac{4!}{4! \times 0!} = \binom{4}{4}$

TABLE C.3 Different Choices to Obtain $X = k$ for General n

$k = 0$	$k = 1$	$k = 2$
$1 = \dfrac{n!}{0! \times n!} = \binom{n}{0}$	$n = \dfrac{n!}{1! \times (n-1)!} = \binom{n}{1}$	$\dfrac{n \times (n-1)}{2} = \dfrac{n!}{2! \times (n-2)!} = \binom{n}{2}$
...	$k = n-1$	$k = n$
...	$n = \dfrac{n!}{(n-1)! \times 1!} = \binom{n}{n-1}$	$n = \dfrac{n!}{n! \times 0!} = \binom{n}{n}$

MULTINOMIAL COEFFICIENT

The *multinomial coefficient* is defined as

$$\binom{n}{n_1 \; n_2 \; \ldots \; n_k} = \frac{n!}{n_1! \cdot n_2! \cdot \ldots \cdot n_k!} \quad \text{(C.3)}$$

for $n_1 + n_2 + \ldots + n_k = n$.[2]

Assume we have some population of balls with k different colors. Suppose n times we draw some ball and return it to the population such that for each trial (i.e., drawing), we have the identical conditions. Hence, the individual trials are independent of each other. Let Y_i denote the color obtained in the i-th trial for $i = 1, 2, \ldots, n$.

How many different possible samples of length n are there? Let us think of the drawings in a different way. That is, we draw one ball after another disregarding color and assign the drawn ball to the trials Y_1 through Y_n in an arbitrary fashion.

First, we draw a ball with any of the k colors and assign it to one of the n trials, Y_i. Next, we draw the second ball and assign it to one of the remaining $n-1$ possible trials i as outcome of Y_i. This yields

$$n \times (n-1)$$

different possibilities. The third ball drawn is assigned to the $n-2$ trials left so that we have

$$n \times (n-1) \times (n-2)$$

possibilities, in total. This is continued until we draw the nth (i.e., the last), color, which can only be placed in the last remaining trial Y_i. In total this yields

$$n \times (n-1) \times (n-2) \times \ldots \times 2 \times 1 = n!$$

different possibilities of drawing n balls.

The second question is how many different possibilities are there to obtain a sample with the number of occurrences n_1, n_2, ..., and n_k of the respective colors. Let red be one of these colors and suppose we have a sample with a total of $n_r = 3$ red balls from trials 2, 4, and 7 so that $Y_2 = Y_4 = Y_7$ = red. The assignment of red to these three trials yields

---

[2] Sometimes, the multinominal coefficient is referred to as the *polynomial coefficient*.

Binomial and Multinomial Coefficients

$$3! = 3 \times 2 \times 1 = 6$$

different orders of assignment. Now, we are indifferent with respect to which of the Y_2, Y_4, and Y_7 was assigned red first, second, and third. Thus, we divide the total number $n!$ of different samples by $n_r! = 3!$ to obtain only nonredundant results with respect to a red ball. We proceed in the same fashion for the remaining colors and, finally, obtain for the total number of nonredundant samples containing n_1 of color 1, n_2 of color 2, ..., and n_k of color k

$$\binom{n!}{n_1 \; n_2 \; \ldots \; n_k} = \frac{n!}{n_1! \times n_2! \times \ldots \times n_k!}$$

which is exactly equation (C.3).

APPENDIX D

Application of the Log-Normal Distribution to the Pricing of Call Options

In Chapter 11, we described the log-normal distribution and applied it to the return distribution for a stock's price. In this appendix, we illustrate the application of this distribution to price a derivative instrument. More specifically, we illustrate an application to the pricing of a European call option.

CALL OPTIONS

A call option is a contract that entitles the holder to purchase some specified amount of a certain asset at a predetermined price some time in the future. The predetermined price is called the strike price or exercise price. The date when the option expires is called the expiration date or maturity date. A call option has an exercise style that means when the option can be exercised by the holder of the option. If the call option can be exercised any time up to and including the maturity date, it is called an *American call option*. If it can only be exercised on the maturity date, it is referred to as a *European call option*. A call option is said to be a Bermuda call option if it exercised on designated dates throughout the option's life. In this appendix, our focus is on European call options.

The underlying asset for a call option can be a stock, bond, or any other financial instrument. Our focus in this appendix is a European call option on a stock. We will assume that the stock in this illustration is stock A.

The buyer of an option must pay the seller (or writer) of that option a price. That price is referred to as the option price or option premium. All that the seller of the option can gain is the option price. In contrast, the buyer of a call option benefits from a favorable price movement for the underlying stock.

DERIVING THE PRICE OF A EUROPEAN CALL OPTION

As just explained, a call option (which we simply refer to as call hereafter) is only valid for a certain specified period of time. At the end of this period, it expires. Let this time of expiration be denoted by $t = \tau$ and is what we referred to as the maturity date or simply maturity. The maturity τ is usually given as fraction of one year (which is assumed to have 360 days).[1] So if the option expires in say 30 days and the stock returns are measured daily, then $\tau = 30/360$. The strike price is denoted as K. Let the stock price at time $t = 0$ be equal to S_0 and at any time T in the future S_T.

Furthermore, in our model world, we assume today's (i.e., $t = 0$) value of some payment X in the future, say at $t = T$, is equal to $X \cdot e^{-r_f \cdot T}$ which is less than X if r_f and T are positive, where r_f is the risk-free interest rate over the relevant time period. In other words, the future payment is discounted at the risk-free rate. Here, we use compounded interest with the constant risk-free rate r_f which implies that interest is paid continuously and each payment of interest itself is immediately being paid interest on.[2]

At maturity $t = \tau$, the call is worth

$$C_\tau(S_\tau) = (S_\tau - K, 0)^+$$

where the expression on the right side is the abbreviated notation for the *maximum* of $S_\tau - K$ and 0.[3]

To understand this final value, consider that the option provides the option buyer with the right—but not the obligation—to purchase one unit of stock A at time τ. So, when the stock price at τ, S_τ, is no greater than K, the option is worthless because we can purchase a unit of stock A for only S_τ on the market instead of having to pay K. And, hence, the option provides a worthless right in that case. If, however, the stock A's price at maturity should be greater than the strike price (i.e., $S_\tau > K$), then the option buyer saves $S_\tau - K$ by exercising the option instead of purchasing stock A in the market. Consequently, in that case the call should be worth exactly that difference $S_\tau - K$ at maturity.

Suppose we are interested in the value of the call at some point in time prior to expiration. However, we do not know what the stock price S_τ will be. Consequently, the entire terminal value of $C_\tau(S_\tau) = (S_\tau - K, 0)^+$ is treated as random prior to expiration. However, we can compute the expected value

---

[1] Sometimes, one finds that a year is assumed to have 250 days, only, instead of 360 if the weekends are disregarded.

[2] The continuously compounded interest r translates into an annual interest payment R according to $R = e^{360 \cdot r_f} - 1$.

[3] Recall that, for $S_\tau > K$, the maximum is $S_\tau - K$ and 0, else.

Application of the Log-Normal Distribution to the Pricing of Call Options

of $C_\tau(S_\tau)$, that is, $E((S_\tau - K,0)^+)$, if the distribution of the stock price at maturity is known. Now, suppose we are at $t = 0$. Then today this $E((S_\tau - K,0)^+)$ is worth simply $e^{-r_f \tau} \times E((S_\tau - K,0)^+)$.

In our model, we assume that the terminal price for stock A, S_τ, is given as

$$S_\tau = S_0 \cdot e^{\left(r_f - \frac{1}{2}\sigma^2\right)\tau + \sigma W_\tau} \tag{D.1}$$

where W_τ is a $N(0,\sigma^2\tau)$ random variable such that the exponent in equation (D.1) becomes a random variable with mean $(r_f - 1/2\sigma^2)\tau$ and variance $\sigma^2\tau$. The exponent is referred to as the *stock price dynamic* and the parameter σ is referred to as the *stock price volatility*.

Consequently, the price of the call is computed as

$$C_\tau\left(S_0, K, \tau, r_f, \sigma^2\right) = r_f e^{-r_f \tau} \cdot E\left((S_\tau - K, 0)^+\right)$$

$$= e^{-r_f \tau} \cdot \int_{-\infty}^{\infty} (s - K, 0)^+ f_S(s) ds \tag{D.2}$$

such that we compute the expected value over all real numbers $-\infty \leq s \leq \infty$ even though S_τ may only assume positive values. The function $f_S(s)$ in equation (D.2) denotes the probability density function of S_τ evaluated at the outcome s of S_τ.

We can explicitly compute the maximum operator in equation (D.2) and obtain

$$C_\tau\left(S_0, K, \tau, r_f, \sigma^2\right) = 0 \cdot r_f P(S_\tau \leq K) + e^{-r_f \tau} \cdot \int_K^{\infty} (s - K) f_S(s) ds$$

which, because we can split the integral of $(S_\tau - K)$ into two integrals with S_τ and $-K$, respectively, simplifies to

$$= e^{-r_f \tau} \cdot \int_K^{\infty} s f_S(s) ds - e^{-r_f \tau} \cdot \int_K^{\infty} K f_S(s) ds$$

and, because K is constant, can be reduced further, to

$$= e^{-r_f \tau} \cdot \int_K^{\infty} s f_S(s) ds - e^{-r_f \tau} \cdot K \cdot P(S_\tau > K) \tag{D.3}$$

Since we know that the exponent in equation (D.1) is a normal random variable, the ratio S_τ/S_0 becomes log-normal with the same parameters as the exponent, that is,

$$S_\tau/S_0 \sim Ln\left(\left(r_f - 1/2\sigma^2\right)\tau, \sigma^2\tau\right)$$

To compute $P(S_\tau > K)$ in equation (D.3) we consider that $S_\tau > K$ exactly when $S_\tau/S_0 > K/S_0$. This again is equivalent to $\ln(S_\tau/S_0) > \ln(K/S_0)$ where, on the left side of the inequality, is exactly the exponent in equation (D.1), which is normally distributed.[4] Thus, the standardized random variable

$$\frac{\ln(S_\tau/S_0) - \left(r_f - 1/2\sigma^2\right)\tau}{\sigma\sqrt{\tau}} \tag{D.4}$$

is greater than

$$\frac{\ln(K/S_0) - \left(r_f - 1/2\sigma^2\right)\tau}{\sigma\sqrt{\tau}}$$

if and only if $S_\tau > K$, which, consequently, occurs with probability[5]

$$P(S_\tau > K)$$
$$= P\left(\frac{\ln(S_\tau/S_0) - \left(r_f - 1/2\sigma^2\right)\tau}{\sigma\sqrt{\tau}} > \frac{\ln(K/S_0) - \left(r_f - 1/2\sigma^2\right)\tau}{\sigma\sqrt{\tau}}\right)$$
$$= 1 - P\left(\frac{\ln(S_\tau/S_0) - \left(r_f - 1/2\sigma^2\right)\tau}{\sigma\sqrt{\tau}} \leq \frac{\ln(K/S_0) - \left(r_f - 1/2\sigma^2\right)\tau}{\sigma\sqrt{\tau}}\right)$$
$$= 1 - \phi\left(\frac{\ln(K/S_0) - \left(r_f - 1/2\sigma^2\right)\tau}{\sigma\sqrt{\tau}}\right)$$
$$= \phi\left(\frac{\ln(S_0/K) + \left(r_f - 1/2\sigma^2\right)\tau}{\sigma\sqrt{\tau}}\right) \tag{D.5}$$

Now, let us turn to the computation of the integral, in equation (D.3). Recall that $Y = S_\tau/S_0$ is log-normally distributed, so that using the rule for change of integration variables we can express the density of $S_\tau = Y \cdot S_0$ by

---

[4]The ln denotes the logarithm to the base $e = 2.718$.
[5]The use of the standard normal cumulative distribution function in the last two of the following equations results from the fact that the random variable in equation (D.4) is standard normal. The last equation results from the symmetry property of the normal distribution explained in Chapter 11 and computational rules for the logarithm.

Application of the Log-Normal Distribution to the Pricing of Call Options

$$f_S(s) = f_Y\left(\frac{s}{S_0}\right) \cdot \frac{1}{S_0}$$

so that the integral, in equation (D.3), changes to

$$\int_K^\infty s \cdot f_S(s)\, ds = \int_K^\infty s \cdot f_Y\left(\frac{s}{S_0}\right) \cdot \frac{1}{S_0} \cdot ds = \int_{K/S_0}^\infty S_0 \cdot y \cdot f_Y(y) \cdot dy \qquad (D.6)$$

where in the last step we substituted the integration variable s with $y = s/S_0$. Note that we have to consider the altered integral limits in equation (D.6) since our new random variable is $Y = S_\tau/S_0$. Due to the log-normal distribution of Y, we can further specify the integral as

$$\int_K^\infty s \cdot f_S(s)\, ds = \int_{K/S_0}^\infty \frac{S_0 \cdot y}{\sqrt{2\pi\sigma^2 \tau} \cdot y} \cdot e^{-\frac{\left(\ln y - \left(r_f - \frac{1}{2}\sigma^2\right)\tau\right)^2}{2\sigma^2 \tau}}\, dy$$

We, next, introduce the additional helper random variable $Z = \ln Y$. Reapplying equation (D.4) and using the rule for the change of integration variables, we obtain

$$\int_K^\infty s \cdot f_S(s)\, ds = S_0 \int_{\ln(K/S_0)}^\infty \frac{1}{\sqrt{2\pi\sigma^2 \tau}} \cdot e^{-\frac{\left(z - \left(r_f - \frac{1}{2}\sigma^2\right)\tau\right)^2}{2\sigma^2 \tau}} \cdot e^z \cdot dz$$

$$= S_0 \int_{\ln(K/S_0)}^\infty \frac{1}{\sqrt{2\pi\sigma^2 \tau}} \cdot e^{-\frac{\left(z - \left(r_f - \frac{1}{2}\sigma^2\right)\tau\right)^2 - 2\sigma^2 \tau z}{2\sigma^2 \tau}}\, dz \qquad (D.7)$$

With the intermediate steps

$$-\frac{\left(z - \left(r_f - \frac{1}{2}\sigma^2\right)\tau\right)^2 - 2\sigma^2\tau \cdot z}{2\sigma^2\tau}$$

$$= -\frac{z^2 - 2z\left(r_f - \frac{1}{2}\sigma^2\right)\tau - 2\sigma^2\tau \cdot z + \left(\left(r_f - \frac{1}{2}\sigma^2\right)\tau\right)^2}{2\sigma^2\tau}$$

$$= -\frac{z^2 - 2zr_f\tau + \sigma^2\tau \cdot z - 2\sigma^2\tau \cdot z + \left(\left(r_f - \frac{1}{2}\sigma^2\right)\tau\right)^2}{2\sigma^2\tau}$$

$$= -\frac{z^2 - 2zr_f\tau - \sigma^2\tau \cdot z + \left(\left(r_f - \frac{1}{2}\sigma^2\right)\tau\right)^2}{2\sigma^2\tau}$$

$$= -\frac{z^2 - 2z\tau\left(r_f + \frac{\sigma^2}{2}\right) + \left(r_f + \frac{\sigma^2}{2}\right)^2\tau^2}{2\sigma^2\tau} - \frac{-\left(r_f + \frac{\sigma^2}{2}\right)^2\tau^2 + \left(\left(r_f - \frac{1}{2}\sigma^2\right)\tau\right)^2}{2\sigma^2\tau}$$

$$= -\frac{z^2 - 2z\tau\left(r_f + \frac{\sigma^2}{2}\right) + \left(r_f + \frac{\sigma^2}{2}\right)^2\tau^2}{2\sigma^2\tau} \ldots$$

$$\ldots - \frac{-r_f^2\tau^2 - 2r_f\frac{1}{2}\sigma^2\tau^2 - \left(\frac{1}{2}\sigma^2\tau\right)^2 + r_f^2\tau^2 - 2r_f\frac{1}{2}\sigma^2\tau^2 + \left(\frac{1}{2}\sigma^2\tau\right)^2}{2\sigma^2\tau}$$

$$= -\frac{z^2 - 2z\tau\left(r_f + \frac{\sigma^2}{2}\right) + \left(r_f + \frac{\sigma^2}{2}\right)^2\tau^2}{2\sigma^2\tau} - \frac{-4r_f\frac{1}{2}\sigma^2\tau^2}{2\sigma^2\tau}$$

$$= -\frac{z^2 - 2z\tau\left(r_f + \frac{\sigma^2}{2}\right) + \left(r_f + \frac{\sigma^2}{2}\right)^2\tau^2}{2\sigma^2\tau} + r_f\tau$$

the exponent of equation (D.7) can be simplified according to

$$-\frac{\left(z - \left(r_f - \frac{1}{2}\sigma^2\right)\tau\right)^2 - 2\sigma^2\tau \cdot z}{2\sigma^2\tau} = -\frac{\left(z - \left(r_f + \frac{1}{2}\sigma^2\right)\tau\right)^2}{2\sigma^2\tau} + r_f\tau$$

Inserting this exponent into equation (D.7), we obtain

$$\int_K^\infty s \cdot f_S(s)\,ds = S_0 \cdot e^{r_f\tau} \int_{\ln(K/S_0)}^\infty \frac{1}{\sqrt{2\pi\sigma^2\tau}} \cdot e^{-\frac{\left(z - \left(r_f + \frac{1}{2}\sigma^2\right)\tau\right)^2}{2\sigma^2\tau}} \cdot dz$$

The integral is equal to the probability $P(Z > \ln(K/S_0))$ of some normally distributed random variable Z with mean $(r_f + 1/2\sigma^2)\tau$ and variance $\sigma^2\tau$. Hence through standardization of Z, equation (D.7) becomes[6]

---
[6] Again, the last of the following equations results from the symmetry property of the normal distribution and computational rules of the logarithm.

$$\int_K^\infty s f_S(s)\,ds = S_0 \cdot e^{r_f \tau} P\left(Z > \frac{\ln(K/S_0) - (r_f + 1/2\sigma^2)\tau}{\sigma\sqrt{\tau}}\right)$$

$$= S_0 \cdot e^{r_f \tau}\left(1 - \phi\left(\frac{\ln(K/S_0) - (r_f + 1/2\sigma^2)\tau}{\sigma\sqrt{\tau}}\right)\right)$$

$$= S_0 \cdot e^{r_f \tau} \phi\left(\frac{\ln(S_0/K) + (r_f + 1/2\sigma^2)\tau}{\sigma\sqrt{\tau}}\right) \quad \text{(D.8)}$$

Inserting equations (D.5) and (D.8) into equation (D.3), we obtain the well-known *Black-Scholes option pricing formula* for European call options[7]

$$C_\tau(S_0, K, \tau, r_f, \sigma^2) = S_0 \phi\left(\frac{\ln(S_0/K) + (r_f + 1/2\sigma^2)\tau}{\sigma\sqrt{\tau}}\right)$$

$$-K \cdot e^{-r_f \tau} \cdot \phi\left(\frac{\ln(S_0/K) + (r_f - 1/2\sigma^2)\tau}{\sigma\sqrt{\tau}}\right) \quad \text{(D.9)}$$

From this formula, all sorts of sensitivities of the option can be determined. That is by differentiating with respect to some parameter such as, for example, today's stock price, we obtain the amount by which the call price will change if the parameter is adjusted by some small unit amount. These sensitivities, often referred to as *Greeks* since they are denoted by Greek symbols, are beyond the scope of our book, however.

ILLUSTRATION

In the following, let us assume that today the stock price of interest is \$100 (i.e., $S_0 = \$100$). Furthermore, we consider a European call option with strike price $K = \$90$ and expiration in 30 days (i.e., $\tau = 30/360 = 0.0833$). Moreover, the variance parameter in equation (D.5) of the daily stock price return is assumed to be 20% (i.e., $\sigma^2 = 0.2$). Finally, suppose that the compounded risk-free interest rate of is 1% (i.e., $r_f = 0.01$). Then, we have everything to compute the price of the call option following the Black-Scholes option pricing formula given by equation (D.9) as follows:

---

[7] This formula is named after its developers Fischer Black and Myron Scholes who were awarded the 1977 Nobel Prize in Economic Science, along with Robert Merton, for their work in option pricing theory. See Black and Scholes (1973).

$$C(100, 90, 0.0833, 0.01, 0.2) = \$100 \times 0.8125 - \$90 \times e^{-0.01 \times 0.0833} \times 0.7758$$
$$= \$100 \times 0.8125 - \$90 \times 0.7751$$
$$= \$11.49$$

In this illustration, we did not explicitly state the technical assumptions allowing the call option price at any time to equal its discounted expected value at maturity because they are beyond the scope of this book. Standard option textbooks provide all of the details.

The pedagogical aspect of the illustration in this appendix is the demonstration of the interplay of the normal and log-normal distributions. Furthermore, we chose this illustration with its continued use of quantiles to familiarize the reader with the one-to-one relationship between quantiles and the related probabilities. For example, in equation (D.3), we bound the integral by the $P(S_\tau \leq K)$–quantile K of the probability distribution of S_τ.[8] Then, we saw how quantiles changed as we replace one random variable with another. This was displayed, for example, in the second equation of equation (D.6) where K was the $P(S_\tau \leq K)$–quantile of S_τ, which translated into the $P(S_\tau \leq K)$–quantile K/S_0 of the new random variable $Y = S_\tau/S_0$.

---

[8] We used the identity $P(X > q) = 1 - P(X \leq q)$ for any quantile q of a continuous random variable X.

References

Bachelier, Louis. [1900] 2006. *Louis Bachelier's Theory of Speculation: The Origins of Modern Finance*. Translated by Mark Davis and Alison Etheridge. Princeton, N.J.: Princeton University Press.

Barndorff-Nielsen, Ole E. 1997. "Normal Inverse Gaussian Distributions and Stochastic Volatility Modelling." *Scandinavian Journal of Statistics* 24 (1): 1–13.

Black, Fischer, and Myron Scholes. 1973. "The Pricing of Options and Corporate Liabilities." *Journal of Political Economy* 81 (3): 637–654.

———. 1987. "A Conditionally Heteroscedastic Time-Series Model for Security Prices and Rates of Return Data." *Review of Economics and Statistics* 69: 542–547.

Bollerslev, Tim. 2001. "Financial Econometrics: Past Developments and Future Challenges." *Journal of Econometrics* 100: 41–51.

Chow, Gregory C. 1960. "Tests of Equality Between Sets of Coefficients in Two Linear Regressions." *Econometrica* 28: 591–605.

Dijkstra, T. K. 1995. "Pyrrho's Lemma, or Have it Your Way." *Metrica* 42: 119–225.

Embrechts, Paul, Claudia Klüppelberg, and Thomas Mikosch. 1997. *Modelling Extremal Events for Insurance and Finance*. New York: Springer-Verlag.

Enders, Walter. 1995. *Applied Econometric Time Series*. New York: John Wiley & Sons.

Engle, Robert F. 1982. "Autoregressive Conditional Heteroscedasticity with Estimates of the Variance of U.K. Inflation."*Econometrica* 50: 987–1008.

Fabozzi, Frank J. 2008. *Bond Markets, Analysis and Strategies*. Hoboken, N.J.: John Wiley & Sons.

Fan, Jianqing. 2004. "An Introduction to Financial Econometrics." Unpublished paper, Department of Operations Research and Financial Engineering, Princeton University.

Humpage, Owen F. 1998. "The Federal Reserve as an Informed Foreign Exchange Trader." *Federal Reserve Bank of Cleveland*, Working Paper 9815, September.

Jacod, Jean, and Phillip E. Protter. 2000. *Probability Essentials*. New York: Springer.

Mandelbrot, Benoit B. 1963. "The Variation of Certain Speculative Prices." *Journal of Business* 36: 394–419.

Markowitz, Harry M. 1952. "Portfolio Selection." *Journal of Finance* 7 (1): 77–91.

———. 1959. *Portfolio Selection: Efficient Diversification of Investments*. Cowles Foundation Monograph 16 (New York: John Wiley & Sons).

McNeil, Alexander J., Rudiger Frey, and Paul Embrechts. 2005. *Quantitative Risk Management*. Princeton, N.J.: Princeton University Press.

Rachev, Svetlozar T., Stefan Mittnik, Frank J. Fabozzi, Sergio M. Focardi, and Teo Jasic. 2007. *Financial Econometrics: From Basics to Advanced Modeling Techniques*. Hoboken, NJ: John Wiley & Sons.

Samorodnitsky, Gennady, and Murad S. Taqqu. 2000. *Stable Non-Gaussian Random Processes*. Boca Raton, FL: Chapman & Hall/CRC.

Savin, N. Eugene, and Kenneth J. White. 1977. "The Durbin-Watson Test for Serial Correlation with Extreme Sample Sizes or Many Regressors." *Econometrica* 45: 1989–1996.

Index

Absolute convergence, 303, 308
   condition, 308–309
Absolute cumulative frequency, 41–43
Absolute data, 20–21
Absolute deviation, 61–63
Absolute frequency, 75
   denotation, 26
   density, inclusion, 86f
   obtaining, 84
Acceptance region, 485
   determination, 508
Accumulated frequencies, computation, 42
Adjusted goodness-of-fit measure, 529
Adjusted R-squared (adjusted R^2), 529
Aggregated risk, behavior simulation, 386
Aitken's generalized least squares (GLS) estimator, 579–580
Algorithmic trading strategies, growth, 367
Alpha-levels (α-levels), chi-square distribution, 571
Alpha-percentile (α-percentile), 56
   concept, 57–58
   obtaining, 58–59
   value, 86, 88
Alpha-quantile (α-quantile), 380
   denotation, 296
Alpha-stable density functions (α-stable density functions), comparison, 287f
Alpha-stable distribution (α-stable distribution), 8–9, 285–292
   assumption, 309
   decay, 286–287
   second moment, non-existence, 308–309
   variance, 311–312
Alpha-stable random variables, parameters, 290
Alternative hypothesis, 482
   null hypothesis, contrast, 483f
   representation, 506
American call option, 625
American International Group (AIG), stock price, 292
Analysis of variance (ANOVA), 528
   component pattern, 528t
AR. *See* Autoregressive process
ARCH. *See* Autoregressive conditional heteroskedasticity
Area approximation. *See* Integral
   rectangles, usage, 594–595
Arithmetic mean, 56
ARMA. *See* Autoregressive moving average

Asian currency crisis, 286
Asset returns
   cross-sectional collection, 380
   heavy tails, 379
   joint distribution, 334–335
   modeling, log-normal distribution application, 272–274
Assets
   dependence, 342
      structures, complexity, 379
   duration estimate, 532
   portfolio, 346–347
Asymmetric confidence interval, construction, 468–469
Asymmetric density function, 243f
Asymmetry, modeling capability, 8
Atoms, 174
Augmented regression, 535–536
Autocorrelation, 156. *See also* Negative autocorrelation; Positive autocorrelation
   detection, 577–579
   presence, modeling, 579–580
Autoregressive conditional heteroskedasticity (ARCH) model, 574–576
   generalization, 576
Autoregressive moving average (ARMA) models, 580–581
Autoregressive of order one, 156
Autoregressive process (AR) of order p, 580
Axiomatic system, 168–170

Bar chart, 78–81
   usage, 79
Bayes, Thomas, 371
Bayes' formula, 9–10
Bayes' rule, 361, 371–374
   application, 372–374
Bell-shaped curve, real-life distribution, 3
Bell-shaped density, 248
Bell-shaped yield, 251
Benchmark, derivation, 504
Berkshire Hathaway
   pie charts, comparison, 79f
   revenues, third quarter (pie chart), 77f
   third quarter reports, 76t
   third quarter revenues
      bar charts, 80f, 81f
      pie chart, 78f
Bermuda call option, 625

635

INDEX

Bernoulli distributed random variable, mean/variance, 193
Bernoulli distribution, 192–195
  generalization, 195
  p parameter, estimation, 425–426
  random variables, relationship, 192–193
Bernoulli population, sample mean (histogram), 426f
Bernoulli random variables, 250
  distribution, 615–616
Bernoulli samples, examination, 426
Bernoulli trials, link, 196
Bessel function of the third kind, 283
  defining, 597–598
Best linear unbiased estimator (BLUE), 445–446
Beta distribution, 247, 269–270
  distribution function, 270f
Beta factor, 143
Beta function, 597
Bias, 428–429. *See also* Sample mean; Sample variance; Squared bias
  determination, 435
  examination, 430
Bias corrected sample variance
  efficiency, 444–445
  usage, 435
Bias-corrected sample variance, 431
Biased estimator, 429
Binary variables, 12, 549
Binomial coefficient, 196, 615–621
  definition, 615
  derivation, 615–621
  general case, 621
  n=3, special case, 616–617
  n=4, special case, 617–621
  usage, 207–208
Binomial default distribution model, application, 203
Binomial distribution, 7, 195–203
  approximation, 449
    normal distribution, usage, 251–252
  mean, 303
  P parameter, confidence interval, 475–479
Binomial interest rate model, binomial distribution application, 202–203
Binomial probability distribution, 199
  example, 224–226
Binomial stock price model, 182
  application, 198–202
  Bernoulli distribution, generalization, 195
  extension, multinomial distribution (usage), 213
  periods, example, 200f
Binomial tree, 200–201
Biometrics, 2
Biometry, 2
Biostatistics, 1–2
Bivariate distribution function, 387
Bivariate normal density function, correlation matrix integration, 390
Bivariate normal distributions (2-dimensional normal distributions), 348–349, 354–355
  joint density function, 350

  mean vector, example, 351f
Bivariate regression, 132
  model, 568
Bivariate student's t density function, contour lines, 356f
Bivariate student's t distribution, density plot, 355f
Bivariate variables, correlation/dependence (relationship), 122f
Black Monday, 286
Black-Scholes option pricing
  formula, 631
  model, 8
  reliance, 4
BLUE. *See* Best linear unbiased estimator
Bond default, probability, 264
Bond portfolio, 265
  credit risk modeling, Poisson distribution application, 218–219
  data, 451
  defaults, time, 305
  observations, 498
Bond spread application, regression data, 557t–559t
Borel-measurability, requirement, 234
Borel-measurable function, 233–234
Borel σ-algebra, 239
  construction, 175–176
  definition, 176
  origins, 234
Box plot, 91–95
  counterclockwise movement, 92
  display, 93f–95f
  generation, procedure, 92–93
Breusch-Pagan-Godfrey test (Lagrangian multiplier test), 573
British Pound-USD Exchange Rate, returns (contrast), 98f
Broad-based market index, 568

Call options, 625
  illustration, 631–632
  predetermined price, 625
  price dependence, 7
  pricing, log-normal distribution (application), 625
  value, 184–185
Capital Asset Pricing Model (CAPM), 8
Casualties, number, 217
Ceiling function, 48–49
Center, 46–58
  measures, 69–70
    examination, 72t
Central Limit Theorem, 250
  requirement, 425
Central moment, 309. *See also* Fourth central moment; Second central moment; Third central moment
Central moment of order k, 309
Central tendency, measures, 55
Certain event with respect to P, 178
Characteristic exponent, 286
Characteristic function, 349–350, 598–599
Characteristic line, 142–147
  goodness-of-fit measure, 147

Index

Cheapest to deliver (CTD), 148–149
Chebychev inequalities, 251, 312, 441
Chi-square density function, degrees of function, 255
Chi-square distributed
  degrees of freedom, 473
  n-k-1 degrees of freedom, 530
  random variable, 254
Chi-square distribution, 247, 254–256
  asymmetry, 507
  degrees of freedom, 256–257
  density functions, 255f
  statistic, 517
Chi-square statistic, 571
Chi-square test statistic, 124–125
Chow test, 550
Circle, size/area, 76
Class bounds, frequency distributions (determination), 43f
Class data
  quantiles, computation, 88
  relative frequency density, 233
Classified data
  interaction, 47
  IQR, defining, 61
  median
    determination, 52–53
    retrieval, interpolation method (usage), 54f
Class of incident, 53
Clayton, David, 399
Clayton copula, 398–399
Closed under countable unions, 174–175
Closed under uncountable unions, 174–175
Closure under convolution, 357
Closure under linear transformations, 357
Coefficient estimate, standard error, 530
Coefficient of determination (R^2), 138–140
  correlation coefficient, relationship, 140
  definition, 139
Coefficient of partial determination, 563–564
Coefficient of variation, 67–69
Collateralized debt obligations, 305
Column vector, 601, 613
  matrix, comparison, 603
  usage. *See* Matrix
Common stock, sale price (example), 531
Common terms, 550
Company value, random variable function, 587f
Complement, 171
Completeness, data class criteria, 33
Complex numbers, 598
  graphical representation, 599f
  representation, two-dimensional space, 599
  set, symbolization, 599
Complex values, random variable assumption, 350
Component random variable, marginal density function, 337f
Component returns, standard deviations, 347
Component variable, value set, 120–121
Component vectors. *See* Linearly dependent component vectors

Composite model, 550
Conditional distribution, 113–114
Conditional expectation, 374–376
Conditional frequencies, 116t
  distribution, 113
Conditional parameters, 114–117, 374–377
Conditional population variance, 117
Conditional probability, 9–10, 361–365. *See also* Unconditional probabilities
  computation, 362–363
    illustration, 364–365
  determination, 366
  expression, 365–366
  formula, 363–365
  knowledge, 373–374
Conditional relative frequency, 114
Conditional statistics, 114–117
Conditional value-at-risk, 9–10
Conditional variance, 376
Conditioning variable, discrete setting, 375
Confidence interval, 11, 463–466
  concept, 548
  construction, 474
  definition/interpretation, 465–466
  obtaining, 473, 478
  realizations, 467f
Confidence level, 463–466
  definition, 465
Consistency criterion, demonstration, 496, 496f
Consistent test, 494–496
  definition, 495
  illustration, 496f
Constant derivative, 591
Constant estimator, 429
Constant intercept, 522
Constant risk-free rate, implication, 626
Constant variance, tests, 573–576
Contingency coefficient, 124–126. *See also* Corrected contingency coefficient; Pearson contingency coefficient
Continuity criterion, 585f
Continuous distribution function, 592
  growth, marginal rate, 235
Continuous function, 583–586
  continuity, 592
  illustration, 584f
Continuously compounded return, 253
Continuous probability distributions, 7–8, 228
  description, 229–230
  discrete probability distribution, contrast, 231
  extreme events, 277
  statistical properties, 247
Continuous random variables, 7, 237–238, 309
  covariance, 343
  density function, 340
  values, assumption, 237–238
Continuous random vector, components, 336
Continuous variable, 32–33
Contour lines, 10, 332–333
  example, 334f

Contours, ellipsoids (comparison), 355–356
Convergence characteristics, 436
Convergence in probability, 436–438
  expected value, 438
Convolution
  concept, 340
  integral (h), impact, 340
Copula (copulae), 379–405. *See also* Gaussian copula; Independence; *t* copula
  bivariate distribution function, 388
  bounds, 382, 383–384
  computation, 390
  construction, 380–381
  contour lines, 391
  density, 382, 383
  estimation, usage, 401–405
  expression, 386
  function, 381–382
  increments, 382, 383
  invariance, 385–386
  plots, 395
  properties, 382–386
  specifications, 381–382
  two dimensions, 387–399
  usage, 386–387
Corporate bond issuer, observation unit, 555–556
Corporate bond market, high-yield bond sector, 556
Corporation, default, 194
Corrected contingency coefficient, 125–126
Corrected sample variance, 431
  obtaining, 63
Correlated random variables, 341–342
Correlation, 123–124, 345–347
  covariance, relationship, 341–347
  criticism, 347
  measure, insufficiency, 10
  role, 129–131
  statistical measure, 3
Correlation coefficient, 10, 123, 345
  absolute value, 563
  coefficient of determination, relationship, 140
  denotation, 391
  matrix, 346
Correlation matrix
  bivariate normal density function, integration, 390
  denotation, 385
  setting, 399
Countable additivity, 177
Countable sets, distinction, 7
Countable space
  mapping, 181–182
  random variables, 181–183, 188–189
    definition, 182
Coupon rate, 556
Coupon securities, 537
Covariance, 120–123. *See also* Transformed random variables
  aspects, 343–345
  change, 343–344
  computation, 343, 344

continuous case, 342–343
correlation, relationship, 341–347
  criticism, 347
  discrete case, 342
  linear shift invariance, 344
  matrix of X, 349
  measure, 341–342
  scaling, 345
  symmetry, 345–346
  weighted sum, 348
Covariance matrix, 344–345
  aspects, 343–345
  correlation structure, 354
  obtaining, 456
  presentation, 344–345
  value, 352–353
Coverage ratio, 556
Cramér-Rao lower bounds, 415, 453–457
Credit migration tables, 369
  illustration, 370t
Credit ratings, 18
Credit risk management approach, 305
Credit risk modeling, 194
  Poisson distribution, application, 218–219
Critical region, 485
Cross hedging, 148–149
Cross sectional data, time series (relationship), 21
Cumulative distribution function. *See* Joint cumulative distribution function
  example, 189f
  usage, 331
  values assumption, 56–57
Cumulative frequency distribution, 27, 41–43
  evaluation, 30
  formal presentation, 30
Cumulative probability distribution function, 230
Cumulative relative frequency
  computation, 42
  distribution, 51–52
Cyclical terms, randomness, 155

Data. *See* Absolute data; Rank data
  accumulated frequencies, computation, 42
  analysis, 17
  center/location, 46–58
  class centers, usage, 62
  collection, process, 17–18
  comparison, problem, 68
  counting, 22, 24
  distribution, kurtosis, 317
  graphical representation, 75
  information, examination, 18–19
  interval scale, 20
  levels, 19–21, 67
  lower class bound, value selection, 38
  measurement, 20
  median computation, 69–70
  ordered array, usage, 39
  points, distance, 67
  range, usage, 39–41

Index

ratio scale, 20
reduction methodologies, 5
requirements, 47
scale, 19–21
scatter plot, 135f
sorting, 22, 24
standardization, 68
transformation, 70
types, 17–21
Data classes, 32–42
frequency density, 84
frequency distribution, determination, 43f
range, obtaining, 59
Data classification
formal procedure, 33–34
procedures, example, 34–41
reasons, 32–33
Sturge's rule, usage, 37
Data-dependent statistics, 238–239
Data level qualification, 72t
Datasets. *See* Multimodal dataset; Unimodel dataset
comparability, 24
minimum/maximum, 59
size, 48
symmetry, 84
variation, 67–68
DAX
closing values, plot, 153, 154t
index values, 155f
d-dimensional elliptical random variable, 357
Debt obligations, credit ratings, 18
Decision rule, 481, 483–485
Decomposition, 492f
form, 154–155
Default intensity, 265
Default number, null hypothesis (impact), 498
Default probabilities, 270
Default rate, 265
Degrees of freedom
estimate, computation, 403, 405
indication, 385
parameter, 354
reduction, 259
Delta, 588
Density, Fourier transformation, 599
Density function, 232–237
contour lines, 332
definition, 235
display, 261–262, 486
example, 267f
histogram, comparison, 234f
means, yield, 241f, 242
presentation, 288
requirements, 233–237
stability, 289f
symmetry, 247–248
usage, 238–239
Dependence, 338–341
alternative measures, 405–412
continuous case, 339–341

discrete case, 338–339
measures, 379
structures
detection, 124
determination, 349
zero covariance, concept, 121–122
Dependent components, 338
Dependent variable, 6, 132
Derivative, 233, 587–591
computation, 595–596
construction, 590–591
function, illustration, 588f
gains/losses, 76–77
integral, relationship, 595–596
setting, 450–451
Derivative instruments, valuation, 3–4
Descriptive statistics, 5–6
Determinants, 609–612
computation, 613–614
Deutsche Bank, absolute deviation, 64
Diagnostic statistics, usage, 535
Difference equations
components, 159
usage, 159
Difference quotient, 590–591
Differentiable, term (usage), 591
Differential calculus, usage, 525
Discontinuous function, 583
illustration, 584f
Discrete cumulative distribution, 188
Discrete distributions, list, 223
Discrete law, 187–192
Discrete probability distributions, 187
continuous probability distribution, contrast, 231
examples, 326–327
usage, 7
Discrete random variables, 230
Discrete random vectors, 326
marginal probability, 334
Discrete uniform distribution, 7, 219–221
independent trials, consideration, 220
mean/variance, 219
Discrete variables, 32–33
Dispersion matrix, 386
Dispersion parameters, 239–244
Distributed random vector, dependence structure, 385–386
Distribution
Chebychev inequalities, 251
connectivity, 304–305
decays, 288
equality. *See* Equality of distribution
exponential family, 457–458
mean, definition, 302
skewness statistic, 572
symmetry, objective measurement, 314–315
tails, 3
VaR computation, 299–300
yields, 194–195

Distribution function, 230–232, 266
    definition, 179
    example, 267f
    mean/variance, 266
    parameters, 322–323
    value, assumption, 387
Distribution of X, 184
Disturbance terms
    linear functional relationship, 134f
    randomness, 155
Diversification, strategy, 348
Dividend payments, variable, 21
Dot-com bubble, 286
Dow Jones Global Titans 50 Index (DJGTI), 22, 24
    list, 25t–26t
    relative frequencies, comparison, 28t
Dow Jones Industrial Average (DJIA)
    components
        industry subsectors, frequency distribution, 24t
        list, 23t
    component stocks analysis, 22
    relative frequencies, comparison, 28t
    stocks
        empirical relative cumulative frequency distribution, 31t, 32f
        share price order, 29t
    value
        consideration, 416
        realizations, 416
Down market beta, 551
Down movement, 198–199, 221
    binomial random variable, impact, 303
    occurrence, 252
Drawing with replacement, 196
d-tuple, requirements, 382
Dummy variables
    design, 549
    incorporation, independent variable role, 548–561
    trap, 549
    usage, 549–550. *See also* High-yield corporate bond spreads; Mutual fund characteristic lines
    value, 551
Dummy variables, usage, 12
Duration, 531. *See also* Empirical duration; Regression-based duration
Durbin-Watson critical value table, 579
Durbin-Watson d statistic, 578
Durbin-Watson test, 578

Earnings
    logarithm, 556
    measures, 556
Earnings before interest, taxes, depreciation and amortization (EBITDA), 556
Earnings before interest and taxes (EBIT), 556
Econometrics, 2
Efficient estimator, 443–444
Efficient frontier, 353
Eigenvalue, 608, 612–614
    multiplicity, 613

Eigenvector, 612–614
    derivation, 612–613
Electronic communication networks (ECNs), 367
Electronic trading, importance, 367
Elementary events/atoms, 174
Elements
    combination, 170
    marginal distributions, 10
Elliptical class, properties, 357–358
Elliptical distributions, 356–358
    closure, guarantees, 357
Empirical cumulative frequency, 27
    distribution, 27–32, 402
        function, computation, 514–515
Empirical cumulative relative frequency distribution, 90f, 91f
    joint observations, 403f
Empirical duration, 531
    data, 533t
    estimated regression coefficient, relationship, 532
    estimation, 531–536
        regression parameters, estimation
            multivariate linear regression, usage, 536t
            simple linear regression, usage, 534t
Empirical relative cumulative frequency distribution, 31t, 32f
Empirical returns/theoretical distributions (comparison), box plots (usage), 98f
Empirical rule, 67, 250–251
Engineering, risk analysis, 1
Engle, Robert, 2
Equality, denotation, 171
Equality of distribution, two-tailed Kolmogorov-Smirnov test (usage), 513–516
Equality of means, test, 509–513
Equal means, null hypothesis, 510
    rejection, impossibility, 511
Equal tails, 472
    probability of errors, creation, 478
Equal tails test
    test size partition, 508f
    usage. *See* Normal distribution
Equation system, 602
Equidistance, data class criteria, 33
Equivalent test rule, 497–498
Erlang distributed statistic, 500
Erlang distribution, 247, 269
    α quantiles, 478
Erlang random variable, creation, 478
Error correction, 161–163
    model, 163
Error probability, asymmetric distribution, 471
Error terms
    constant variance, tests, 573–576
    distribution, 133
    standard deviation, 574
    statistical properties, assumption, 570
Error types, 485–490
Estimate
    empirical distribution, 432

Index

population parameter, contrast, 429
Estimators, 415, 420–421. *See also* Biased estimator; Point estimators
    asymptotic properties, 436
    basis, statistics (impact), 460
    consistency, 438
    convergence characteristics, 436–438
    denotation, 421
    expected value, 441
    large-sample properties, 436
    properties, 436
        assessment, Central Limit Theorem (usage), 436
    quality criteria, 428–435
    random sample dependence, 421
    understanding, 421
    usage, 432–433
    variance, Cramér-Rao lower bounds, 415
Euler constant, 217
Euro-British pound exchange rate returns, 97
European banks
    investment banking revenue ranking, 64t
    revenue, sample standard deviation (obtaining), 65
European call option, 625
    Black-Scholes option pricing formula, 631
    expected value, computation, 627
    maximum operator, computation, 627
    price, derivation, 626–631
    standardized random variable, 628
    stock terminal price, 627
EUR-USD exchange rate
    daily logarithmic returns, box plot, 93f
    QQ plot, contrast, 97f
    returns, box plot, 95f
Event-driven hedge fund, 372
Events, 7, 173–176. *See also* Independent events
    correspondence, 390f
    defining, 368
    marginal probability, 334
    no value, 177
    origin, determination, 183
    probability, 231, 367
        obtaining, 389
    rectangular shape, 388f
Excess kurtosis, 286–287, 318
Excess return, 142–143
Execution slippage, 367
Execution time, 367–368
Exemplary datasets, usage, 84
Exemplary histograms, 85f
Exercise price, 625
Expected inflation, monthly data, 538t–542t
Expected shortfall, 9–10, 300
Expected tail loss, 376–377
Expected value, interpretation, 303
Exponential data, linear relationship, 141–142
Exponential density function, 263f
Exponential distribution, 8, 262–265
    characterization, 262
    exponential family, 458

λ parameter
    confidence interval, 477–479
    one-tailed test, 499–501
λ parameter, MLE (impact), 450–451
    Cramér-Rao bound, 455
    mean, 262, 304–305
    no memory property, 264
    parameter space, computation, 500f
    Poisson distribution, inverse relationship, 264
    skewness, 315
    variance, 262, 310–311
Exponential family, 457–461. *See also* Exponential distribution; Normal distribution; Poisson distribution
Exponential function, value yield, 271
Exponential functional relationship, linear squares regression fit, 141f
Exponential random variable, observation, 450
Extreme events, continuous probability distributions, 277
Extreme value distribution. *See* Generalized extreme value distribution
Extreme value theory, 278

Factorial, 596–597
    operator, usage, 615
Factorization theorem, 460
Fan, Fianqing, 2–3
F-distribution, 247, 260–262
    density function, 261f
Feasible portfolios, 353
Federal Reserve, intervention, 209–211
Finance
    correlation, 130
    exponential distribution application, 265
    multivariate linear regression model application, 531–543
    probability/statistics dependence, 3
    regression analysis applications, 142–149
Financial asset returns, modeling, 9
Financial econometrics, 2
Financial quantities, research, 18
Financial returns, 290
    modeling, 282
    simulation, copula (usage), 386–387
First derivatives, usage, 447–448
First moment, 239, 302–306
Fisher's F distribution, 247
Fisher's Law, 537
Fitch Ratings, 369
    credit ratings, 18
Foreign exchange (FX) markets, electronic trading (importance), 367
Formal derivation, 585–586
Forward price, 161
    innovation, 163
Fourier transformation. *See* Density
Fourth central moment, 318
Franklin Templeton Investment Funds, 12-month returns, 34, 35t–36t
    ordered array, 37t–38t

Fréchet distribution, 278, 280
Fréchet lower bound, 383
Fréchet type distribution, application, 281
Fréchet upper bound, 383
Freedman-Diaconis rule, 33–34
  application, 39
  distribution, fineness, 40
Frequencies
  accumulation, 27
  formal presentation, 26–27
Frequency density, 82, 84
Frequency distributions, 22–27. *See also* Empirical cumulative frequency
Frequency histogram, 82–88
Frisch, Ragnar, 2
F-statistic, 521
  computation, 528
F-test
  application. *See* Lehman U.S. Aggregate Bond Index
  inclusion, 530–531
  usage, 550
Full rank, 605
Full rank $n \times n$ matrix, inverse (definition), 607
Fully pooled data, regression equation (estimation), 560
Function
  decrease, absence, 592
  features, 583, 596
  formal expression, 30
  integral, 595
  slopes, measurement, 589f
Functional relationship, assumption, 141
Fund returns. classes
  centers, 47t
  Freedman-Diaconis rule, usage, 40t
  Sturge's rule, usage, 39t

Gamma function, 255, 268–269, 354
  density function, 268f
  illustration, 598
  mean/variance, 269
  non-negative values, 597
GARCH. *See* Generalized autoregressive conditional heteroskedasticity
Gauss, C.F., 247
Gauss copula, specification, 385
Gaussian copula, 383, 385
  concretizing, 390
  contour lines, 395
  d=2 example, 390–393
    contour plots, 393f–394f
    illustrations, 391f, 392f
  d=2 simulation, 399–405
  data, fit, 412
  degrees of freedom, 401f
  example, 409–410
  marginal normal distribution functions, 404f–405f
  returns, computation, 403, 405
  tail dependence, 408–409
  usage, 404f

values, generation, 400f
Gaussian distribution, 97, 247
Gaussian random variables, 418
General Electric (GE)
  daily return data
    consideration, 439
    sample means, histogram, 424f
    sample variance, histogram, 442f–443f
  daily returns
    bivariate observations, 402f
    empirical cumulative relative frequency distribution, joint observations, 403f
    generated returns, 405f–406f
    non-normal distribution, Central Limit Theorem (impact), 511
  daily stock returns, 418, 468, 509
    distribution, 423
    mean parameter, confidence interval, 468
  data
    simulation, estimated copulae (usage), 401–405
    Spearman's rho, 410–412
    tail dependence, 410–412
  marginal normal distribution functions, generated values, 404f
  marginal student's t distribution functions, generated values, 404f
  monthly returns, 131t
  return data
    observation, 441
    sample mean, 434f
  returns
    conditional frequency data, 120
    conditional sample variances, 117
    examination, 510–511
    usage, 126
  sample means, histograms (comparison), 440f
  stock price
    observations, 419t
    variable, 168–169
  stock returns, 434
    conditional sample means, 118t
    indifference table, 127t
    marginal relative frequencies, 115t
General Electric (GE) daily returns, 317f
  kurtosis, 318–319
  left tails, comparison, 320f
  modeling, 319f
  skewness, 315–316
Generalized autoregressive conditional heteroskedasticity (GARCH) model, 574–576
Generalized central limit theorem, 290
  application, 425
Generalized extreme value (GEV)
  density function, 279f
  distribution, 277–281, 286
  function, 279
Generalized Pareto distribution, 8–9, 281–283
  function, parameter values, 280f, 282f
General multivariate normal distribution, density function, 349–350

Index

Glejser test, 573
Global minimum-variance portfolio, 353f
Global minimum-variance portfolio (global MVP)
    denotation, 353
GLS. *See* Aitken's generalized least squares
Goldfeld-Quandt test, 573
Goodness-of-fit denotation. *See* Regression
Goodness-of-fit measures, 6, 138–140. *See also*
    Adjusted goodness-of-fit measure
    coefficient of determination, relationship, 527
    improvement, 530
    usage, 147, 521–522
Granger, Clive, 2
Growth, marginal rate, 232–235
Gumbel, Emil Julius, 398
Gumbel copula, 398–399
Gumbel distribution, 278
    distribution function, 278, 280

Half-open intervals, 327
Hazard rate, 265
Heavy-tailed distribution, 572
Heavy tails, 357–358
    modeling capability, 8
    presence, 379
Hedge funds, 372
    data, 372–373
Hedge ratio, refinement, 148
Hedging. *See* Cross hedging
    instrument
        shorting amount, 147
        usage, decision, 147
        regression analysis, application, 147–149
Heteroscedastic error term, 570
Heteroskedasticity
    advanced modeling, 574–576
    modeling, 573–576
Higher dimensional random variables, 326–327
    continuous case, 327
    discrete case, 326–327
Higher order, moments. *See* Moments of higher order
High-yield bond, 555–556
High-yield corporate bond spreads
    prediction, 555–561
    regression results, 561t
        credit rating classification, dummy variable
            (usage), 561t
        time, dummy variables (usage), 562t
Histogram. *See* Exemplary histograms; Frequency
    histogram
    absolute frequency density, inclusion, 86f
    density function, comparison, 234f
    skew, 572
Homoskedastic error term, 570
Homoskedasticity, 573–576
Hypergeometric distribution, 7, 204–211
    application, 209–211
    approximation, Poisson distribution (usage), 218
    random variable, 208–209
    representation, 417

Hyperplane, 524. *See also* Regression; Vector
    hyperplane
Hypothesis (hypotheses), 482–485
    setup, 482–483
    test
        alternative use, 484
        size, 486–488
    testing, 11, 480
        examples, 496–518

IBM
    daily returns
        bivariate observations, 402f
        empirical cumulative relative frequency
            distribution, joint observations, 403f
        generated returns, 405f–406f
        non-normal distribution, Central Limit
            Theorem (impact), 511
    data
        simulation, estimated copulae (usage), 401–405
        Spearman's rho, 410–412
        tail dependence, 410–412
    marginal normal distribution functions, generated values, 404f
    marginal student's t distribution functions,
        generated values, 404f
    returns, examination, 510–511
    stock return, 509–510
Identically distributed Bernoulli random variables,
    615–616
Identity matrix, 608
Imaginary number, 349–350, 598
Imaginary signs, matrix (setup), 610–612
Incident class, 53
Independence, 117–120
    assumption, 196
    copula, 384
        obtaining, 398–399
    definition, 119
    discussion, 10
    formal definition, 338
Independent components, 338
Independent distributed random variables, 616
Independent drawings, n-fold repetition space, 417
Independent events, 365–366
    probability, multiplicative rule, 368–369
Independent random variable, distribution, 406
Independent standard exponential random variables, summation, 478
Independent variable, 6, 132
    calculation, 551
    covariance, 121
    determination, 562
    role, 548–561
    significance, testing, 529–530
    space formation, 524
    usage, excess, 529
Indicator function, 237, 374, 586–587
    expression, 587
    usage, 238–239

Indifference table, 126
Inductive statistics, 10–11
Industry Classification Benchmark (ICB), usage, 22
Inferences, 416. *See also* Statistical inference
   making, 4
Infinitum (lowest bound), 296
Inflation
   monthly data. *See* Expected inflation
   rate, information (availability), 537
Influence, concept, 119
Information matrix, 454
   inverse, 456
Inner product
   computation, 606
   definition, 606
Insurance premiums earned, 77
Integral, 592–596
   area approximation, 594f
   computation, 628–629
   derivative, relationship, 595–596
   left-point rule, 595
   limiting area role, 594–595
   right-point rule, 595
Inter-arrival times, 478–479
   distribution, 264, 269, 304–305
   observation, 500–501
Interest
   independent variable, 549
   random variable, 214
Interest rate
   changes, 203
   index, independent variable, 532
Intermediate statistics, computation, 511
Interpolation method, usage, 54
Interquartile range (IQR), 34, 60–61
   computation, 92–93
   data representation, 91–92
   defining, 61
Intersection operator, usage, 171
Interval scale, 20
Invariance. *See* Strictly monotonically increasing transformations
Investment banking revenue
   data, standardized values, 69t
   ranking, 68t
Investment-grade bonds, consideration, 369
IQR. *See* Interquartile range
Irregular term, 155–156
   coefficient, 158
Isolines, 332

Jarque-Bera test statistic, 571–572
Joint behavior, knowledge, 329
Joint cumulative distribution function, 328–329
Joint density function, 332, 339–340
   expression, 340
Joint distribution
   copula, 406
   function, display, 380–381
   governance, 409
   marginal distribution, comparison, 335–336
   reduction, absence, 382
Joint frequencies
   average squared deviations, measurement, 125
   payoff table, 122t
Jointly normally distributed random variables, tail dependence (absence), 409
Joint probability, 366–367
Joint probability density function, 330
   contour lines, 333
Joint probability distributions, 9, 325, 328–338
   continuous case, 330–332
   discrete case, 328–329
   understanding, importance, 329
Joint random behavior, measures, 325
Joint randomness (measure), correlation/covariance (criticism), 347
j-th independent variable, 547
Junk bond, 556

k components, pairwise combinations, 346
k-dimensional elements, 327
k-dimensional generalization, 340
k-dimensional random variable, 354
k-dimensional random vector, 327
   covariances, 346
   density function, 355–356
k-dimensional real numbers, general set, 421
k-dimensional volume, generation, 330
Kendall's tau, 407
K events, Bayes' rule, 372
k independent variables
   regression coefficients, 522
   test, 529–530
Kolmogorov, Andrei N., 168–169
Kolmogorov-Smirnov (KS) statistic, 515
Kolmogorov-Smirnov (KS) test
   application, 11
   usage. *See* Two-tailed Kolmogorov-Smirnov test
k-th moment, 307
k-tuple, 326
   representation, 330
Kurtosis, 317–318, 572. *See also* General Electric
   excess, 286–287, 318

Lag, size, 428
Lagged correlation, 577
Lagrangian multiplier test, 573
Lambda (λ) parameter, 427–428, 448–450
   one-tailed test, 499–501
   test, 496–499
Large capitalization mutual funds, characteristic line estimation (data), 144t–146t
Large capitalization stock funds, 143, 147
Large sample criteria, 10, 435–446
   consistency, 436–438
Law of Large Numbers, 426, 437
Least squares
   provision, 136
   regression fit, exponential relationship values, 142t

Index

result, 138f
Left-continuous empirical distribution, 514–515
Left-point rule, 595
Left-skewed data, indication, 66
Left skewed indication, 244
Left-skewed returns, 316
Left tails, comparison, 301f
Lehman U.S. Aggregate Bond Index
   asset duration estimate, 532, 534
   F-test, application, 536
   regression, analysis, 535–536
Leptokurtic distribution, 318
Light-tailed distribution, 572
Likelihood functions, 446–447
   computation, 517
   obtaining, 448, 450
Likelihood ratio, 516
   random sample dependence, 516
   test, 11, 516–518
Limiting distribution, 425
Limiting normal distribution, 427–428
   Poisson population, contrast, 427f
Limit probability, 235
Linear estimators, 424–425
   lags, inclusion, 428
Linear functional relationship, 134f
Linearity, tests, 568–569
Linearly dependent component vectors, 613
Linear regression, 6. *See also* Simple linear regression
   creation, 569
Linear transformation
   measures, 69–71
   sensitivity, 72t
Linear unbiased estimators, 445–446, 570
Location, 46–58
   measures, 45, 59, 69–70
   examination, 72t
Location parameters, 9, 249, 295. *See also* Right-skewed distribution
   mode, relationship, 301
Location-scale invariance property, 249
Location-scale invariant, 149
Logarithmic return
   normal distribution, 252–254
   probability, 231
   usage, 254
Log-likelihood function, 446–447
   computation, 448, 450
   first derivatives
      obtaining, 452
      usage, 447–448
   obtaining, 451–452
   setup, 447
Log-normal distribution, 8–9, 247, 271–274
   application. *See* Call options
   asymmetry, 314
   closure, 272
   density function, 273f
   impact, 629
   second moment, 308

Lognormal distribution, density function, 272f
Log-normally distributed random variable, 305
Log-normal mean, presentation, 305–306
Log-normal random variable, mean/variance, 271
Long-term corporate bond, risk manager hedge position, 147–148
Long-term interest rates, short-term interest rates (correlation), 148
Look-up reference, 611
Lower limit, 590
   defining, 92
Lower quartile, 296–297
Lower tail dependence, 407–408
   coefficient, 409–410
   expression, 408
   measure, 408
λ parameter, estimation, 427–428

MA. *See* Moving average
Marginal density function, 336–337
   concept, illustration, 337
   denotation, 417
   integration, 338
   value, 336
Marginal distributions, 10, 333–338. *See also* Elements
   continuous case, 336–338
   discrete case, 334–336
   function, 399
   heavy tails, 357–358
   joint distribution, comparison, 335–336
   values, 339, 382
Marginal increment, 590
   limit analysis, 590–591
Marginal probabilities, 362
Marginal probability distribution, 333
   obtaining, 335
Marginal rate of growth, 232–235
Marginal variances, weighted sum, 348
Market capitalization, 19
Market environments, mutual fund characteristic lines (testing), 551–555
Market excess return, 143
Markowitz portfolio selection, 353
   theory, 8
Matrix (matrices), 601–602. *See also* Square matrix
   algebra, usage, 525
   determinant, 609
   determination, 609–612
   direction (3×3 matrix), column vectors (usage), 604
   display, 602
   examination, 603–614
   extension, 603
   full rank $n \times n$ matrix, inverse (definition), 607
   inverse, 608
      algebraic derivation, avoidance, 608
   multiplication, 602–603
      feasibility, 603
   $n \times n$ matrix

INDEX

Matrix (*Cont.*)
    component vectors, 606
    determinant, computation, 609
    operations/concepts, fundamentals, 601
    reduction, 611–612
    sizing, 607–608
    2 × 2 matrices, determinants, 609
Matrix notation
    analogue, 523–524
    usage. *See* Minimization problem
Maximum likelihood estimate, 448
    determination, 452
Maximum likelihood estimator (MLE), 446–457
    examples, 448–453
    illustration, 449
    reference, 448
    sample size, 455
Maximum operator, computation, 627
Mean absolute deviation (MAD), 61–63
    analysis, 70–71
    computation, 62
    contrast, 66
    definition, 61–62
    standard deviation, comparison, 65
Mean (μ), 46–48, 189–191. *See also* Weighted mean
    computation, original data (usage), 48
    defining. *See* Sample mean
    estimator, 421–423
    example, 306f
    expected value interpretation, 303
    first moment, 302–306
    interpretation, 48–49
Mean parameter, confidence interval, 476–477
Mean squared of errors (MSE), 528
Mean-square error (MSE), 432–433
    MSE-minimal estimators, unavailability, 433
    square root, 567
Mean squares of regression (MSR), 528
Mean-variance portfolio optimization, 353
Measurable function, 237–238
    definition, 180
    relationship, 181f
Measurable space, 176–177
    definition, 176
    formation, 230
Measurement level, 19
Measures, summary, 71–72
Median, 48–54
    ambiguity, avoidance, 57
    calculation, 51
    determination, linear interpolation (usage), 53
    example, 306f
    feature, 49
    percentiles, 58
Minimization problem, matrix notation (usage), 524–525
Minimum-variance estimator, discovery, 433
Minimum variance linear unbiased estimator (MVLUE), 445–446
Minimum variance portfolio (MVP), 353

Minimum-variance unbiased estimator, 433
Mode (M), 54–55, 301–302
    class, determination, 55
    distribution shape, 317
    example, 306f
Moment of order k, 307
Moments of higher order, 240–244, 307–309
Monotonically increasing function, 383, 592
    illustration, 593f
Monotonically increasing value, 230
Monotonic function, 591–592
Monthly return data, usage, 140
Moody's Investors Service, 369
    credit ratings, 18
Moving average (MA). *See* Autoregressive moving average
    method, 158
Moving average (MA) process of order q, 580
MSE. *See* Mean squared of errors; Mean-square error
MSR. *See* Mean squares of regression
Multicollinearity. *See* Near multicollinearity
    indicators, 547
    mathematical statement, 546
    problem, 545–548
    procedures, mitigation, 547–548
Multimodal dataset, 55
Multinomial coefficient, 214–215, 615, 622–623
Multinomial distribution, 7, 211–216
Multinomial random variable, probability distribution, 212
Multinomial stock price model, 213–216
    discrete uniform distribution application, 220–221
    example, 216f
Multiple events, total probability (law), 371
Multiplicative return (R), 252–253
Multiplicative rule. *See* Probability
Multiplicity, 613
Multivariate distribution function, expression, 381
Multivariate distributions, 6, 325, 328
    application, 350–353
    introduction, 384
    selection, 347–358
Multivariate linear distribution, 12–13
    models, building process, 12
Multivariate linear regression
    analysis, estimates/diagnostics, 521
    usage, 536t
Multivariate linear regression model, 522–523
    assumptions, 523
        testing, 567
    construction, 545
        techniques, 561–565
    dependent variable, impact, 531–532
    design, 526, 545
    diagnosis, 526
    diagnostic check, 526–531
    error term, 567
    fitting/estimating, 526
    independent variables, consideration, 546
    parameters, estimation, 523–525

Index

significance, 526–531
  testing, 527–529
  specification, 526
  standard stepwise regression model, 564–565
  steps, 526
  stepwise exclusion method, 564
  stepwise inclusion method, 562–564
Multivariate normal distribution, 347–353
  characteristic function, 350
  properties, 348–349
Multivariate normally distributed random vector, 399
Multivariate normal random vector, 349
Multivariate probability density function, 330
Multivariate probability distributions, 9
Multivariate random return vector, density function, 333
Multivariate t distributions, 354–356
Municipal bond futures, availability, 148
Mutual exclusiveness, data class criteria, 33
Mutual fund characteristic lines
  estimation data, dummy variable (usage), 552t–554t
  regression results, dummy variable (usage), 555t
Mutual fund characteristic lines, testing, 551–555
Mutually exclusive events, 368–369
MVLUE. *See* Minimum variance linear unbiased estimator

Natural logarithm, usage, 451–452, 497, 583
Near multicollinearity, 546
Negative autocorrelation, 577
Negatively correlated random variables, 346
Nested models, 516
Neutrality property, 607–608
New York Stock Exchange (NYSE) stocks, categorization, 18
Neyman-Pearson test, 499
NIG distribution
  normal distribution, contrast, 320f
  scaling property, 285
NIG random variables, 283, 285
  scaling, 285
  skewness, 316
NIG tails, 318–319
No default probability, 265
No memory property, 264
Nominally scaled data, 19
Non-decreasing function, 172–173
  hypothesis, 173f
Nonemptyness, data class criteria, 33
Noninvestment grade bond (junk bond), 556
Nonlinear regression, 569
Non-linear relationship, linear regression, 140–142
Non-negative integer
  usage, 207
  value, assumption, 217
Non-negative probability, 178
Non-negative real numbers, 254
Non-negative values
  assumption, 261
  gamma function, 597

Non-redundant choices, usage, 208
Non-redundant options, usage, 206–207
Non-redundant outcomes, 205
Non-redundant samples, 212
Non-speculative-grade rating, unconditional probability, 371
Normal distribution, 3, 247–254
  assumption, 291
  exponential family, 459
  finance usage, 8
  mean, 305–306
  NIG distribution, contrast, 320f
  normal inverse Gaussian distribution, contrast, 285
  parameter components, MLE, 451–453
  parameter µ
    equal tails test, usage, 506–508
    one-tailed test, usage, 501–503
    two-tailed test, usage, 505–506
  parameters
    confidence intervals, 475t
    MLE, Cramér-Rao bounds, 456–457
  properties, 248–251
  skewness, 315
  t distribution, difference, 410
  usage, 251–252
  variance (σ^2), 311–312
    one-tailed test, usage, 503–505
Normal inverse Gaussian density function, parameter values, 284f
Normal inverse Gaussian distribution, 8–9, 283–285
  function
    normal distribution, contrast, 285
    parameter values, 284f
Normalized sums, asymptotic properties, 436
Normally distributed data, sample mean (consistency), 438–440
Normally distributed random variables, tail dependence, 411f
Normally distributed stock return, 456
Normal random variable
  mean, confidence interval, 466–469
    unknown variance, 469–475
  variance, confidence interval, 471–473
Null hypothesis, 482. *See also* Equal means; True null hypothesis
  acceptance, 494
  accounting, 497
  alternative hypothesis, contrast, 483f
  autocorrelation, relationship, 579
  avoidance, 488
  formulation, 509–510
  probability, value, 501–502
  rejection, 485, 505–506, 508
    impossibility, 501, 512
    probability, 492, 495f, 500f
    test statistic, impact, 510
$n \times n$ matrix. *See* Matrix

Observation pairs, scatter plot, 137, 138f

Observed values, frequency, 32–33
Off-diagonal cells, zero level, 445
Ogive diagrams, 89–91
One-day stock price return, 306f, 311, 313
　distribution, bounds, 314f
　rule of thumb, 313f
One-dimensional probability distributions, 9
One-tailed test, 483. *See also* Exponential distribution
　usage. *See* Normal distribution
Open classes, 33
Open intervals, 327
Optimality, determination, 135
Ordered array, length equality, 96
Ordinally scaled data, 19
Ordinary least squares (OLS), 135–136
　method, usage, 577–578
　regression method, 524–525
　usage, 141–142, 574
Outcomes, 7, 173–176
　influence, 250
　likelihood, null hypothesis (impact), 497
　occurrence, 231
　probability, 210–211, 231
　realization, 196
Outliers, 92
Over-the-counter (OTC) business, data inaccessibility, 18

Pairwise disjoint, 171, 177
Parallelepiped (three-dimensional space), 609
Parameters
　components, MLE, 451–453
　definition, 45
　estimation, 257
　invariance, 466
　statistics, contrast, 45–46
Parameters of location. *See* Location parameters
Parameters of scale. *See* Scale parameter
Parameter space, 416
　computation, 500f
　interval, selection, 464
　probability computation, 502f
Parameter values
　benchmark, 505f
　normal density function, 248f
　normal distribution function, 249f
Parametric distributions, 239
Pareto distribution, 281
　β/ξ parameters, 282
　density function characterization, 281–282
Pareto tails, 286–287
Park test, 573
Partial determination, coefficient. *See* Coefficient of partial determination
Partial integration method, 305
Path-dependence, 202
Payoff table, 122t
Pearson contingency coefficient, 125
Pearson skewness, 244
　definition, 65–66

Percentage returns, empirical distribution, 49t, 51t
Percentage stock price returns, empirical distribution, 50f
　even number companies, 52f
Percentile, 56, 296
Performance measure estimates, results (estimation), 147
Pie charts, 75–78
Platykurtic distribution, 318
P-null sets, 232
Point estimators, 415
Poisson distributed random variable, 307
Poisson distribution, 7, 216–219
　delta parameter, 217
　exponential distribution, inverse relationship, 264
　exponential family, 458
　λ parameter
　　estimation, 427–428
　　MLE, 448–450
　　sufficient statistic, 461
　　test, 496–499
　mean, 304
　parameter, 304, 310
　probability, insertion, 497–498
　second moment, 307–308
　variance, 310
Poisson parameter, MLE (illustration), 449
Poisson population, limiting normal distribution (contrast), 427f
Poisson probability distribution, 227
Poisson random variable
　mean/variance, 217
　modeling, 499
Population
　alpha-percentile, 56–59
　mean, 114
　　knowledge, 431
　multivariate linear regression model, 522
　standard deviation, definition, 64
Population parameter
　estimate, contrast, 429
　generation, *n*-dimensional sample (usage), 419–420
　information, 416
　statistic, contrast, 419–420
Population variance
　computation, 431–432
　definition, 63
　estimator replacement, 430–431
　examination, 71
　unbiased estimator, obtaining, 431
Portfolio
　assets, currency denomination, 346–347
　diversification, 408
　return, 343–344
　　computation, 352
　　mean/variance, 352
　selection, multivariate distribution application, 350–353
　theory, risk measure (role), 241
　value, representation, 351–352
Positive autocorrelation, 577

Index

Positive semidefinite matrices, 614
Positive-semidefinite matrices, 444–445, 454
Power law, 286–287
Power of a test, 490–493
   definition, 491
   probability line, 491f
Power set, example, 174t
Power tails, 287–288
p parameter, estimation, 425–426
Price/earnings (P/E) ratio, 364–365
   list, 364t
Price process
   efficiency, 160
   time series application, 159–163
Probabilistic dependence, 10
Probability
   alternative approaches, historical development, 167–170
   computation, 197–198
      density function, usage, 238–239
   concept, problem, 170
   congruency, 201–202
   convergence, 436–438
   determination, 196, 373
   double summation, 342–343
   expression, 408
   law, 184, 187
   mass, 241–242
   models, 4
   multiplicative rule, 366–371
      application, 368
      illustration, 367–368
   plot, 96
   relative frequencies, 167–168
   statistics, contrast, 4–5
   studying, 4–5
   theory, 7–10
      concepts, 167
   university course, offering, 2–3
Probability density
   function, 233
   understanding, 592
Probability distributions, 7, 187, 194
   continuousness, 330
   moments, 9
   obtaining, 203
   values, 198
Probability measure, 177–179
   continuous distribution function, 232
   definition, 177
   density function, 233–234
Probability of error, 486
   creation, 478
Probability space, 178
   random variable, definition, 375–376
Probability-weighted values, sum, 239
Procter & Gamble (P&G)
   par value, 148–149
   trading day, yield/yield change, 150t–152t
Psychometrics, 2

p-value, 11, 489–490
   illustration, 490f
   reduction, 571
Pyrrho's lemma, 562

Quadratic forms, 614
Qualitative data, 18
   usage, 22
Quality criteria, 428–431, 490–496
Quality grades, consideration, 19
Quantile-quantile plot (QQ plot), 96–99
Quantiles, 56–58, 296–301. *See also* 0.25 quantile; 0.75 quantile
   one-to-one relationship, 632
Quantitative variable, 19
Quartiles, 58
   computation, 297–298
   example, 306f
Queue theory, 1

Radius, attributed meaning, 75–76
Randomness, inheritance, 259
Random return vector, 337
   joint probability density function, contour lines, 334f
   probability density function, 333f
Random sample, joint probability distribution, 446
Random shock, 160
Random variables, 179–185. *See also* Continuous random variables; Higher dimensional random variables
   Bernoulli distribution, relationship, 192–193
   boundary k, 503–504
   company value function, 587f
   consideration, 384
   correlation coefficient, 345
   countable space, 181–183, 188–189
      definition, 182
   density function, 422
      skew, 471
   deterministic function, 180
   distribution, 196
   exponential distribution, 265
   independence, definition, 338
   linear transform, 278
   link, correlation inadequacy, 3
   location, parameters, 295
   mean, 190, 208–209
      value, 252
   measurability, 376
   measurable function, 181
   moments, 309
   multiplication, 477–478
   n degrees of freedom (chi-square distribution), α-quantiles (position), 472f
   negative correlation, 346
   obtaining, 504
   outcomes, 192
   parameters, 238–239
   positive correlation, 346
   probability law, 325
      relationship, 217

Random variables (*Cont.*)
  scale, parameters, 295
  uncorrelation, 346
  uncountable space, 183–185
  usage, 189, 212
  values, assumption, 389
  variance, 509–510
    definition, 191
Random vectors, 10, 326
  density function, 350
  joint distribution, 348
    function, 381
  multivariate distribution function, 383
  probability, 328
Random walk, 160–161
Range
  obtaining, 59
  variation measure, 59–60
Rank correlation measures, 406–407
  symmetry, 407
Rank data, 19
Ratio scale, 20
Real numbers
  open/half-open intervals, 327
  three-dimensional space, 605
Real rate, monthly data, 538t–542t
Real-valued parameter (λ), 262
Real-valued random variable
  usage, 240, 242, 244
Real-valued random variables
  gamma distribution, 268
Real-valued variable, continuous function, 583
Recession, restricted space, 375
Rectangular distribution, 266–267
Rectangular event, 388–389
Redundant components, 605–606
Regime shift, occurrence, 561
Regression
  analysis, 129
    finance applications, 142–149
    hedging application, 147–149
  equations, examination, 550
  errors, 570
  goodness-of-fit denotation, 530
  hyperplane, 524
  method. *See* Standard stepwise regression method
  modeling, 129
  theory, 12
Regression-based duration, 531
Regression coefficients, 524
  estimates, distribution, 570
  estimation, 147, 532–534, 556
  unbiased estimates, 577
Regression model, 131–132
  distributional assumptions, 133–134
  estimation, 134–138
  evaluation
    goodness-of-fit measures, usage, 521–522
  extension, 6
  industry sectors, consideration, 549

  misspecification, 570
Regressor, 132
Reichenbach, Hans, 167–168
Rejection region, 485
Relative frequency, 27, 236
  comparison, 28t
  computation, 24
  density, 233
    histogram, usage, 88f
  dependence, 169
  distributions, usage, 120
  probability, relationship, 167–168
  usage, 84, 124–125
Replacements, 206–208
Rescaled random variables, generation, 257
Rescaling effect, 70
Residuals, 525
  analysis. *See* Standardized residuals
  autocorrelation, absence, 576–581
  correlation, importance, 577
  distribution, 133
  observation, 569
  tests, 570–572
Residual term, 132
Restricted model, 516
Retransformation, 141–142
Return
  alternative set, 60t
  attributes, 56
  data, sample, 69t
  density function, 331–332
  joint distribution, 334–335
  standard deviations, 347
  vector, two-dimensional pdf, 332f
  0.25 percentile, 57
  0.30 percentile, 57f
Rho. *See* Spearman's rho
Right-continuity, demonstration, 172f
Right-continuous empirical distribution, 514–515
Right-continuous function, 172–173
Right-continuous value, 230
Right-point rule, 595
Right-skewed distribution, location parameters, 316f
Right skewed indication, 244
Risk analysis, 1
Risk-free asset, excess, 142–143
Risk-free interest rate, 143
  composition, 344
Risk-free position, addition, 344
Risk-free rate
  implication. *See* Constant risk-free rate
  market return, relationship, 551
Risk-free return, 143
Risk management, variables consideration, 6
Risk manager, hedge position, 147–148
Risk measures, 9–10
  role, 241
Risky assets, 343
Rounded monthly GE stock returns, marginal relative frequencies, 115t

Index

Row vector, 601
 form, 606
R-squared (R²). *See* Adjusted R-squared
Russian ruble crisis, 286

Sample, 415–419
 criteria. *See* Large sample criteria
 length, increase, 432
Sample mean
 analysis, 429–430
 bias, 429–430
 computation, 437–440
 defining, 48
 efficiency, 444
 histogram, 423f, 424f, 427f
 mean-square error, 433–434
 sampling distribution density function, 422f
 standardization, 466–467
Sample outcome, impact, 428–429
Sample size, 417
 Central Limit Theorem, impact, 476
 sufficiency, 425
Sample skewness, definition, 65–66
Sample standard deviation, definition, 64
Sample variance
 bias, 430–432
 definition, 63
 histogram, 442f
 obtaining. *See* Corrected sample variance
Sampling distribution, 420
 density function, 422f
Sampling error, 429
Sampling techniques, 417
Scale parameter (σ), 288, 295, 306–319
Scaling invariant, 123–124
Scaling property, 285
Seasonal components, 157
Second central moment, 309
Second moment, 240. *See also* Log-normal distribution; Poisson distribution
 non-existence. *See* Alpha-stable distribution
Securities, types, 536–537
Security, return, 143
Seller, option payment, 625
Serial correlation, 577
Serial dependence, 428
Set countability, 170–171
Set operations, 170–177
Set value, assumption, 328
 Shorthand notation, usage, 329
Short hedge, 147
Short rates, 255–256
 dynamics, 256
Short-term interest rates
 modeling, application, 255–256
Short-term interest rates, long-term interest rates (correlation), 148
Sigma algebra (σ-algebra), 175, 327, 420
 definition, 176
Significance level, 488

determination, 510, 513
Simple linear regression, 6, 132, 521
 model, 532
 regression parameter estimation, 534t
Simultaneous movements, observation, 402
Skewness, 65–67, 242–244, 314–317. *See also* Exponential distribution; General Electric; Normal distribution
 computation, 315
  probability distribution, knowledge, 316
 definition, 65, 66. *See also* Pearson skewness; Sample skewness
 formal definition, 315
 indication, 244, 288
  parameter β, impact, 288
 parameter, 244
 statistic, 572
Sklar's Theorem, 381
Slippage, occurrence, 368
Space (Ω), 173–179. *See also* Countable space; Measurable space; Uncountable space
 mapping, 217
 occurrence, 180
Spearman's rho, 407
Special Erlang distribution, 247
Speculative-grade rating, 370
 unconditional probability, 370–371
Spherical distribution, 355
Spread measures, 45
Spurious regression, 130
Squared bias, 433
Squared deviations, variance (impact), 576
Square matrix, 603–604
 eigenvalues/eigenvectors, 612–614
SSE. *See* Sum of squares errors; Unexplained sum of squares
SSR. *See* Sum of squares explained by the regression
SST. *See* Total sum of squares
Stability property, 290
Stable distributions, 285
 characterization, 286
Stand-alone probabilities, 362
Standard deviation (std. dev.), 191–192, 242, 312–314
 definition. *See* Population standard deviation; Sample standard deviation
 outcomes, relationship, 193f
 usage, 68
 variance, contrast, 63–65
Standard error of the coefficient estimate, 530
Standard error of the regression, 530. *See also* Univariate standard error of the regression
Standard error (S.E.), 432–433
Standardization, 67–69
Standardization of X, 249
Standardized data, 278
Standardized random variable, value, 628
Standardized residuals, analysis, 572
Standardized transform, 385
Standard normal distribution, 247–248, 298
 distribution function, 271

Standard normal distribution (*Cont.*)
  mode, 302f
  α quantile, 510
  returns, contrast, 98f
  0.2 quantile, determination, 299f
Standard normal population, observations, 515–516
Standard & Poor's 500 (S&P500)
  daily returns, distribution test, 517
  equity index, value, 375
Standard & Poor's 500 (S&P500) index
  asset correlation, 501
  commercial bank sector, asset duration estimate, 532, 534
  daily returns, 503
    zero mean, determination, 505–506
  electric utility sector, asset duration estimate, 532, 534
  prices, 83t
    stem and leaf diagram, 82f
  returns, 568
    examination, 506–507
    time series application, 157–158, 161
Standard & Poor's 500 (S&P500) logarithmic returns, 177
  class data, 87t
  daily returns (2006), 89t
  frequency densities, 87t
  probability, 179
Standard & Poor's 500 (S&P500) returns, 375–376
  empirical cumulative relative frequency distribution, 90f
  QQ plot, contrast, 97f
Standard & Poor's 500 (S&P500) stock index, 534–536
  closing prices, consideration, 82
  daily logarithmic returns, 86
  daily returns, 157f
  logarithmic returns, box plot, 94f
  marginal relative distributions, 114
  monthly returns, 131t
  prices, changes (vertical extension), 158f
  usage, 143, 147
  values/changes, 162t
Standard & Poor's 500 (S&P500) stock index returns
  box plot, 95f
  classes, 90t
  conditional, 116t
  empirical cumulative relative frequency distributions, 91f
Standard & Poor's (S&P), 369
  credit ratings, 18
    assignation, 49
    empirical distribution, 50f
  index, cross sectional data (obtaining), 21
  ratings
    attributes, 56
    empirical distribution, 49t, 51t
    even number companies, 53f
Standard stepwise regression method, 12, 562, 564–565

State space (Ω'), 237–238
  probability, 188
Statistic, 415, 419–420
  definition, 45
  degrees of freedom ratio, 512
  empirical data, 4
  parameters, contrast, 45–46
  population parameter, contrast, 419–420
  probability, contrast, 4–5
  university course, offering, 2–3
Statistical independence, 159
Statistical inference, 416
Statistical noise, 133
Statistical physics, 1
Statistical significance, determination, 527
Stem and leaf diagram, 81–82
Stepwise exclusion method, 12, 564
Stepwise inclusion method, 12, 562–564
Stepwise regression, usage, 12
Stochastic terms, 155–156
Stochastic variable, 180
Stock price
  changes, joint movements, 409
  dynamic, 627
  evolution, 199–200, 291
  expected value, relationship, 163
  pricing process, evolution, 200–201
  probability, 253–254
    correspondence, 221
  random variable, 221
  ratio, 308
    bounds, 313
  variation, 21
  volatility, 198–199
    parameter, 627
Stock price to free cash flow per share (PFCF), ratio, 372–373
Stock returnillustration, 340–341
  modeling, 418
  properties, application, 251
  regression, application, 6, 130
  regression model, application, 136–138
  skewness, example, 315–316
  student's t distribution application, 259–260
Stocks
  characteristic line, 143, 147
  index futures, availability, 148
  logarithmic return, 229
  performance, 364
  portfolio
    benchmark, 361
    consideration, 251
  relative performance, 364t
Strictly monotonically increasing transformations, invariance, 382, 384–386
Strictly monotonic function preserves, natural logarithm, 447f
Strictly monotonic increasing function, 592
  illustration, 593f
Strike price (exercise price), 625

Index

Student's *t* distributed
   n-1 degrees of freedom, 469–470
   statistic, 512
Student's *t* distributed random variables, tail dependence, 411f
Student's *t* distribution, 247, 256–262
   assumption, 291
   cumulative distribution function, 260
   degrees of freedom, 470–471
   density function, 258f
   α-quantiles, n-1 degrees of freedom, 470
   returns, 300
   shape, 257
Student's *t* observations, generation, 401f
Student's *t* random variable, mean, 259
Sturge's rule, 33, 47
   usage, 37
Subprime mortgage crisis, 286
Subsets, 174
Sufficiency, 457, 459–461
Sufficient statistic, 11, 461
   distribution, 476
Summation, infeasibility, 595
Sum of squares errors (SSE), 139
Sum of squares explained by the regression (SSR), 13, 527
Support, term, 237
Supremum
   comparison, 515
   selection, 513–514
Survival time, 264
Symmetric confidence intervals, advantages, 11
Symmetric density function, 243f
   zero skewness, 244
Symmetric matrix, 608, 614
Symmetric $n \times n$ matrix, 608

Tail dependence, 347, 407–412
   absence, 409
   measure, 407
Tail index, 286
Tails, 3
t copula, 383, 385–386
   d=2 example, 394–398
      contour plots, 397f–398f
      illustration, 395f–396f
   degrees of freedom, estimate (computation), 403, 405
   joint distribution, 410
   tail dependence, 409–410
t distribution, 247, 256. *See also* Multivariate *t* distributions
   density function, 258f
   normal distribution, contrast, 410
   obtaining, 394–395
   probability mass, attribution, 395
t-distribution, tabulation, 530
10-year Treasury note, 537
10-year Treasury yield
   forecasting, regression results, 543t

   monthly data, 538t–542t
   prediction, 536–537
Test power. *See* Power of a test
Test quality criteria, 490–496
Test size, 486
   determination, 487f
   partitioning, 508f
   value, 501–502
Test statistic
   basis, 495
   computation, 513
   density function, 487f
   obtaining, 509–510
   usage, 506–507
Theoretical distributions/empirical distributions (comparison), box plots (usage), 98f
Third central moment, 314
30% quantile (determination), histogram (usage), 88f
Three-dimensional real numbers, span, 606
Time
   deterministic functions, 155
   variance, 575
Time series, 6
   analysis, 153, 576–577
      serial correlation/lagged correlation, 577
   cross sectional data, relationship, 21
   data, 21
      model, description, 575
   decomposition, 154–158
      components, 156f
   definition, 153–154
   obtaining, 55–56
   representation, difference equations (usage), 159
Tinbergen, Jan, 2
Total probability, law, 369
   illustration, 369–371
   multiple events, 371
Total sum of squares (SST), 139, 527
Transformed data
   median computation, 69–70
   range, computation, 70–71
Transformed random variables, covariance, 343–344
Treasury bill futures, availability, 148
Treasury bond futures, availability, 148
Treasury yield
   changes, 535
   prediction. *See* 10-year Treasury yield
True null hypothesis, 527–528
t-statistic, 521
   application, 550
t-test
   n-k-1 degrees of freedom, 529–530
   results, 548
Turned vectors, 607f
2-dimensional normal distributions, 349
Two-dimensional copulas, 387–399
   maps, 387
Two-dimensional random variable, usage, 331
Two-dimensional space, complex number representation, 599

Two-dimensional variable, 524
Two-period stock price model, probability distribution, 215t
Two-tailed Kolmogorov-Smirnov distance, computation, 514f
Two-tailed Kolmogorov-Smirnov test, usage. *See* Equality of distribution
Two-tailed test, 483
  usage. *See* Normal distribution
Type I errors, 11, 485–486
Type II errors, 11, 485–486
  probability, 490–491

UMP test. *See* Uniformly most powerful test
Unbiased efficiency, 443–444
Unbiased estimator, 453–454
  obtaining, 431
Unbiased test, 494
Unconditional probabilities, 362
  product, 368
Uncorrelated random variables, 341–342, 346
Uncorrelated variables, regression, 137f
Uncountable real-valued outcomes, 184
Uncountable sets, 170–171
  distinction, 7
Uncountable space, 229–230
  random variables, 183–185
Unexplained sum of squares (SSE), 528
  degrees of freedom, determination, 530
Uniformly most powerful (UMP) test, 491–493
  size α, 493
    comparison, 493f
Unimodal dataset, 55
Univariate data, 17
Univariate distributions, 6, 325
Univariate normal distribution, density function, 349
Univariate probability distributions, 9
Univariate random variable, probability, 328
Univariate regression, 521
Univariate standard error of the regression (s^2), 530
Unknown mean (normal random variable), variance (confidence interval), 473–475
Unknown variance (normal random variable), mean (confidence interval), 469–475
Up market beta, 551
Up movement, 198–199, 221
  binomial random variable, impact, 303
  occurrence, 252
Upper limit, 590
  defining, 92
Upper quartile, 296–297
Upper tail dependence, 408
u-quantile, 388
U.S. dollar, appreciation, 209–211
U.S. Treasury bond, comparison, 555

Value-at-risk (VaR), 298–299, 376–377
  computation, 299–300
  equivalence, 300
  measure, 295

Variables
  inclusion, *F*-test (usage), 530–531
  linear functional relationship, 131–132
Variance inflation factor (VIF), 547–548
Variance (σ^2), 189, 191, 240–242, 309–312. *See also* Analysis of variance
  contrast, 66
  definition. *See* Population variance; Sample variance
  elimination, 433
  estimator
    comparison, 431
    consistency, 441–443
    mean-square error, 435
  knowledge, 501
  non-existence, 290
  one-tailed test, 503–505
  parameter, estimation, 255, 631–632
  proxy risk, 574
  reduction, 443–444
  square root, 192, 312
  standard deviation, comparison, 63–65
  yield, 311
Variation, 59–69. *See also* Coefficient of variation
  measures, 59, 67, 70
  examination, 72t
Vector, 601–602. *See also* Column vector; Row vector; Turned vectors
  dimensionality, 601
  linear composition, 605f
  three-dimensional space, 604
Vector hyperplane, 525f
Vector notation, usage, 522
Venn diagram, 362
  example, 363f
VIF. *See* Variance inflation factor
Volatility clustering, 574–575
von Mises, Richard, 167–168

Waiting time, 269
Weibull distribution, 280
Weighted least squares (WLS) estimation technique, 573–574
Weighted least squares (WLS) estimators, 574
Weighted mean, 55–56
  mean, form, 55
Whisker ends, 93
White's generalized heteroskedasticity test, 573
WLS. *See* Weighted least squares
Writer, option payment, 625

Yield beta, 148–149
Yield spread, difference (test), 560

0.2 quantile, 297
  determination, 297f, 299f
0.25 quantile, 34
0.75 quantile, 34
Zero-coupon securities, 536–537
Zero covariance, concept, 121–122
Zero-one distribution, 193

Printed in the United States
By Bookmasters